DESIGN OF LANDFILLS AND INTEGRATED SOLID WASTE MANAGEMENT

DESIGN OF LANDFILLS AND INTEGRATED SOLID WASTE MANAGEMENT

Third Edition

AMALENDU BAGCHI

JOHN WILEY & SONS, INC.

0.0ΛΓΔ

For general information on our other products and services or for technical support, please contact our Customer Care Department within the United States at (800) 762-2974, outside the United States at (317) 572-3993 or fax (317) 572-4002.

Wiley also publishes its books in a variety of electronic formats. Some content that appears in print may not be available in electronic books. For more information about Wiley products, visit our web site at www.wiley.com.

Library of Congress Cataloging-in-Publication Data:

Bagchi, Amalendu.
  Design of landfills and integrated waste management / Amalendu Bagchi.—3rd ed.
      p. cm.
  Rev. ed. of: Design, construction, and monitoring of landfills. 2nd ed. 1994.
  Includes bibliographical references and index.
  ISBN 0-471-25499-1 (cloth)
    1. Sanitary landfills. 2. Integrated solid waste management. I. Bagchi, Amalendu. Design, construction, and monitoring of landfills. II. Title.

  TD795.7.B36 2004
  628.4′4564—dc21

                                                              2003057179

10 9 8 7 6 5 4 3 2 1

To my Parents and Parents-in-law

CONTENTS

PREFACE

The first two editions of this book were well accepted both within the United States and internationally. The solid waste profession is becoming increasingly demanding. Since waste reduction is undertaken in most communities, professionals involved in landfill design are finding it necessary to have a basic understanding of the waste reduction issues. Because of this expanded role of the solid waste professionals, the scope of the book has been revised to include fundamentals of integrated solid waste management (ISWM). Discussion on redevelopment of contaminated land has also been included. Part I focuses on ISWM issues and redevelopment of contaminated land, and Part II focuses on landfill design, construction, and monitoring. Professionals from several disciplines (such as environmental science and engineering, hydrogeology, toxicology, microbiology, planning, and mass communication) are involved in ISWM and usually work as a team. Although the book is written primarily for engineers and technical professionals involved in ISWM and landfill design, it also provides fundamental knowledge for professionals from the above disciplines engaged in ISWM programs. As one can see, it is difficult to provide in-depth information from all the disciplines and about all the issues related to ISWM. Therefore the book focuses on reduction, utilization, and disposal of solid waste, and redevelopment of contaminated land. The book may be used for teaching undergraduate courses in solid waste management and contaminated land remediation, as well as by practicing professionals.

Integrated solid waste management involves actions at two levels: national and local. At the national level laws are formulated to foster technological development to reduce the volume and toxicity of the waste. At the local level the emphasis is primarily on reduction of the volume and toxicity of waste destined for land disposal or incineration. The book is written primarily to help professionals working at the local level. However, ideas regarding total integration techniques have been discussed in the book. Although models are required for optimization, no model is included in the book. Models, along with the necessary instructions, are usually stand-alone documents, which is beyond the scope of the book.

Part I includes Chapters 1 through 10. The different types of waste generated within a community and their management approaches are included in Chapters 1 through 8. Issues related to redevelopment of contaminated lands are discussed in Chapters 9 and 10. This is a topic of significant interest for solid waste professionals and has some bearing on ISWM. Redevelopment of contaminated lands may be considered recycling because it is a use of resources after processing.

Part II, consisting of Chapters 11 through 26, concentrates on landfill-related issues. In addition to information regarding design, construction, and monitoring of landfills, remediation of landfills and compensatory wetland development are included in Part II. Fundamentals of health and safety and economic analysis are also included in this part.

The opinions expressed in this book are mine and do not necessarily reflect the views and policies of the Wisconsin Department of Natural Resources.

I am very much thankful to the following individuals who provided suggestions, comments, or critical review during the manuscript preparation: Ajit Chowdhuri (RMT Inc., Madison, WI); Bruce Ramme (We Energies, Milwaukee, WI); Barb Derflinger, Chris Carlson, Daniel Kolberg, Dave Siebert, and Gretchen Wheat (Wisconsin Department of Natural Resources, Madison); John Reindl (Dane County Public Works, Madison, WI); Gene Kramer (Onyx, Burlington, WI); Sonia Newenhouse (Madison Environmental, Madison, WI); Craig Benson (University of Wisconsin, Madison); Edward Kavazanjian (Geosyntec Consultants, Huntington Beach, CA); and Leo LaRochelle (Performance Technology Inc., Lewiston, ME). Dipak Maitra, Goutam Chakraborty, and Rina Chakraborty provided secretarial help including word processing. Their help is greatly appreciated.

I thank my daughter, Sudeshna, for her comments and help in preparing the manuscript. Finally, I express my deepest gratitude to my wife, Sujata, for her patience, encouragement, and help.

AMALENDU BAGCHI

Madison, Wisconsin
May 2003

PART I

INTEGRATED SOLID WASTE MANAGEMENT

1

INTRODUCTION

In the past, solid waste management primarily included collection, land disposal, and incineration of household waste. Industrial waste disposal did not receive much attention. Environmental awareness by the general public increased over time because of various reasons such as advancements in environmental science and technology and interest in pollution-related health problems. Attention was also drawn toward the fact that Earth's material and energy resources are finite (Meadows et al., 1972). It became apparent that landfilling and incineration have significant environmental impacts and that landfilling and incineration are not enough to deal with the huge volumes of solid waste generated by communities and industries. All these issues and the emergence of the *sustainable development* concept (WCED, 1997) helped to foster the idea of recycling as a way to reduce waste volume. It became apparent that besides landfilling and incineration, solid waste can be managed by other means (e.g., composting of putrescible waste). Subsequently, the concept of integrated solid waste management (ISWM) emerged. The goal of sustainable solid waste management is the recovery of more valuable products from that waste with the use of less energy and a more positive environmental impact (McDougall et al., 2001). The practice of the three R's (reduction, reuse, recycle) fits very well within the sustainable development concept. Rather than relying on a waste reduction hierarchy (Fig. 1.1), integrated solid waste management suggests optimization of the system. Although the concept of integrated waste management can be applied to both hazardous and nonhazardous wastes, discussion in this book will be restricted to municipal and industrial nonhazardous wastes.

Integrated solid waste management lacks a clear and widely accepted definition. Although many prefer to use a hierarchy (Fig. 1.1) in defining ISWM,

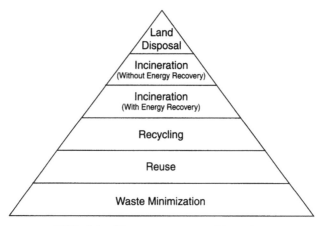

FIG. 1.1   Waste management hierarchy.

it may not always provide an optimum choice between cost and environmental impacts. The current consensus is to allow ISWM planners flexibility to choose from different elements of waste management options, which will result in minimum energy use, minimum environmental impact, and minimum landfill space at a cost affordable to the community. This goal can be achieved by various means such as recycling of certain types of waste, beneficial reuse of industrial by-products, and segregating waste type (e.g., infectious from noninfectious waste).

Figure 1.2 is a simplified diagram showing the various sources of solid waste generation. Source reduction is an important part of an ISWM program. On a global (or perhaps national) scale industries, governmental bodies, and citizenry need to work together for a comprehensive program to reduce the volume and toxicity of waste. On a local level, a designer of an ISWM program has very little control over the volume and toxicity of the waste generated by a community. Thus, the strategies for an ISWM program at a local level are significantly different from those at the global level. From a practical standpoint, at a local level an ISWM program essentially means reduction of volume and toxicity of waste destined for disposal or incineration. The planner also needs to ensure that the chosen elements have low environmental impacts and are energy efficient. All these have to be done at a cost acceptable to the community. On a local level a designer of an ISWM program needs to identify sources and generation rates of different waste types, characteristics and total volume of each waste stream, and so on. Based on this information, the planner needs to optimize management of all the waste streams generated in a community with an aim of minimizing environmental impact at an affordable cost.

To integrate a solid waste management program within a community, the program should address the needs of the community as a whole. In other

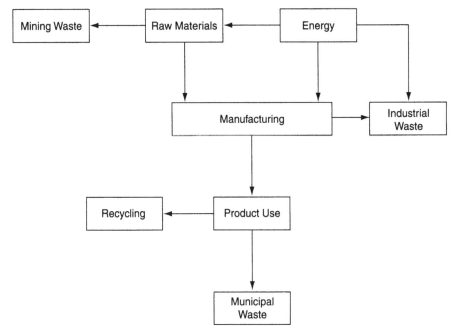

FIG. 1.2 Simplified diagram showing solid waste generation sources.

words waste generated from individual homes and apartments, public places, businesses, and industries located within a community should be taken into consideration for efficient management of all types of solid waste generated within the community. The program must satisfy the regulatory requirements and address the economic parameters set by the community. Enough flexibility should be built into a program so it can protect the environment in a variable marketplace. Educating the public (including managers of industrial and commercial institutions) regarding the benefits of an ISWM program is key to the success of the program in the long term. Willing participation of the community as a whole (which includes both industrial and nonindustrial sectors) in reducing waste is essential. Thus, apart from management practices, due consideration should be given to educating the public regarding the waste reduction program.

The goal of the book is to provide information regarding ISWM pertinent to the local level. The various wastes generated within a community, their characteristics, ideas regarding reduction, reuse, and recycling of these wastes, and the management options are included. ISWM should not be viewed as a hierarchical scheme but as an optimization tool for maximum environmental benefit at an affordable cost. There can be many sequence combinations for an ISWM program. Rather than discussing the combination of sequences in detail, attempt has been made to provide basic information regarding the characteristics and management options for the various waste types. This information will help the planner to choose the needed elements and sequence them to fit the needs of a specific community.

1.1   BASIC CONCEPTS OF AN INTEGRATED SOLID WASTE MANAGEMENT PROGRAM

Integrated solid waste management is a complex task. An ISWM program involves various disciplines. A successful program must include both short-term and long-term goals. It must also provide a balance between three main factors: environmental regulations, cost of running the program, and community needs. To develop a program one needs to comprehend the basic principles involved in managing each component and their effect on one another. For instance, if the ash generated by incineration of municipal waste tests out to be hazardous, then either the ash must be detoxified or it must be disposed in a specially designed landfill/landfill cell. Therefore, prior to including incineration in the program, one needs to ascertain the characteristics of the incinerator ash so that correct disposal practice is included in the program. On a local or community level, ISWM programs essentially consists of the following five steps:

1. Waste source identification and characterization
2. Efficient waste collection
3. Reduction of volume and toxicity of the waste to be discarded
4. Land disposal or incineration of the waste past the reduction goal
5. Optimization of the first four steps to reduce cost and environmental impact

The actions on the first four steps are interdependent. For instance, method of waste collection depends on the method chosen for reduction and disposal of waste. Several distinct issues are associated with each step and several choices are available to optimize the solid waste management program.

Although public comments are reflected in environmental regulations, there are a rigid set of rules for the ISWM planner. A thorough knowledge regarding the rules and necessary environmental permits is very crucial for a successful program. The total cost of running an ISWM program may vary primarily due to the cost of achieving the waste reduction goal mentioned in step 3. ISWM programs are planned on a long-term basis. However, it should be flexible enough to accommodate a change in the waste stream. For example, a paper mill located within a community may decide to close its own landfill and request permission to dispose of its waste in the community-owned landfill. This change in waste profile and volume may significantly impact the community-owned landfill operation and future landfill needs. An ISWM program should be flexible enough to accommodate this community need.

The main goal of a community-based ISWM program is to reduce the volume and toxicity of waste being disposed or incinerated. Many believe that recycling is an additional financial burden on a community. They judge land disposal based solely on the current tipping fee, which in most cases is lower than recycling cost. However, the total cost of siting future landfills and the hauling cost is overlooked. A second group of citizenry may think that if recycling is adopted, there will be no need for landfills in the future. This is also a wrong perception because the entire municipal waste stream cannot be reduced to zero through a waste reduction program. In summary an ISWM program attempts to optimize a solid waste program so that the community will benefit both in the short and long term while providing environmental benefits as well.

1.2 SCOPE AND ORGANIZATION OF BOOK

This book is divided into two parts. Part I includes various components of ISWM with an emphasis on principles and practices for reuse, reduction, and recycling of waste. Part II includes principles and practices of waste disposal. Until about late 1980s, solid waste professionals were involved mainly in managing land disposal of municipal and industrial solid wastes. However, over the last decade the job responsibilities of these professionals expanded and changed. Currently, many of them are involved with various other types of issues such as recycling, reuse of both municipal and industrial wastes, infectious waste, and redevelopment of contaminated land. Because of this change, Part I is being added to the original book to reflect current professional needs. The book is expected to provide fundamental concepts regarding ISWM to students and working professionals so that they can meet today's challenge.

The book provides information regarding various options for reduction and reuse on a local level. Although federal/state regulations followed in the United States are cited occasionally, the book concentrates on principles and practices for developing a good understanding of an ISWM program. The book is written primarily for professionals and students interested in learning the subject. However, others such as lawyers, government officials, and industry managers involved in solid waste management issues will also find the book helpful in gaining a basic understanding of the principles of ISWM and landfill technology.

Both theories and practices of an ISWM program have been discussed in a comprehensive manner in Part I. The following is a brief summary of the chapters in Part I:

This chapter has given a brief history of ISWM and basic concepts of an ISWM program.

Chapter 2 discusses the basic concepts of ISWM. Information regarding collection and transportation of waste, transfer stations, and waste composition studies are also included.

Chapter 3 considers the basic approaches for reduction of both the volume and toxicity of waste destined for land disposal or incineration.

Chapter 4 discusses the management of waste generated in health-care-related institutions. Storage, transportation, treatment technologies, and ideas regarding source reduction of this waste stream are also included.

Chapter 5 discusses incineration. Various types of incinerators, environmental issues related to incineration, and issues related to siting an incinerator are included.

Chapter 6 considers composting. Facility design as well as issues related to marketing and economics are included.

Chapter 7 discusses reuse and recycling of municipal solid waste. Information on various recyclable materials, quality control, and data collection are included. Information regarding material recovery facilities (MRF), refuse-derived fuel (RDF) processing plant, and land spreading of sludge are also included.

Chapter 8 considers information on reuse of industrial by-products in civil engineering projects. Comments on engineering characteristics and testing protocols of industrial by-products are included.

Chapter 9 discusses information on both *in situ* and *ex situ* remediation technologies. Methods involving physical, chemical, and biological remediation methods are discussed. Basic concepts of contaminant transport, project life, cost estimation, and maintenance of wells used for remediation are included.

Chapter 10 discusses primarily the redevelopment of lands contaminated by industrial activities. Fundamentals of risk assessment are included in this chapter.

Part I includes information on issues related to ISWM but does not consider land disposal. Part II is dedicated to issues related to land disposal of solid waste. The range of issues in ISWM is quite wide. Knowledge is required in various disciplines such as planning, mass communication, marketing, economics, chemistry, hydrogeology, civil engineering, and so on. Therefore, running a successful ISWM program requires professionals from various disciplines to work together. As the list implies, some of the issues require background in science-related fields whereas for others knowledge in the liberal arts is required. Although the book focuses mainly on issues related to science and technology, comments on nontechnical issues are also included. The primary purpose is to provide well-round information for professionals involved in ISWM. While environmental engineers will benefit from both Parts I and II, Part I is written for all professionals involved in solid waste management.

Currently in most organizations, both government and private, job duties no longer focus on just one discipline; employees and managers are involved in multidisciplinary tasks. While in-depth knowledge in one discipline is required, it is also becoming essential to have functional knowledge in other related disciplines. The book is written with this idea in mind. Professionals involved in ISWM will find it very helpful in performing their day-to-day job duties. The book may also be used in undergraduate environmental science and engineering courses. Attempts have been made to provide information on various topics in sufficient detail, but justice could not be done in some cases. References are cited for readers interested in follow-up reading.

2

INTEGRATED SOLID WASTE MANAGEMENT

In addition to the discussion on integrated solid waste management (ISWM) issues, this chapter includes information on collection and transportation of waste, transfer stations, and waste composition study. Although waste management principles started evolving in the late 1800s, the concept of ISWM emerged in the late 1980s (see Chapter 1). Ideally, the goal of ISWM is to optimize all aspects of solid waste management to achieve the maximum environmental benefit at least cost.

As mentioned in Chapter 1, a well-accepted definition of ISWM is still absent. Initially various waste management alternatives were combined to form a hierarchy (Fig. 1.1). However, current studies indicate that the hierarchical scheme is not the correct option in all cases. The waste management hierarchy is not based on any scientific principle. It cannot help choose the best combination for a community. For example, it cannot help compare the following two options in terms of overall environmental impact and cost: (a) recycling plus landfilling or (b) composting plus incineration. Additional analysis is needed to find the cost and the environmental impact. If the cost is compared, which is done in most cases, then it will be a step toward integration. Since the hierarchy is not a tool to compare cost, or environmental impact, its use cannot lead to optimization of the two factors. Choosing the best combination by optimizing the two factors is the goal of ISWM. Instead of following the hierarchy, ISWM aims at optimizing environmental impact and cost while choosing the best combination from the five broad areas of current waste management practices: reduction, reuse, recycling, incineration (with or without energy recovery), and land disposal. ISWM helps in recovering valuable products from waste with less energy and environmental impact. Social acceptability for the chosen option should also be included in the decision-making process.

Solid waste planners need more data about the environmental impact associated with the five waste management options mentioned above. Life-cycle inventory is a tool that can provide the data needed for choosing the best combination from an environmental standpoint. Life-cycle assessment is a compilation and evaluation of the inputs, outputs, and potential environmental impacts of a product throughout its life cycle. Input refers to raw materials, resources, and energy. Output refers to emissions to air, water, and solid waste. Since a quantitative methodology is followed, life-cycle assessment provides an objective basis for decision making. It should be noted that life-cycle analysis does not predict actual impact, assess risk, safety, or whether a threshold may be exceeded by choosing an option. Although risk assessment can predict the most likely effect, it cannot take into account overall environmental impact due to an action.

White (1996) argues that an overall approach is important for ISWM. An ISWM program should manage all waste in an environmentally and economically sustainable way. Environmental sustainability can be achieved by optimizing resource consumption and the generation of air emissions. Economic sustainability means that the cost of the program is acceptable to all sectors of the community. However, Hickman (1996) and Ham (1996) think that sustainability in waste management is an impractical idea.

Integration of waste management involves actions at both the global (or more realistically national) and local levels. Industries, governmental bodies, and citizenry need to work together at the national level to reduce volume and toxicity of the waste before its generation. Actions at the national level should include identification of the sources where reduction are possible, development of technologies to reduce toxicity and increase recyclability of the waste, fostering the idea of minimizing resource consumption, development of laws and standards for implementation of the goal, provide incentives for implementation of the goal, and so on.

The ISWM planners at the local level have very little control on the waste generated by the communities. Therefore, the actions at the local level need to combine multiple waste management techniques to minimize volume and toxicity of the waste destined for disposal or incineration. Rather than following a rigid hierarchy, the ISWM program for each community should include alternatives that will minimize both environmental impacts and cost. Kreith (1994) has suggested 9 strategies for managing municipal solid waste. McDougall et al. (2001) summarized 13 case studies where principles of ISWM were followed (Table 2.1). It does not appear that any of the cited communities used any computer-based integrated model for choosing the solid waste management options.

Several researchers have discussed the use of computer-based models for integrating solid waste management program (Pugh, 1993; Sundberg et al.,

TABLE 2.1 Summary of Integrated Waste Management (IWM) Case Studies

Program	Treatment Technology	Key IWM Characteristics
Brescia, Italy	Composting, recycling	Economy of scale, stability, long-term perspective
Copenhagen, Denmark	Composting, recycling, incineration	Economy of scale, stability, control of all waste arisings, continuous technology improvement, enabling legislation
Hampshire, UK	Composting, recycling	Economy of scale, flexibility, public support
Helsinki, Finland	Composting, recycling	Economy of scale, control of all waste arisings
Lahn-Dill-Kreis, Germany	Composting, recycling	Flexibility
Madras, India	Composting, recycling (informal)	Vision, public support
Malmö, Sweden	Composting, recycling, incineration	Economy of scale, control of all waste arisings, continuous technology improvement
Pamplona, Spain	Recycling	Economy of scale
Prato, Italy	Composting, recycling	Control of all waste arisings
Seattle, USA	Composting, recycling	Control of all waste arisings, public support
Vienna, Austria	Composting, recycling, incineration	Stability, public support
Zürich, Switzerland	Composting, recycling, incineration	Economy of scale, stability, Long-term perspective, continuous technology improvement, polluter pays

From McDougall et al. (2001).

1994; Hokkanen et al., 1995; Anex et al., 1996; Haith, 1998; Katugampola et al., 1999; Fabbricino, 2001; McDougall et al., 2001; Abou Najm et al., 2002). Not all models cited in these publications are commercially available. At present, use of a computer-based model for integrating the solid waste management system at the local level is not common.

In summary, although several models are available, the current practice at the local level is to choose from five broad areas of waste management (namely reduction, reuse, recycle, incineration, and land disposal) to develop an economically viable program that has low environmental impact. Most communities, if not all, follow a mandated reduction goal for developing a solid waste management program.

Most communities have to manage both hazardous and nonhazardous wastes. The definition and characteristics of hazardous waste, as adopted in the United States are discussed in Chapter 14; it may be noted that the definition of hazardous waste is not the same in all countries. The management approach of these two types of waste is different. The only exception is the inclusion of household hazardous waste (HHW) in the solid waste stream. This becomes necessary because HHW is, in most cases, discarded at the curbside and gets mixed up with otherwise nonhazardous waste. It is hoped that in the future citizens will understand the environmental risk of disposing HHW with nonhazardous waste whereby this waste stream can be separated from household solid waste. As part of pollution control programs, many industries are reducing the toxicity and waste volume through in-house measures. This approach is also helping in separating solid waste generated in industries, which is being beneficially reused in many cases.

In addition to traditional municipal solid waste, there are other types of solid waste whose management needs to be addressed so as to provide an integrated approach. Probably the largest contributor to the waste stream is construction and demolition (C&D) waste. It is estimated that in the United States, 38% of the total solid waste generated is C&D waste (University of Florida, 1998). This book addresses both the recycling and disposal of this material into the waste stream. Another type of waste normally managed by solid waste program personnel is the waste generated in health-care facilities, commonly known as medical waste. Management of this waste should also be addressed as part of an integrated approach. Agricultural waste, which includes unused herbicides and pesticides and animal waste, is also a significant contributor to the waste stream especially in rural areas. Normally, solid waste program personnel do not manage this waste stream. Unused herbicides and pesticides may be collected through "clean sweep" programs (see Chapter 3). Although many communities use animal waste as manure, environmentally acceptable management program for this part of the waste stream is yet to be developed in some communities.

As indicated before, the main goal of ISWM is to optimize management options for minimizing land disposal. Although recycling has been practiced for almost two decades in many countries, it is still being debated by many. Many see recycling as an extra expense or a "fashionable thing" to do, not a necessary practice. Many fail to realize that landfill sites are not easy to obtain; while regulations in most cases put restrictions on landfill siting, the biggest obstacle for landfill siting comes from residents living near the proposed site. Because of difficulty in siting landfills in new locations, even in the United States where population density is very low, many landfill owners are expanding existing landfills both horizontally and vertically. In addition to disposal of municipal waste, landfill space is needed for disposal of industrial by-products. Reuse and recycling of these by-products are necessary. So, in the long run there will still be a landfill space problem that can only be addressed through reduction of solid waste volume disposed in landfills.

It must be pointed out that emissions of air pollutants from municipal landfills are significant and the cost of long-term care of landfills is quite high. Therefore, both from an environmental and an economic standpoint, land disposal of waste must be minimized by adopting the principles of reduction, reuse, and recycling.

A new concept of zero waste is also emerging. The concept comes primarily from the concerns regarding global warming. According to the concept of zero waste, even 100% recycling creates quite a volume of waste in the process of bringing the goods to the consumer. It is estimated that for each load of waste discarded by individual homeowners, 71 loads of waste were generated during manufacturing, distribution, and marketing of the product. The concept of zero waste looks at the overall picture of consumption and disposal. Because of government subsidies in various forms such as tax break, low-cost electricity to manufacturing units, and so on, the producers of consumer goods do not pay serious attention to minimize in-house waste generation (i.e., waste generation within the factories). In addition, long-term planning by governing bodies is inherently impeded because of immediate gains favored by business leaders, who mostly have relatively short tenure in office. The recent mandate in European Union countries to buy back the computer products after useful life by the manufacturer is an example, which shows that the concept of zero waste is gaining ground. Table 2.2 shows a list, partial at best, of various communities and industries that are working toward a zero waste goal. The following five key elements for achieving zero waste has been suggested by Wood and Tarman-Ramcheck (2002):

1. Investment in community waste reduction and recovery systems
2. Citizens' participation in recycling, which will create more jobs locally
3. Product redesign to make it nontoxic and reusable after useful life

TABLE 2.2   List of Communities and Industries That Adopted Zero Waste Plan

Community/Industry	Zero Waste Action
Del Norte County, California, USA	Adopted zero waste plan in 2000.
Seventy-four communities in New Zealand	Adopted 2015 as the target year for achieving zero waste.
Fetzer Winneries	Adopted 2009 as the target year for achieving zero waste.
Xerox Corporation	Adopted a worldwide solid waste reduction by 87% and beneficial management of 94% of hazardous waste.

Adapted from Wood and Tarman-Ramcheck (2002).

4. Extension of producer's responsibility beyond initial sale, which will provide incentive to buy back
5. End of subsidies to enterprises that uses virgin resources only

As is obvious from these statements, elements 3, 4, and 5 are far beyond the control of individual communities. Nevertheless, communities, individually or collectively, need to work toward the zero waste goal by discussing these ideas with business leaders and government officials whenever possible. Such interactions will eventually help in educating everyone involved in the decision-making process. "Darkness of mind comes from the lack of education" is an old saying that may be utilized to reach the goal of zero waste in the long run. Prolonged efforts to educate citizens regarding the danger of unnecessary resource use are needed to minimize waste generation. Discussion on various issues, ideas, and currently available technologies for minimizing solid waste are included in this book. Each community needs to develop a customized plan for integrating its solid waste management program.

2.1   PLANNING

Planning is important for implementing an idea. One important issue in efficient and realistic ISWM planning is to define the unit for which the planning is to be done. It could be just one small community (e.g., village) or a group of communities (e.g., county). While a realistic ISWM plan is extremely difficult for a small community, the issues for a group of communities are complex. In addition, both short-term and long-term planning are important, which adds to the degree of complexity in the process.

For small communities, the volume of recyclables (e.g., paper) is so low that finding a buyer is not easy. For countywide planning, the control of waste flow in and out of the county cannot be totally controlled by county government, especially if waste hauling is done by private contractors and privately owned landfills are present. However, if all these issues are considered during a planning process, then a realistic compromise can always be obtained. Each community must realize that to develop an efficient ISWM plan, it is essential to work with other neighboring communities, the regulatory agency, and private farms. For instance, a village's recyclable waste volume may be low, but it can either join with other neighboring communities and form a bigger unit or can contract directly with a private vendor who is willing to buy from such small communities.

Usually a long-term contract needs to be negotiated in such cases so that the community gets the benefit of a buyer and the private vendor is assured of a steady supply of goods for setting up a business. The state's solid waste regulatory agency may help set up these arrangements initially by providing grants.

For proper planning it is essential to have an up-to-date estimate of volume or weight of various recyclable waste streams (e.g., paper) and various components within municipal solid waste (MSW). For this purpose, a proper waste classification study (also known as waste sort) should be undertaken. Details regarding waste sort are discussed in Section 2.7. For each waste stream within the MSW the three R's (reduce, reuse, and recycle) should be implemented to minimize the need for solid waste disposal. Ideas regarding solid waste reduction at the source are discussed in Chapter 3 and reuse and recycle are discussed in Chapter 7. Although the planning process can be complex, it is possible to develop an ISWM plan for any size of community if the basic concept of the three R's is utilized properly.

Reduction refers to minimizing the waste generation. Reuse refers to using the discarded waste without going through any process. An example could be the reuse of office furniture or computers through stores specializing in the sale of used items. Recycle refers to utilizing a waste after running it through a process. An example of recycling is the remanufacture of a glass bottle after melting glass items. It should be mentioned that these definitions are not strictly adhered to when using the terms reduce, reuse, and recycle.

There may be several levels at which program planning is done. In many situations, the planning is done at national, state, and local (county/city/ village/town) levels. The planning process is different at each of these levels. A plan at the national level may provide the goals and ideas of implementation, whereas the planning at the local level needs to emphasize how to implement the solid waste program and funding sources. In general, the technical guidelines and rules are developed at national and/or state levels. Once the various waste streams in a community are identified, the combined management of all of these streams needs to be planned. The main features of local planning should include:

1. *Goal*   It defines the desired outcome of the ISWM program. It is very likely that goals are already mandated by national or state regulatory authority.
2. *Objective*   It specifies how the set goals are to be evaluated for success or failure. A stepped objective, that is, measuring achievements at specified time intervals or for each waste type, may be used for evaluating success or failure of the program.
3. *Assessment of Resources*   Both solid waste program personnel needs and funding needs should be assessed. In addition, possible help from outside the local government (e.g., state, industry) should also be assessed. Comments and recommendations regarding additional personnel and funding needs should be included in the planning. The program structure needed for implementation of the goal must be assessed for achieving the stated goals and objectives.

4. *Study of Alternatives* All possible alternatives should be studied and resource (both human and financial) requirements for each alternative should be specified. Depending on the complexity of the alternatives, it is better not to choose one alternative. However, if from the analysis of alternatives, one option obviously stands out, then it may be recommended. While choosing from the alternatives, the planner must ensure that the chosen alternative meets the goal.

5. *Recommendations for Implementation* The plan should provide recommendations regarding implementation of each option. The responsibilities of each person and/or group should be clearly mentioned. A management structure for integrating all waste streams should be identified.

6. *Public Participation and Education* The plan should indicate how the public input for the ISWM program is to be obtained. It also should suggest how to assess public opinion and awareness regarding ISWM. Specific recommendations regarding public awareness and education campaign should be included.

In general, municipal and commercial waste streams are managed by community government whereas industrial waste stream is managed by the individual industry. However, a joint session of the representatives from community government, industry(ies), and other interested parties can help identify the disposal needs and beneficial reuse projects. Efforts should be made so that all parties can work together to reach the community's goal(s).

The following are additional ideas that may be used for ISWM planning:

1. All plans must look at both short-term and long-term goals. It is very possible that certain actions may provide a gain in the short term, but it may not be beneficial in the long term. A compromise should be chosen in such cases.

2. While analyzing recycling, attention must be given toward price fluctuation of items. This is very crucial when beginning a program. Long-term market trends should be analyzed before estimating the cost of the program.

3. Regulatory agencies should be contacted to find out about available grants. In many cases, demonstration grants to try new ideas are awarded by the state or federal government.

4. Funding from other departments of local government may be available for particular projects. For instance, a city transportation department may fund a road leading to a recycling facility.

5. Permit requirements for different facilities must be researched thoroughly. Some of the permitting processes are time consuming (e.g., landfill permit) and may need expertise in developing the permit proposal. The plan must indicate the total time, expertise, and expenses needed for different permits.

6. The plan must identify all waste streams generated within the community and the party responsible for managing each waste stream. For instance, a paper mill may generate waste that is landfilled within its own property. The industry should be invited to take part in the ISWM planning process. It is possible that during the course of meeting(s), the representative will learn about recycling options or joint venture options. For example, the paper mill may agree to incinerate a waste type in its boiler (e.g., tire chips) generated by the community or the community may agree to disposal of paper mill sludge in the landfill owned by the community.

7. The plan must provide estimates for funding requirements for each segment (e.g., landfill, incineration) and how such expenses will be met (e.g., general revenue, general obligation bond). The expected revenue from each segment should also be estimated.

8. The plan should identify all the waste streams generated within the community, the management options for each waste stream, the steps to be taken for implementing the concept of the three R's for each waste stream.

9. The plan should discuss the public education strategy to be used for implementing the ISWM. Both short-term (e.g., organizing informational meetings) and long-term steps regarding public education should be included. This issue is discussed further in Section 2.4.

10. The plan should include more than one option to achieve maximum waste reduction. The trade-offs for each of the options should also be indicated clearly so that meaningful debate can occur within the community. The cost estimates for each option should be included in the plan. It should be mentioned that the cheapest option may not be the best option. Socioeconomic and environmental impact of each option must be identified. The ranking of each option based on the ISWM concept should also be identified.

11. The planner should assess the economic value of the environmental impact.

12. The plan should include ideas for optimization of the action plan. For regional planning, models using a system analysis approach may be used for optimization.

2.2   BENEFITS OF ISWM

The main benefit of ISWM is to meet the desired goal in an optimal fashion. Although there are short-term benefits, the startup costs of a recycling program are rather high. In the past, solid waste management (SWM) (which mainly included municipal solid waste and commercial waste) focused on disposal. Slowly aesthetic issues were included for disposal sites; then actions to minimize environmental impacts in the immediate surroundings were addressed. As areas for disposal are becoming scarce and knowledge about long-term effects of discarding waste without proper methods for protecting the environment is increasing, attention is given toward minimizing waste volume. Since the 1980s, several types of solid waste (e.g., infectious waste) that needed proper management were added to the solid waste programs at all levels of government—local, state, and federal. The addition of nonhazardous industrial waste to SWM programs expanded the complexity of management significantly. However, these additions helped the growth of the concept of *waste reduction.* For industries, minimizing cost of disposal was the driving force for researching waste reduction techniques. Although the practice of reduction of municipal solid waste volume through incineration and the beneficial use of waste (e.g. land spreading of wastewater sludge) are quite old, an organized holistic approach toward waste management is rather new. Thus, the current approach of ISWM, developed partly due to environmental issues and partly due to need, emerged over the last two decades of the twentieth century.

Many communities that do not have enough disposal space are using ISWM for better management of wastes. For these communities the available disposal space is so low that waste reduction is not an option, but it is the only choice. Even in the countries where the MSW generation rate is already low (around 1 lb/person/day), attempts are being made to further minimize disposal volume through composting and recycling (CPHEEO, 2000). In the United States, the number of communities participating in recycling, and the total volume of waste being recycled, has increased significantly (Goldstein, 1997; EPA, 1998).

An integrated approach for managing waste helps in identifying the waste streams, which can be further reduced to minimize environmental impact both in the short term and in the long term. Although the market for recyclables fluctuates over time, an ISWM approach can make necessary adjustments to fund allocation for running the program in the long run. Rather than looking at recycling as an option, ISWM fosters newer ideas of reuse and recycling.

2.3   REGULATORY ISSUES

As indicated in Section 2.1, it is very important to take into consideration the various regulations that are applicable to various solid waste facilities (e.g., landfill). Knowledge about regulations under consideration is also important because it is cheaper to design a facility according to the rule than to retrofit it later. There may be regulatory mandates regarding reduction of a solid waste item (e.g., paper). If the target cannot be achieved, then the issue should be discussed with the regulatory agency as early as possible. Based on similar comments from other communities, the regulatory agency may allocate funds for demonstration grants or revise the target. Sometimes a regulatory mandate may be faulty. For instance, currently all types of plastics cannot be recycled. So a rule banning all types of plastics from landfills would not be able to be implemented.

It should be mentioned that planning an ISWM sets an optimal goal. However, in the initial years a lower target may be adopted. It is better to start with a lower yearly target and slowly work up to the goal. Because of fluctuations in recyclable markets, experience in coping with low markets must be gained by the program managers. It is a good idea to take a small step, consolidate the position before taking the next step. Risk of adopting the optimal goal of waste reduction in the initial years of the program is rather high and may even lead to a catastrophic failure. A failure at the beginning will receive adverse public reactions, which may be extremely difficult to overcome in the future.

2.4 PUBLIC PROMOTION OF ISWM

For a successful ISWM program, educating the public about the program is very important. Both short-term and long-term measures should be taken for sustaining the program. The public needs to be informed about the facts so individuals can form their own opinion. Both the short-term issues (e.g., regulatory mandates) and long-term issues (e.g., an overall picture of solid waste disposal) should be considered in developing public education campaign.

Traditional marketing techniques for selling a product may not be effective in promoting a sustainable change in behavior (Constanzo et al., 1986). Community-based social marketing tools are to be used for waste reduction campaigns. For an effective public education program, barriers and benefits of the recommended activity must be identified. There are two types of barriers: internal and external. Lack of knowledge regarding how to perform an activity/task (e.g., how to compost) is defined as an internal barrier. Physical obstacles for conveniently performing an activity/task (e.g., availability of a low-cost composting bins) are defined as external behaviors. Barriers are activity specific ((McKenzie-Mohr et al., 1995; Okamp et al., 1991). So, the barrier(s) for the proposed ISWM-related activity (e.g., recycling) must be studied carefully. There are many internal barriers regarding recycling or re-use such as lack of knowledge regarding disposal issues, apathy, and lack of knowledge regarding the importance of individual actions in achieving a community goal. Information regarding barriers and benefits of an activity can be found from a literature review, observational studies, and survey of the target community. (Note: Observational studies involve watching one or more individuals perform an activity without being told about the study.) Surveys can be done through the mail, personal interviews, and telephone interviews. Although personal interviews provide the most reliable data, it is expensive and time consuming. Developing a survey is a difficult task. A focus group consisting of randomly chosen residents, who should be paid for the task, may be employed for developing the survey under the supervision of a person knowledgeable about the job. Once a survey is developed, it should be used for a pilot study using a small group (10–15 persons). Such pilot studies help in improving the survey questionnaire and the length of time required for finishing the survey questionnaire. The survey questions should be simple and nonambiguous. The survey data should be analyzed by knowledgeable person(s) using proper statistical tools. Success of developing a sustainable behavior depends on correct identification of the barriers and benefits related to the chosen activity.

Once the barriers and benefits are identified for the chosen activity, the proper strategy to implement sustainable behavior changes needs to be adopted. Public information is the first step for promoting the behavior change.

2.4.1 Public Information

Informing the public about an activity is essential for fostering change in behavior. Vivid, concrete, and personalized information is the most effective way to ensure attention (McKenzie-Mohr and Smith, 1999). The attitude, beliefs, and behavior of the target population are studied either through surveys or focus groups. For a campaign to be successful in achieving a permanent behavior change, the target audience must pass through the following 12 steps (McGuire, 1989):

1. Get exposed to the communication.
2. Pay attention.
3. Become interested in the message.
4. Understand the message.
5. Learn what and how the activity is to be performed.
6. Change attitude toward the activity.
7. Agree with the message.
8. Learn more about the activity.
9. Decide to take action.
10. Change behavior.
11. Reinforce the idea of behavioral change.
12. Assimilate the activity and make it part of one's lifestyle.

While the public information campaign manager may make every effort to bring the message to the target audience, the target audience must have at least a leaning toward adopting new waste management practices. The target audience may select to avoid exposure to a message and even if exposed make no effort to assimilate the message. Thus it means that vigorous and repetitive messages are necessary at the beginning of a public information campaign.

Proper selection of the media (such as television, radio, newspaper, and posters) is important. Vividness through dramatic colors, loud music, use of familiar or credible person(s) or organization(s), familiar themes, and values are useful tools in drawing public attention (Andreasen, 1995; Eagley and Chaiken, 1975). Although both positively and negatively framed messages (e.g., positive frame: Home composting reduces waste collection costs; negative frame: Waste collection cost will increase if home composting is not done) are useful, negativly framed messages are more persuasive (Davis, 1995).

Messages that are easy to remember and that include specific actions are very effective (Heckler, 1994). Messages that include specific goals or targets for a household or community are more effective for reaching waste reduction goals (Folz, 1991).

Public information campaigns should be a long-term effort. Feedback to the community regarding impact of the newly adopted activity helps in wider acceptance and promotion of sustainable behavior (McKenzie-Mohr and Smith, 1999). For example, the amount of recycled paper increased by 26% when households were informed each week regarding the amount of paper recycled by the community (DeLeon and Fuqua, 1995). Public information campaigns should not create expectations of the participants that are too high. For example, a campaign message should not state that new landfills are not required at all if recycling is undertaken. The message should also include negative consequences of an activity. For example, the message should include that recycling may increase expenditures for waste management initially.

2.4.2   Social Marketing Tools

A social marketing approach appears to be very useful for fostering a change in behavior. Social marketing is a relatively new concept that evolved out of techniques such as public communication, social mobilization, and social advertising (Andreasen, 1995). There are a number of community-based social marketing tools that can be utilized for promoting change of behavior. For an effective outcome more than one tool should be used.

2.4.2.1   Commitment   Individuals are more likely to follow up on an activity if they agree to undertake the activity verbally or in writing. Written commitments are probably more effective than verbal ones (Pardini et al., 1983). Commitments from cohesive groups (e.g., players of a neighborhood soccer team) can also be effective. Commitments should be requested only from individuals who express interest in an activity.

2.4.2.2   Slogans and Prompts   These are helpful in fostering sustainable changes in behavior. Slogans such as "Don't be Fuelish" are designed to promote sustainable change in reduction of fuel consumption. Prompts are effective for both one-time and repetitive actions. Installing a low-flow shower head is an one-time action that has an ongoing positive environmental impact. Prompts should be short and direct and should be delivered at convenient locations. While negatively framed media messages are more persuasive, prompts should direct toward a positive action (e.g., "Turn off the light when leaving the office" and should be placed next to the light switch). Long indirect prompts embedded with environmental consequences due to lack of the suggested actions (e.g., "If lights are not switched off when leaving office, valuable natural resources will be wasted") are not effective.

2.4.2.3   Financial Incentives   Financial incentives are effective in motivating people to undertake an activity. For example, a waste disposal fee based on waste volume collected from individual homes increases recycling (McKenzie-Mohr and Smith, 1999). Incentives should be directly linked with the target activity and reward positive behaviors. Incentives have little effect if people are unaware of them and the benefit is very low.

While a financial incentive can have positive effects in promoting behavioral changes, it may also encourage prospective participants to undertake unwanted activity. For instance, a waste disposal fee based on waste volume can prompt individuals (especially living in remote places) to dispose of waste illegally in highway dumpsters or in nearby woods. Clear statements regarding consequences of violation and organized enforcement is essential for financial incentives to work. Nonmonetary forms of incentives, such as public recognition of individuals or organizations, can also be used for motivation especially for commercial and industrial establishments.

2.4.2.4   Norms   The behavior of neighbors and the community influences the behavior of individuals. For example, if many members of a community compost food waste, then chances are high that a new person coming to the neighborhood will start composting. Norms have substantial impact on behavior. Norms affect behavior in two different ways: compliance and conformity (McKenzie-Mohr and Smith, 1999). In many situations people undertake a behavior not just because it is the right thing to do but to receive favorable remarks from others and to avoid punishment. Conformity occurs by observing others doing an activity and then changing one's behavior. To be effective norms must be made visible, at least initially. For example, a city may give stickers (indicating that the household composts food waste) to all those who compost and request that the sticker be put on their waste container(s).

2.4.2.5   Removal of External Barriers   Any external barriers to a certain activity are to be identified through survey or other means. Removing external barriers associated with an activity greatly increases participation in the activity. For example, participation in composting can be increased by supplying compost bins to interested households. It should be mentioned that inconvenience is to certain extent a matter of perception. This type of perception changes once the activity is undertaken. Although strongly perceived inconvenience is difficult to overcome, commitment and norms can be used to overcome moderate perception of inconvenience (McKenzie-Mohr and Smith, 1999).

2.5   COLLECTION AND TRANSPORTATION OF WASTE

The solid waste and recyclables are usually stored over a period of time (e.g., week) by the generators, which is collected at a set frequency and transported to a proper facility for disposal or recycling and so forth. The storage and collection strategies for large communities with high population densities are different from those suitable for relatively small communities with low population densities. In addition, funding levels for the program will dictate the level of mechanization and frequency. Collection frequency also depends on solid waste volume discarded per day and the accessibility to the dumpster, weather, and availability of storage space. For instance, even in cold climatic regions, it is essential to collect putrescible solid waste outside of an apartment on a daily basis, whereas for single-family homes, putrescible solid waste may be stored outside of the building for an entire week. A different strategy may be adopted for storage, collection, and transportation of nonputrescible solid waste. Collection and transportation from commercial facilities, health facilities, and industries (where applicable) are different from individual apartments or homes. A discussion on various strategies followed for storage, collection, and transportation is included in the following sections. Collection and transportation of medical waste are discussed in Chapter 4.

2.5.1   Collection of Putrescible Waste

Collection of putrescible waste from apartments, individual homes, and commercial establishments is one of the largest and most complicated operation managed by solid waste personnel of all communities—large and small. In apartment buildings and most commercial establishments, the solid waste is usually collected in dumpsters outside of the buildings. Residents are informed about items that are banned from land disposal with the hope that these items are not discarded into the dumpsters. It is very difficult to enforce the disposal of banned items in these dumpsters. Usually, separate containers for the banned items are kept along side the dumpsters or at a different location within the apartment building. In commercial establishments such as restaurants, the majority of the waste is food waste. However, the items discarded in different commercial establishments are very diverse and many times nonputrescible. In cities with a high population density, because of lack of space, placing a dumpster outside of apartment buildings is difficult. So the putrescible waste is stored overnight within the apartment or house. A door-to-door hand collection is done every day of the week. The waste is taken to a truck stationed at a central point within the locality. After completion of one building or a group of buildings, the truck moves to another location and finally to a landfill when the truck is full.

In individual homes, the waste is first discarded in small containers, which are transferred to a bigger container placed outside of the house (often in garages). The waste is stored for usually a week before being collected by a waste collection truck. In communities where the yard waste (e.g., grass clippings) are composted, it is bagged separately or are taken to a central location within the community; the collected yard waste may be composted by the community at a location other than where it is collected.

In some communities, a fee is charged for waste collection from individual homes. It is popularly known as pay-as-you-throw (PAYT). The fee may be based on volume or weight (unit) of the waste. Where a volume-based fee is charged for collection, it is based on the volume of the container [usually between 30 and 90 gal (113 to 340 liters)] that a resident chooses. A periodic bill is sent to the homeowner by the city or the company that collects the waste. In a weight-based system, the container is weighted at collection. Reduction in waste generation as high as 59% has been achieved in communities where a unit-based collection fee is used (DiMartino, 2000). Vehicles used for unit-based collection need to be more mechanized because the containers tend to weigh more, leading to more worker injuries. A survey conducted by the U.S. Occupational Safety and Health Administration (OSHA) indicates that worker injury is very high among workers in the solid waste industry.

Solid waste is also transported via river or railroad. If the landfill is located on an island or the waste needs to be transported on waterways, the waste transportation is done using boats. In the United States, substantial volume of solid waste is transported via rail. Not only municipal solid waste but construction and demolition (C&D) waste, ash, and sludge are also transported via rail (Woods, 1995). For rail transportation of waste to be successful, the daily waste generation quantity needs to be very high. In addition, for a fault in the engine or track, a whole train may be canceled, which may create a health hazard at the point of generation or collection. This is not the case with road transportation. There may be a breakdown of one truck, but all trucks do not breakdown simultaneously. However, inclement weather may impact both road and rail transportation. Rail transportation does not provide the flexibility of road transportation because a long-term contract (e.g., 5 years or more) is essential for the former transportation mode. While transporting nonputrescible waste (e.g., C&D waste) fits well with railroads' main business of hauling bulk goods, transporting putrescible waste does not. Putrecible waste must be hauled away as soon as a rail car is full. Special loading and unloading docks are necessary for rail hauling. Rail hauling is economical for long distances [100 miles (160 km)]. It can be attractive if goods can be transported during the return trip. Rail transport is a good option if the highway access is poor or a community wants to reduce truck traffic in a congested area. All these issues must be considered before opting for rail transportation of waste.

The collection of solid waste from apartments, single-family homes, and commercial establishments are done by trucks. There are two issues related to collection: vehicles used and vehicle routing.

2.5.1.1 Vehicles Used   Vehicles used for collecting solid waste door-to-door should be chosen based on load limits on roads (due to bridges and culverts), design load of the roads, and available funding. The full weight of vehicles should be checked against the road limits before choosing one. While checking the weight limits on the bridges and culverts, the entire route must be scrutinized. The collection vehicle may terminate at a transfer station or it may travel all the way to the place of disposal (incinerator or landfill). Mainly three types of vehicles are used for residential collection: top loading, rear loading, and side loading. Usually the waste is dumped manually in these vehicles. However, these vehicles may be fitted with robotic hands to grab the container, a practice that is steadily growing. This type of mechanical dumping is necessary in communities where the collection is unit price based. The available design and capacity of collection vehicles vary widely. Manufacturer's catalogs, the experiences of the surrounding communities, and load tests should be used to choose the most suitable vehicle for a community. Worker safety, cost, ease of collection, weight of vehicles, ease of maintenance, fuel consumption rates, and availability of spare parts and tires should be considered when choosing a vehicle. In many communities, collection and transportation services are contracted to private haulers.

2.5.1.2 Vehicle Routing   Proper routing of vehicles leads to substantial cost savings. Much data are needed before a routing can be done. The goal of vehicle routing is to find out the travel path of a vehicle so that pickup can be completed at minimum cost. The time spent on each pickup, travel time between pickup points, street layout, topography of the area, and capacity of the vehicle greatly influences the efficiency of collection.

The routing of vehicles for collecting waste from each source is complex. The main goal of efficient routing is to minimize traveling on a segment that has no pick up or a second time without pick up (also called dead distance). In most cases, houses are located on two-way streets whereby both sides have to be serviced. This means that the collection vehicle must travel on the same street twice, collecting waste from either side of the street. Although computer programming concepts can be adapted to develop collection routes (Ford and Fulkerson, 1962; Quon et al., 1965; Wathane, 1972), a manual development of collection vehicle routing is preferable (Liebman et al., 1975; Vesiland et al., 2002). Manual routing is still in use in many communities. A common-sense approach (heuristic approach) for vehicle routing developed by Schuster and Schur (1974) provides useful guidance for developing vehicle routing networks manually. Calculating the number of houses is easy for communities that use unit-based fees because the volume "purchased" by each house on the route is known. However, this number has to be estimated for communities that are not unit based. A new routing map needs to be adjusted several times before finalizing it. In addition, since ISWM is expected to reduce MSW waste generation rates, the routing map should be revisited if waste sort studies indicate a significant reduction in waste generation rates.

There are three different routing needs: macrorouting, route balancing, and microrouting (Fig. 2.1). The purpose of macrorouting is to optimize the col-

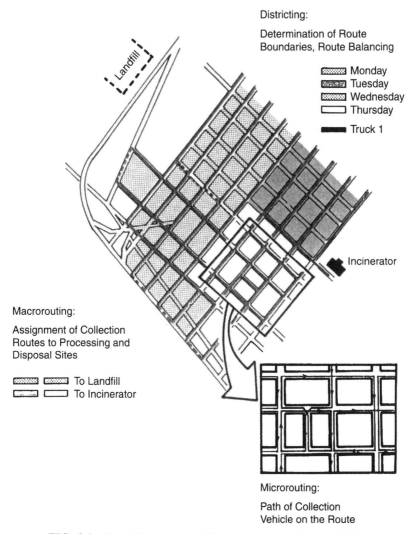

Districting:

Determination of Route
Boundaries, Route Balancing

▨ Monday
▨ Tuesday
▨ Wednesday
☐ Thursday

▬ Truck 1

Incinerator

Macrorouting:

Assignment of Collection
Routes to Processing and
Disposal Sites

▨ ▨ To Landfill
☐ ☐ To Incinerator

Microrouting:

Path of Collection
Vehicle on the Route

FIG. 2.1   Heuristic routing. After Schuster and Schur (1974).

lection routes based on location of disposal or processing facilities (i.e., landfill, incinerator, transfer station). Macrorouting may also be used to determine cost-effective location(s) of several disposal or processing facilities (or garage). Macrorouting exercises can be undertaken if more than one option or site are available for dealing with the collected waste. If all collected waste must be delivered to one landfill or one incinerator, then macrorouting is not needed. Route balancing on the other hand aims at distributing workload for collection crews so that workers in each route or collection district spend more or less the same time on the job. It also helps in determining the total collection time required in each collection district. Thus route balancing can be used to determine district size. The purpose of microrouting is to lay out exact travels paths for a collection vehicle so that collection time is minimized. Although deterministic models can be used for microrouting, the heuristic approach is most suitable, probably the best way, to develop travel path for a collection vehicle within a district. Manual heuristic approach for microrouting is faster, less costly, and more flexible compared to deterministic computer models. However, with the advent of geographical information systems (GIS), computerized microrouting may become attractive in the future.

Microrouting should be undertaken whenever there is a change in frequency of collection, point of collection, size and type of collection vehicles, type and size of containers, separate collection of recyclables, disposal size, or location of garage for equipment. Collection policies and methodologies influence collection efficiency. Districting and route balancing are important for efficient collection. A sound estimate of the average time spent to service a pickup point, the average amount of waste collected at each pickup point, and the travel time between the last pickup point and the waste delivery point (e.g., transfer station) are taken into account in estimating the length of a workday. The number of pickup points served in a day by a truck depends on the length of the workday.

The following heuristic rules are used for developing microrouting (Schuster and Schur, 1974):

1. Each route should consist of streets clustered in the same geographical area.
2. The length of a workday should be reasonably the same for each route.
3. Collection point should start as close to the garage as possible.
4. Routing on one-way streets or segments should start where the one-way begins.
5. Collecting dead-end streets when they are at the right side of the truck minimizes left turns.
6. Whenever possible, both sides should be collected on streets with steep slope.
7. Collection should start at higher elevations.
8. It is best to use clockwise turns when routing collection on one side of the street.

9. For collection from both sides of the street at the same time, it is generally best to route with long straight paths across the grid before looping clockwise.
10. Heavily traveled streets should not be collected during rush hours.

2.5.2 Collection of Recyclables

Recyclables can be collected separately or along with other waste. The following three alternatives are available for a recycling program:

1. Collection of recyclable items and waste separately
2. Collection of commingled recyclable items and waste separately
3. Collection of all recyclable items and waste together

The white goods (e.g., refrigerators, stoves) are usually collected separately and taken either to a material recovery facility (MRF) or to a dismantling/recovery shop.

The experience of various communities regarding the effectiveness of each of the options stated above should be studied to choose the best option. An understanding of "what happens after pick up" will help in choosing an option. First of all, if the community has a refuse-derived fuel (RDF) plant or an incinerator, then there is normally no need to separate the recyclables from the putrescible waste at source. However, source separation of recyclables from waste increases the value of recyclables whereby the earnings from recyclables will be higher. Processing of waste from which metal and glass have been removed in a RDF is easier and provides better fuel. In terms of sorting of recyclables before selling, the best option is to have all items separately bagged. The second option is to have the paper bagged/bundled separately from the plastics, glass, and metal. The third option is to have all recyclables bagged together. One of the problems of the third option is contamination of the paper from broken glass. This type of contamination leads to reduction in the value of the paper. Of all these options, the second option is a good compromise. With proper public information campaigns and participation, it is possible to choose the option that is suitable and economically viable. A system analysis may identify the third option as the best because it reduces the collection cost.

There are several choices of vehicle types available for collection of recyclables. The choice of a vehicle type depends on the way the recyclables are handled. If the waste mixed with recyclables is delivered to an RDF plant, then a packer truck can be used. If the recyclables are separated, then it has to go to an MRF. If the MRF is en route to the landfill or incinerator, then a dual-compartment packer truck may be used. However, if the MRF is not on the way to a disposal site, then separate vehicles for collection of recyclables and waste need to be used. Often a transfer station and MRF are sited together. Multicompartment trucks for collecting separated recyclables are a good choice if each type of recyclables is bagged separately. As for the collection of white goods or bulky goods from home improvement projects, it is normally done separately because these materials do not fit into a regular collection vehicle and cannot be loaded onto the vehicle by a one-person collection crew. In many communities, special fees are charged for disposing white goods. In most cases white goods are demanufactured and recyclables are salvaged. Usually collection truck drivers inform the community solid waste personnel regarding white goods sighted at a property.

2.6   TRANSFER STATIONS

At a transfer station, wastes from relatively smaller trucks are transferred to larger trucks or trailers for transporting the waste to landfills or incinerators located at a great distance. The purpose of using a transfer station is to consolidate waste from relatively smaller trucks to a larger trailer so that the waste collection and transportation cost is minimized. The overall cost is reduced due to better utilization of collection trucks and their crews, lower maintenance cost for collection vehicles due to a significant reduction in tire maintenance, and lower total waiting time at landfills. All these costs should be considered (for all collection vehicles combined) for calculating transportation cost (Walsh et al., 1993). A transfer station may be used in cities with narrow streets where larger trucks cannot be used. They are also constructed in thinly populated communities where door-to-door collection is not needed or economical. A transfer station can be used for recovering recyclables. If a transfer station is not tied to a dedicated landfill or incinerator, then the owner can take advantage of lower rates at competing facilities. The cost effectiveness of a transfer station depends primarily on the total volume of waste handled that is affected by recycling (Ficks, 2001). Therefore, if the cost of running a transfer station is based on the tipping fee of the waste tonnage brought to the facility, then the effect of waste reduction due to ISWM should be taken into consideration when setting the tipping fee.

The site of a transfer station must satisfy location criteria set by the regulatory and local authority. The location should be such that waste collection trucks need not travel long distances for dumping the waste. At the same time, it should be located at a place that is easily accessible to the outgoing transport route. The size of the lot should be big enough to provide queuing space for incoming and outgoing vehicles. It should also have enough space for future expansion or redesign. A transfer station should include enough floor space for segregating recycling materials in addition to space for waste processing. It is a good idea to include unloading space for small vehicles of individual citizens. If the transfer station is to be used for collecting household hazardous waste, then a separate space (preferably with a separate entrance) should be included in the floor plan. A bigger space is need for handling bulky waste (e.g., C&D waste) (Dempsey, 2000).

Based on waste loading, there are basically two types of transfer stations: direct dumping and reloading. In direct-dumping transfer stations, the collection trucks drive up a ramp and unload directly into a trailer or compactor. In a reloading-type transfer station the collection trucks unload the waste on the floor. The waste is reloaded to the trailer using a wheeled loader. The transfer station may be designed either to accommodate trailers directly into the loading bay or to back up to the loading bay. Depending on the waste volume transferred per day, a drive-through type of transfer station may require more space than a back-up design. The transfer station capacity depends on the length of the loading day (e.g., 8 hr), technology used for transfer operation, material density, and type of transport used (road, rail, or river) for delivery or hauling. However, in most cases, the capacity of a transfer station is dictated by the availability of transferable materials, capacity of the transport vehicles, and availability of loading equipment. The storage volume at transfer stations should be more than the volume needed in a normal day because if hauling is delayed due to temporary failure of the transport system (e.g., bad road due to inclement weather, rail track failure) the whole day's load may need to be stored.

Transfer stations for rail cars and barges are costlier to build. Additional arrangement for loading into a trailer should be made to address emergency situations when the rail traffic gets shut down for an extended period due to rail track or engine failure. The following issues should be taken into consideration while planning a transfer station: capacity, site, building design, safety, equipment to be used, and support facilities (Ficks, 2001).

2.7 WASTE COMPOSITION STUDY

A waste composition study, commonly known as waste sort, is needed to estimate the fraction of various waste materials or items present in a waste stream. A waste sort is done for various projects such as designing of recycling programs or finding out whether a waste is suitable for incineration. For a proper waste sort, the purpose of the waste sort must be identified at the beginning. Usually waste generated by households, commercial establishments (e.g., restaurants, banks), and institutions (e.g., hospitals, schools) are managed by municipal authorities, whereas wastes generated due to manufacturing processes are managed by the generating industries. Normally, a waste sort is not undertaken for industrial wastes because they are composed of a single type of waste. To estimate an overall composition of municipal waste, an idea regarding the ratio of waste from households and from commercial and institutional sources must be established. The following major issues need to be resolved prior to undertaking a waste sort:

1. Purpose of the study
2. Sorting location
3. Number of samples to be analyzed
4. Waste sort design
5. Sample sorting
6. Statistical analysis of the data.

2.7.1 Purpose of Study

The purpose of the study is important for developing a list of items that will be sorted out of the waste. For instance, the list of items to be identified for incineration of waste will target the items based on heating value or pollution potential. The heating value of paper is high compared to food waste. On the other hand, if the purpose is to identify recyclable items present in waste, then the list needs to include items that can be recycled.

2.7.2 Sorting Location

The sorting location is important because it will help identify the target(s) to concentrate on further action. Sorting can be done at the point of generation or at the point of processing. For household wastes, the point of generation is the curbside, whereas for commercial or industrial waste the point of generation could be the dumpster or the storage room of a restaurant. The points of processing are a transfer station, incinerator tipping floor, or landfill. A waste sort done at the point of generation can identify the difference in percentage of recyclables between single-family and multifamily households. In a waste sort in Washington State where recycling was mandatory, it was found that multifamily household (greater than four units) has about 50% more recyclables compared to single-family household. A high percentage difference between these two types of household can help in doing further surveys regarding the cause of the difference and then take necessary steps to improve the situation. It is difficult to find data regarding such differences if the waste sort is done at a landfill. However, such data can be obtained through a study performed at a landfill only if routes consist entirely of multifamily households and the trucks carrying such waste are targeted at the landfill. Such data will be completely lost if the waste is consolidated first in a transfer station. Random selection is very important for arriving at a good estimate regarding percentages of various items present in the waste. Targeting or preselecting trucks will skew the data due to selection bias if done improperly.

2.7.3 Number of Samples

The total number of samples that should be sorted depends primarily on the allowable error and confidence level. Statistical analysis, discussed in Section

2.7.6, is done to arrive at the number of samples needed to provide the desired accuracy of the percentages of various items. However, a rough estimate of each item is needed to estimate the number of samples. This estimate may be obtained from an earlier study or from a waste sort done in a community with similar socioeconomic background and population. If such data is not available, then a large number of random samples should be sorted at various locations representing 80–90% of the waste. There are items [e.g., items related to household hazardous waste (HHW) such as paint thinner] that are not normally distributed. The HHW distribution data usually have a right skew, which are known as J distribution. For proper estimate of an item with J distribution, data from more samples need to be analyzed. Thus, the percent distribution of an item will influence the number of samples to be sorted. For instance, to measure the success of a "clean sweep" program (see Chapter 3), an extremely high number of samples must be sorted from materials collected separately at curbside because the HHW is expected to have a J distribution.

2.7.4 Waste Sort Design

Location of sorting and method for selecting samples must be finalized before undertaking a field sorting operation. An initial estimate regarding the number of samples should be done as part of the design. These issues are discussed in the following sections.

If sampling is to be done at curbsides, the addresses of houses where sampling is to be undertaken must be finalized. If sorting is to be done for a state, then the landfills where sorting is to be undertaken must be finalized. A randomly chosen geographical location is the correct way of collecting representative samples. However, there are practical problems associated with this type of choice. The owner(s) of the geographical location (e.g., the house or the landfill) may not permit the sorting operation at his or her premises. If an owner refuses to allow sorting operation, then another location should be chosen randomly. However, if sorting of a significantly high number of samples is undertaken, then choosing representative geographical locations is not expected to compromise the quality of data. It should be mentioned that the randomness of samples is lost as soon as the owner or the waste collector is notified regarding the date(s) of sampling or sorting.

If the waste sort is done at a landfill or transfer station, then the trucks entering the premises are chosen randomly. The driver is interviewed to find out the route characteristics. The number of trucks chosen for waste sort should have approximately the same ratio as that of the ratio of waste from households and commercial and institutional sources.

The ratio of the waste from single- and multifamily homes, commercial, and institutional sources is estimated to decide the percentage of samples from each source. Usually, waste from multifamily buildings with more than five units have dumpsters. Since the type of vehicles used for collection from single-family households is different from those used for collection from dumpsters, it is possible that waste from multifamily homes and from commercial and institutional establishments are collected by the same trucks. A study of the microroutes helps to identify the percentage of such mixed collection. If the number of such mixed routes is high, and if the purpose of the waste sort is to study the effectiveness of an existing recycling program in households, then it is better to undertake the waste sort at curbside.

For obtaining unbiased and representative estimates, waste sorts should be done in different seasons (e.g., once in winter and once in summer). Waste stream composition has a seasonal variability. For instance, the number of soda containers is high in summer. A waste sort done right after a holiday or in a day following a day of local celebration is very likely to have different results than a waste sort done at a time when there is no such celebration.

2.7.5   Waste Item Sorting

The selected trucks are directed to dump the load on a turf placed on a flat and level surface. The waste is dumped from the truck in the shape of a loaf of bread. There are two methods by which a representative sample is collected from a truckload—*cone-and-quarter* method and *grid-and-pull* method. In the cone-and-quarter method half of the waste is removed longitudinally from entire side. The waste is then mixed, coned, and quaterd using a back hoe. One quarter is chosen randomly for sorting. In the grid-and-pull method a 3-ft (1-m) height of the "loaf of waste" is maintained while dumping. The waste is divided into several 3 × 3 ft (1 × 1 m) cells, either physically using tapes or visually. One cell is pulled for sorting. The weight of waste used for sorting should be 200–300 lb (91–136 kg).

The waste to be sorted is spread over a 10 × 12 ft (3 × 3.6 m) turf. From the turf the waste is scooped out and placed over a sorting table using a wide-mouth shovel. Each listed item is sorted and deposited into cans, positioned around the table, by the sorting crew. The cans are weighed before and after the sorting. Weighing is done on a portable scale placed on level ground. The scale should be calibrated at the site following manufacturer's recommendations. The accuracy of the scale should be checked using a known weight. To minimize moisture loss, the weighing should be done as soon as the sorting of a sample is finished. The weight of each sorted sample is recorded in a separate sheet for each sample with a given sample identification number or description.

The sorting crew should be knowledgeable about the purpose of the waste sort and be detail oriented. Sorting is done by either experienced professional sorters or by paid college students. The sorting crew should wear protective gear such as heavy leather gloves, hard hats, dust masks, safety glasses, and safety boots. The sorting crews are instructed to be careful to avoid injury from sharp objects such as hypodermic needles and broken glass. Crews should not force their hands into the waste pile for pulling objects. The crew should be current on immunization, especially against hepatitis B and tetanus. In tropical climates the crew should receive immunization against typhoid and cholera or any other vaccinations recommended by the local medical professionals. The crew should take proper protection against climatic conditions (e.g., proper clothing in cold weather). The crew must wash hands with germicidal soap before eating.

2.7.6 Statistical Analysis

The number of samples (n) required to achieve the desired precision level (p) is computed by using the following formula (ASTM, 1992):

$$n = (ts/px)^2 \qquad (2.1)$$

where t = Student's t statistic corresponding to the desired confidence level
$\quad s$ = estimated standard deviation
$\quad x$ = estimated mean

The values of t are given in Table 2.3. The values of s and x for various waste items (based on waste sort studies in the United States) are given in Table 2.4. Ideally, prior to starting a composition study, the standard deviation and mean values of municipal solid waste components should be determined by sampling a few waste loads

TABLE 2.3   Values of Mean ($\bar{x}$) and Standard Deviation(s) for Within-Week Sampling to Determine MSW Component Composition[a]

Component	Standard Deviation(s)	Mean ($\bar{x}$)
Mixed paper	0.05	0.22
Newsprint	0.07	0.10
Corrugated	0.06	0.14
Plastic	0.03	0.09
Yard waste	0.14	0.04
Food waste	0.03	0.10
Wood	0.06	0.06
Other organics	0.06	0.05
Ferrous	0.03	0.05
Aluminum	0.004	0.01
Glass	0.05	0.08
Other inorganics	0.03	0.06
		1.00

[a] The tabulated mean values and standard deviations are estimates based on field test data reported for MSW sampled during weekly sampling periods at several locations around the United States.
After ASTM (1992).

TABLE 2.4   Values of t Statistics (t^*) as a Function of Number of Samples and Confidence Interval

Number of Samples, n	90%	95%
2	6.314	12.706
3	2.920	4.303
4	2.353	3.182
5	2.132	2.776
6	2.015	2.571
7	1.943	2.447
8	1.895	2.365
9	1.860	2.306
10	1.833	2.262
11	1.812	2.228
12	1.796	2.201
13	1.782	2.179
14	1.771	2.160
15	1.761	2.145
16	1.753	2.131
17	1.746	2.120
18	1.740	2.110
19	1.734	2.101
20	1.729	2.093
21	1.725	2.086
22	1.721	2.080
23	1.717	2.074
24	1.714	2.069
25	1.711	2.064
26	1.708	2.060
27	1.706	2.056
28	1.703	2.052
29	1.701	2.048
30	1.699	2.045
31	1.697	2.042
36	1.690	2.030
41	1.684	2.021
46	1.679	2.014
51	1.676	2.009
61	1.671	2.000
71	1.667	1.994
81	1.664	1.990
91	1.662	1.987
101	1.660	1.984
121	1.658	1.980
141	1.656	1.977
161	1.654	1.975
189	1.653	1.973
201	1.653	1.972
∞	1.645	1.960

After ASTM (1992).

The number of samples (n_0) is computed first by using the s and x values for a waste item from Table 2.3. The value of t for the desired confidence level is obtained from Table 2.4 assuming $n = \infty$. A new value of n_0 (n_1) is calculated by using the t value corresponding to n_0 and the desired confidence level. The process is repeated until the value of n_1 is within 10% of n_0.

Example 2.1 Calculate the number of samples required for newsprint with a confidence level of 90% and a precision level of 10%. From Table 2.3,

$$s = 0.07 \qquad x = 0.1$$

From Table 2.4,

$$t = 1.645 \quad \text{for} \quad n = \infty$$

From Eq. (2.1)

$$n_0 = (1.645 \times 0.07/0.1 \times 0.1)^2$$

$$= 133$$

The value of n is to be computed using $n_0 = 133$:

$$t = 1.6568 \quad \text{for} \quad n_0 = 133$$

$$n_1 = (1.645 \times 0.07/0.1 \times 0.1)^2$$

$$= 135$$

Since the value of n_1 is within 10% of n_0, the number of samples (n) to be sorted for newsprint with a confidence level of 90% is 135.

The values of n for all listed items are computed. The number of samples sorted is the highest of all the values of n.

After the waste sort is completed, the mass fraction (M_i) of each item is computed using the following formula:

$$M_i = \frac{w_i}{\sum_{i=1}^{j} w_i} \tag{2.2}$$

where w_i = weight of the item i
       j = number of items

The percent of component i (P_i) is computed using the following formula:

$$P_i = M_i \times 100 \tag{2.3}$$

The mean of M_i (m_i) of all samples is computed using the following formula:

$$m_i = \cfrac{1}{n\left[\sum_{i=1}^{k} (M_i)_k\right]} \tag{2.4}$$

The mean of P_i (p_i) of all samples (n) is computed by using the following formula:

$$p_i = \cfrac{1}{n\left[\sum (P_i)_k\right]} \tag{2.5}$$

The above formulas are valid only if the item has a normal distribution. The following formula is to be used for transforming items with J distribution (RecyleWorlds Consulting, 1994):

Transformed value $= 2 \times$ arcsin $\sqrt{\text{Fraction of the material in the sample}}$

$$\tag{2.6}$$

3

SOURCE REDUCTION

The aim of source reduction is to reduce both the volume and toxicity of waste. The activities related to source reduction needs the involvement of governmental bodies, industries, and consumers. Source reduction (also known as waste prevention) programs need to target changes in behavior of both industry and consumers. The goal of source reduction is to reduce waste generation, whereas the goal of waste management is to find solutions for proper handling of already-generated waste so that environmental impact is minimized. In other words source reduction eliminates the environmental impact that would have been caused in the handling of waste once generated. While source reduction is also applicable to reduction of hazardous industrial waste (which is the goal of pollution prevention), this chapter focuses mainly on municipal solid waste (MSW). Source reduction primarily includes reuse activities, but it can also include activities that reduce waste generation further up the chain (e.g., packaging). For instance, a consumer can reuse plastic grocery bags for lining small waste cans. On the other hand, use of a canvas bag by a shopper to carry groceries eliminates the need for a new plastic grocery bag entering the waste stream. Technically both are reuse, but the second choice is far better than the first reuse choice. The second choice is an example of source reduction because it eliminates the grocery bags (plastic or paper). Source reduction aims at reduction of generation of waste rather than reuse, for instance, the elimination of a plastic wrapper on a juice box.

3.1 ROLE OF GOVERNMENTAL BODIES

As mentioned, source reduction must involve government, industry, and consumers. The role of national, state, and local governments are different in source reduction. The role of national governmental agencies responsible for environmental protection is to develop source reduction policies and goals for both industries and consumers. Such an agency works with state agencies to implement the policies at a state level. It also works with other national agencies to implement the policies and goals. Research and planning on waste reduction activities and ideas are also generally carried out by the national agency. In some cases comprehensive laws for the entire nation may be developed by the federal government.

The role of state environmental protection agencies is to develop laws, rules, and guidelines for implementing the waste reduction goals developed by the national agency. State government also helps develop the framework for implementing source reduction goals at local levels. The laws developed by the state should address waste reduction for all generators, namely the residential, commercial, institutional, and industrial generators. The state also helps foster regional planning by local governments and provides support for educating the citizens.

The role of local government is to implement laws and rules promulgated by different levels of government. The local governments may sometime take additional initiative to go above and beyond the national and state mandates.

3.2 PLANNING

Planning is key to developing proper source reduction programs. Appropriate organizational structure and funding is necessary for implementing such programs.

Specific policy statements including definitions of terms must be developed first for any effective program. In addition, both the short-term and long-term goals and assessment criteria need to be clearly stated in the planning document. A baseline year and a target year must be identified when setting goals. There are several methods by which the achievements can be judged. There are direct methodologies such as a waste composition study and sampling and indirect methodologies such as material flow analysis. It is a good idea to opt for detail waste sort within the first or second year of a program and at every 5–6 years thereafter until the set reduction goal is reached. A random sampling may be undertaken at selected location every 1–2 years. Material flow method, though involved at the beginning, is somewhat easy after 1 or 2 rounds. In addition to representatives from the industrial sector, representatives from the other three major sectors—residential, commercial (e.g., restaurants), and institutional (e.g., hospitals)—should be consulted for developing source reduction goals. Specific separate goals should be developed for each item (e.g., glass). These itemized goals can be different for different sectors.

Proper administrative structures and funding are important aspects of planning. Proper estimates of personnel requirements based on scope of work, job descriptions, and required knowledge must be done before implementing a program. A well-organized program includes both field staff and administrator(s). The funding for source reduction programs for governmental bodies needs to be budgeted quite early in the process. While local governments can expect a broad-based direct long-term payoff from source reduction initiatives, national and state governmental bodies cannot expect such direct payoff except for the in-house initiatives (e.g., double-sided printing and copying leading to savings in paper cost). If source reduction is compared directly with the cost savings from management of the "not generated waste," then a long-term tangible benefit becomes obvious. However, in many instances this tangible benefit may not be obvious. A population not knowledgeable about environmental issues does not appreciate the benefits of source reduction. Many consider all these waste recycling efforts as needless and think that reduction is optional. So, there needs to be a well-organized effort in informing and educating citizenry regarding the importance of source reduction.

Source reduction is closer to resource management than waste management. A planned approach involving the producers, distributors, and consumers needs to be undertaken jointly by national, state, and local governmental bodies. If necessary, laws are promulgated to achieve the target. Local governments may issue ordinances to implement source reduction.

3.3 SOURCE REDUCTION IDEAS

There are numerous examples and ideas of source reduction (Fishbein, 1993). Many countries have laws regarding source reduction. These laws are aimed mainly at reducing packaging, which is a big source of waste (Fishbein, 1994). A ban on a waste item may lead to change in packaging. Appliances and automobiles, if manufactured for a longer life and maintained properly, will remain serviceable for a longer time. According to one estimate, extending the service life of all household appliances in the United States by one third will slow the discard rate by about 25%. The total number of appliances thus saved will add up to 12 million (Stein, 1997). There are many publications that provide ideas and strategies for source reduction (Goldbeck and Goldbeck, 1995; Durning, 1992; Young, 1991).

3.4 REDUCING TOXICITY

The toxicity of municipal solid waste (MSW) can be reduced by reducing the volume of hazardous chemicals in garbage. There are many chemicals that can be found in household products. A partial list of chemicals usually found in MSW landfill leachate and their potential contributors can be found in Table 15.6. The toxic constituents in MSW include many heavy metals (e.g., lead), organic chemicals (e.g., methylene chloride), pesticides, and so on. A study by the U.S. Environmental Protection Agency (EPA) found over 100 hazardous substances in household products (USEPA, 1980b).

Toxicity of MSW can be reduced by two different methods: direct collection of hazardous waste from households by local waste management authority (also known as clean sweep program) and changing chemical constituents of household products. Pollution prevention during manufacturing processes also helps in reducing toxicity of industrial waste.

3.4.1 Clean Sweep Program

The purpose of clean sweep programs is to collect household hazardous waste (HHW) generated from hazardous chemicals used in household products (e.g., oven cleaner) and small quantities of hazardous agricultural waste. A clean sweep program reduces the HHW disposed in landfills. Such reduction in HHW disposal helps reduce the toxicity of the landfill leachate, whereby leachate treatment cost can be reduced. So, the program should be viewed as complimentary to landfill operation. Citizens are encouraged to bring HHW to a collection site. There are several issues connected with a clean sweep program: organizing and managing the collection site, arranging proper management (management may include either or all of the following: recycling, detoxification, or incineration) of the collected waste, and funding. The program must be well coordinated and well publicized. Usually these sites are run for households and small businesses.

To run a clean sweep program one or more collection site(s), convenient to residents, must be chosen. The site should be big enough so that it can accommodate the estimated maximum number of visitors at any one time. Since hazardous waste will be collected and stored at the site, all regulatory requirements for these two activities must be satisfied. Depending on the volume of hazardous waste collected and the length of time the waste is expected to be stored on site, the site may need a hazardous waste storage license. Necessary arrangements must be made to transport the hazardous waste to a licensed facility capable of managing the waste. A contractor may also be hired for proper management of the waste. The list of items to be collected depends on the capabilities of the facility(ies) or the contractor responsible for managing the waste.

The location and hours the site remains open must be well publicized. Brochures, posters, press releases, and a Web site are good avenues for publicity. In sparsely populated rural areas, mailing of brochures to farm units and agricultural businesses may need to be done. Maintaining a clean sweep site at the same location for a long time helps to increase users.

The personnel working at the site should be familiar with types of materials to be collected. They must be well trained in handling the accumulated hazardous waste and must follow necessary health and safety procedures both for themselves and the visitors. Records regarding visitor's addresses and type of materials brought to the site prove to be helpful in the long run. This type of information is also helpful in assessing the program. Comments and suggestions from visitors should be collected either on a regular basis or through surveys. The clean sweep site may also be used for product exchange.

The hazardous wastes and materials collected at a clean sweep program site must be managed properly. The cost of management depends on the existence and proximity of facilities capable of managing the waste. The need for management of collected waste can be minimized by product exchange. If arrangements for managing a particular waste or material cannot be made, then that item should not be collected. The purpose of clean sweep is to avoid disposal of HHW in landfills. If the collected items are landfilled, then the purpose of clean sweep program will be defeated.

Funding for running a clean sweep program is needed for publicity, administrative and personnel costs, and management of collected waste and materials. Proper estimates and funding must be budgeted for running the program. Reduction of toxicity of the waste disposed in a landfill also reduces toxicity of the leachate. It is extremely difficult to provide proper treatment to a leachate with high concentrations of toxic compounds. Pretreatment of MSW leachate may become necessary if the concentration of toxic compounds increases beyond the capabilities of the wastewater treatment plant where the leachate is usually treated. So running a clean sweep program can be justified by directly relating to the increased cost of leachate treatment. Because of the reduction in leachate treatment cost, it is in the best interest of the landfill owner to provide funding for running a clean sweep program in the community.

3.4.2   Changing Chemical Constituents of Products

Although there are few success stories about the change of chemical constituents of products due to community pressure, involvement of national or state government is needed for an effective program. Government may impose bans or restrictions on the production, sale, use, or disposal of such products. For instance, use of mercury in dry cell batteries has been banned by quite a few state governments in the United States. This is a ban on use of a material, which reduces the toxicity of the product. A land disposal ban on polystyrene foam made with chlorofluorocarbons (a chemical that depletes ozone) eventually will lead to a change in manufacturing processes. This type of disposal ban can be adopted by local government but is likely to face opposition from businesses and trade groups. Because of growing environmental awareness of consumers, industries and retailers are also opting for product substitution. For instance, Gillette Corporation no longer uses trichloroethylene in the liquid typewriter correction fluid, and Dow Chemicals eliminated the use of chloroethylene in spot lifters. Product labeling programs in several countries have led to certifying organizations, which identify "Ecofriendly" products based on recyclability, toxicity, and so on (Fishbein et al., 1994).

3.4.3 Pollution Prevention

Pollution prevention programs not only reduce the toxicity of industrial waste they also aim at designing products that have no toxic materials, last longer, are easily repairable, and can be recycled. For instance, elimination of metal-based ink used for printing is an example of pollution prevention. Computers and automobiles can be designed so that they can be bought back and recycled (U.S. Congress, 1992). If designed and managed properly, pollution prevention programs can reduce both production cost as well as toxicity of the waste.

4

MEDICAL WASTE MANAGEMENT

Medical waste refers to all waste generated in hospitals and clinics (for both humans and animals), nursing homes, and laboratories that use microorganisms. All these institutions are considered "medical institutions" in this book. All wastes generated in medical institutions are not infectious. The various types of waste generated in these and similar institutions are pathological waste (e.g., body tissues and parts removed during surgery), radiological waste (e.g., waste containing low-level radionuclides), sharps (e.g., needles, scalpels, pipettes, glass slides), chemical waste, chemotherapy waste (chemical wastes generated from drugs used for the traetment of cancer), laboratory waste (e.g., wastes containing microorganisms), items soaked or contaminated with blood and body fluids, infectious municipal solid waste from isolation rooms or wards, and noninfectious municipal solid waste. These waste types fall into the following four categories: noninfectious and nonhazardous solid waste (considered municipal solid waste), infectious waste, hazardous waste, and low-level radioactive waste. Infectious waste may be generated during home-based health care. Source separation of these various categories of wastes greatly reduces management cost as well as risk to hospital workers. The treatment methods and disposal of these various waste types are different. With proper knowledge regarding transmission of disease, the volume of actual infectious waste can be identified and thereby the volume of actual infectious waste can be reduced. In addition, all types of waste generation within these institutions can be minimized with proper attention, which in turn can reduce cost of managing the waste. Note that discussions in this chapter do not concentrate on management of medical waste within the medical institutions. The emphasis is rather on treatment and disposal of medical waste. Discussions regarding in-house policies and procedures for managing medical waste can be found elsewhere (Reinhardt and Gordon, 1991; Garvin, 1994).

4.1   DEFINITION OF INFECTIOUS WASTE

Various terms such as biological waste, biohazardous waste, hospital waste, and so on are used to refer to infectious waste. Infectious waste may be defined as solid waste, which contains pathogens with sufficient virulence and in sufficient quantity that exposure of a susceptible human or animal to the solid waste could cause the human or animal to contract an infectious disease. Based on our definition, infectious waste may include (but is not limited to) waste soaked in blood and/or body fluid, microbiological waste, bulk blood and body fluids, contaminated or used sharps, pathology waste, contaminated animal carcasses, body parts and bedding exposed to infectious waste, contaminated surgery or autopsy waste, laboratory waste with microorganism(s), and dialysis waste. Items generated from patient care rooms may be infectious. If the items are grossly soiled with blood, body fluids, or excretion, then they should be treated as infectious waste. However, partially soiled items may be infectious in some situations. When in doubt, guidelines issued by infection control professionals should be followed. Plastic tubes, except the front portion containing a needle, used for administering intravenous medications may not be considered as infectious waste. Infectious waste may also be generated while providing first aid in schools, factories, offices, and so on. A blood-soaked towel generated during managing an accident should be treated as infectious waste. However, a blood-tinted adhesive strip from a minor cut or bruise need not be considered as infectious waste unless the patient is known to be a carrier of a major blood-borne pathogen (e.g., hepatitis B). While medical caregivers need to be cautious in all situations, the blood-tinted materials left untouched for a long time do not pose a health risk for waste handlers. However, since microorganisms grow very fast in blood and body fluids, facilities that generate a significant quantity of such materials (e.g., clinics where wounds are treated) should manage these wastes as infectious waste. When in doubt, the opinions of health experts or infection control practitioners and regulators should be sought. While utmost care should be exercised in identifying and managing infectious waste, unnecessary caution only increases the infectious waste volume and thereby the cost of management. Infectious waste commingled with noninfectious waste should be managed as infectious waste. All used and unused discarded sharps should also be managed as infectious waste.

4.2   SOURCE SEPARATION AND MANAGEMENT

Source separation helps in efficient management of various waste types mentioned before. Usually bags with a particular color are used to collect infectious waste. In the United States the color of these bags is red. Both noninfectious and infectious waste may be generated in patient rooms, examination rooms, and so on. In such room/areas, the infectious and noninfectious waste should be separated by keeping two different types of containers—one for the infectious waste and one for the noninfectious municipal solid waste. Bags or containers of the designated color are used for storing infectious waste. Health-care workers need to be trained in proper use of these containers. Figure 4.1 shows the various waste types generated in medical institutions and how they are managed. The following sections discuss management of each waste type separately.

4.2.1 MSW Type of Waste

Both recyclable and nonrecyclable wastes are generated in medical institutions. These wastes can be managed as municipal solid waste. Ideas on minimizing municipal solid waste are discussed in Chapters 3 and 7. Ideas on minimizing waste in medical institutions are discussed in Section 4.8.

4.2.2 Infectious Waste

If properly isolated, infectious waste constitutes a small percentage (10–15%) of total waste generated in medical institutions. There are primarily two types of waste that constitute this stream: sharps and waste contaminated with microorganisms and pathogens. Although pathological waste is also infectious waste by definition, it should not be commingled with infectious waste because storage and treatment of pathological waste is different from infectious waste. While several treatment options are available for infectious wastes, pathological wastes are incinerated. The focus of this chapter is the management of infectious waste, which is discussed in subsequent sections.

4.2.3 Chemotherapy Waste

The drugs used for chemotherapy are known as antineoplastic drugs. Several of the antineoplastic drugs are listed as toxic waste by the USEPA (Table 4.1). Most chemotherapy drugs are considered mutagenic, teratogenic, and/or carcinogenic to both human and animals (Reinhardt and Gordon, 1991). Chemotherapy waste, generated during patient treatment, includes tubing and bottles used for intravenous drug administration, gloves, aprons, empty vials, and coverings used in biological cabinets. These wastes are chemically contaminated and hence are not handled as municipal solid waste. Sharps used for administering the chemotherapy drugs are considered as infectious waste.

The other source of chemotherapy waste is unused portions of drugs, drugs with expired use life, and surplus mixtures. In major medical institutions, handling and disposal of these drugs is a significant management issue (Vaccari et al., 1984).

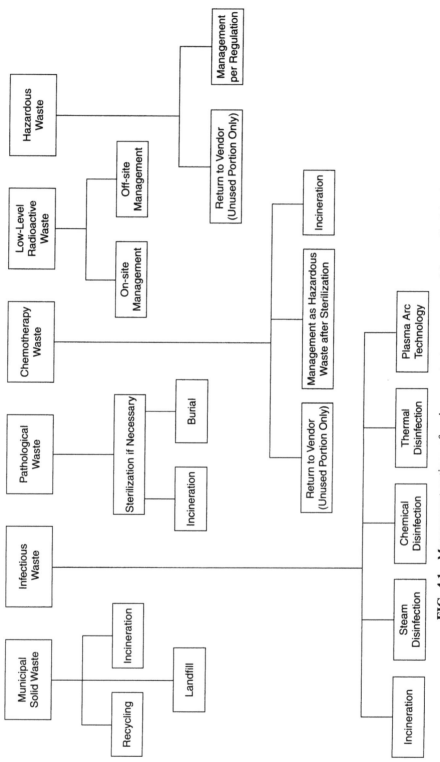

FIG. 4.1 Management options of various waste types generated in medical institutions.

48

TABLE 4.1   Antineoplastic Drugs Listed by USEPA as Toxic Wastes[a]

Drug	Hazardous Waste Number
Mitomycin C	U010
Chlorambucil	U035
Cyclophosphamide	U058
Daunomycin	U059
Malphalan	U150
Streptozotocin	U206
Uracil mustard	U237

[a]This pertains only to discarded commercial chemical products and spills resulting from these materials.

Reprinted with permission from Reinhardt and Gordon (1991). Copyright Lewis Publishers, an Imprint of CRC Press, Boca Raton, Florida.

The unused drugs may be sent back to the vendor if the buyer negotiates a buy-back policy with the vendor. If the needle is detached from the tubing, then the needle can be treated as infectious waste and the rest of the tubing and bottles may be treated as chemically contaminated waste and/or hazardous waste as the case may be. Depending on the amount of chemical remaining in the tubing and bottle, hazardous waste regulators may allow disposal of this waste as nonhazardous waste. Hazardous waste incinerators may be used for treating noninfected chemotherapy waste.

4.2.4   Low-Level Radioactive Waste

Low-level radioactive waste is generated in medical institutions when providing health care and doing research. Although most of the radionuclides used in medical institutions have half-lives of 90 days or less, a few of them have significantly long half-lives (Evdokimoff, 1987; Vetter, 1987).

The waste with short half lives, generated mostly from diagnosis and therapy, may be stored in a secured place until the radioactivity reduces to background level. After this period it may be disposed with solid waste if permitted by the regulatory authority. The volume of the waste can be reduced through evaporation under restricted conditions. Necessary permit for storing on-site should be obtained. Burial of these waste in specially designed vaults or cells may be permitted in some instances. Management of low-level radioactive waste containing radionuclides with long half-lives should be discussed with the regulatory agency for proper disposal.

4.2.5   Hazardous Waste

Although hazardous waste is generated in many medical institutions, it is generated mainly in medical research laboratories. Many unused hazardous waste products may be returned to vendors provided the contract allows for buy-back. In most instances special permits are needed for storage, transportation, and treatment of hazardous waste. A list of hazardous waste in the facility should be developed first. The issue of managing the identified hazardous wastes should be discussed with the regulatory agency. Based on volume and/or characteristics or both, parts of the waste may be exempted from regulations.

4.3 STORAGE

Regulatory mandates dictate storage requirements of infectious, hazardous, and radioactive wastes. Regulations may also specify the length of time these wastes can be stored on-site. There may be exemptions from regulatory requirements based on volume. These issues should be discussed with the regulatory agency before developing a management plan. Chemotherapy wastes that are identified as hazardous waste should be considered as hazardous waste for storage purposes also. Usually, tubing and other similar items attached to sharps, which are essentially chemically contaminated infectious wastes, can be considered as infectious waste for storage purposes, unless dictated otherwise by the regulatory agency. Usually there are no specific regulatory requirements regarding storage of municipal solid waste.

In addition to the storage of hazardous and radioactive waste, storage of infectious waste in medical institutions also requires special attention. Putrescible infectious waste should be sent for treatment as soon as possible for both aesthetic and health reasons. Microorganisms grow rapidly in this type of waste, which leads to odor problems and increases health risk during subsequent handling. There may be regulations regarding a time limit for storing putrescible infectious waste at room temperature. All containers of putrescible infectious waste should be closed tightly or sealed. These waste should be stored in a separate room or at least in a separate area with refrigeration. Access to these wastes should be restricted to designated personnel only. The floor of the storage area should be impermeable to liquid. It is recommended that wooden floors or carpeting is not used in the storage area. The waste must be labeled properly, indicating date and time of generation. If the infectious waste is stored for a length of time, then it should be refrigerated below 42°F (5.5°C).

Nonputrescible infectious wastes are separated into two streams—sharps and nonsharps. All nonsharp infectious waste is stored in separate bags or containers. All infectious waste should be stored in reusable containers whenever possible. The bags or containers should have sufficient impact and tear resistance. Impact and tear resistance can be tested by American Society for Testing and Materials (ASTM) publications D1709-91 and D1922-89, respectively (Rutala and Sarubbi, 1983). Liquid infectious wastes are best contained in rigid containers (e.g., bottles) with a secured cap. The bottles are placed in another secondary container (e.g., box) to avoid spillage during movement within the facility and while transporting to a treatment facility. The containers for liquid wastes should be compatible with the treatment process. A metallic container is best for steam sterilization whereas a non-metallic container is best for incineration.

Sharps should be collected directly into reusable or recyclable containers to minimize handling. Sharps should be stored in special puncture-resistant containers for protecting waste handlers. Specially designed sharps containers should be used in medical institutions.

4.4 TRANSPORTATION

All types of wastes generated within a medical institution needs to be transported to the storage area within the building and subsequently to an on-site or outside treatment facility. Both issues must be addressed properly to minimize risk to waste handlers and the environment. Usually carts are used to move these wastes from the point of generation to the storage area. These carts must be cleaned and disinfected regularly. It is preferable that only dedicated carts be used for moving waste within the building. Waste should be moved within the building during less busy hours and through routes with the least possible exposure risk. If gravity chutes or pneumatic tubes are used for waste transport, then proper care should be taken to ensure that the containers remain intact at the collection point. It is better to avoid compaction of infectious waste. However, if a compactor is used, it must be operated under negative pressure to retain aerosols and dust within the compactor.

Specially marked vehicles should be used for the transportation of infectious waste from medical institutions. There may be special license requirements for these transportation vehicles. Regulations regarding transportation of hazardous and low-level radioactive wastes are usually different from infectious waste. The regulatory agency needs to be contacted to find out about transportation regulations for these types of wastes. Vehicles transporting infectious waste must be cleaned periodically. All vehicles transporting infectious waste should carry a written contingency plan for spills and accidents and should carry necessary tools for cleanup. The driver(s) of the vehicle should be trained to handle all spill situations. Infectious waste transported from an individual's home should be kept in a secured location (e.g., car trunk). All vehicles used for infectious waste transportation should be maintained properly to avoid mechanical failure on the road. Transportation of infectious waste from medical institutions to treatment facilities should be manifested to ensure proper documentation of disposal. The waste manager of the medical institution (generator) should keep one copy of the manifest. The driver should carry at least two copies, for handing over to the treatment facility manager at the time of delivery. The treatment facility manager must verify and sign off on the third copy, which should be returned to the generator. If the transportation is done by an waste handling company, then a fourth copy of the manifest should be kept by the handling company for its records. Instead of a manifest a log book signed by all parties involved in dispatching and receiving of waste may also be used to document proper disposal and/or treatment.

4.5 TREATMENT

The goal of infections waste treatment is to render it disinfected not sterilized. Sterilization is commonly defined as the complete elimination or destruction of all forms of microbial life, including highly resistant bacterial endospores. Disinfection refers to the reduction of microbial population to a level such that it will not spread disease. There are many technologies available for the treatment of infectious waste, which can be grouped into the following five categories:

1. Incineration
2. Steam disinfection
3. Chemical disinfection
4. Thermal disinfection
5. Plasma arc technology

Each of the technologies are discussed in the following sections. While gas (ethylene oxide and formaldehyde) can be used for disinfection, there is significant health risk associated with its use. Both ethylene and formaldehyde are suspected human carcinogens (USDL, 1984, 1987). Because of this problem, currently gas disinfection as a treatment technology is seldom used. It is important to establish the efficacy of a treatment process prior to its regular use.

4.5.1 Incineration

Information regarding infectious waste incinerator design, ash disposal, and air emission issues can be found in Section 5.5. Medical waste incinerator effectiveness for the destruction of infectious microorganisms are tested. Several studies have indicated that microorganisms may remain in the ash or can be found in air emissions (Barbeito and Shapiro, 1977; Barbeito et al., 1968; Peterson and Stutzenberger, 1969). These problems can be avoided by using improved technology. However, efficacy testing should be done for all newly installed incinerators.

Ash containing a high amount of uncombusted waste indicates poor burnout. Causes of poor burnout include excessive waste feed rate, lack of heat generation due to particle size of waste, and insufficient burning time. The completeness of burning can in most situations be detected by visual observation of the ash. Both design and operation of an incinerator are important to ensure proper burning of waste. The efficacy of an incinerator for the destruction of infections agents is verified by using *Bacillus subtillis* var. *niger* spores. Simply explained, spores are like peanuts inside the shell. Vegetative bacteria is like the peanut without the shell. Spores are more difficult to destroy. The biological indicators are added to the waste and monitored by testing the ash and the stack emission. If there is a discharge of wastewater, then the effluent should also be monitored. The efficacy is verified by testing many samples over a period of time. Once the efficacy is established, then a routine monitoring is done to ensure proper destruction of the bioload.

4.5.2 Steam Disinfection

Disinfection using steam is done using autoclaves. In an autoclave, saturated steam (e.g., steam with very small water droplets, usually less than 3%) is used to disinfect the infectious waste. If saturated steam is further heated, then the steam becomes superheated. Figure 4.2 shows a cross section of a typical autocalve. If the outer jacket is not included, it is called a retort. Condensation forms without a preheating outer jacket. Such condensation, although not desirable for reusable medical instruments, is not an issue for treating infectious waste. Usually, pressurized steam [about 30 psi (0.01 kPa)] is used in retort for increasing efficiency. Typically, infectious waste containers are placed inside the chamber and treated for sufficient time (up to 90 min) with steam temperature of 250–270°F (121–132°C).

Several studies have indicated that type of container, moisture content, amount, and density of the waste influences the steam penetration into the waste. Destruction of pathogens depends on thorough penetration of steam into the waste (Rutala et al., 1982; Lauber et al., 1982). Steam temperature and residence time of waste are established by studying the kill efficiency of *Bacillus stearothermophilus*. Detail discussion regarding efficacy testing can be found in Section 4.5.6.

Compared to incinerators, operation and testing of autoclaves are less complex. In addition, environmental releases from incinerators contain a broader range of constituents (e.g., dioxin) than autoclaves (U.S. Congress, 1988). Autoclaves are cheaper and require less space. Although the bags shrink and look curly, they remain intact. Landfill operators may consider these as untreated waste and deny disposal. In addition, the limited capacity of autoclaves for treating waste and longer residence time make them a less common waste treatment method. A survey of commercial infectious waste treatment facilities in the United States indicated that the ratio of incinerator to autoclave used for infectious waste treatment is more than 2 to 1 (Malloy, 1995a). However, the ratio may be different in different countries and may change as technology develops.

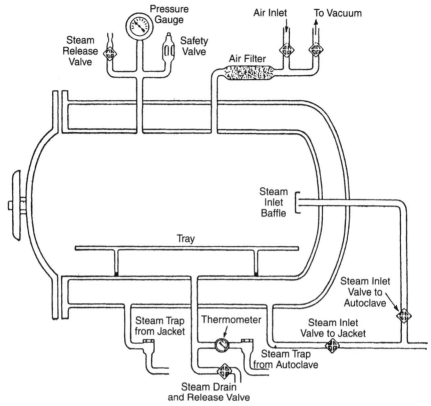

FIG. 4.2   Autoclave shematic. Reprinted with permission from Reinhardt and Gordon (1991). Copyright Lewis Publishers, an Imprint of CRC Press, Boca Raton, Florida.

4.5.3   Chemical Disinfection

Chemical disinfection is normally used to treat liquid infectious wastes that are uniform in characteristics. Even if the composition is not uniform, tests can be done easily to find the effectiveness of the process. Chemical disinfection is also used for treating sharps, pathological wastes, and infectious waste in bags.

Usually chlorine compounds, iodine compounds, alcohol, glutaraldehyde, and so on are used as the reagent (USEPA, 1982; Scott and Gorman, 1983). Quality of disinfectant, contact time, pH, and temperature are important parameters requiring adjustment for effective treatment. In addition, mixing of the chemical with the waste, especially for large volume of liquid and shredded solid waste, must be thorough.

For liquid waste, effectiveness of the process can be tested by measuring the reduction of most resistant microorganisms expected to be present in the waste. For solid infectious waste *Bacillus subtilis* spores may be used for assessing effectiveness of the method (STA, 1994). Toxicity of the disinfectant dictates whether the posttreatment waste is hazardous. If the chemical used for treatment is hazardous, then the spent chemical may need to be treated as hazardous waste as per regulatory mandate.

4.5.4 Thermal Disinfection

The process uses either electric heating or dielectric heating for raising the temperature of the waste. Currently, use of dielectric heating is gaining wider acceptance. Dielectric heating is achieved by agitating polar molecules such as water. Dipolar molecules have asymmetric electronic structure that align themselves with an imposed electric field. If the polarity of the electrodes are reversed, the molecules rotate to realign with the new polarity (Fig. 4.3). If the reversal of the applied field is changed rapidly, the molecules agitate at a high frequency giving rise to heat. Since all the molecules are agitated simultaneously, the heat is produced throughout the material. Thus dielectric heating is more efficient and quicker compared to conventional heating, which starts at the surface.

To increase efficiency of the system, the waste is shredded before dielectric heating. The shredded waste is fed into a screw conveyor where the waste is heated to about 200–220°F (93–104°C). Dielectric heating can be achieved by using either radio frequency (below 300 MHz) or microwave frequency (300–30,000 MHz). The energy absorbed by a given material is directly proportional to its dielectric properties. Infections waste is typically a mixture of materials with low dielectric constant (3–6) such as paper and plastics and high dielectric constant (80 or more) such as organic material. Because of its high dielectric constant, organics absorb the majority of the energy and get heated quickly. The heat generated raises the temperature more than 90°C, high enough to destroy the microorganisms. Since the microorganisms absorb energy at a high rate, the liquid molecule inside the membrane gets heated, causing the cells to die. In addition, the high voltage used for the process weakens the cellular membrane, causing it to rupture (known as cell lysis). Thus in dielectric heating, destruction of microorganisms is achieved in two ways by high heat and cell lysis.

Figure 4.4 shows a typical flow diagram for a high-volume [1000 lb (455 kg) per hour or more] infectious waste processing unit. Figure 4.5 shows a

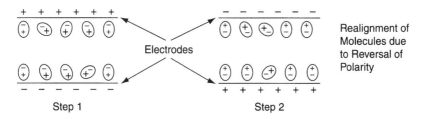

FIG. 4.3 Rotation of polar molecules due to change in electric field.

FIG. 4.4   Flow diagram for a typical high-volume waste processing facility using dielectric heating.

typical cross section for a low-volume [500–600 lb (227–272 kg) per hour] infectious waste processing unit.

Since the waste is shredded prior to treatment, the design of the units should be such that they do not release microorganisms into the surroundings. This is necessary to ensure the safety of operators and the environment. To prevent this type of release, the shredder housings are equipped with high-efficiency particulate air (HEPA) filters. HEPA filters can arrest up to 0.3-μm particle size with a 99.97% efficiency. The size of microorganisms normally found in infectious waste is bigger than 0.3 μm. Hence, release of microorganisms is not expected from shredder housings equipped with HEPA filters. The HEPA filters should be tested before installation. These filters must also be monitored regularly to ensure efficiency. Usually, a HEPA filter housing comes with a gauge indicating whether it is operational. In addition to the microorganisms, volatile organic carbon (VOC), associated with various chemicals used in medical institutions, may also be released from the unit. So VOC concentration should also be monitored near the dielectric treatment units. A detail discussion on efficacy testing and monitoring is included in Section 4.5.6.

4.5.5   Plasma Arc Technology

Plasma arc generates a temperature of 6000–12,000°F (3315–6630°C) or more. Plasma, commonly described as a fourth state of matter, is essentially a gaseous state capable of conducting electricity. Plasma torches operate in

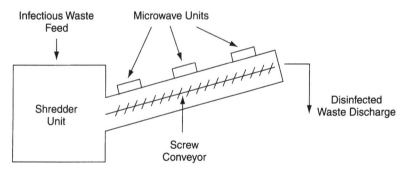

FIG. 4.5   Cross section of typical low-volume infectious waste processing facility using dielectric heating.

the 100-kW to 10 MW-power range. Plasma systems do not need a high volume of air, which reduces effluent gas volume (Malloy, 1995b). Plasma arc technology can be used to build small units as well as large units [up to 2000 lb (900 kg) per hour]. The ash gets vitrified to a inert glasslike material. Regulations may require that the plasma units be treated like incinerators, in which case, necessary testing of stack emissions and ash must be done. This being a heat treatment unit, microbiological testing similar to incinerators should be undertaken.

4.5.6 Efficacy Testing

Efficacy testing involves establishing the ability of the method to destroy microorganisms present in the infectious waste. Terms such as decontamination, sterilization, disinfection, and so on are used to indicate level of treatment. However, these terms are nondescriptive and do not reflect a quantitative level of microbial destruction. Although most infectious waste treatment technology uses one of the methods described in the previous sections, a combination of two methods may also be used for destruction of microorganisms (e.g., chemical/thermal). The total process by which the infectious waste is treated will influence the validation process. The difference between sterilization and disinfection is as follows:

1. Sterilization indicates complete elimination of all forms of microbial life including resistant microbial spores. Since complete elimination of microbial life is difficult to prove, sterilization of infectious waste refers to a 6 $\log_{10}$ reduction or a one-millionth (0.000001) survival probability (i.e., 99.9999% reduction) of the most resistant microorganisms relative to the process.
2. Disinfection of infectious waste refers to 4 $\log_{10}$ reduction (i.e., 99.99% reduction) of most resistant microorganisms relative to the process.

Table 4.2 provides a list of microbial indicators that can serve as surrogates of the group. Table 4.3 provides a list of levels of microbial inactivation, which should be used for complete efficacy testing of a method. To validate a method for treating infectious waste, all levels should be tested to assess the capability of the method. Although traditionally *Bacillus subtilis* var. *niger* spores are used to determine resistance to dry heat and *Bacillus stearothermophilus* spores are used to determine resistance to wet heat, either of the above spores may be used for demonstrating microbial inactivation in all types of infectious waste treatment processes (State and Territorial Association, 1994). Opinions of microbiologists and epidemiologists should be sought for assessing the efficacy of an infectious waste treatment process. Regulatory mandate, if any, regarding chemical parameters, level of inactivation, and type of surrogate microorganism must be followed for addressing efficacy of a treatment system.

TABLE 4.2   Recommended Biological Indicators

Vegetative bacteria
    Staphylococcus aureus (ATCC 6538)
    Pseudomonas aeruginosa (ATCC 15442)
Fungi
    Candida albicans (ATCC 18804)
    Penicillium chrysogenum (ATCC 24791)
    Aspergillus niger
Viruses
    Polio 2, Polio 3
    MS-2 Bacteriophage (ATCC 15597-B1)
Parasites
    Cryptosporidium spp. oocysts
    Giardia spp. cysts
Mycobacteria
    Mycobacterium terrae
    Mycobacterium phlei
    Mycobacterium bovis (BCG) (ATCC 35743)

After State and Territorial Association (1994).

4.6   TRACKING

Because improper treatment and disposal of infectious waste may spread diseases, it is essential to keep track of the generated waste until final destination (either disposal or recycling facility). A manifest or tracking form or log book should be used to document the flow of the waste from the point of generation to the final destination via the treatment facility. Table 4.4 shows the essential information that should be included for tracking. It is good practice to use tracking, even if the treatment facility is in the same premises as the medical institution where the waste was generated. If mandated by regulations, a copy of the completed manifest must be sent to the regulatory agency. The treat-

TABLE 4.3   Levels of Microbial Inactivation

Level I	Inactivation of vegetative bacteria, fungi, and lipophilic viruses at a 6 $\log_{10}$ reduction or greater
Level II	Inactivation of vegetative bacteria, fungi, lipophilic/hydrophilic viruses, parasites, and mycobacteria at a 6 $\log_{10}$ reduction or greater
Level III	Inactivation of vegetative bacteria, fungi, lipophilic/hydrophilic viruses, parasites, and mycobacteria at a 6 $\log_{10}$ reduction or greater; and inactivation of *B. stearothermophilus* spores or *B. subtilis* spores at a 4 $\log_{10}$ reduction or greater
Level IV	Inactivation of vegetative bacteria, fungi, lipophilic/hydrophilic viruses, parasites, mycobacteria, and *B. stearothermophilus* spores at a 6 $\log_{10}$ reduction or greater

After State and Territorial Association (1994).

TABLE 4.4   Essential Information for Infectious Waste Tracking

1. Generator:
   a. Name and Address
   b. Contact Phone No.
2. Type of Generator:
   a. Hospital
   b. Clinic
   c. Nursing Home
   d. Other
3. Weight of infectious waste (should indicate whether measured/estimated)
4. Number and type of containers with identification number (if any)
5. Name, signature, and date of authorized agent with a statement indicating accuracy of information
6. Transporter:
   a. Name and Address.
   b. Contact Phone No.
   c. Vehicle License No.
   d. Name and dated signature of driver indicating receipt of items described by the generator
7. Treatment Facility:
   a. Name and Address.
   b. Contact Phone No.
   c. License no. issued by the regulatory agency (if any)
   d. Name and dated signature of agent indicating date of treatment completion received from the transporter
8. Disposal or recycling facility:
   a. Name and Address.
   b. Contact Phone No.
   c. License no. issued by the regulatory agency (if any)
   d. Name and dated signature of the agent indicating disposal of waste received from the treatment facility.
   e. License no. of the vehicle and the total tonnage received for disposal

ment facility may need to inform the regulatory agency periodically (e.g., every year) the total amount of various waste types (e.g., pathological waste) received and treated from various generators. If the infectious waste is taken to a transfer station before treatment, then the manifest should indicate that intermediate stop. In the absence of any regulatory mandate, tracking form should be reconciled periodically (e.g., each month) and retained for 2–3 years after the treatment date.

4.7 COLLECTION OF INFECTIOUS WASTE GENERATED FROM HOMES

Normally infectious waste is not generated in homes unless health care is provided at home. However, used needles (commonly referred to as sharps) are generated at home due to self-administration of insulin by diabetic patients and other drugs. If significant quantity of infectious waste is generated in homes, then arrangements needs to be made to transport it to clinics or hospitals for proper treatment. Regulations may allow discarding of small quantities of infectious waste along with household waste. Sharps generated in homes must not be disposed with regular garbage. Sharps, if disposed with regular garbage, may injure waste haulers. Risk of sharps related injury of waste haulers and recycling facility workers must be eliminated to protect their health. Liability from this type of injury can be very high if injured worker develop any communicable disease [e.g., acquired immune deficiency syndrome (AIDS)]. To protect all parties involved in infectious management, additional efforts must be made to collect sharps from individual homes.

There are many ways in which sharps can be collected from individuals for proper treatment and disposal. Arrangements can be made with the company that sells injection devices to take them back for a nominal fee or free of charge. In some communities, sharps collection stations are set up where individuals can drop off the needles free of charge. In some cities, pharmacies set up sharps collection station where individuals can drop off needles free of charge or for a nominal fee. Such a program of sharps collection stations is being run successfully in many places in Wisconsin (Derflinger, 2002).

Sharps generated in homes can be stored in puncture-resistant plastic bottles (e.g., thick-walled laundry detergent bottles) until it is full. The plastic bottles should be labeled to avoid disposal with household waste. A full bottle may then be handed over to the sharps collection station. Specially designed sharps collection containers may also be used to avoid double handling. The public's cooperation is an important element in running a successful program. Periodic education of generators should be undertaken to ensure proper collection and disposal of sharps generated in individual homes.

4.8 WASTE REDUCTION

If properly planned and implemented, waste reduction not only reduces the amount of waste generated in medical institutions it also saves money. It helps to identify wastage and misuse of products and equipment. The waste reduction plan should look at both short-term and long-term benefits. Waste minimization also reduces occupational and environmental risks.

4.8.1 Waste Audit

An audit of the current generation amount of various types of wastes in a medical institution should precede development of a waste reduction plan. A waste audit should list all areas where waste is generated and the type of waste generated in each area. The waste may be divided into the following categories, which will help both in identifying proper management issues and developing future reduction plans: infectious waste, hazardous waste, hazardous and nonhazardous chemotherapy waste, radioactive waste, solid waste, and recyclable waste materials. The audit should identify whether the present practice involves collection of two separate types of waste that, if separated, can be managed in a different way. For instance, if solid waste and infectious waste is mixed in patient examination rooms, arrangements can be made to collect each type separately. The disposal cost of solid waste is less than infectious waste. The audit should also identify the total quantity of waste type generated in an area, management practice (e.g., collection, storage, transportation, and treatment/disposal) for each waste type.

In the absence of any regulatory mandate, the following schedule may be used. To account for seasonal variation of waste types, the initial audit should be repeated 2–3 times within a 12-month period for obtaining a good estimate. The audit should be done at least once every year, after implementation of the waste reduction plan (discussed in the following section), for 5–6 years. Subsequently, the audit should be done every second or third year to ensure continuation of the waste reduction plan.

4.8.2   Waste Reduction Plan

The waste reduction plan should address reduction, reuse, and recycling of waste generated in a medical institution. Reduction can be accomplished in many ways such as reducing the amount of packaging, reducing disposable items, change in purchasing policy to reduce waste generation rate, change in housekeeping practices, and so on. Since in many instances infectious waste is mixed with solid waste because of wrong collection practices, a significant reduction in infectious waste amount can be achieved simply by collecting the wastes separately right at the point of generation. Reuse of supplies can be done by adopting proper practices for disinfection or sterilizing. Recycling of paper products, glass, plastics, wood, and so on can be done by collecting these items in a separate container and then sending these off for recycling. While developing a waste reduction plan, it is a good idea to consider reduction in power, water usage, and steps to reduce environmental impacts (e.g., reducing release of air contaminant by installing a better incinerator).

When developing a reduction plan, the following issues should be considered and addressed:

1. Compliance with regulatory mandates from different governmental agencies—federal, state, and local
2. Probable adverse effects on patient care and workers safety
3. Cost of implementation and savings/benefit from the proposed actions
4. Recycling rules and recycling options available for the identified recyclables

5. Availability of equipment or product needed for implementation of the proposed action and their cost

Note that in most instances waste reduction leads to net cash savings. However, there may be instances where capital investment is needed up front for upgrading an equipment or collection method. These investments pay off in the long run and also provide environmental benefits.

The plan should set a reduction target both in terms of quantity and time. The results of the initial audit will provide the baseline to judge the success or failure of the plan. A detailed discussion regarding available alternatives and how the proposed alternative was chosen should be included in the plan. This narrative will help in adjusting the plan in the future, if needed.

4.8.3   Education

The plan must identify training needs for staff as well as for patients and visitors. One of the biggest challenges is to educate patients and visitors regarding source separation of recyclables from waste (especially in the cafeteria) and avoiding mixing of infectious wastes with noninfectious wastes in patients' rooms. However, in general, the patients and visitors are not involved in disposing infectious wastes in containers.

The staff must be informed regarding the goals and objectives of the waste reduction plan and how to implement them. It is always helpful if the staff is provided with a summary of the overall plan. A detailed staff training should be undertaken initially and should be repeated once every year initially. New employees must be trained about the waste-reduction-related work ethics during their orientation.

The training plan should include the regulatory as well as environmental issues associated with the waste reduction plan. Employee inputs regarding practicality of various activities should be considered carefully to come up with a sustainable plan. One or more employees should be designated for proper development, implementation, and continuation (through monitoring/ audit) of the waste plan. While outside consultant(s) may be hired to carry out the audit, development of the plan including evaluation of alternatives and staff training should be delegated to one or more employees. A progress report should be developed annually showing whether the institution is meeting the objectives and goals set in the initial plan. If necessary, the plan should be revised to raise or lower a goal or change the objective(s) based on the experience gained in the past years.

4.8.4   Helpful Ideas

There are many ways in which waste reduction can be achieved. Ideas regarding packaging needs to be discussed with the vendor or supplier so that there is a source reduction in packaging materials. The following are just few ideas:

1. Replacing disposable items with reusable items is easy in many situations.
2. Launderable cloth items can be used in place of paper or plastic products.
3. Clean and reusable containers can be used for sharps and nonsharp infectious waste.
4. Unlined trash cans can be used in administrative areas.
5. Liquid cleaning products can be bought in big refillable type containers.
6. Disposable water pitchers, glasses, and bed pans can be replaced with reusable ones.
7. Suppliers can be asked to deliver supplies in reusable crates.
8. Suppliers of custom surgical packs can be asked to exclude items not used frequently.
9. Reusable liquid-proof surgical gowns can replace single-use gowns.
10. Sterilized bowls and instruments (whenever possible) should be used in operating rooms.
11. Outdated pharmaceuticals can be returned by negotiating a contract with the supplier. Timely delivery of medicinal admixtures can reduce the volume of waste drugs.
12. Drug inventory management can be improved by implementing centralized purchasing, leading to reduction of outdated pharmaceuticals.
13. In the cafeteria and in patient's rooms, washable plates, eating utensils, glasses, and cups can be used replacing disposable items. In some instances kitchen and food waste may be composted on site, which would reduce the disposal cost of solid waste.
14. Water use in a hospital can be reduced by adopting new technology and common sense. Dry developer for X-ray film can be used to reduce water used for wet developing. Sink and shower faucets can be fitted with water reducers leading to less water use.

The list is not comprehensive but provides some ideas regarding waste reduction. There is a significant scope for waste minimization without increasing cost, compromising quality of health care, or occupational safety. On the contrary, apart from the environmental benefits, properly administered waste reduction plans lead to net savings and reduced occupational risk. There are many organizations with Web sites that provide useful ideas and case histories regarding medical waste reduction. Addresses of a few of these Web sites are included in the reference list under Health Care (2002).

5

INCINERATION

In some communities the entire mass of municipal solid waste (MSW) is incinerated; in other communities only the waste fraction that cannot be recycled is incinerated. Incineration reduces the waste volume and helps generate heat for commercial use; in many instances the heat is further utilized to generate power. Although waste-to-energy (WTE) facilities are not popular in the United States, they are widely used in Europe and Japan. Japan incinerates 50%, and Switzerland and Sweden incinerate 75 and 60%, respectively (Kreith, 1994), of their MSW.

In the United States, the incineration of solid waste declined due to several reasons; stricter air emission standards combined with the rise in the cost of incineration forced some incinerators to shut down. Another issue that lead to the shutdown was the decline in waste flow. Each facility is designed for a minimum waste volume. However, due to lack of proper planning in many communities, the waste entering a facility was much less than the minimum required design volume. In the late 1980s, incinerator ash from many facilities was found to contain high concentrations of lead and cadmium (Sopcich and Bagchi, 1988; U.S. Congress, 1989). Thus special attention was required for disposal of incinerator ash, which added to the cost of incinerator operation. All these issues influenced public opinion regarding incineration of solid waste. However, in recent years, public opinion in the United States seems to be changing in favor of WTE facilities.

A successful integrated solid waste management (ISWM) program may include incineration. In addition to MSW, sludge-type wastes (e.g., wastewater treatment plant sludge) are also incinerated. Thus incineration is a management option that should be considered during ISWM planning for environmental reasons. In terms of greenhouse gasses, a WTE is better than a landfill.

In a modern WTE facility, because of complete combustion and savings in fossil fuel (due to energy recovery), a ton of MSW burnt will generate about one-twelfth of greenhouse gas compared to the disposal in landfills without methane recovery. The ratio lowers to one-sixth for landfills with energy recovery (Taylor, 1991).

5.1 INCINERABLE WASTE

For MSW, incinerability will depend on two main factors: moisture content and heating value. Both of these factors have seasonal variability, which must be considered while designing an incinerator—especially a WTE facility.

Moisture Content Additional fuel requirements increase with increase in moisture content. Usually the moisture content of MSW is 15–20%. However, the average moisture content depends on the composition of the waste. For a community with a successful recycling program, the percentage of low-moisture-content waste components (e.g., paper, plastics) as well as the percentage of high-moisture-content waste components (e.g., vegetable waste, yard waste) are expected to be low. For sludge-type waste the moisture content is normally about 40%. However, if the sludge is dewatered, the moisture content can be as low as 25%. So, rather than assuming a moisture content of the waste proposed to be incinerated, it is better to determine moisture content using proper reprensative samples. Waste with a high moisture content would require a high volume of additional fuel, which will increase the operational cost of the incinerator.

Heating Value Heating value refers to the heat generated due to burning a pound/kilogram of waste. For an economically viable WTE facility, the heating value of the waste should, in general, be around 1000 Btu/lb (0.55 kcal/kg). While the heating value of paper, wood, and so on is high (about 8000 Btu/lb or 4.4 kcal/kg), the heating value of vegetative waste is low (about 2000 Btu/lb or 1.1 kcal/kg). The overall heating value of MSW for a community will depend on the percentage of various waste components. Table 5.1 shows four major components of MSW in various U.S. cities. Table 5.2 shows five major components of MSW in various other countries. As shown in the Tables 5.1 and 5.2, the waste composition can be different for different communities. There is also a seasonal variation in waste components. The following seasonal variations have been reported in the literature: paper 35%, yard waste 98%, and food waste 27% (Niessen and Alsobrook, 1972). Variation of waste composition due to both location and season should be considered when designing a WTE facility for a community.

While the incinerability of a waste can be assessed from moisture content and heating value estimates, the presence of contaminants (e.g., household hazardous waste) may heavily influence the air emission and ash from the

TABLE 5.1 Percentages of Four Major Components of Municipal Solid Waste in Various U.S. Cities (1988 and 1990 Averages)

Waste Component	National Average (estimate)	Indianapolis, IN	Ann Arbor, MI	Portland, OR	San Diego, CA
Paper	34	38	29	29	26
Plastics	9	8	8	7	7
Yard Debris	20	13	8	11	21
Miscellaneous Organics	20	22	39	33	23

Reproduced by permission of the publisher, ASCE, from Gay et al. (1993).

facility. It is a good idea to undertake waste sort more than once prior to adopting incineration as part of an ISWM program. Waste sort will provide a better picture of the waste stream(s) proposed to be incinerated. The success of an WTE facility depends not only on efficient design of the incinerator but also on the pollution load both in air emission and in ash. While the presence of undesirable wastes is unavoidable, every effort should be made to minimize their percentages.

5.2 WASTE LOAD ESTIMATE

The quantity of waste destined for incineration must be estimated properly. Proper long-term planning guaranteeing a range of waste quantity is very critical for the design and operation of a WTE facility. Although waste load

TABLE 5.2 Percentages of Five Major Components of Municipal Solid Waste in Various Countries

Country	Paper	Metal	Glass	Food	Plastics
Australia	38	11	18	13	0.1
Bangladesh	2	1	9	40	1.0
Columbia	22	1	2	56	5.0
France	30	4	4	30	1.0
Hong Kong	32	2	10	9	11.0
India	3	1	8	36	1.0
Italy	31	7	3	36	1.0
Japan	21	6	4	50	6.0
New Zealand	28	6	7	48	0.1
Saudi Arabia	24	9	8	53	2.0
Sweden	50	7	8	15	8.0
United Kingdom	37	8	8	28	2.0
United States of America	29	9	10	18	3.0

After Niessen (1995).

estimate is important, it is not critical for design or operation of an incinerator without heat recovery. In ISWM, the goal is to reduce the volume and pollution load of the waste ultimately disposed in a landfill. So, the waste to be incinerated should be as free of unwanted materials (e.g., waste with low Btu and waste that will increase pollution load both in air emission and in ash) as possible. When estimating the waste load for an incinerator, the planner needs to take into consideration possible reduction in load due to source reduction (e.g., recycling) and the flow of waste. It is somewhat difficult to predict or even estimate the volume of waste reduction until the source reduction program is well established. Thus it is probably not prudent to opt for incineration at the initial phase of starting an ISWM program, especially if the goal is to recover energy from waste incineration. Once the source reduction part of the program is well established both in terms of funding and organizational structure, steps should be taken to investigate the viability of incineration.

For ensuring enough waste for a WTE facility, waste flow must be estimated properly. Sometimes waste flow ordinances are adopted by communities to ensure minimum waste volume for an incinerator. For bigger communities guaranteeing a minimum volume of waste is easier. Smaller communities can ensure such a minimum volume only through a regional approach.

For the design of WTE facilities both percentages of each waste type (e.g., food waste, paper) and the total weight of the waste entering the facility are important. Since source reduction can alter the waste composition significantly, a waste sort should be done to develop the range of waste composition prior to designing a WTE facility. In most cases the commercial waste stream is considered part of MSW. So if a waste sort is performed on the combined waste, then the waste composition will automatically include commercial waste. However, if the commercial waste is collected separately or if the waste sort is performed at residential curbside, only then can actual waste composition entering a WTE facility be estimated properly. The composition of commercial waste may include a higher percentage of incinerable waste (e.g., paper) and hence can influence the design of an incinerator.

If nonhazardous waste (e.g., paper mill waste) from one or more industries in the community is considered for combined incineration with MSW, then proper characterization of the waste must be done. A careful study of the flow diagram of a production line(s) should be done to identify desirable and undesirable waste components regarding incinerability and pollution load. Niessen and Alsobrook (1972) have provided major waste components of various industries in the United States. Similar waste component analyses of the participating industries should be developed for designing an incinerator.

5.3   TYPES OF INCINERATORS

The goal of incineration dictates the type of incinerator chosen. If the goal of incineration is merely to reduce the waste volume, then open burning or burning in closed chambers may be utilized. The effectiveness of combustion depends on furnace temperature, resident time, and exposure of waste to the flame achieved through some arrangement of turning the waste, supply of sufficient quantity of oxygen as dictated by the incinerator design (note: oxygen requirements are different for starved-air vs. non-starved-air units), and proper combustion of atomized particles in the gas phase to reduce contaminants in the effluent air.

5.3.1   Open Burning

This is a very old method of burning waste. In this method, combustible waste is piled up and burned in an open area. This type of burning has significant fire hazards, and the emission may contain significant amounts of air pollutants depending on the characteristics of the waste materials. In many communities burning waste has a long history. In sparsely populated areas open burning in a metalic barrel, commonly known as a "burn barrel" (Fig. 5.1), is practiced to avoid waste collection and disposal. The danger of waste burning, from fire hazard (especially in forested areas), air, and groundwater pollution are well established. Smoke generated from leaf burning contains high concentration of particulate, which can reach deep into lung tissue causing coughing, wheezing, and shortness of breath. In addition to a particulate problem, it may contain many hazardous chemicals such as carbon monoxide and benzo(a)pyrene at significantly high concentration. Benzo(a)pyrene is known to cause cancer in animals and believed to be a major factor in lung cancer. Open burning of plastics, asphalt, rubber, and so on can generate hazardous air pollutants. A USEPA study showed that each pound of garbage burnt in

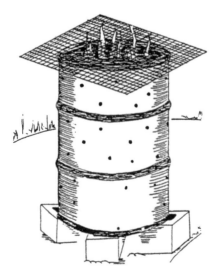

FIG. 5.1   Burn barrel.

a burn barrel emits twice as much furans, 20 times more dioxin, 40 times more particulates than burning one pound in an incinerator with air pollution control (WDNR, 1996). Burning garbage in a burn barrel often generates ash with high levels of metals (e.g., lead, cadmium). It is a common practice to dump the ash on open lands. Such open dumping will cause the metals to leach out and contaminate the groundwater.

Open burning of automobile tires emits carbon monoxide, sulfur dioxide, and significantly high concentrations of organic compounds, mainly aromatics. The ash contains high concentrations of zinc and other metals that can be leached out (Ryan, 1989). The reclamation process for retrieving ferrous and nonferrous metals from scrap automobiles generates a nonmetallic waste product, commonly known as shredder fluff. The components of shredder fluff are plastics (e.g., polyethylene, polyurethane foam), rubber, glass, wood products, cloth, paper, dirt, and electrical wiring. Open burning of shredder fluff can emit a significant amount of benzene, chlorobenzene, ethylbenzene, and so on. The particulate emission includes, among other things, high concentrations of copper, lead, and zinc; it may be postulated that these metals are also present in the ash (Ryan and Lutes, 1993).

Thus open burning of garbage, tire chips, and shredder fluff can pose significant health risks. While open burning of garbage from individual homes should not be undertake, small quantities of leaves, plant clippings, paper cardboard, and untreated wood from individual homes located in sparsely populated areas may be allowed. It should be noted that burning a large volume of these materials can create a health hazard.

5.3.2 Open-Pit Incinerators

This type of incinerator (Fig. 5.2) is usually built below grade. The walls and floors may be lined with refractory. Air is injected from the top at an angle of 25–35° using several closely placed nozzles that force the flame to bend down, creating a rolling action of high-velocity air. Because of long resident time and high flame temperature, complete combustion is achievable. A screen is placed over the pit to control emission of large airborne particles; the screen also serves to control the entry of rodents into the pit when not in use. The pit should be oriented in such a way that wind does not create a fire hazard during charging of solid waste. Thus, the loading ramp should be located on the upwind side, provided the direction of wind is the same most of the time. Open-pit incinerators should be designed for 100–300% of ideal air supply at a rate of 850 st ft^3/min (24 kL/min) at 11 in. (28 cm) of water column per foot of pit length (Brunner, 1991).

Although virtually visible smoke is absent, this type of incinerator cannot meet current air quality standards even for particulate emission rate. Comments regarding open burning of garbage, tire chips, and shredder fluff, mentioned in Section 5.3.1, are also applicable to open-pit incinerators. In addition, if the solid waste contains even small amounts of hazardous material,

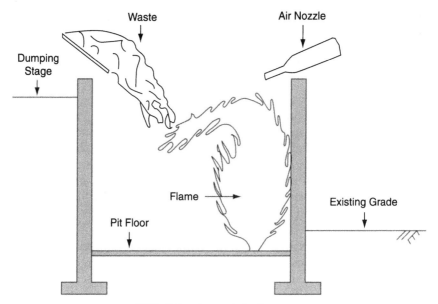

FIG. 5.2   Open-pit incinerator.

the air quality around the incinerators can pose real health risk. An open-pit incinerator located in a remote place may be used for burning a larger quantity of the allowable materials mentioned in Section 5.3.1. However, air quality in the immediate vicinity must be studied for air emission to provide proper health protection to incinerator operator(s) and people living in the vicinity of the open-pit incinerator.

5.3.3   Single-Chamber Incinerators

These incinerators are essentially an enclosed incinerator with a stack (Fig. 5.3). The waste may be fed sidewise or from the top. A variation of the single-chamber incinerator is a jug-type incinerator. The solid waste is placed on the grate. In some incinerators, an afterburner is added to control air emission. Particulate emission improves significantly with the addition of the afterburner (Macknight, 1990). However, these incinerators cannot meet current air regulation standards in the United States.

5.3.4   Multiple-Chamber Incinerators

There are two types of multiple-chamber incinerators: retort type and in-line type. These incinerators provide significant improvement in air emission and provide better combustion.

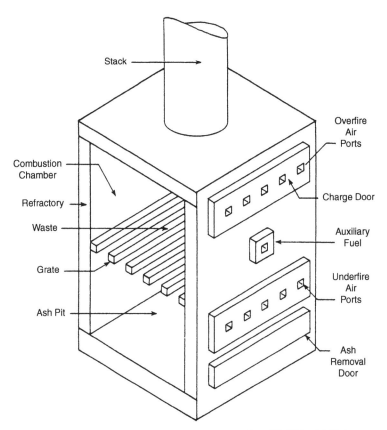

FIG. 5.3   Single-chamber incinerator. After USEPA (1980a).

5.3.4.1 Retort Type   This is a compact unit (Fig. 5.4) with multiple bafles, which provide a longer flow path for the flue gases before exiting the chamber. Both underfire and overfire are provided in the primary chamber. Under proper operation this type of incinerator may meet air emission standards particularly the particulate emission. It is possible to add air pollution control devices to this type of incinerator.

5.3.4.2 In-Line Type   These units are bigger than retort-type incinerators. A moving grate and automatic ash discharge conveyor allow continuous burning. Natural gas is used as supplemental fuel both in the primary and secondary chambers. Use of supplemental fuel in the secondary chamber helps in complete combustion of the solid waste. Figure 5.5 shows a cross section of a typical in-line type of incinerator. This type of incinerator can also be used for destruction of pathological waste. Figure 5.6 shows a cross section of a typical in-line incinerator with a pathological waste retort.

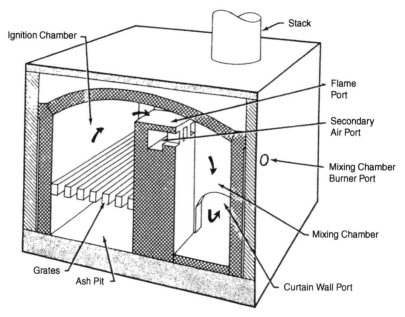

FIG. 5.4   Retort multiple-chamber incinerator. From Brunner (1991). Reproduced with permission of The McGraw-Hill Companies.

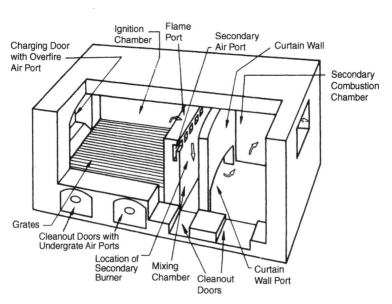

FIG. 5.5   In-line multiple-chamber incinerator. From Brunner (1991). Reproduced with permission of The McGraw-Hill Companies.

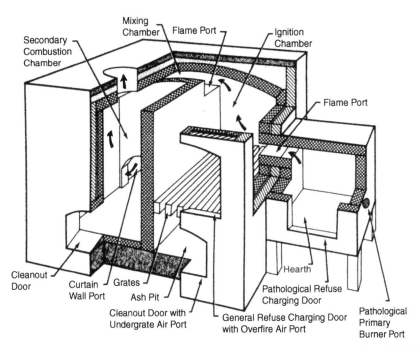

Mixing
Chamber

Flame Port

Ignition
Chamber

Secondary
Combustion
Chamber

Flame Port

Cleanout
Door

Curtain
Wall Port

Grates

Ash Pit

Cleanout Door with
Undergrate Air Port

Hearth

Pathological Refuse
Charging Door

General Refuse Charging Door
with Overfire Air Port

Pathological
Primary
Burner Port

FIG. 5.6 Multiple-chamber incinerator with pathological waste retort. From USEPA (1973).

5.3.5 Types of Incinerators Used for Burning of Solid Waste

There are several types of incinerators used for burning of MSW. Usually these incinerators are designed as WTE facilities. Figure 5.7 indicates different WTF facilities based on their method of combustion and construction. In MSW burning facilities, waste may be introduced using continuous feed or batch feed. For complete combustion of waste and for removing contaminants from the flue gas, usually a primary and a secondary chamber is used in most incinerators. The major types of incinerators used for the burning of MSW include rotary kiln, controlled-air incinerator (using pyrolysis), fluidized bed, and moving-grate furnaces. The major types of incinerators used for sludge incineration include multiple-hearth incinerator, conveyer furnace, and cylonic furnace. The choice of a furnace type depends on incinerable waste content and economics.

In rotary kilns (Fig. 5.8) the waste is fed into a rotating drum that is lined with refractory. The advantages of rotary kiln are: (1) it can incinerate a wide variety of waste streams, (2) the residence time of waste in the kiln can be controlled very easily, and (3) uniform contact of waste with air leading to better combustion. The disadvantages of rotary kiln are: (1) the particulate concentration is high in the emission, (2) the air requirement is high, and (3) heat loss is high (Brunner, 1991).

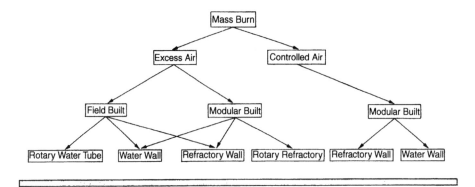

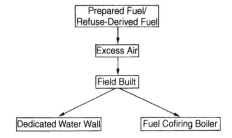

FIG. 5.7   Type of municipal solid waste combustion facilities. From Hegberg et al. (1991b).

Another technology used for combustion of MSW is controlled air systems or pyrolysis. In pyrolysis, the waste is burned in the absence of theoretically required (stoichiometric) oxygen. This is a destructive distillation process in which heat must be applied for completing the reactions. After removal of glass and metal, the waste is shredded before charging the waste to the pyrolytic converter. Although the system has merits, currently pyrolysis is not a popular system for MSW incineration.

Another type of controlled air incinerators is the starved-air unit (SAU). In these units only 70–80% of stoichiometric air is introduced in the primary chamber. The off gas is burnt in a secondary chamber using additional air. Figure 5.9 shows a cross section of a typical SAU. The exhaust gas from SAUs is relatively clean (Brunner, 1991).

In fluidized-bed furnaces, air at high pressure is forced through a sand bed. The sand gets suspended in the air and behaves like a fluid. Shredded MSW or sludge cake (for burning sludge-type waste) is fed in small quantities, which burn rapidly. A variation of the fluidized bed is the circulating fluid bed (CFB) furnace. In CFB, the air velocity is much higher because of partially burnt solid waste particles that get thrown to the top of the chamber. The unburnt particles are captured and sent back to the bed. The burning efficiency of the CFB furnaces is higher than in fluidized-bed furnaces. The furnaces may be lined with a water wall for heat recovery.

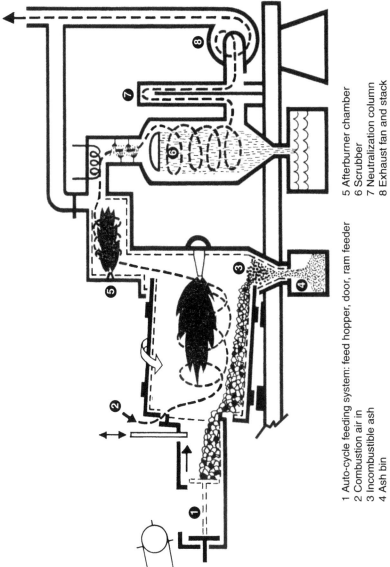

1 Auto-cycle feeding system: feed hopper, door, ram feeder
2 Combustion air in
3 Incombustible ash
4 Ash bin

5 Afterburner chamber
6 Scrubber
7 Neutralization column
8 Exhaust fan and stack

FIG. 5.8 Rotary kiln. After C. E. Raymond Combustion Engineering, Inc.

75

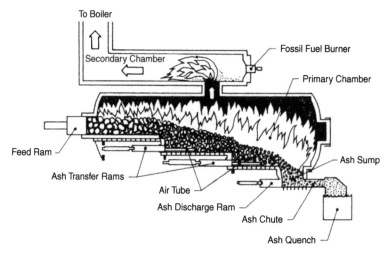

FIG. 5.9   Starved-air incinerator. After Brunner (1991). Reproduced with permission of The McGraw-Hill Companies.

The multiple-hearth incinerators (Fig. 5.10) can handle sludge with a wide range of moisture content. The furnace has hearths at different levels. Each hearth has openings so that the sludge drops from the top to the next hearth. The vertical shaft with arms rotates slowly, turning the waste and facilitating the downward movement. The amount of excess air is 75–125% of stoichiometric air requirement (Brunner, 1991). The furnace requires trained and experienced operators to ensure proper burning efficiency.

In a conveyor furnace, the sludge is placed on a conveyer belt that passes through a furnace. The sludge is burnt either by fuel-fired or electrically heated radiant tubes. A simple air pollution control system (e.g., a cyclonic scrubber) is used for cleaning the flue gas.

The cyclonic furnace (Fig. 5.11) is a single-hearth furnace in which sludge is introduced at the periphery of a rotating hearth. Air and supplemental fuel is introduced tangentially, which creates a cyclonic funnel making the sludge airborne. The sludge burns as it moves up. The flue gas exits through the top. These units, although simple, can handle relatively low feed rates.

5.3.5.1   Grate Design   Currently, stationary grates for batch-type processing are rather uncommon for WTE facilities except for old existing incinerators. Grates used for continuous flow uses some type of rocking or reciprocating mechanism (Fig. 5.12) for advancing the waste. All grates are sloped downward toward the exit point. Depending on the length of traverse, several sections are installed in series. In addition to these variations in grate design, other grate designs are also used in incinerators (USDC, 1977).

5.3.5.2   Furnace   Both static and rotating burner or kilns (also called hearths) are used for burning MSW. The burners can be designed with or without heat recovery systems. Table 5.3 shows a comparison of systems with

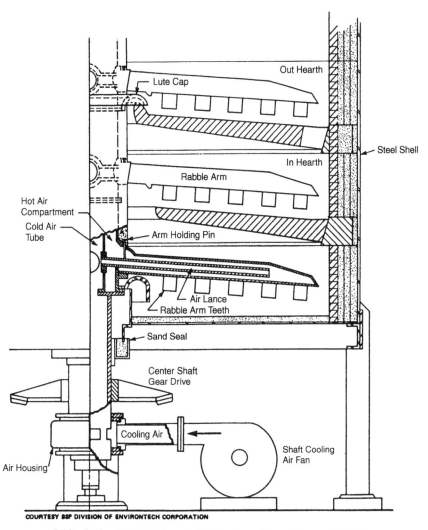

FIG. 5.10 Multiple-hearth incinerator. From Niessen (1995).

and without heat recovery. However, these furnaces can include sophisticated air pollution control devices. Currently, furnaces are designed with two chambers: primary and secondary. Waste is burnt in the primary chamber and flue gas is further burnt in the secondary chamber (or afterburners) using additional fuel. There are many areas that include design variations, but the following three are most common: air and fuel addition, cooling of flue gas, and heat recovery.

Air and Fuel Addition Air is injected to supply additional oxygen for enhancing combustion. Airflow rate is controlled to supply the stoichometric oxygen for complete combustion of the refuse. Air may be injected from the

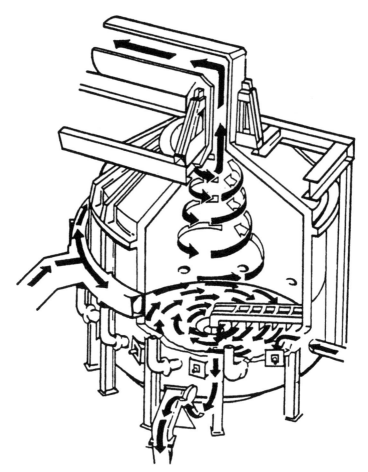

FIG. 5.11 Cyclonic furnace. From Brunner (1991). Reproduced with permission of The McGraw-Hill Companies.

top or bottom of static furnaces or from both. In general, additional fuel in gaseous form is added where needed. The need for additional fuel in the primary chamber for burning the waste depends on the moisture content of the waste. Fuel is also added in the secondary chamber to burn the flue gas for burning contaminants (especially organic).

Cooling of Flue Gas Flue gas is cooled by adding water or air or by heat removal. There are two methods of cooling—wet-bottom and dry-bottom methods. In wet-bottom systems water is sprayed on the flue gas. The excess water falls to the bottom, which is collected and treated. The advantages of this method are that it is reliable, simple, and inexpensive. In addition, removal of particulate is easier from moist gas in electrostatic precipitators. The disadvantages of this system are that it creates significant volume of water

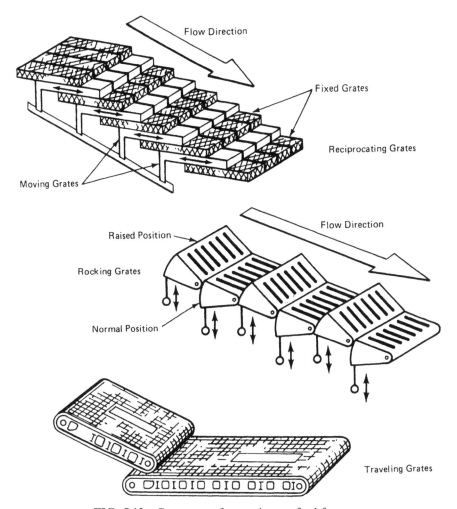

FIG. 5.12   Grate types for continuous-feed furnaces.

needing further treatment and if recirculated causes increased maintenance due to corrosion. At present new incinerators are usually not designed with a wet-bottom system (Niessen, 1995). In the dry-bottom system fine atomized water (created using pressure) is sprayed to cool the flue gas to an optimum temperature. The advantages of dry-bottom systems are it minimizes excess water treatment and produces a relatively dry flue gas. The disadvantages of this system are that both design and installation are expensive, power consumption is high, and requires relatively high maintenance. Cooling the gas through dilution with air is simplest and most reliable. However, the operating cost of this system is high. Like the wet system, air cooling is usually not used in new incinerators. The flue gas can also be cooled by recovering heat,

TABLE 5.3 Comparison between Incinerators with Heat Recovery and without Heat Recovery

With Heat Recovery	Without Heat Recovery
Reduced gas temperatures and volumes due to absorption of heat by heat recovery system	Hotter gas temperatures
Moderate excess air	High excess air required to control furnace temperatures
Moderate-size combustion chamber	Large refractory-lined combustion chamber to handle high gas flow
Smaller air and induced-draft fans required for smaller gas volume	Higher gas volumes due to higher temperatures, requiring larger air and gas flow equipment
Steam facilities including integral water wall, boiler drums, and boiler auxiliary equipment required	No steam facilities required
Operations involve boiler system monitoring, adjustments for steam demand, etc.	Relatively simple operating procedures
Steam tube corrosion is possible as well as corrosion within exhaust gas train	Corrosion possible in the exhaust gas train
Licensed boiler operators are required to operate incinerator	Conventional operators satisfactory
Considerable steam credits possible, including in-plant energy savings in addition to salvage	Only credits are possible salvage of the equipment after its useful life

After Brunner (1994). Reproduced with permission of The McGraw-Hill Companies.

by passing it through a water boiler. The gas temperature is used to generate steam. The advantages of this system are heat recovery and high shrinkage of flue gas, which is helpful for air pollution treatment. The disadvantages of heat recovery systems are that both design and installation are expensive and require relatively high maintenance due to corrosion problems.

Heat Recovery   Heat is recovered either from the flue gas or from the burning waste directly. To recover heat from the flue gas, the heated gas is passed through a system of water-filled tubes, which heats the water. To recover heat from the burning waste, water wall technology is used. A water wall consists of a number of metal tubes filled with water; instead of refractory, the tubes are used to line the furnace. The hot water or steam generated from the furnace is used to generate electricity. On an average 400–600 kWh of electricity can be generated by burning a ton of MSW (Brunner, 1994). If hot water is generated, then it is supplied to a nearby building or industry (e.g. laundry) for use. In some cases, heat recovered from a flue gas is used to heat air that is injected into the furnace. The hot air helps to raise the temperature or reduce moisture of the waste; thus the energy recovered is utilized to reduce additional fuel consumption.

5.4 ENVIRONMENTAL ISSUES

The choice between incineration and recycling is an issue related to environmental impact. In addition, there are two major environmental issues directly related to incineration: air emission and ash disposal. It is essential to address all issues before an incinerator is installed.

5.4.1 Choice between Incineration and Recycling

The heating value of MSW must be high enough for incineration. Both paper and plastic has high heating values. So for good-quality incinerable waste, the percentages of these two items should be high, where as the goal of recycling is to minimize these two items in the solid waste stream. If a community is successful in recycling, the composition of its MSW will have a low quantity of paper and plastics. However, if ISWM is implemented, the MSW will also be low in compostable waste (e.g., yard waste) and household hazardous waste. Thus the overall composition of the MSW destined for incineration, although low in paper and plastic, will also be low in low heating value materials and in hazardous material. For most communities, infectious waste may also be part of the nonhazardous solid waste stream, which will need some form of treatment for disinfection; part of it may even need to be incinerated (e.g., pathological waste). If the infectious waste is incinerated in a separate facility, then that must be taken into consideration. All these issues must be sorted out properly before undertaking an incineration project. In general the environmental benefits of reuse, recycling, and composting of MSW is much higher than reducing it through incineration. While for bigger communities incinerability of MSW even after recycling is expected to remain high, it may be a critical issue for smaller communities.

5.4.2 Air Emissions Control

The air emissions from an incinerator must meet the regulatory criteria for air quality. The allowable air emissions from incinerators are different in different parts of the world. Both the parameters and the allowable discharge concentration can be different. The emission from incinerators include a wide variety of chemicals, which depend on the incinerator type and waste composition. In addition to many metals (e.g., arsenic and mercury), the emissions include HCl, SO_2, NO_x, CO, dioxins and furans, polycyclic aromatic hydrocarbons (PAHs), and particulate. Dioxins, furans, and their cogeners, although not present in the waste, are formed during combustion. Probable emissions from the chosen incinerator type should be determined by test burning of the candidate waste. It should be noted that because of inherent variation in feed waste, the concentration of emissions of different parameters varies widely. Air emissions control equipment can be grouped into three systems: dry, semidry, and wet (Teller, 1994). In dry systems, the gas is (cooled, if necessary) passed through equipment(s) to achieve desired reduction in emission. Electrostatic precipitator, dry venturi, and fabric filter are some of the dry-type equipment. In a semidry system, a lime slurry is used first to reduce the gas temperature and to neutralize it. The gas is then passed through a dry system such as electrostatic precipitator and dry venturi. In a wet system, the gas is sprayed with a reagent within a tower-type equipment. Table 5.4 indicates the capabilities of different equipment for removing air pollutants.

5.4.3   Ash Management

There are mainly three types of ash generated from an incinerator: bottom ash, fly ash, and scrubber ash. The waste characteristics of each ash type are different. Since fly ash and scrubber ash may test out to be hazardous, their disposal strategy may be different. This section deals with disposal only. Beneficial use of incinerator ash is discussed in Chapter 8.

Bottom ash is the residual product after combustion of the waste. Bottom ash also includes ash fallen through furnace grates and ash that gets attached to boiler walls (known as boiler ash). Bottom ash consists mostly of inert residue and noncombustible fraction (e.g., metals, glass). The range of particle size of bottom ash is 0.15–10 mm and above. It may be wet if the residue is quenched in water, or it may be dry. Of the total ash generated from an incinerator 85–90% is bottom ash. It is nonhazardous.

Fly ash is essentially the particulate generated during combustion. This ash is discharged from the particulate control devices (e.g., electrostatic precipitator). The particle size of fly ash is much finer than bottom ash; it is like a powdery material. The range of particle size of fly ash is 0.05–0.1 mm, although a small percentage of agglomerates as big as bottom ash particles are also found. Because of the finer size, the fugitive dust potential is high during handling. Fly ash may be sprinkled with moisture to reduce fugitive potential. In general, fly ash is dry or semidry. However, it may be wet if wet scrubbing is used in the air pollution control system. Fly ash contains significantly high levels of contaminants and the leachate may turn out to be hazardous with lead or cadmium (Sopcich and Bagchi, 1988).

The scrubber ash is generated during reactions of flue gas with chemicals (e.g., lime) used in air pollution control systems (e.g., spray dryer). In most cases, it is collected along with the fly ash. The combined fly and scrubber may turn out to be hazardous and thus may need special handling and a disposal strategy.

If each of the ash components is collected separately, then appropriate leachate test should be done on each component separately. However, if the

TABLE 5.4 Emission Control System Types[a]

Dry	Semidary	Wet	Particulate	Acid Gases	Heavy Metals	Mercury	Organics	NO$_x$
Electrostatic precipitator			X		X			
Fabric filter			X		X			
Furnace injection + FGD		Tray or packed scrubber	X	X	X			
		Condensing scrubber	X	X	X		X	
Dry reactor or dry venturi	Quench reactor							
+ ESP	+ ESP	ESP, wet scrubber	X	X	X		X	
+ FF	+ FF	FF, wet scrubber	X	X	X		X	
Dry reactor or dry venturi	Quench reactor (+ DV)							
+ ESP + carbon injection	+ ESP + carbon injection	ESP or FF + wet scrubber inclusive of oxidation	X	X	X	X	X	
+ FF + carbon injection	+ FF + carbon injection	SNCR or SCR with wet scrubber inclusive of oxidation	X	X	X	X	X	
SNCR or SCR with dry reactor or dry venturi	SNCR or SCR with quench reactor (+ DV)							
+ ESP + carbon injection	+ ESP + carbon injection		X	X	X	X	X	X
+ FF + carbon injection	+ FF + carbon injection		X	X	X	X	X	X

Materials Addressed[b]

[a]ESP = electrostatic precipitator, FF = fabric filter, DV = dry scrubber, SNCR = selective noncatalytic reduction, SCR = selection catalytic reduction, and FGD = flue gas desulfurization.

[b]Merely addressing specific materials is not reflective of the efficiency achieved.

From Teller (1994). Reproduced with permission of The McGraw-Hill Companies.

83

three ashes are collected together, then appropriate leachate tests should be done on composite samples. Leachate quality for municipal incinerator is discussed in Chapter 13. In general, bottom ash has less leachable constituents compared to the fly ash. Because of low leachability, disposal or beneficial reuse of bottom ash is less problematic. Hence, incinerators with arrangements for separate collection of bottom ash and fly ash are preferable.

Since the composition of waste fed into a boiler varies significantly with time, the chemical composition of ash also varies widely (Andrews, 1991). It is important to perform leachate tests on representative samples. A representative sample should be obtained by collecting a sample at regular intervals. USEPA (1986a) and ASTM (1989) provide guidelines for sampling waste with variable characteristics. The representative concentration of various parameters may be obtained through statistical analysis of leachate data collected over a long period of time (e.g., month).

Comments regarding leaching tests used for MSW incinerator ash can be found in Chapters 13 and 14. Leachate characteristics will dictate the best strategy for ash management. The ash may be chemically treated to reduce toxicity. There are three options available for managing MSW incinerator ash: land disposal, beneficial reuse, and vitrification. The ash may be co-disposed in an MSW landfill or disposed in a monofill cell. Discussion of land disposal is included in Chapter 23. Options regarding beneficial reuse of untreated or chemically treated ash is discussed in Chapter 8. Vitrification of ash produces glassy frits with low leachability constituents. Technology for vitrification using electric arc furnaces (Decesare and Plumley, 1992) and cold crown glass furnaces (Wexell, 1993) has been reported in the literature.

5.5   INCINERATION OF OTHER NONHAZARDOUS WASTE TYPES

In addition to MSW there are other types of wastes generated by communities that may need to be incinerated. It is possible that some of these wastes may be managed without being incinerated. For instance, nonhazardous wastewater treatment sludge can be managed by other means (e.g., land spreading). However, there are certain waste streams for which incineration currently seems to be the only option. Some of these are nonradioactive wastes generated in laboratories, chemotherapy and anatomical wastes generated in hospitals, anatomical waste from pet clinics, and so on.

For ISWM, the incineration issue is to be addressed according to the size of the community. While most communities may generate all of the above-mentioned waste type(s) for which incineration is the only option, it may not be economical to install incinerators in each community. Several communities may install a jointly owned incinerator for incinerating these wastes. The issue is more important for smaller communities than bigger communities. Hazardous waste generated in research laboratories and in hospitals (e.g., chemotherapy) must be incinerated in hazardous waste incinerators. Many hospitals incinerate anatomical waste in-house. So the volume of waste that actually needs incineration must be assessed properly before installing an incinerator.

Many hospitals incinerate anatomical waste in modular units (e.g., in-line type of multiple-chamber incinerators). However, the current trend is to use starved-air units, which can provide better air emissions control. In many communities these units are regulated. It may be mentioned that air emissions from these units may not meet current standards. However, if a new unit is planned, it is better to opt for a unit capable of meeting current air emissions standards even in the absence of any regulatory requirements; it is highly likely that regulatory mandates for individual hospital units will become the same as other incinerators. Rotary kilns are used for incinerating nonseparated infectious waste if the waste volume is high. However, if proper waste reduction policy is adopted in a hospital, then the waste will contain very little amount of paper or other waste components, which have high Btu values. All these issues should be taken into account before finalizing incinerator installation for these types of waste.

Sludge from wastewater treatment plants or other sources (e.g., paper mills) may need to be incinerated. As mentioned in Section 5.3.5 there are four types of incinerators available for burning sludge-type waste (e.g., wastewater treatment plant sludge). The volume of sludge to be incinerated dictates the choice of furnace. In general, the air pollution control from these furnaces is not as complicated as MSW incinerators. However, test burn must be done to find out about emission quality.

5.6 SITE SELECTION, INSTALLATION, AND FINANCING

Both technical and social issues need to be considered for siting an incinerator. The type of incinerator to be installed must be identified before starting the site selection process.

5.6.1 Data Collection

The technical data needed for siting an incinerator will depend on the type and size of the incinerator. The acceptability of the site to the residents around the site also needs to be ascertained. Initially more than one site should be investigated for an incinerator. In addition to data identified in Sections 12.1.1–12.1.8, the following data must be obtained for selecting a site.

5.6.1.1 Air Quality The air quality of the area should be such that an additional industrial-type facility should not be in violation of air quality standards for the location. The preliminary air emissions data from the incinerator should be obtained from the vendor. The expected release concentration of various air parameters should be judged against the allowable limit imposed on new facilities for the target area. The location with the maximum allowable air release parameters should be given priority.

5.6.1.2   Critical Habitat Area   Any site identified as critical habitat for one or more endangered species must be excluded from the siting process.

5.6.1.3   Geotechnical Data   Preliminary subsoil investigation to assess the suitability of the foundation for a major structure should be undertaken. If possible, data for subsoil investigation of neighboring major buildings or industry should be obtained to find out whether a costly foundation (e.g., pile foundation) will be needed for the incinerator. Detailed subsoil data necessary for foundation design must be obtained for the top two favorable sites.

5.6.1.4   Wetlands   Wetland map of the targeted areas should be obtained. In general, it is better to avoid sites that may impact wetlands. If disturbance of wetlands for a chosen site is unavoidable, then the issue of wetland compensation must be discussed with the regulatory agency.

5.6.1.5   Utilities   The targeted areas must be close to electric and telephone lines. Proximity of a fire station is helpful for protecting the plant from fire if needed. Necessary arrangements for fire protection should be made within a plant if a fire station is not close by. Availability of water, either through supply lines or from a surface water body, is essential for siting an incinerator. Proximity of sewer lines is helpful for an incinerator because there will be a significant amount of process water that will need treatment.

5.6.1.6   Assessment of Public Reaction   The Public's initial reaction to any new industrial-type facility is always negative. This is because of concerns primarily regarding health and property value. Although loss of property value is often thought of as a major concern for residents near incinerators, this fear is not well founded. The literature indicates that the impact of incinerator sitting on property value is not significant. Residents reject property value guarantees. They prefer that emphasis be given on pollution prevention (Zeiss, 1991). Proposal to site an incinerator in an area identified as an industrial zone will create less public reaction compared to a location for which zoning needs to be changed. Public ownership draws less opponents than private ownership. To ease antisiting public sentiment an informational meeting should be organized to discuss the need, environmental impact, socioeconomic impact, and so on at potential sites. Public opinion regarding siting should be surveyed at the sites with highest potential.

5.6.1.7   Energy Market   The incinerator type and design will dictate whether heat can be recovered from the burning process. The design will also indicate whether energy is in the form of steam or hot water. Steam can be utilized for generating electricity. However, if steam is generated, then the incinerator must be located preferably within a 2-mile (3.2-km) radius of the user. This would restrict the siting process. This issue should be considered while searching for a potential site. Excess electricity can be sold to the electric utility. However, because of the unsteady nature of generation from an incinerator, the electric utilities tend to pay less because they need to backup generating capacity. The power sale price should be negotiated carefully because the revenue generated is normally used to finance the incinerator (Rogoff, 1987).

TABLE 5.5   Typical Cost Items for Incinerator Installation

I. Incineration system
  A. Waste conveyance
     (1) Open or compaction vehicles and commercial containers
     (2) Special design containers
     (3) Piping, ducting, and conveyors
  B. Waste storage and handling at incinerator
     (1) Waste receipt and weighing
     (2) Pit and crane, floor dump, and front-end loader
     (3) Holding tanks, pumps, and piping
  C. Incinerator
     (1) Outer shell
     (2) Refractory
     (3) Incinerator internals (grates, catalyst)
     (4) Burners
     (5) Fans and ducting (forced and induced draft)
     (6) Flue gas conditioning (water systems, boiler systems)
     (7) Air pollution control
     (8) Stacks
     (9) Residue handling
     (10) Automatic control and indicating instrumentation
     (11) Worker sanitary, locker, and office space
II. Auxiliary systems
  A. Buildings, roadways, and parking areas
  B. Special maintenance facilities
  C. Steam, electrical, water fuel, and compressed air supply
  D. Secondary pollution control
     (1) Residue disposal (landfill, etc.)
     (2) Scrubber wastewater treatment
III. Nonequipment expenses
  A. Engineering fees
  B. Land costs
  C. Permits
  D. Interest during construction
  E. Spare parts inventory (working capital)
  F. Investments in operator training
  G. Startup expenses
  H. Technology fees to engineers and vendors

From Niessen (1995).

5.6.2 Data Analysis and Permit

After collecting the data each site may be ranked using a numerical or alphabetical system. Detailed investigation regarding air quality, transportation needs, public sentiment, and subsoil investigation should be undertaken at the top two to three sites. If necessary a proposal should be submitted to the regulatory agency for obtaining necessary permits. In most cases permits regarding air emissions and ash disposal are necessary. In addition, a permit may be necessary for the sale of electricity.

5.6.3 Installation

Once a site is selected, a detailed soil investigation for foundation design purposes is undertaken. The incinerator design is also finalized at this stage. The design and installation can be contracted out to an engineering firm. If the solid waste management department of the local government does not have personnel to provide necessary technical advice, then a consultant is hired to advise the solid waste department. In such cases, the consultant usually prepares the bid documents and overseas the design and construction done by an outside contractor. Sometime a turn key approach is taken in which design, construction, and initial startup is handed over to the same consulting firm. It is also possible to select a private firm to install, operate, and own the incinerator completely. Usually private firms are interested only in WTE facilities because of long-term revenue potential. Although this type

TABLE 5.6 Typical Cost Item List for Incinerator Operation

I. Fixed costs (credits)
  A. Repayment of debt capital
  B. Payment of interest on outstanding capital
  C. Tax credits for depreciation
II. Semivariable costs
  A. Labor (including supervision) with overheads
  B. Insurance
  C. Operating supplies
  D. Maintenance and maintenance supplies
III. Variable costs (credits)
  A. Steam usage (or credits)
  B. Electricity
  C. Water supply and sewerage fees
  D. Oil or natural gas fuels
  E. Compressed air
  F. Chemicals (catalysts, water treatment)
  G. By-product credits
  H. Disposal fees

From Niessen (1995).

of arrangement cannot be made for incinerators without energy recovery, it is possible to contract out the operation of such units to outside contractors. Based on business potential WTE incinerators may also be built and owned by a private firm.

5.6.4 Financing

If a WTE incinerator is handed over to a private firm, then financing of the project is done by that firm. If the WTE incinerator is to be financed by the community, then usually the community government issues a bond to obtain money for the project. Either a general obligation bond or project revenue bond are issued for such purposes. Discussion regarding funding solid waste projects can be found in Chapter 26. A detailed financial analysis should be done to compare the best option for installing an incinerator. Whether the community will own the facility or it will be handed over to a private firm should be decided on the basis of short-term and long-term cost and benefits. Cost analysis for design, construction, startup, and operation should be undertaken at an early stage of decision making, which will help identify the best option for the community. Tables 5.5 and 5.6 list typical cost items for installation and operation of an incinerator.

6

COMPOSTING

Composting plays an important role in integrated solid waste management (ISWM). It is essentially recycling of readily biodegradable materials into their basic components of water, carbon dioxide, energy, and a composted matter. Composting can thereby reduce the municipal solid waste volume destined for land disposal or incineration. Composting yields a valuable product that can be used for soil amendment and mulch. Properly processed and cured compost product improves soil nutrients, reduces erosion, and helps suppress plant diseases. The potential of composting cannot be realized through empirical practices (Finstein and Hogan, 1993). A composting operation must be designed to satisfy the needs and resources (both material and funding) of a community. For a small community with limited resources, a simple composting facility using windrows is sufficient, whereas for a large community a specially designed facility capable of handling several tons of biodegradable waste will be needed. Proper attention must be given during planning a siting so that it is sufficiently funded and does not run into public opposition due to odor or other composting-related problems.

6.1 FUNDAMENTALS OF COMPOSTING

Composting is a biodegradation process that transforms organic matter into water, carbon dioxide, energy, and a composted matter. Composting involves two major stages. In the first stage microorganisms decompose the feedstock into simpler chemical compounds. The metabolic activities generate heat. In the second stage the composting pile is cured. Depletion of food source slows the activities of the microorganisms. As a result, heat is reduced and the compost becomes dry and crumbly resulting in a soil-like material rich in nutrients.

Composting consists of a series of microbial activities (Fig. 6.1). Different types of microorganisms are active in different phases of the process. Macroororganisms such as nematodes, mites, sow bugs, and earthworms reduce the size of the feedstock by foraging or chewing. These actions physically break down the feedstock, creating greater surface area, which helps microbial activities. The bacteria and fungi, important in decomposing the material, can be classified into mesophilic (those that grow best at temperatures from 77 to 113°F (25–45°C) and thermophilic (those that grow best at temperatures from 113 to 158°F (45–70°C). Mesophites are dominant initially when the temperature is low. These microorganisms use the available oxygen in transforming carbon to obtain energy, and the process produces carbon dioxide, water, and heat. Inside the compost pile, most of the heat remains trapped (without turning) raising the temperatures to near 113°F (45°C). At this high temperature mesophites either die or become dormant, and the thermophiles become active. Thermophiles multiply rapidly and raise the temperature, killing most pathogens and weed seeds. The thermophiles continue to decompose the feedstock as long as nutrient and energy sources are plentiful. With the depletion of nutrients and energy sources, thermophiles die, causing the lowering of the temperature of the pile. At this time the dormant mesophiles become active again until all readily available energy sources are utilized. After the active composting is completed, the biodegradable waste is reduced to a soil-like material. At this phase the microbial activities subside, and the temperature decreases. Additional stabilization is usually done for several weeks allowing slow decomposition of cellulose and lignin. This process is called curing.

6.2   INFLUENCE OF VARIOUS PARAMETERS ON COMPOSTING

Microbial activities are influenced by several parameters such as particle size of the feedstock, nutrient level, temperature, and so forth . These parameters are interdependent; a change in one often causes a change in another.

6.2.1   Particle Size

Particle size of the feedstock influences the composting process because smaller particle size provides greater surface area for microbial activities. Smaller particles helps to maintain optimum temperature within the composting pile (USEPA, 1994d). However, very small particle size may compact the feedstock, which will reduce void size and thereby oxygen availability. Ideally the particle size should be between 0.5 and 3 in. (1.25 and 7.5 cm). However, for highly decomposable materials, the particle size can be as high as 6 in. (15 cm).

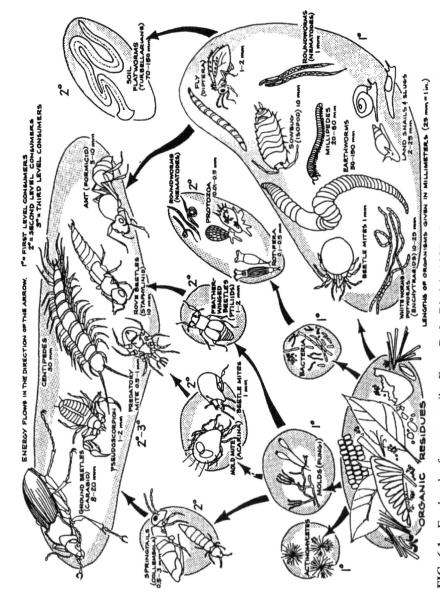

FIG. 6.1 Food web of compost pile. From D. L. Dindal (1984). Ecology of Compost. State University of New York, College of Environmental Science and Forestry, Syracuse, New York.

6.2.2 Moisture Content

Microorganisms require moisture to assimilate nutrients and increase colony size. They also produce water as part of the decomposition process. Accumulation of the water due to decomposition will result in oxygen depletion, which will lead to anaerobic conditions (Gray et al., 1971a). Water helps in the transport of nutrients within the mass and creates a physical and chemical environment for assimilation of nutrients by the microbes. The optimum moisture content for microbial activities is between 40 and 65%. Microbial activity almost stops if the moisture content falls below 20% (Haug, 1980).

Usually moisture content of grass is above the optimum range of 40–65% but that of dry leaves is below the range. Although moisture content of municipal solid waste varies widely, it usually requires additional moisture. The moisture content of a feedstock can be determined by drying it at about 219°F (104°C) for 8 hr (USEPA, 1984). The moisture content is the ratio of weight lost to the weight before drying. The necessary amount of water should be added initially to the feedstock and later to the composting materials if the moisture content is below the optimum range. The following approaches are available to overcome high moisture content in a feedstock:

1. Use of previously composted dry material
2. Dry amendments (e.g., sawdust)
3. Bulking agents (e.g., wood chips)
4. Aeration through turning or mechanical agitation.
5. Temporary spreading (when precipitation is not expected)

Air or heat drying can reduce the moisture content of dewatered sludge cakes. Aerobic composting leads to the generation of carbon dioxide, water, and heat, which influences the moisture content of the feedstock. Too high a moisture content will inhibit the temperature rise of the pile; too low a moisture content will inhibit microbial activities, which will also inhibit the temperature rise of the pile.

6.2.3 Temperature

The rate of decomposition is dependent on temperature. Aeration, cooling at the surface, and loss of heat due to evaporation of moisture affect the temperature of the compost pile. As indicated earlier microbial activities are temperature dependent. Microorganisms tend to be more active at the higher end of their respective survival range (i.e., 25–45°C for mesophiles and 45–70°C for thermophiles). The rate of decomposition increases up to an optimum temperature. As shown in Fig. 6.2, because of microbial activities, the temperature in a compost pile tends to change with time. Thus in a composting process attempts are made to maintain the optimum temperature to ensure high microbial activity (USEPA, 1994d).

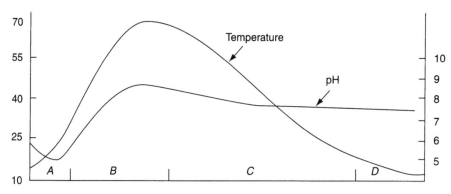

FIG. 6.2 Influence of temperature and pH on microbial activities: *A*, mesophilic; *B*, thermophilic; *C*, cooling; and *D*, maturing. After USEPA (1994).

Temperature control is also critical for destroying weed seeds and disease-causing pathogens. A temperature of 131°F (55°C) should be maintained for several days to destroy pathogens. A temperature between 149 and 160°F (65 and 71°C) needs to be maintained for at least 12 hr to destroy weed seeds. Temperatures between 120 and 150°F (49 and 66°C) are ideal for destroying weed seeds and pathogens but cool enough for the microorganisms to be active. The temperature of a compost pile should not be allowed to rise above 160°F (71°C). At temperatures above 175°F (79°C) the pile will stop composting (Minnesota Extension Service, 1993). Temperature readings should be taken at different surface locations and at various depths in the same surface location. The hottest zone is at the center of the pile. The temperature of a pile depends on its size. Piles should be turned or aerated to maintain temperature and microbial activities. The temperature may be monitored using a long-stem digital or dial-type thermometer.

6.2.4 pH

The pH is a function of the hydrogen ion concentration. The pH indicates the acidic (below 7) or alkaline (more than 7) condition of a substance. For bacteria, a near neutral pH (6.5–7.5) is best, whereas fungi tolerate a wider range of pH (5.5–8). Microbial activities are inhibited significantly if the pH falls below 5.5 or rises above 9. As indicated in Fig. 6.2 microbial decomposition is high between pH range of 5.5. and 9. The pH level in a compost pile changes with time (Fig. 6.2). Initially, the pH is lowered to about 5 due to the formation of organic acids. Fungi, which can tolerate a low pH, are active in this stage of decomposition. As the pH gradually rises due to the breakdown of the acids, bacteria become more active. The rise of pH in a compost pile is a good indicator of the maturity of the pile (USEPA, 1994d).

A near neutral pH is best for composting. pH testing kits are widely available and are easy to use. Acidic materials such as lemon juice or pine needles may be added if the pH is 8 or above, and lime (calcium carbonate) may be added if the pH is 6 or below.

6.2.5 Nutrient Levels

Nutrients in available form and in adequate concentration are required for the desired level of microbial activities. Nutrients needed by microorganisms include carbon (C), nitrogen (N), phosphorus (P), and potassium (K). The microorganisms also need micronutrients and trace elements such as boron, calcium, chloride, cobalt, copper, iron, molybdenum, zinc, and so on. High levels of these and other micronutrients can be toxic to the microorganisms. High carbon-to-nitrogen (C : N) ratios (i.e., high C and low N concentration) reduce microbial growth whereas low C : N ratios (i.e., low C and high N concentration) accelerate microbial activities. For aerobic decomposition, C : N ratio influences composting time. The composting time for a C : N ratio of 20 is about 12 days, which increases to 14 days for a C : N ratio between 20 and 50 (Haug, 1980). A very high rate of decomposition depletes oxygen rapidly, creating anaerobic conditions. Anaerobic conditions in compost piles produce foul smells. Excessive amounts of N in a feedstock may produce ammonia at a high concentration, which can become toxic to the microbial population. Such a toxic condition reduces microbial activities. The chemical form of nutrients is important for microbial uptake. Some microorganisms cannot break down certain chemical bonds. For example, lignin (a chemical compound found in wood) is a large chemical compound with complex bonds. Because of this complexity, it takes longer for microorganisms to break it down. It is for this reason that wood decomposes slowly (USEPA, 1994d).

Accurate calculation of the C : N ratio is difficult. In addition to the unavailability of nutrients described above, microbial decomposition itself may deplete nitrogen (Gray et al., 1971a). The following relationship with an accuracy of 2–10% (Gotaas, 1956) has been proposed:

$$\%C = (100 - \text{percent ash})/1.8 \tag{6.1}$$

Percent carbon (%C) in Eq. (6.1) is obtained by burning the feedstock at 550°C (1022°F) for 1 hr. The equation is not applicable for feedstocks with a high percentage of nonbiodegradable materials (e.g., plastic), which will disappear at 550°C (1022°F) (Polprasert, 1989). Normally the C : N ratio should be 20–25. However, if the feedstock contains a significant amount of carbon, which is not readily available for microbial activities, a C : N ratio of 40 will produce good-quality compost within a reasonable time period.

If the C : N ratio of the feedstock is above or below the optimum range of 20–25, the C : N ratio of a feedstock must be adjusted. The C : N ratios of various organic materials are listed in Table 6.1. Although the C : N ratio of a single item can be determined through laboratory analysis, the C : N ratio

TABLE 6.1   Carbon-to-Nitrogen Ratio of Various Materials

Type of Feedstock	Ratio
High Carbon Content	
Bark	100–130 : 1
Corn stalks	60 : 1
Foliage	40–80 : 1
Leaves and weeds (dry)	90 : 1
Mixed MSW	50–60 : 1
Paper	170 : 1
Sawdust	500 : 1
Straw (dry)	100 : 1
Wood	700 : 1
High Nitrogen Content	
Cow manure	18 : 1
Food scraps	15 : 1
Fruit scraps	35 : 1
Grass clippings	12–20 : 1
Hay (dry)	40 : 1
Horse manure	25 : 1
Humus	10 : 1
Leaves (fresh)	30–40 : 1
Mixed grasses	19 : 1
Nonlegume vegetable scraps	11–12 : 1
Poultry manure	15 : 1
Biosolids	11 : 1
Weeds (fresh)	25 : 1
Seaweed	19 : 1

After USEPA (1994d).

of a mixture (e.g., municipal solid waste) can be calculated. Various materials can be mixed to arrive at an optimum level.

6.2.6   Time

Time is an important factor for producing good compost. Composting can be completed in 6 months or it may take up to 4 years. The time to complete the biodegradation process depends on the composition of waste. Protein and fats (e.g., meats, oils), cellulose, and lignin (e.g., wood, paper) take longer to break down. Curing is important for producing good-quality compost. High-tech composting methods may take about 6 months whereas low-tech methods may take up to 4 years to complete composting. Usually a long curing process results in better quality compost.

6.3 FACILITY DESIGN

Composting facilities vary in size and complexity. While composting in a single-family home may consist of a wooden box or an enclosed area, specialized equipment and a larger area are needed for the composting operation for a big community. A facility can be designed with windrow piles requiring equipment for turning or it may consist of a rotating drum composter. The end use also influences acceptable feedstock and facility design. The entire composting operation may be divided into three parts: preprocessing, composting, and postprocessing.

6.3.1 Preprocessing

The purpose of preprocessing is to separate the materials that cannot be composted from the waste through sorting. Subsequent operations include reduction of particle size of the feedstock and conditioning of the feedstock. A significant effort needs to be put in preprocessing if the waste is a mixed source (e.g., municipal solid waste). Although yard waste or other such single-source feedstock may also require sorting, an elaborate sorting operation may not be needed for such wastes. There is no need to sort if only biodegradable materials are used as feedstocks.

6.3.1.1 Sorting   As indicated, the sorting effort is high if there is no source separation. The end use of the composted material also influences the sorting operation. For example, compost used in landfill final cover may include some contaminants (contaminants in compost refers to objects such as broken glass pieces or chemical contaminants, which may have environmental impacts) compared to compost used for agricultural fields or distributed to the public. Visual inspection and hand sorting may be done for yard trimmings or such single-source wastes. Single-source waste (e.g., yard trimmings) without a significant amount of unwanted objects can be hand sorted more efficiently with the help of a mechanical conveyor belt. The unwanted materials removed from the waste are collected in storage containers. Usually the collected materials do not contain significant amounts of recyclables and are therefore landfilled. Workers involved in hand sorting must follow necessary health and safety measures. Discussion on health and safety issues can be found in Chapter 25.

Since municipal solid waste is extremely heterogeneous, it requires more machinery and labor for sorting materials that are not readily biodegradable. If recyclables are not source separated at curbside, the municipal solid waste will contain a significant amount of recyclables. Like sorting of yard trimmings, municipal solid waste can also be manually sorted with much labor and the help of a mechanical conveyor belt. Mechanical sorting equipment uses physical and magnetic properties of the waste and is generally expensive to purchase, operate, and maintain. Considering labor costs involved in sorting materials after collection, it is much more efficient to sort materials at the point of generation.

Screens Screens separate small materials (e.g., food scraps, glass) from the bulky oversize fraction. The type of screen used depends on the moisture content, heterogeneity, particle size and shape, and density of the materials. Trommel screens are commonly used for screening municipal solid waste (Fig. 6.3).

Magnetic Separators Using electric current, a high-strength magnetic field is generated that attracts ferrous metals (Fig. 6.4). The efficiency of magnetic separators depends on the feed rate, belt speed, size and shape of the ferrous scraps, and the separation distance between the magnet(s) and the belt.

Eddy-Current Separators These machines separate aluminum and other non-ferrous metals from municipal solid waste. An electromagnetic field is created that induces electrical charge in nonferrous metals. This charge causes a repelling force between conductors (e.g., aluminum, copper) and nonconductors (e.g., wood, paper) that causes these items to separate (Fig. 6.5). These machines should be placed after the magnetic separators. Efficiency of the machine for separating conductors from nonconductors depends primarily on the feed rate and belt speed. The strength of the induced charge should be adjusted to provide a high separation efficiency of around 90%.

Ballistic Separators The density and elasticity difference of incoming waste materials are utilized for separating the items. The waste is dropped on a rotating drum. High-density materials such as glass, metals, and stone are thrown at a longer distance compared to organic materials (Fig. 6.6).

Air Classifiers In air classifiers differences in the weight of various materials are utilized to separate them. Thus air classifiers separates heavier fractions (e.g., glass) from lighter fractions (e.g., plastics). Waste is fed into an air column, which is fitted with a blower. The lighter materials are sucked up

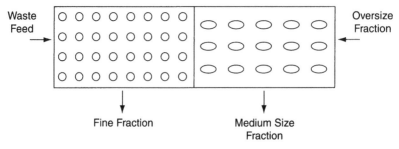

FIG. 6.3 Trommel screen.

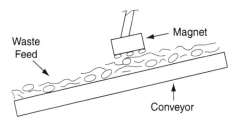

FIG. 6.4   Magnetic separator schematic.

and are collected using a cyclone separator. Heavier materials fall at the bottom of the column. Although most of the heavier materials separated cannot be composted, a fraction of the material that can be composted (e.g., wet paper) will also fall with the heavier materials. If used after size reduction, air classifiers can reduce metal contaminants significantly.

Wet Separators   In wet separators, the waste is fed into a tank of water often with a circulating water stream. The heavy fraction drops to the tank bottom while the lighter fraction floats near the tank surface. Each fraction is removed using stationery or moving bars or rotating screens. Wet classifiers are usually used after size reduction, for separating heavier fraction. One disadvantage of this method is that a significant area may be required for the tank. Also, the feedstock becomes wet and may require dewatering prior to composting.

6.3.1.2 Size Reduction   The primary purpose of size reduction is to increase the surface area, which facilitates composting. However, a very small particle size of feedstock reduces overall porosity of the pile, which inhibits aerobic conditions. Size reduction of woody material is very essential because wood decomposes at a very slow rate. Size reduction increases available surface area for bacterial activities. Normally either hammer mills or shear shredders are used for size reduction. Hammer mills have two counterrotating

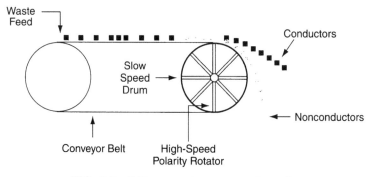

FIG. 6.5   Eddy-current separator schematic.

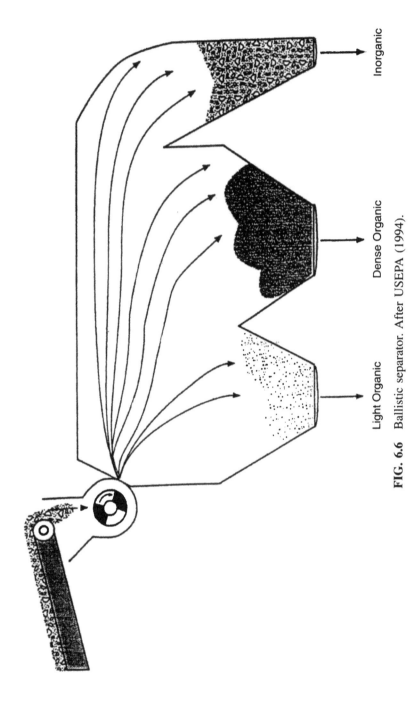

FIG. 6.6 Ballistic separator. After USEPA (1994).

Inorganic

Dense Organic

Light Organic

100

hammers that shred the waste into smaller pieces. The shredded feedstock exits the mill after passing through a grate. In a shear shredder, two counter-rotating knives or hooks do size reductions. A shear shredder cuts through the materials, which enhances decomposition.

6.3.1.3 Pretreatment Pretreatment of feedstock may be needed to enhance composting. Moisture content, C : N ratio, and pH of the feedstock may need to be adjusted to achieve desired results. If adjustment of any of these parameters is done, a thorough mixing of the feedstock is essential. In general, mixing of the pile after size reduction and before starting of the compost pile enhances composting. However, the mixing equipment is costly and requires maintenance. Mixing can be done using a rotating drum, pug mill, or auger mixer. Pug mill or auger-type mixers can compress the feedstock, which inhibits aeration. So, if these mixers are used, care should be taken to loosen the material before or during a windrow formation. Moisture content, C : N ratio, pH, and temperature are adjusted as necessary before starting a pile. Discussion on these topics are included in Section 6.2.

6.3.2 Composting Methods

There are many compost systems, each of which has its advantages and disadvantages. Selection of a composting system depends on volume and nature of feedstock, desired speed of production, funding, and end use. Composting can be done using simple equipment and completely subject to the elements or using complex equipment to create a controlled environment. Aeration may be provided by turning the materials regularly or air may be forced through the materials. In either case, the desired optimum temperature should be maintained at 60°C (140°F) to reduce composting time (Finstein and Hogan, 1984).

6.3.2.1 Passive Composting Passive composting (Fig. 6.7) is usually used only for feedstock mixes that have good porosity and high C : N ratio, such that the microbial process proceeds very slowly and frequent turning is not necessary to aerate the materials. For example, leaves mixed with chipped wood can normally be composted successfully using primarily passive composting, turning the materials perhaps only once or twice per year to mix the materials and eliminate preferential air channels that have formed.

Materials are commonly formed into elongated rows (windrows) with a height and width of about 8 ft (2.5 m), depending on climate and type of equipment used to form the windrows. During the winter in cold climates, windrows will freeze if they are not large enough to produce and maintain sufficient heat to keep the microbes active. Also, to prevent freezing materials the windrows should not be turned after the weather begins to turn cold. A processing pad and roof can improve control and access, as well as reduce environmental impacts, especially in cold or wet climates. Depending on available space and desired rate of composting, winter access and freezing

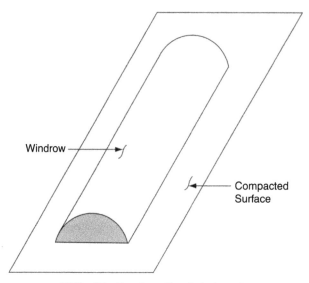

Windrow

Compacted
Surface

FIG. 6.7 Passive piles (windrows).

may or may not be of concern. In hot and dry climates, and where water may not be available, a roof alone may be desirable. A layer of finished compost or wood chips is spread over the top of the pile to minimize loss of heat and moisture, discourage flies, and provide odor control.

6.3.2.2 Active Composting This is similar to passive composting except that the turning is more frequent and a higher moisture content is needed. Active composting is commonly used for feedstock mixes that have low porosity (e.g., municipal solid waste). The windrows may require tuning every 2–3 days depending on the feedstock and state of maturity. Figure 6.8 shows a typical windrow site layout. If no set back distance is mandated by the regulatory agency, then a set back distance of 200 ft (61 m) from surface water bodies, water supply wells, and residence should be used (USEPA, 1994d).

The pile dimensions depend on the type of equipment used for turning, moisture content of the feedstock, weather, and how frequently the material can be turned. Usually the maneuvering spaces for turning equipment is about 3–8 ft (1–2.5 m). If the pile width is w, then the processing pad width for a single pile should be at least $2w + 2s$, where s is the required maneuvering space for the equipment. The length of the pad should be 25 ft + l (7.5 m + l), where l is the length of the pile. The width for multiple piles should be such that the piles can be turned easily from either side. The height and width of windrows may range from 6 to 12 ft (1.8 to 3.6 m). If municipal solid waste is composted, then preprocessing will be necessary to produce a good-quality compost.

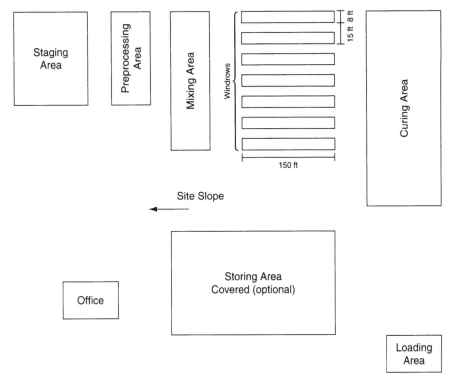

FIG. 6.8 Typical windrow site layout (not to scale).

Figure 6.9 shows a typical forced aerated static pile configuration; forced aerated static piles are essentially static windrows with forced aeration. Air can be supplied to the pile either through suction or forced air. Wood chips are usually spread over the pipes to maintain airflow. A layer of finished compost or wood chips is spread over the top of the pile to minimize loss of heat and moisture, discourage flies, and provide odor control. It may take about 5–6 months to complete composting using this method. Composting time depends on factors such as the C : N ratio and temperature. Aerated static piles are used for granular-type feedstock such as biosolids.

6.3.2.3 In-Vessel Systems   In-vessel composting is a mechanized system. Common designs include horizontal or vertical cylindrical vessels that can be either be flow-through or batch systems. A flow-through horizontal rotating drum composter (Fig. 6.10) continuously mixes the composting materials by a tumbling action. The drum is rotated slowly (10 revolutions/min). Air is forced into the drum through nozzles. The feedstock is composted up to 6 days in the vessel and then cured up to 3 months (USEPA, 1984). In-vessel detention time depends primarily on vessel size and rotation speed. Curing time depends on feedstock type and final use.

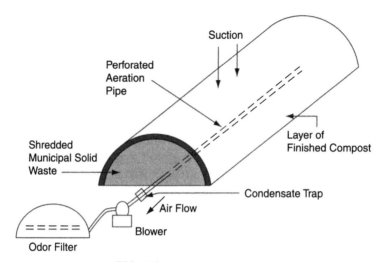

FIG. 6.9   Aerated static pile.

A flow-through vertical tank composter (Fig. 6.11) continuously mixes the composting materials with mixing equipment inside the tank. Air is forced through the holes in the mixing screw. This is a continuous-flow process with a detention time of 5 days. The process can be used for dewatered sludge composting (Haug, 1980). Table 6.2 provides a list of additional in-vessel processes.

6.3.3 Postprocessing

Postprocessing may include curing (for organic stabilization), special treatment to destroy pathogens and remove undesirable materials, and bagging (for product sale). Quality control for meeting the specifications is important to ensure customer satisfaction.

6.3.3.1 Quality Control   Compost should be tested routinely for quality control purposes. Maturity refers to the biological stability. Respirometry and analysis of biodegradable constituents and biochemical parameters assess biological stability (Morel et al., 1984). Seed germination (usually using pea and beans) and plant growth trails are commonly used to confirm maturity. Concentrations of nutrients and toxic compounds are also checked to ensure quality. Long-term monitoring data is useful for process adjustment and marketing.

6.3.3.2 Product Specification   The following eight quantifiable parameters, which represents basic chemical, physical, and biological data of a compost, may be used for product specification: pH, soluble salt content, nutrient content (nitrogen, phosphorus, and potassium at a minimum), water-holding ca-

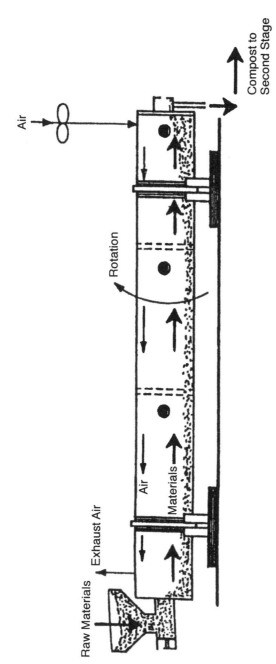

FIG. 6.10 Rotating drum composter. After USEPA (1994d).

105

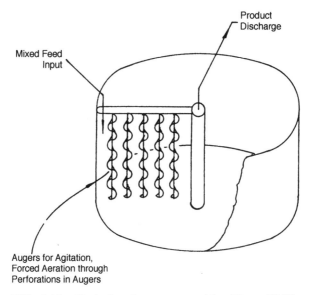

Product
Discharge

Mixed Feed
Input

Augers for Agitation,
Forced Aeration through
Perforations in Augers

FIG. 6.11   Vertical tank composter. After Haug (1980).

pacity, bulk density, moisture content, organic matter content, and particle size (Composting Council, 1995). Table 6.3 includes a range of values of these parameters for various applications. Recommended test methods for these parameters can be found elsewhere (Composting Council, 1994). The finished compost should be stored properly to ensure quality. Storage under the roof should be provided for unsteady or seasonal markets.

6.4   ENVIRONMENTAL ISSUES

Composting can cause detrimental impacts to groundwater, surface water, soil, and air. Proper design and operation is necessary to prevent such impacts. Compost quality and use applications must also be considered to prevent such impacts.

6.4.1   Leachate

Leachate from a composting operation can contain high levels of biological oxygen demand (BOD), nitrate, and phenols. In addition, the leachate may contain high levels of toxic metals, synthetic compounds such as polychlorinated biphenyls (PCBs), and persistent herbicides. The age of the pile also affects the composition of the leachate. Table 6.4 provides a list of toxic chemicals that may be found in compost. It may be noted that currently the database of leachate quality from composting operations is not large. Both total organic carbon and total nitrogen in the leachate is reduced with the age

TABLE 6.2 Various In-Vessel Composting Processes

Serial No.	Process Name	Description
1	Earp Thomas	Composting is done in a silo with eight decks stacked vertically. Agitation is done using a shaft at the center, which also helps in downward movement of feedstock from deck to deck. Composting done for 2–3 days in the silo, followed by windrowing.
2	T. A. Crane	Composting is done in two cells, each with three horizontal decks. Horizontal ribbon screws extending the length of each deck recirculate the feedstock from deck to deck. Air is introduced through the bottom of the cells. Composting is done for 3 days followed by a 7-day curing.
3	Varro	Feedstock is placed in eight deck digesters. A plow moves the feedstock downwards. Air is introduced in each deck. Composting time is 40 hr.
4	Triga	Composting is done in a concrete tower, which is divided into four separate vertical components. A screw-extractor agitates the materials and removes compost from bottom. Extracted material is reintroduced three to five times during composting. Air is introduced at the bottom of the tank, which is allowed to exhaust from top. Composting time is 4–10 days, with curing time of 2–4 months.
5	Euroma	The reactor is similar to Triga process with an agitation mechanism at the bottom of tank. Composting time is 6 days. After composting the material is windrowed. This is a batch process.
6	Kneer/BAV	The composting is done in a tower. Feedstock is introduced at the top and removed from the bottom using a mechanical scrapper. Air is introduced through the bottom and exhausted from the top. Composting time is 7–12 days.
7	Geochemical Eweson	Composting is done in slightly inclined rotating drum. The drum is 11 ft (3.4 m) in diameter and 110 ft (34 m) long. The drum consists of three compartments. Feed stock is transfered from one to the next compartment every 1–2 days. The total digestion time is 3–6 days. The output is screened and cured in open piles.
8	Agitated box	Composting is done in rectangular tanks for about 7 days. A compost turning machine mounted on tracks on top of each wall of a box turns the feestock. In addition, aeration is done through the floor of the boxes. Curing is done for 2–4 weeks after composting in the box. The boxes are 6–20 ft (1.8–6.1 m) wide, 3–10 ft (0.9–3 m) deep, and 200 ft (61 m) or more in length.

Adapted from Haug (1980).

TABLE 6.3 Recommended Range of Parameters for Various Applications

Parameter	Soil Amendment					Media Component		Media	Surface Application	
	Turf	Vegetable Crop	Silviculture	Marginal Soils	Planting Beds	Horticultural Substrate	Blended Topsoil	Sod Production	Landscape Mulch	Erosion Control
pH	5.5–8.0	5.0–8.0	5.5–8.0	5.5–8.5	5.5–8.0	5.5–8.0		5.0–8.0	5.5–8.0	
Soluble salt content	Must report, 4 dS/m maximum for soil blend	Must report, 6 dS/m maximum for soil blend	Must report		Must report, 2.5 dS/m maximum for soil blend	Must report, 3 dS/m maximum for soil blend	Must report, 6 dS/m maximum for soil blend	Must report, 3 dS/m maximum	Must report	
Nutrient content			Must report			Must report		Must report	Must report	
Water-holding capacity			Must report			Must report		Must report	Must report	
Bulk density			Must report			Must report		Must report	Must report	
Moisture content			33–55%			35–55%		35–55%	35–55%	Must report
Organic matter content			Must report			Must report		Must report	Must report	
Particle size	1 in. minus				1 in. minus	½ in. minus	Must report	⅜ in. minus	Must report	
Stability	Stable highly stable		Moderately to highly stable		Stable to highly stable	Highly stable	Moderately to highly stable	Stable to highly stable	Moderately to highly stable	Must report
Growth screening			Must pass			Must pass		Must pass	Must pass	Test not required
Trace elements/ heavy metals	Must meet USEPA Part 503 Exceptional Quality Concentration Limits		NEED NOT meet USEPA Part 503 Exceptional Quality Concentration Limits	Must meet USEPA 503 Exceptional Quality Concentration Limits		Must meet USEPA Part 503 Exceptional Quality Concentration Limits		Must meet USEPA Part 503 Exceptional Quality Concentration Limits	Must meet USEPA Part 503 Exceptional Quality Concentration Limits	

From Composting Council (1995).

TABLE 6.4 Possible Toxic Chemicals in Yard Trimming Composts

Chemical	Range (mg/kg)
Pentachlor phenol	0.001–0.53
Chlordane	0.063–0.37
apDDT	0.004–0.006
ppDDT	0.002–0.035
DDE	0.005–0.019
Aldrin	0.007
Dursban	0.039
Dieldrin	0.019

After Hegberg et al. (1991a).

of the pile. Leachate generation can be minimized or prevented by monitoring the moisture content of compost piles and making sure adequate oxygen is available throughout the pile. If the feedstock contains excess moisture, it will generate leachate initially even without the addition of water.

Control of leachate from in-vessel systems is relatively easy. Placing the piles under a roof can minimize leachate generation from windrows. Composting should be done on a low-permeability base pad or concrete pad so that leachate cannot seep into groundwater from the base of the compost piles. Any leachate accumulated on the compost pad should be collected and may be reintroduced into the pile. However, if the pile is already past the high-temperature phase, then the reintroduction will add harmful microorganisms from the leachate, which were destroyed by the high temperature.

If a processing pad is used, it should provide a good working surface capable of withstanding the weight of the equipment. A low-permeability pad and leachate collection may be required by the regulatory agency. To construct a low permeability pad, clay with a hydraulic conductivity of 1×10^{-7} cm/sec or less is considered suitable (see Section 18.1 for clay specifications and Section 20.2 for construction practices). To prevent rutting and maintain the integrity of the low-permeability clay layer, a geotextile should be placed over the clay, and then 4–6 in. (10–15 cm) of stone should be placed over the geotextile. Other suitable construction material for the pad include asphalt, concrete, and soil admixtures.

6.4.2 Runoff and Runon

Runoff from a compost area can contaminate surface water bodies nearby. Waste contact liquid from municipal solid waste is considered as leachate and can be generated from the tipping floor or from the process floor. The facility design should include provision for collection, treatment, or disposing of waste contact liquid. Like runoff, runon can also create problems for a site. Runon of water will add too much water to the piles and make working conditions difficult. The design should include provisions such as diversion channels and dikes to divert runon water around compost areas.

6.4.3   Air Quality Including Odor

Dust, bioaerosols, and odor are major sources of air pollution from composting facilities. Although volatile organic compounds (VOCs) such as chloroform and trichloroethylene can be found in municipal solid waste compost facilities, the off-site migration of VOCs is not significant. Air pollution from vehicular traffic used for delivery of materials for composting and pick up of compost can be of concern for large composting facilities. At such facilities proper scheduling should be done to minimize air pollution from vehicular traffic. In addition all on-site equipment should be maintained properly to meet emission standards (USEPA, 1994d).

6.4.3.1   Dust   Dust can be a problem, especially in dry weather. Dust is generated from the tipping floor, screening and shredding operations, and by placing feedstock in windrows. Dust generation can be minimized by keeping feedstock moist, constructing asphalt or concrete driving surfaces, watering the unpaved surfaces, and providing engineering controls for dust suppression in enclosed operations.

6.4.3.2   Bioaerosols   A variety of biological aerosols (bioaerosols) can be generated during composting. Bioaerosols are suspension of particles containing microorganisms. Actinomycetes, bacteria, viruses, molds, and fungi can be found in bioaerosols. Bioaerosols can also carry endotoxins, which are released upon destruction of a microorganism. Since bioaerosols are carried by dust, proper dust control from the process area will minimize bioaerosol levels.

6.4.3.3   Odor   Odor from a composting facility can cause serious problem, which may even lead to the closing of a facility due to public opposition. Odor from a composting facility can be produced from the staging, preprocessing, and mixing areas. In addition, if the composting is done in open piles, then odor can be produced from the process area as well. Odor from a composting facility depends greatly on the type of feedstock. Odor from composting of leaves is less compared to municipal solid waste or wastewater sludge. Odor is generated due to improper composting design and operation. Anaerobic decomposition generates odor due to the formation of various odorous compounds. A common problem is inadequate drainage, such that standing water contacts the compost, causing an anaerobic condition at the base of pile. Table 6.5 lists compounds known or suspected to generate odor. Rate and extent of microbial action influences odor from a composting facility (Finstein and Hogan, 1993). Moist and nitrogen-rich materials, such as grass

TABLE 6.5   Compounds either Specifically Identified or Implicated in Composting Odors

Sulfur Compounds

Hydrogen sulfide	Dimethyl disulfide
Carbon oxysulfide	Dimethyl trisulfide
Carbon disulfide	Methanethiol
Dimethyl sulfide	Ethanethiol

Ammonia and Nitrogen-Containing Compounds

Ammonia	Trimethylamine
Aminomethane	3-Methylindole (skatole)
Dimethylamine	

Volatile Fatty Acids

Methanoic (formic)	Butanoic (butyric)
Ethanoic (acetic)	Pentanoic (valeric)
Propanoic (propionic)	3-Methylbutanoic (isovaleric)

Ketones

Propanone (acetone)	2-Pentanone (MPK)
Butanone (MEK)	

Other Compounds

Benzothiozole	Phenol
Ethanal (acetaldehyde)	

After USEPA (1994d).

clippings or food waste, will become anaerobic quickly and may be a source of odor upon delivery.

Therefore, both the age of the material upon delivery the and nitrogen content are important. If odor develops even after proper care is taken in design and operation of the facility, then the following actions should be taken to assess the problem:

1. Identify the source(s) of odor. Facility or process design and practice should be changed as needed to correct the situation.

2. Develop a database indicating the intensity (qualitative), frequency, and odor type (e.g., rotten egg), the wind direction, time of day, and month(s) when odor is considered to be high. It is a good idea to enlist volunteers or designated persons from the neighboring community who will work in developing the database.

3. Monitor the air to find out whether any compound with air standards is associated with the odor. Subjective odor quantity known as effective dilution (ED) may be used to assess odor. ED_{50} is the number of effec-

tive dilution required so that 50% of a panel of 10 people can detect the odor. Odor standards can be based on odor measurements, number of complaints, or existing legal standard, if any (USEPA, 1992b; Walker, 1993).

Careful monitoring and process control to avoid development of anaerobic conditions will help reduce odor production. To minimize odor, attempts should be made first to change the process. The following practices helps to reduce odor:

1. Bulking agents should be added to already odorous incoming feedstock to increase the C : N ratio to 30 (Glenn, 1990).
2. Feedstock should be placed in windrows as soon as possible, preferably by the same day.
3. The windrow dimensions should be such that the oxygen can penetrate easily from outside.
4. Windrow turning must be done frequently enough to prevent formation of an anaerobic core.
5. Odorous piles should be broken down and spread out for drying.
6. Mixing of dry compost with odorous piles is helpful.
7. Constructing a roof can help control moisture and temperature, if needed.

Engineering controls are available for odor control. The choice of odor control depends on odor sources, degree of reduction sought, and the compound causing odor. Instituting engineering controls for odor control from open windrow type of facilities are impractical since the volume of free air is virtually unlimited around the piles. Control technologies for treating odorous air can be used effectively for forced air and in-vessel systems. Available technologies include biofilters, wet scrubbers, adsorption, dispersion enhancement, and combustion. While combustion is the most effective method for odor control, it is expensive (Ellis, 1994).

Biofiltration   In a biofiltration system, the odorous gas is forced through a moist filtration medium such as compost, soil, or sand. As the gases pass through the medium, odor is removed by physical, chemical, and biological processes. The process is largely self-sustaining because of microbial activities that oxidize the odorous gasses (Williams and Miller, 1992). The removal of sulfur compounds such as H_2S and dimethylsulfide was enhanced by seeding fresh peat with *Thiobacillus* and *Hyphomicrobium* 155, respectively (Shoda, 1991). In an open biofilter design, the filter medium is placed below grade and the odorous gas is introduced at the bottom of an 1-m-thick biofilter (Fig. 6.12). In a closed biofilter air is introduced at the bottom of a biofilter placed on top of a perforated support plate (Fig. 6.13). The filter is housed in a concrete tower. The moisture content of the biofilter must be appropriate

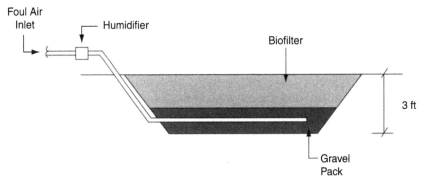

FIG. 6.12   Open biofilter.

for microorganism activity. Usually soil and compost are mixed to create a porous biofilter with large-particle surface area. The pH of the filter should have sufficient buffering capacity and the foul air must be evenly distributed for desired odor control. Organic biofilters must be replaced periodically as decomposition reduces organic fraction.

Air Scrubber   In air scrubbers, the foul gas is "cleaned" primarily through oxidation and absorption. Although several types of air scrubbers are available, a multistage air scrubber (Fig. 6.14) is required to achieve comprehensive odor removal (Ellis, 1994).

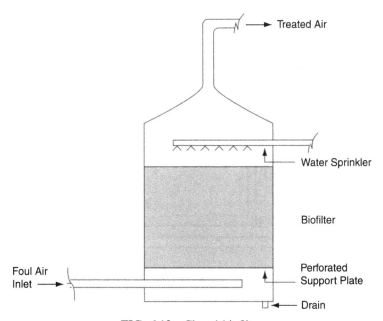

FIG. 6.13   Closed biofilter.

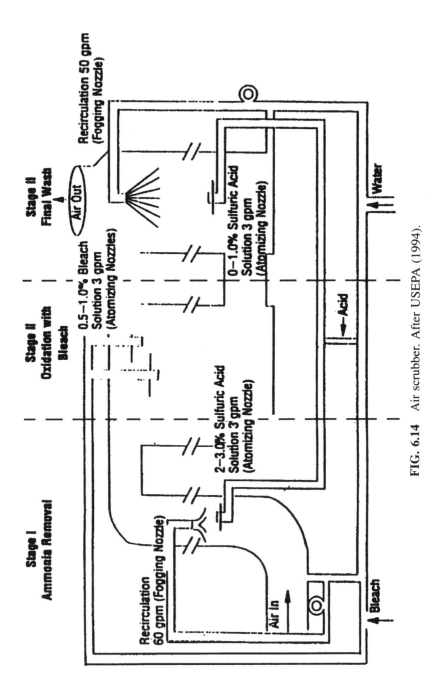

FIG. 6.14 Air scrubber. After USEPA (1994).

114

6.4.4   Environmental Contaminants in Compost

The primary benefit of compost is to enrich the organic matter in the soil with nutrients. If the municipal solid waste contains metals and toxic chemicals, they will contaminate the compost. Such metals and toxic chemicals may eventually bioaccumulate in animal directly (through ingestion of soil) or indirectly via the food chain. The presence of lead (up to 500 mg/kg) in municipal solid waste compost is of concern to many (USEPA, 1994d). Municipal solid waste compost may also contain a substantial level of boron, which can be phytotoxic if applied at a high rate to crops such as beans and wheat. Toxic chemicals such as PCBs, polycyclic aromatic hydrocarbons (PAHs), and polychlorinated aromatics (PCAs) may be present in municipal solid waste. Although PCBs are sometimes found in compost, their concentrations are below the level of concern. PAHs may be phytotoxic. Although PCAs may bind to organic fraction of compost, sufficient information regarding their bioavailability is not available. Additional studies are necessary to fully understand the risk of the presence of these chemicals in compost (USEPA, 1994d).

Source separation prior to composting is necessary to reduce contamination of municipal solid waste compost. Concentration of metals such as zinc, lead, and cadmium in compost was found to be significantly reduced when source separation was undertaken. (Oosthnoek and Smit, 1987). In practice, it is difficult to remove many of the materials containing heavy metals from municipal solid waste.

6.4.5   Vectors

Vectors are small animals or insects that can carry diseases. Mice, rats, flies, and mosquitoes are often found at compost facilities. The vector nuisance can be minimized by running a clean facility and maintaining aerobic conditions within the windrows.

6.4.6   Litter

Paper, plastics, and other lightweight materials create littering problems around composting facilities. Litter from a compost facility can be caused by the incoming vehicles if they are not covered and during tipping and preprocessing of compost materials. To minimize litter, the incoming vehicles should be covered. Additional discussion on litter control can be found in Chapter 23.

6.4.7   Noise

There are many sources of noise at a composting facility. The incoming and outgoing vehicles, and preprocessing equipments such as hammer mills create noise. Noise related regulations may specify a noise level [expressed as decibels (dB)] at property boundary. To prevent hearing loss, workplace noise levels should not exceed 85 dB (USEPA, 1994d). Table 6.6 provides noise levels of various equipments and locations related to composting facilities. The following measures may be taken to reduce noise levels at a composting facility (USEPA, 1994d):

1. Provision for adequate buffer zone around the facility
2. Inclusion of noise-reducing features such as mufflers and noise hoods
3. Proper maintenance of equipments
4. Proper coordination with fleet operation to control traffic flow
5. Imposition of restriction on noise levels of vehicles

6.4.8 Fire

There is a potential of fire hazard in improperly maintained composting facilities. Organic material can ignite spontaneously if the moisture content is between 25 and 45% and the temperature is above 199°F (93°C). This type of condition may develop in piles 12 ft (4 m) or more in height (USEPA, 1994d). Routine monitoring of piles or composting vessels should be done to ensure the temperature and moisture content stays out of the danger levels. In addition, the windrows and other aspects of the facility should be designed with fire protection and fire fighting in mind. The facility layout should ensure enough space for fire fighting and fire fighting devices to put out small fires, and water supply should be provided on-site. The site manager should be trained in spontaneous combustion issues and dealing with small fires. Emergency contacts for bigger fires should be readily available. Size of windrows, moisture content, and temperature of windrows should be monitored for both

TABLE 6.6 Noise Levels of Various Equipment and Locations

Equipment/Location	Weighted Sound Pressure Level in dBA
Tipping floor	85–90
Shredder infeed	85–90
Primary shredder	96–98
Magnetic separator	90–96
Secondary shredder	91–95
Air classifier fan	95–120
Shop	78
Control room	70

Adapted from USEPA (1994d).

compost quality and fire prevention. Fire drills should be done on a regular basis. The local fire department should be contacted at the time of site design to ensure that all fire codes are implemented. Periodic tours by local fire department personnel can be very helpful. Any suggestion(s) and/or advice for fire protection arising out of such a tour should be implemented to ensure fire safety. It may be noted that insurance companies may deny payment of fire damages if fire code violation(s) is(are) found.

6.5 HEALTH AND SAFETY

Workers in composting facilities may be exposed to bioaerosols, toxic chemicals, and noise, in addition to general physical safety and equipment-related accidents. Adequate worker training on health and safety issues will help minimize injuries. Proper training regarding safe operation of equipment is essential. The windrow turning equipment usually have high-speed rotating mixing flails, which can throw stones or wood chips a long distance. Employees and site visitors need to maintain safe distances from such equipment during operation. The presence of gas cylinders and other ignitable liquids in municipal solid waste, if not sorted out prior to size reduction, can explode in equipment such as a shredder. This type of equipment should be located in a separate room with adequate structural protection. Proper safety design for such equipment is essential for worker safety. Additional discussion on health and safety issues can be found in Chapter 25.

6.6 MARKETING

Market assessment should be done at the early stages of the planning process. It should be noted that while composting is not highly profitable, it can generate revenue to offset the cost of running an operation. The marketing can be done directly by the community or rely on private companies that specialize in composting business. The private company may collect compost from several city-owned operations in a region and market it directly.

Compost price depends on quality, transportation cost, production cost, and cost for implementing regulatory requirements. Like any other new product on the market, the price of compost should be competitive. Although some communities provide compost free of charge, such a practice can create an impression that the compost has no value.

Cost of transportation is a key item in the compost price. In general, the market should be within a 20-mile radius. Bagging can increase the sale of compost and increase the market area, but bagging requires additional investment in capital and maintenance costs (Razvi et al., 1990). Municipalities may use the compost they generate on their own streets and gardens. In addition, independent expert opinion can help establish the benefits of composting.

In addition to the issues related to marketing of the product, the "idea needs to be sold" to the community. Without community support a composting operation cannot be successful. In public education campaigns, both the environmental benefits and the cost benefit of composting should be emphasized. Compost price may be judged against the cost of disposal of the materials composted. Usually the cost of composting is competitive with land disposal. The environmental benefits of reduced landfill and/or combustion cost is a key issue that must be stressed in public education. There are several hidden cost benefits from composting that should be added to estimate savings for the integrated solid waste management program of a community.

6.7 ECONOMICS

Good financial planning is key to the success of any program, and composting is no exception. Although funds collected from the direct sale of the product may not be significant, composting will reduce solid waste management program costs. The following are indirect economic benefits of composting:

1. Extending the life of an existing landfill, which amounts to a significant reduction in both short-term and long-term cost of land disposal
2. Possible reduction in landfill tipping fees due to extended site life
3. Savings from the purchase of soil amendments used for streets and parks
4. Revenues from selling of the compost
5. Creation of new jobs within the community (Kashmanian and Taylor, 1989)

The cost–benefit analysis must be used while campaigning for composting. Cost for developing a composting program typically include:

1. Capital costs for establishing a facility
2. Cost of operation and maintenance of the program

The capital costs usually include site acquisition, land improvement, equipment procurement, process design, site construction, permit application fees (if any), and training. Operation and maintenance cost include:

1. Collection (if on-site drop off not used), including curbside collection or cost for running a drop-off collection site
2. Labor cost, equipment operation, and maintenance
3. Marketing
4. Education
5. Monitoring, which includes both for quality control and those mandated by any regulatory agency (Goldstein et al., 1989)

7

REUSE AND RECYCLING

Reuse and recycling are no longer optional but essential to address the long-range problems associated with solid waste management. Recycling should not be viewed as environmental activism but as an integral part of integrated solid waste management program. Because of environmental and socioeconomic problems associated with disposal of waste on land, siting landfills is becoming increasingly difficult even in the United States where the land-to-population ratio is high. It is a much greater problem in countries where land availability is rather low. While from an environmental impact standpoint incineration (when properly designed and operated) is better than landfilling (Taylor, 1991), incineration has it's limitations too. Recycling may not eliminate the need for landfills, but it does help extend the life of existing landfills and delays the need for siting new landfills. In many instances recycling programs do not generate income for the community but the same is true with landfills. A community has to pay to throw waste in landfills. Because of environmental issues, public demand, and a growing lack of disposal alternatives for solid waste, governmental agencies are mandating recycling. If the mandates are not implemented, the community is likely to face enforcement actions. Therefore, there are legal issues associated with recycling. Recycling should be viewed as part of an overall environment friendly, low-cost waste management strategy.

Recycling saves resources. It should be noted that throughout human history access to the total amount of natural resources has increased. However, things are different now. Currently, many valuable resources are depleting. Although substitutes are being discovered for many resources, still many others are not replaceable. The depletion rate for most of the resources is very high. This resource depletion is causing communities of the world to collide for resource control. Apart from the energy savings associated with recycling (e.g., manufacturing an aluminum can from a recycled can rather than from raw materials saves half a can of gasoline), there are direct environmental consequences of mining. Because of environmental awareness, communities living near natural resource repositories often vehemently oppose further harnessing. There are many instances, in which resources are harnessed with significant impact to the neighboring ecology and human communities. Thus recycling is the right thing to do, not only for environmental, monetary, and legal reasons but for ethical issues as well.

This chapter will address recycling and reuse of various materials. Reuse of industrial waste is covered in Chapter 8 and reuse of contaminated land is covered in Chapter 10.

For a successful recycling program both planning and marketing are very important.

7.1 PLANNING

To plan a recycling program, the target community must be identified properly. For a recycling business the customers who bring or sell recyclable materials to the business may be considered as the community. For a local government the community refers to the political boundary of the government. Since from an economic standpoint participation of a larger community is advantageous, two or more contiguous local governments may join together to run a recycling program. For proper planning purposes a budget for running the program must be developed first. It is possible to enlist local residents as volunteers. However, for initial planning purposes such free assistance should not be taken into account. In the budget, funding should be set aside for campaigning for volunteer assistance. A proper estimate must be done for expenses related to building, procurement, equipment, maintenance, outreach, and personnel needs. The budget should include both the projected income and expenditure for one year and a long-range prediction for 7–10 years or more. Both the short-term and long-term goals of the program, and the list of materials to be recycled, should be clearly identified. The plan should indicate a baseline year and method to document the baseline against which the success or failure of the recycling program will be judged. The organizational structure should be chosen carefully. A recycling program may be run by a local government, a community-based nonprofit organization, or by a business. If funding is received from outside (either government or private/individual), then proper accounts as mandated by law must be maintained. In addition, the organizational structure should identify person(s) responsible for specific tasks to run the program. If run efficiently, with proper financial records, it is easy to discover problem source(s) and their rectification. Since trained personnel are required for operating most of the recycling equipment, the plan must identify the personnel needs and their training needs. The plan should identify the milestones of success and probable future expansion both in terms of materials and community boundary. Additional comments regarding planning are included in Section 2.1.

7.2 MARKETING

Establishing a steady market for recyclables is a difficult task. Although environmental, regulatory, and ethical reasons have been the prime driving force behind recycling, currently emphasis is being given to develop end-use markets. However, this imbalance is diminishing slowly. Many industries are becoming interested in recycling for various reasons, one of which is economics. Recycling markets have suffered both for financial reasons and bad press. One of the prime reasons for the misperception regarding the nonviability of recycling is that all tangible and nontangible benefits are not accounted for while estimating the cost–benefit ratio of recycling. Another reason was the initial imbalance between material collected and end-use market.

Not only community recycling personnel but governmental agencies, businesses, and consumers also have a role in developing a recycling market. The role of each of these entities is discussed in the following sections.

7.2.1 Role of Governmental Agencies

Both state and federal agencies have significant roles in developing recycling programs. Apart from promulgating recycling-related laws and administrative rules, governmental agencies also provide financial incentives, in the form of grants, for initiating programs. Laws banning recyclable items from municipal solid waste landfills also help promote recycling programs. Such laws draw attention to environmental issues related to solid waste management. However, if banning of recycling items from landfills is not done gradually, then selling of recyclable materials becomes a problem due to the lack of time for market development.

Governmental agencies also need to provide incentives to end users for the development of processing methods and technology and for replacing virgin materials with recycled materials. Governmental agencies provide funding to research institutes and universities for the advancement of recycling technology. There are several ways of providing incentives to businesses to increase markets for recyclable materials. Loan or loan guarantees and tax incentives for businesses involved in recycling are common. Grants for startup costs and short-term subsidies are also provided to help develop markets. Local and state governments can also provide tax credits for properties owned by recycling businesses and business investments.

7.2.2 Role of Community Recycling Personnel

The market development issues discussed in this section are applicable to both local government officials involved in recycling programs and nonprofit recycling organization personnel. Establishing markets for recyclable items collected in a community is a difficult task. The price of recyclable items often fluctuates significantly, which makes it difficult to run a recycling program. So part of the revenues generated in a good market should be saved for "rainy days." Bigger communities collect a high enough volume of recyclables whereby direct marketing to manufacturers becomes possible. However, smaller communities often need to contract with a broker for selling the materials. Knowledge about the manufacturers specifications regarding raw materials is essential when designing a program. For instance, there are several grades of plastics [e.g., polyethylene terephthalate (PET), polypropylene (PP)], each of which must be sorted separately for marketing purposes. Some, not all, of these can be blended for recycling purposes. Contamination of recyclable items (e.g., broken glass pieces mixed with paper), reduce their market value. These issues should be discussed clearly with the broker when developing the contract. Rather than picking up materials from the community warehouse, the contract may specify delivery at the broker's or manufacturer's warehouse. Since cost of transportation reduces recycling revenues, every effort must be made to reduce this cost. Transportation cost can be reduced by packing more materials in a truck and by reducing the number of trips of trucks. To pack more materials in a truck, a baler may be used to reduce the volume of recyclable items. To minimize the number for trips in a week/month, each truck should be packed to the maximum capacity. If the total quantity of materials required for a full truck load is not generated within a week/month, then sufficient space should be available at the community recycling center to store the materials over the time period. In general, paper makers prefer baled rather than loose newsprint. The installation of a baler and creation of more storage space will increase both the capital and operational cost of the recycling program. This increase in cost should be compared with the transportation cost to optimize the program cost. All these issues should be taken into consideration when negotiating a contract with a broker or manufacturer.

Local community recycling program personnel have very few options to provide economic incentives for a recycling business except a reduction in property and sales taxes. It may adopt rules for using recycled products within the local government offices. However, a local government can enhance a market development by informing residents about the benefits of recycling. Details regarding public information campaigns can be found in Chapter 2. It should be noted that recycling does not include only plastics, paper, and glass but many other materials (e.g., oil, tires). Recycling of construction and demolition (C&D) waste is also increasing.

7.2.3   Role of Businesses and Manufacturers

Businesses and manufacturers involved in dealing with recycled materials have a role different from the two bodies discussed in previous sections. It is a well-established fact that the use of recycled materials leads to energy savings and causes less pollution compared to the use of virgin materials. In the late 1970s when recycling was just at the development stage, there was a lot of debate regarding the benefits and viability of recycling from an economic standpoint. However, over the years recycling programs have shown that recycled materials reduce the use of resources. Because of the environmental awareness of consumers and economic benefits of substituting virgin materials with recycled materials, more and more manufacturers are recycling. Because of this benefit, more manufacturers are committing capital for research and development for substituting virgin materials with recycled materials. This type of research is not restricted to the few items recovered from municipal solid waste, but goes well beyond. Businesses and manufacturers ideally should work together with community recycling programs to ensure a steady supply of recycled materials.

Integrating recycling materials in manufacturing processes takes much longer than developing a collection program for recycling. For instance, there are many steps for the installation of equipment to use recycled paper in paper mills. Such installation may need several permits from state (may even include federal) and local governments, allocation of company funding for the project, equipment installation, and so forth. Thus the necessary lead time could be several years. In addition to the time needed for these items, time is also necessary to develop a technology. Since recycling is demanded by many consumers, businesses and manufacturers that want to respond to consumers should adopt recycling. Manufacturers may increase their sales if they incorporate recycled materials in their products and label them.

7.2.4 Role of Consumers

In order to increase recycling markets, consumers must be willing to buy products with recycled content. If consumers avoid buying products with recycled materials, then it will ultimately affect the recycling. However, in general consumers have shown strong preference for using products with recycled material content. Consumers are increasingly scrutinizing the labels. Consumers, working through local governments and institutions, may include mandates to procure products with recycled materials content in the procurement standards. Consistent and strong consumer demand for recycling is essential for long-term viability of recycling program and market development as a whole.

7.2.5 Summary

As discussed in the above sections, recycling market development is not a task to be carried out by any one group. Since recycling of materials compared to the use of virgin materials is less costly, everyone involved needs to support recycling. Contrary to current belief, recycling is a better economic choice than the use of virgin materials. This is not apparent because the hidden costs of using virgin material and landfilling of recyclable materials are not taken into account when comparing their costs with recycling. However, there are situations where use of recycled material may compromise the quality of the end product. This issue should be considered when mandating recycled content. Since clear labeling about the recycled material content is an important factor in market development, there is a need to develop a national or international standard and/or policy for labeling products with recycled material content. While laws and regulations for recycling play a key role in market development, manufacturers and businesses also need to step forward to sustain recycling.

As indicated in Section 7.2.3, adaptation to recycling by manufacturers is time consuming. The apparent lack of viability of recycling program could be attributed in part to this lag time for adaptation. Overlooking consumer's preference for recycling may hurt an industry in the long-run. Recycling program personnel also need to develop alternative uses for recycled materials to avoid a "glut" when the market is down. Although recycling is becoming integrated into the main-stream economy, communities need to develop strategies to address market fluctuations (e.g., more storage space for recyclables).

7.3   INFORMATION ON RECYCLABLE MATERIALS RECOVERED FROM MUNICIPAL SOLID WASTE

There are several recyclable materials recovered from municipal solid waste. Basic information regarding processing of some of these materials is discussed in the following sections. This information will help in collection and sorting of the materials for proper marketing. Details of reclamation technology and industrial processes for incorporation in products are beyond the scope of this book.

7.3.1   Plastics

The following six types of resins are primarily used in disposable packaging: polyethylene terephthalate (PET), high-density polyethylene (HDPE), low-density polyethylene (LDPE), polyvinyl chloride (PVC), polypropylene (PP), and polystyrene (PS). PET is commonly used to manufacture milk containers. Usually pigmented, HDPE is used for manufacturing laundry detergent containers. LDPE is used for making plastic bags used in grocery stores. PVC is extensively used for manufacturing plumbing pipes. Its use as a packaging material is being reduced. Bottle caps are usually made of PP. PS is used for manufacturing plastic spoons, forks, and so on.

Usually sorting postconsumer plastics into different resin types is done by hand in a material recovery facility (MRF). Workers, standing next to the conveyor belt carrying the discarded items, do the separation by hand. Plastics can also be separated using a microsorting technique, which separates chopped bottles with centrifugal force. Currently, this technology is used to separate PET from HDPE (Pearson, 1993). After plastics are separated into the various types, each type is chopped up and cleaned in a detergent bath. Further separation of light and heavy plastic components is done using hydrocyclone. The light component is bagged for sale after drying. The heavy component is further cleaned using electrostatic process (Rankin, 1993).

Commingled plastics, for which sorting is not done, can also be used to manufacture new products (e.g., plastic lumber). Usually fillers and reinforcing agents are mixed when casting a product to enhance physical properties.

While reclaimed PET is cheaper than virgin PET, reclaimed HDPE is not cheaper than virgin HDPE. So all recycled plastics do not have the same economic value. However, with new product ideas and technological advances, it may become economically possible to reuse the reclaimed resins in the future.

7.3.2   Paper

There are three categories of paper salvaged from municipal solid waste stream—newspaper, corrugated paper, and office paper. Newspaper is also directly recycled by newspaper publishers who send their scrap papers directly to mills. Successful newspaper recycling depends heavily on the "cleanliness" of the collected paper. However, contamination (e.g., food products commingled with newspaper) reduces the value of the recycled paper.

First, the paper is deinked before mixing the processed pulp with virgin pulp. Currently paper recycling technology is quite developed whereby the value of recycled pulp and virgin pulp are competitive. Although most categories of paper are recyclable, there are problems with certain types of papers. Papers with holograms are difficult to recycle. Colored paper must be separated from white paper for the deinking process. Glues on paper (e.g., glue binding on a notebook) makes deinking difficult. Currently a different type of glue is being used to improve the situation, which allows adhesion under normal thumb pressure. Paper can be sorted before or after collection. When sorted properly into different categories, recycled paper can provide a revenue stream.

7.3.3   Aluminum Cans

Beverage cans are made of aluminum. Aluminum cans provide high revenues and can be recycled easily. Ninety-five percent less energy is required for manufacturing aluminum cans from recycled materials (Buckholz, 1993). Recovered aluminum cans are melted in a furnace and mixed with other materials to form an alloy suitable for industrial use (Buckholz, 1993).

Aluminum cans are easily separated from a commingled recycle materials using eddy currents (see Section 6.3.1.1). In general, aluminum cans are not sorted by hand. The recycling manufacturers may specify receiving baled or loose flattened cans and may exclude noncan aluminum (e.g., pie pans) from the load. Manufacturers may specify moisture content (usually up to a maximum of 1%), which means the material should be stored inside a building.

7.3.4   Glass

Because glass is an inert material and does not pose a environmental threat directly, it does save energy to recycle glass, which is beneficial to the environment. Nine gallons (34 liters) of fuel oil is saved per ton of recycled glass. Currently most glass used for manufacturing of containers (e.g., jars and bottles) are recycled. Glass used for other items (e.g., window paves, lightbulbs, mirrors) cannot be recycled with container glass. Although the reduction in waste volume due to glass recovery is not significant, glass recovery is beneficial for waste destined for incineration. The presence of glass interferes with the burning process.

Usually three colors are used for glass containers: green (also called emerald), brown (also called amber), and clear (also called flint). The price of glass will increase if different colors are sorted separately. Glass is recovered from mixed recyclable items through hand sorting. Glass containers are fully recyclable. It can be easily remanufactured without any appreciable loss of material (Gilmore and Hayes, 1993).

In addition to remanufacturing, crushed glass can be used for bituminous road paving. Crushed glass can also be used as gas pipe bedding below the final cover of landfills. However, the sharpness of the crushed glass should be considered before such uses. Sharp glass pieces may damage auto tires and pipes. Wine bottles are reused after washing and sterilizing (Papke, 1993).

7.3.5   Tires

Tires pose both disposal and environmental problems. Because of the shape and composition, tires cannot be easily compacted. Thus tires occupy a high volume. Tires also "float" to the top of landfills because of the buoyancy created by the air and landfill gas trapped within the hollow spaces of tires. The "floated" tires can break the final cover of landfills. Such cracks in landfill final cover can attract rodents, allow escape of landfill gases, and increase leachate generation. The environmental problems related with scrap tires stock piles are also a fire hazard and can cause mosquito and rodent problems. Tire fire releases both thick smoke and toxic chemicals. Thus tire recycling is very beneficial.

There are many options for tire recycling. Tire casings can be used for retreading. Cost of retreading truck tires is lower than passenger car tires. Retreading passenger car tires is further reduced because of the advent of all-season radial tires. Whole tires are used for constructing artificial reefs in the sea. Scrap whole tires are bundled together and anchored underwater, which creates an attractive habitat for fish (Blumenthal, 1993). Whole tires are also used as crash barriers in highway bridges. Shredded tires are used for fuel in incinerators. Rubber from shredded tires can be used for the manufacture of rubber goods (e.g., floor mats), thus saving virgin rubber. Pyrolysis of scrap tire is used to recover gas and oil. Tire chips are also used as bulking agents in sludge composting.

7.3.6 Metals and Steel Cans

The percentage of metals and steel cans in municipal solid waste is not high. In addition to municipal solid waste, there are other sources of scrap metals (e.g., automobiles, railway tracks) that are also recycled. Apart from saving landfill space, recycling of each ton of scrap metals and tin cans saves 2500 lb of iron ore, 1000 lb of coal, and 40 lb of limestone. Seventy-five percent less energy is used for making steel from recycled items (Jordan and Crawford, 1993). There is a good market for scrap metal recycling. However, since the volume of scrap metals and steel cans separated from municipal solid waste is not high, the revenue earned from the sale of these items is low, primarily because of transportation cost. Scrap metals and steel cans can be easily separated from municipal solid waste because of its magnetic properties. Steel cans are used as containers for food, paint, aerosol, and so on. Bigger steel cans are used for packaging food for use in restaurants, hospital cafeteria, and so on.

The separated metals and steel cans are flattened and bundled together for transportation to steel mills, de-tiner facilities, or iron foundries. Tin is removed from the tin can using a chemical and electrolytic process.

7.3.7 Construction and Demolition Waste

Although construction and demolition (C&D) waste is not commingled with municipal solid waste, it is disposed mostly in municipal landfills. While a portion of C&D waste can be recycled, the remaining portion needs to be disposed in landfills. Disposal-related issues are discussed in Chapter 23 and recycling-related issues are discussed in this section.

Construction and demolition waste is generated during home improvement projects in single-family homes, during construction of new buildings or structures, and in the demolition of old buildings and structures. The C&D waste generated during home improvement projects in single-family homes is usually picked up by waste haulers in cities. In rural areas usually the owners haul it to a landfill for disposal. In general, contractors are involved in construction and demolition of buildings and other civil engineering structures. The contractors manage the C&D waste generated during these activities. It is estimated that the total volume of C&D waste generated in the United States is approximately 38% of municipal solid waste (University of Florida, 1998). Thus the landfill space needed for this waste stream is significant.

Parts of this waste, especially concrete, have been recycled for quite some time. Interest in recycling this waste stream is increasing primarily because of the rise of disposal costs and environmental issues. In addition, efforts at a construction site indicate a significant part of C&D waste can be recycled effectively, leading to savings in disposal costs. Currently there are consulting companies that specialize in providing advice regarding recycling of C&D waste.

The definition of C&D waste adopted by a regulatory agency must be followed when developing the management approach for this waste stream. Certain items (e.g., uncontaminated concrete, bricks) may be defined as "clean," whereby they may be used as fill materials in and around construction sites or elsewhere. However, there are better uses for much of this material in concrete used for roads, sidewalks, basement floors, and so on. On the other hand there may be restrictions regarding use of treated wood generated during demolition of a structure, such as mulch. A clear knowledge regarding the pollution potential of the of items generated during demolition is essential for setting up a C&D recycling program. Except for a few items (e.g., treated wood), in general, waste generated during construction of new buildings or structures is clean. Waste items generated during a demolition or construction can be reused. Care should be taken when demolishing a building, however, to manage hazardous materials such as mercury thermostats, fluorescent bulbs, and so on.

For a major project, both the general contractor and subcontractors should discuss the recycling process and opportunities. Usable and recyclable items should be removed from an existing building before demolition if possible. It may be necessary to identify items or portions of a building or structure that are contaminated. If these items are removed prior to demolition, then rubble can be easily recycled for construction-related work.

Usually properly labeled containers for each major waste type (e.g., scrap metal, wood) are placed as close to the site as possible. Space needed for keeping the containers may pose a challenge in sites where available open area is restricted. Container placement should be considered in the construction site plan. Containers for collection of recyclable items at workers' lunchrooms should be placed at convenient location(s). For example recycling bins for beverage containers and newspapers should always be placed next to the waste bins. This setup make it easier for the employees and reduces contamination. Waste paper, most of which is discarded drawing sheets, can also be collected and recycled from construction project sites. Although with careful planning many C&D items can be recycled, in most cases the following items are recyclable: wood, concrete, brick and block, scrap metals (both steel and aluminum), copper wire and pipes, pavement, paper, cardboard, shingles, beverage containers, and so on. Currently carpet and drywall recycling is also gaining momentum.

Identifying markets for recyclable items is extremely important for the success of any recycling program. Markets are well developed for certain items of C&D waste (e.g., concrete, scrap metal). The scrap metals (e.g., structural steel, aluminum) can be easily sold to scrap metal dealers.

The recycling program should be discussed with all employees involved in a project. This type of discussion should be undertaken at the beginning of the project and on a regular basis thereafter (weekly/monthly). Information regarding the amount of each item recycled and the target should be shared at each weekly/monthly meetings.

Special equipment is needed for preparing certain items for recycling (e.g., crushing concrete). The owner(s) of such equipment may contract for processing and marketing of the materials. If the owner(s) of such equipment agree only to process the materials, then marketing these materials needs to be arranged by the project owner. A C&D waste reduction and recycling project in Madison, Wisconsin, reported a 22.5% savings in disposal cost over just a 5-month period (Fuller and Newenhouse, 2002); the project is expected to run for another 20 months. Since hauling is a major cost for both landfill disposal and recycling, the cost of hauling should be considered while comparing cost for these two management options. In some cases hauling cost for recycling may be very low compared to landfill disposal. Even if the hauling cost is the same, a tipping fee is needed for the disposal of an item.

7.3.8 White Goods

White goods refer to large appliances such as refrigerators, washers, dryers, and air conditioners. These appliances may contain polycholorinated biphenyls (PCBs) and chlorofluro carbons (CFCs), which have an environmental impact. PCBs are suspected carcinogens. Capacitors containing PCBs may be found in refrigerators, washing machines, televisions, and so forth. CFCs are linked to the depletion of Earth's ozone layer. CFCs may be found in refrigerants for air conditioners, refrigerators, and freezers. The PCB- and CFC-containing parts in the white goods are removed first. In many instances the nonmetallic parts are separated before baling or shredding. Usually white goods are collected from single-family homes by the local authority (directly or through a hauling contractor). Many times the installation contractors or the homeowner hauls the discarded appliance to a scrap yard.

The scrap dealer separates the steel, nonferrous metals (e.g., copper, lead), and nonmetallic components from white goods. Usually the nonmetallic parts are sent to landfills for disposal. Although the price of scrap metals fluctuates, there is relatively steady market for it. Both steel and nonferrous metals are marketed directly to respective industries.

7.3.9 Electronics

With the constant improvement in information technology, large numbers of computers and other electronic items are disposed in landfills every year. In addition to depleting landfill space, the pollution potential of these electronic items is significant. If managed properly many devices can be refurbished for reuse. Valuable materials, including gold, can be retrieved from obsolete computers. Electronic waste primarily consists of television sets, computers and peripherals, audio and stereo equipment, video cassette recorders, telephones, cell phones, copy machines, and video game consoles. Although most of these items can be recycled without significant expense, currently the recycling focus is on computers and peripherals and televisions.

In the United States two items, computers along with the peripherals and televisions, comprise about 2–5% of municipal solid waste. The 1998 estimate (for the United States) of these two items was:

1. About 36 million new personal computers (PCs) were shipped in United States; 20 million PCs were estimated to be at the end of useful life.
2. Only 14% were estimated to have been recycled.
3. Over 50% of households own at least one computer.
4. Over 95% of households own at least one television.
5. Over 75% of old computers are stockpiled.

As mentioned earlier there are many parts in computers and televisions that generate hazardous waste. For instance:

1. Cathode ray tubes (CRTs) and solders in various electronic circuits contain lead.
2. Chip resistors contain cadmium.
3. Batteries, switches, and printed circuit boards contain mercury.
4. Flat liquid-crystal display (LCD) monitors may contain mercury over the toxic characteristic leaching procedure (TCLP) limit.

Many electronic devices can be reused (upgraded or used by another user without upgrading). The ones that cannot be reused can be demanufactured and recycled. The materials that can be recovered are glass, steel, plastic, silver, gold, copper, lead, and zinc.

To create a better atmosphere for recycling, governmental bodies need to develop new regulations and guidelines for businesses and individuals. To reduce pollution potential, disposal and incineration of electronic items containing hazardous waste could be banned. In addition businesses engaged in recycling electronics could be exempted from hazardous waste regulations regarding storing and transporting of electronic items destined for recycling. However, transporting of electronic waste to other countries should not be allowed without ensuring that the recovery process is sound both from human health and environmental standards. Apart from humanitarian and environmental concerns, transporting of waste to other countries where environmental and human health issues are not properly addressed may create legal problems in the future.

7.4   MATERIAL RECOVERY FACILITIES

Material recovery facilities (MRFs) are buildings or facilities where the various recyclable items are housed for sorting and distribution. In an MRF, various recyclable items are separated and prepared for marketing. The commingled recyclable items are usually tipped on to the floor. Paper and cardboard are usually source separated, whereas glass, aluminum can, and plastic items are collected together. The commingled items can be separated by hand or mechanically. If hand sorted, the commingled recyclable items are put on a conveyor belt, which moves past several workers. Each worker is designated to pick up one type of item from the stream. The speed of the conveyor is adjusted to ensure satisfactory recovery of various items.

Trommel screens are used to sort material by size. A trommel is essentially a cylindrical screen, usually 7–10 ft (2.1–3.5 m) in length. The trommel screen is set at a slight angle. Because the slow rotation of the screen, the refuse tumbles forward through the screen. The screens usually have three or four different size openings at different parts of the trommel. The smallest size opening, usually 2 in. (5 cm), is placed at the beginning. The size opening increases toward the "exit end." Thus different size material can be separated at different sections of the same trommel screen. Alternatively, several trommel screens are placed in series, each with different sized openings to collect different sized materials. Air classifiers, magnetic separators, and eddy current machines (see Section 6.3.1.1) are used to separate various materials. The materials are moved from one equipment to the next using conveyor belts. Depending on the extent of source separation, an MRF may include just a hand-sorting operation or all of the equipment mentioned above. More equipment is needed if the recyclable items are mixed with garbage. Increasing the amount of equipment used for recovering recyclable items will also increase the cost. There are three options for recovering recyclable items from municipal solid waste:

1. Collection of recyclable items and garbage separately from the curbside. The residents need to separate each item (e.g., paper, plastics) and place them in separate containers and put them all at the curbside on collection day. The pickup trucks must have separate compartments for each recyclable. The garbage collector puts each type in respective compartments in the truck. In addition to the truck for collection of recyclable items, another truck is used for the collection of household waste. In this option two separate trucks are used. Use of an MRF can be avoided for this option. Only a transfer station (see Section 2.6) is needed for collecting and sending recyclables to buyers.

2. Collection of commingled recyclable items and garbage separately from the curbside. For this option the residents need to put all types of recyclable items in one container and household waste separately in another container and set it out at the curbside on collection day. There are two alternatives for collecting the recyclable. The recyclables and the household waste are collected by the same truck or the contents of the two containers are collected by two separate trucks. Thus two trucks will be involved—one for collecting recyclables and the another for collecting household waste. Use of an MRF will be needed. If paper is bundled up separately and collected in a separate compartment within the truck, then the sorting operation could be less mechanized.

3. Collection of unsorted recyclables and household waste mix (all placed in one container) from the curbside. The residents need not separate the recyclable items at all. Only one truck will be involved in picking up the household waste and recyclables. Several pieces of equipment will be needed in the MRF to separate recyclable items and the organic fraction from the recyclables and household waste mix.

A planner needs to look into the pros and cons of each of the three methods of collecting recyclable items. While the first option will need significant cooperation from the residents, the overall cost could be high. In the second option, the residents need not be as involved as the first option; however, the cost of construction and operation of the MRF should be included in the plan. In the third option, involvement of the residents is the lowest, the cost of construction and operation of an MRF will be the higher than for the second option. All these and other issues (e.g., availability of space, cost of public campaign) should be factored into the decision-making process. The size and layout of an MRF are important to run an efficient operation. The average processing capacity of the facility can be estimated by dividing the expected weight of input in a month with the number of operating days. Since this will provide an average estimate, a reasonable excess capacity (between 10 and 25%) should be assumed to provide redundancy.

The space needed for incoming and outgoing vehicles at the MRF must be estimated properly to avoid unnecessary idle time. For an efficient system, multiple handling of materials should be minimized as much as possible. The system design should be such that most of the sorted materials are directly loaded into the containers in which it will be transported to the buyer(s). In MRFs, materials are either "positive sorted" or "negative sorted." In a positive sort the target material is physically removed from the commingled stream (e.g., removal of ferrous metals using magnets). The negative sort refers to the material, which is not physically removed from the stream. Ideally the negative sort material should be the material with the largest volume.

7.5 REFUSE-DERIVED FUEL PROCESSING PLANTS

Municipal solid waste can be used to derive fuel. The fuel quality depends on the degree of removal of noncombustibles (e.g., metals) from municipal solid waste This type of plant predates current MRFs. If not source separated, the percentages of paper and plastics in municipal solid waste are quite high. The Btu value of these materials is also high. Refuse-derived fuel (RDF) can be a good source of fuel, provided the incinerator is designed to use RDF.

Almost all the equipment used in an MRF (where unsorted municipal solid waste is processed) are also used in an RDF. Figure 7.1 provides a schematic of a typical RDF plant. A simple RDF plant may consist of a shredder and minimum amount of sorting to yield a high-quality fuel. Modern RDF plants are capable of recovering organic compostable fraction and other recyclable iterms from the fuel. The key to the success of an RDF plant is availability of an incinerator capable of burning the RDF.

7.6 QUALITY CONTROL AND DATA COLLECTION

The quality of recyclable items must be high to ensure proper sale value. Quality control measures should be implemented at all stages—from the source separation by generators (e.g., residents) to proper sorting. The contamination may consists of:

1. Household hazardous waste (e.g., paint thinner cans) disposed in recycling container
2. Nonrecyclable items (e.g., window glass)
3. Cross contamination (e.g., broken glass mixed with newsprint)

To maintain quality, it is very important to inform and educate all groups involved in recycling. Generators, primarily the residents of the community, must be informed properly about the items that can and cannot be recycled. While a massive campaign should be undertaken at the beginning of a program, a periodic campaign (e.g., once every 6 months) should also be done. Details regarding such campaigns are included in Chapter 2. The workers involved in hauling recyclable items should be trained in identifying nonrecyclable items. The workers may be instructed to leave containers with recyclable items at the curbside. If a drop-off center is used for collecting recyclable items, then the center staff should be instructed to monitor the materials being dropped off. The tipping floor of MRFs may also be monitored to remove nonrecyclable items. Even with a good education program and quality control measures at various collection points, nonrecyclable items usually find its way into MRFs. Monitoring of the tipping floor and/or a station for removing nonrecyclable items prior to actual operation is very

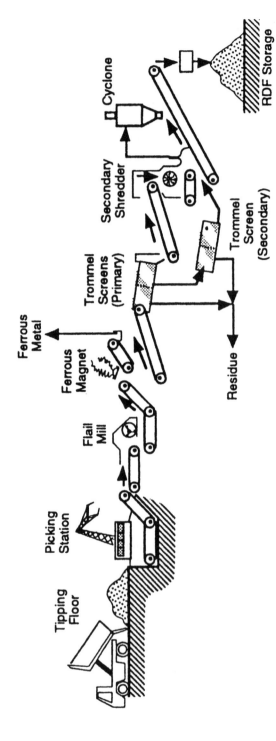

FIG. 7.1 RDF processing system schematic. From Kreith (1994). Reproduced with permission of The McGraw-Hill Companies.

helpful in improving quality. The cost of implementing such a monitoring program will pay off due to receiving a better selling price for the recyclable items.

Although a mechanized MRF can handle more materials, it may not always provide high-quality recyclable items. Hand sorting, in combination with mechanized separation of certain items, may provide better quality recyclable items. A cost–benefit study should be undertaken prior to designing a sorting plant and an overall program. The extent of source separation and the extent of mechanization of the MRF heavily influences the quality of the recyclable items. The program design, education, and training of persons involved with recycling and monitoring are all important for ensuring the quality of recyclable items.

In addition to quality control, data collection at each step of the process is important. Data for a number of program parameters should be collected and studied for improving the cost efficiency and operation of the program. In addition to helping to manage the program, such data may be mandatory by state and/or federal regulations. The objectives for use of data should also be clearly identified. The following is a list of items for which data may be collected:

1. Weight of each recyclable item collected
2. Weight of each recyclable item sold
3. Revenue generated from each recyclable item
4. Number of times a nonrecyclable item were rejected at curbside and the neighborhood in which such an incident occurred
5. Weight of recyclable items collected from single-family homes, multi-family units, and commercial and industrial sources
6. Number of times and the total weight of nonrecyclable items removed from tipping floor or presort station
7. Price of each recyclable items by month/week

In addition, a well-organized survey should be conducted after a year of starting a program, and approximately every other year thereafter, to assess participation and community issues or concern. An annual report is useful to assess overall program success, problem areas, the cost efficiency of the adopted collection method, and so on. Such annual reports can help identify areas where improvements can be made. The report is also useful for preparing the annual budget. The annual report should include all raw data and analysis to assess whether the program objectives are being met. If objectives are not met, the source(s) of the problem can be traced easily by analyzing the data. Thus, data on various issues should be collected and stored properly with an objective of analyzing it to assess the program.

7.7   LAND SPREADING

Only nonhazardous sludge, which is not expected to contaminate surface or groundwater, can be reused as soil conditioner or fertilizer. The various sludge that are land spreadable include wood ash from the combustion of untreated wood, lime sludge (from paper mills and water supply treatment facilities), coal ash, wastewater treatment sludge, and so on. The aim of land spreading is to enhance soil nutrients. It also reduces landfill space needs. Land spreading can be done on agricultural or forest lands. Like landfills, location and performance standards, operation, and monitoring plans should be followed.

Sludge with sufficient organic material is biodegraded by plants and microorganisms. Sufficient time must be allowed to assimilate the organic components. The sludge liquid may be considered as additional water, which is very helpful for crop growth in arid climates. Land spreading improves soil quality by providing nutrients for plants and increases the water retention capacity of soil.

7.7.1   Risk Analysis

Risk analysis should be done if the sludge contains any chemicals that can contaminate groundwater, air, or human/animal through direct contact. Direct exposure pathway analysis for humans and animals should be done in these situations (see Chapter 10). If necessary the sludge should be treated to minimize the contaminant(s). During the active life of a land spreading program, direct exposure to sludge or sludge-amended soil is a risk to personnel involved in the operation. Necessary protective equipment and clothing should be used by the employees to ensure safety. If the crops are to be used as animal feed or grazing, then proper test of the feed should be done to ensure food chain safety.

7.7.2   Preliminary Land Spreading Projects

Issues such as nutrient level in sludge and soil properties are assessed for a land spreading program. A brief discussion of these issues follows.

7.7.2.1   Sludge Parameters   A preliminary determination of sludge application rate is based on the following sludge parameters: total solids content, pH, biodegradable organic matter, nutrients, carbon-to-nitrogen ratio, soluble salts, calcium carbonate equivalent, and pathogens. The significance of these parameters for land spreading projects are listed in Table 7.1.

7.7.2.2   Soil Properties   Physical, biological, and chemical characteristics of soil are important for land spreading projects. Physical properties such as texture, structure, and pore size distribution influence the infiltration rate and

TABLE 7.1 Significance of Sludge Parameters

Sludge Parameter	Significance
Total solids content	Indicates ratio of solids to water in waste and influences application method.
pH	Controls metals solubility (and therefore mobility of metals toward groundwater) and affects biological processes.
Biodegradable organic matter	Influences soils water-holding capacity, cation exchange, other soil physical and chemical properties, and odor.
Nutrients (nitrogen, phosphorus, and potassium)	Affect plant growth; nitrogen is a major determinant of application rate; can migrate to and contaminate groundwater or cause phytotoxicity if applied in excess.
Carbon-to-nitrogen ratio	Influences availability of nitrogen to plants.
Soluble salts	Can inhibit plant growth, reduce soil permeability, and contaminate groundwater.
Calcium carbonate equivalent	Measures a waste's ability to neutralize soil acidity.
Pathogens	May threaten public health by migrating to groundwater or being carried off-site by surface water, wind, or vectors.
Groundwater constituents designated in the risk assessment section including metals and organic chemicals	May present public health risk through groundwater contamination, direct contact with waste–soil mix, transport by surface water, and accumulation in plants. Metals inhibit plant growth and can be phytotoxic at elevated concentrations. Zinc, copper, and nickel are micronutrients essential to plant growth but may inhibit growth at high levels.

After USEPA (1999).

attenuation capacity of soil. Sites with high or low permeability soil or poorly drained soil are not suitable for land spreading projects.

Chemical and biological properties such as pH and organic matter percentage also influence the attenuation properties of soil. The attenuation mechanisms in soil include adsorption, cation-exchange capacity, biological uptake, and precipitation (see Section 15.2).

If the soil is overloaded with nutrients, then rapid oxygen depletion, extended anaerobic conditions, and the accumulation of odorous and/or phytotoxic end products could impair soil fertility and productivity in the long run.

7.7.2.3 Plant Microbial Effects The type of crop to be grown on the land influences the application rate. Plant nutrient uptake affects rate of assimilation. The presence of toxic constituents and pathogens in the sludge may render the crop unsuitable for food or feed. In the absence of existing plant uptake data, greenhouse or field house studies should be done to assess the effect of the sludge and microbes.

7.7.2.4 Climate Climatic conditions also influence land sludge spreading projects. In arid climates use of sludge with high soluble salt concentration is not desirable. Plant uptake of water is inhibited due to osmotic pressure of salts (USEPA, 1999). However, downward migration of contaminants is less likely in arid climates. The types of plants that can grow in a region depend on the climate. Sludge application should be avoided when the soil is frozen or too wet.

7.7.2.5 Agronomic Rate For agricultural benefit, the application rate is governed by the agronomic rate. The agronomic rate is the application rate designed to provide the amount of nitrogen plants need for desired yield. It may be noted that in some cases phosphorus, potassium, or salt content, rather than nitrogen, will be the limiting factor.

7.7.3 Location and Performance Standards

The following location standards should be followed while choosing a land for a land spreading project:

1. The land should not be within 100 ft (30 m) of lakes, river, or other water bodies.
2. The land should not be within 1000 ft (300 m) of private water supply wells or 200 ft (60 m) of private water supply wells.
3. The land should not be within 500 ft (150 m) of any residence.

If the sludge has a potential for attracting birds, then the land should not be located within 10,000 ft (3000 m) of an airport. In addition to these location criteria land spreading projects should be located and designed in such a way that it does not have any adverse impact on wetlands, critical habitat areas, surface water bodies, or groundwater.

7.7.4 Waste Characterization

Prior to undertaking land spreading the characteristics of the waste should be studied carefully. The sources(s), process(es), or treatment system(s) from which the waste originated and the chemicals used during the process are noted. Physical characteristics of the waste, including solid and organic fractions, are tested. Chemical and biological analysis of the waste are done for the following parameters (Wisconsin Department of Natural Resources, 2003; USEPA, 1999):

1. pH
2. Nutrient content: kjeldahl-nitrogen, ammonia nitrogen, nitrate nitrogen, phosphorous, and potassium
3. Salt content: chloride, fluoride, and sulfate
4. Element content: aluminum, barium, boron, calcium, copper, iron, manganese, magnesium, sodium, strontium, and zinc
5. A bulk chemical analysis is performed for the waste, based on chemicals used in the treatment process
6. Total solids content
7. Biodegradable organic matter
8. Carbon-to-nitrogen ratio
9. Soluble salts
10. Calcium carbonate equivalent
11. Pathogens

Appropriate leach tests (see Section 13.3) are done on sludge prior to land spreading, especially for sludge that has never been used in a land spreading program or that has questionable chemicals. Leach tests should be performed on the waste material, the land spreading site soil, and a mixture of waste material and site soil. The leach test should include the following parameters:

1. All priority pollutants listed in Table 14.3
2. All chemicals used in the process that may impact the groundwater
3. Aluminum, barium, boron, calcium, iron, manganese, magnesium, sodium, and strontium

The waste characterization data are analyzed to assess the benefits and adverse effects of the land spreading proposal. The goal of the assessment is to determine whether the waste has value as a soil conditioner or fertilizer, and whether it will have any detrimental effect on public health or the environment.

7.7.5 Design and Operation

The following issues need to be addressed when designing a land spreading facility:

1. The location and size of the property on which the waste will be spread. The amount of waste needed can be calculated based on the necessary nutrient load and the size of the facility. Usually, a 2- to 3-in. (5–7.5 cm) layer of waste is spread over the entire site.
2. The road from the location where the waste is generated to the facility should studied carefully. The road design should be adequate to support

the transportation need. The allowable axle load on the road bridges should be high enough to carry the trucks loaded with the waste.

3. The mode of transportation and type of vehicles to be used should be identified. There may be regulatory requirements regarding vehicle design for such projects.

4. The application rate should be such that both annual and cumulative weight of nitrogen, arsenic, cadmium, copper, lead, mercury, nickel, zinc, and other heavy metals do not exceed the regulatory standards, if any.

5. The waste should be plowed, disked, or otherwise incorporated with the surface soil layer initially and once or twice within the following year. This practice will minimize contaminated surface water run off, leaching contaminants into groundwater, and odor.

6. A vegetative buffer strip should be maintained between the land and any surface water bodies (e.g., river).

7. Waste containing significant amount of pathogenic bacteria should be stabilized prior to land spreading.

8. Waste that received pesticide, persistent organic materials, or PBT should be used with caution in the food chain (both human and animal) crop growing fields. The crop should be tested to ensure that it meets all regulatory requirements regarding allowable contaminant levels.

7.7.6 Monitoring

Monitoring of a land spreading facility is essential for protection of surface water and groundwater. Monitoring will also help to ensure that required amount of nutrients are added to the soil. If multiple sludge application is planned for a site and the sludge contains any parameters listed in Tables 7.2 and 7.3, then a lysimeter (see Section 22.8) should be constructed below the land for monitoring downward migration of contaminants. Groundwater monitoring, using monitoring wells and lysimeters should be undertaken annually. If groundwater results indicate unacceptable levels of contaminants the land application should be suspended.

TABLE 7.2 List of Parameters for Testing Soil Prior to Land Spreading

Cation-exchange capacity	Organic matter content
pH	Extractable calcium
Extractable magnesium	Available phosphorus
Extractable potassium	Total arsenic
Total cadmium	Total trivalent chromium
Total hexavalent chromium	Total copper
Total lead	Total molybdenum
Total nickel	Total zinc

TABLE 7.3 List of Parameters for Testing Soil after Land Spreading

Total Kjeldahl nitrogen	Available nitrogen
Available phosphorus	Available potassium
Total arsenic	Total cadmium
Total chromium	Total copper
Total lead	Total mercury
Total molybdenum	Total nickel
Total zinc	

The soil should be tested both before and after receiving the sludge. The number of sampling points and parameters depends on the size of the land, sludge characteristics, type of crop to be grown, and climate. At a minimum five samples per acre should be tested for the parameters listed in Table 7.2 before receiving the sludge and five soil samples should be tested for the parameters listed in Table 7.3 after receiving the sludge.

8

REUSE OF INDUSTRIAL BY-PRODUCTS

Although the practice of using industrial by-products in construction projects is quite old, until the early 1980s, a significant portion of industrial by-products were landfilled. However, currently attempts are being made to reuse industrial by-products to a greater degree. In many cases the by-products are utilized directly while in others they are utilized in combination with other by-products or materials. For instance, foundry sands are reused for highway embankment construction, and fly ash is mixed with cement for producing concrete. Rather than providing a complete list and processes used regarding the reuse of various industrial by-products, comments on engineering characteristics, basic principles involved in reuse projects, and environmental issues will be discussed in this chapter. Information regarding uses of various industrial by-products are included. References are cited for readers interested in additional information.

The industrial by-products are reused mainly to replace soil, aggregates, and cement. Identification of proper application of a by-product is a crucial step in its reuse. A comprehensive knowledge about the construction requirements, economics, and environmental regulations prevailing at the site of reuse is essential to choose a by-product (Edil and Benson, 1998).

8.1 COMMENTS ON ENGINEERING CHARACTERISTICS AND TESTING PROTOCOLS

The properties needed for a particular application must be studied carefully to choose the appropriate by-product. Properties such as strength, deformability, constructability, long-term compression, and plastic strain are important for use as construction fill (Bosscher et al., 1992, 1997). Although properties such as optimum moisture content for compaction and related compactive effort can be identified in the laboratory, field trials are needed for resolving material handling issues.

The American Society for Testing and Materials (ASTM) standards are widely used for measuring various properties of soil. All these standards have been developed for naturally occurring soil and hence may not be applicable to assess physical properties of industrial by-products. For instance, compaction standards developed for earthen materials (ASTM D698 and D1557) with a maximum particle size of 19 mm are not applicable to tire chips, which ranges in size from 25 to 300 mm (Edil and Bosscher, 1994). Atterberg limits of foundry sands, mixed with dehydrated bentonite, have been found to be dependant on hydration time (Kleven et al., 1998). Thus the standard tests used for earthen materials may not be suitable for determining relevant properties of industrial by-products. Innovative approaches for testing the relevant properties may be necessary for reuse projects.

Proper understanding of engineering behavior of industrial by-products is needed before such by-products can be used in construction projects. Although in many instances existing models for soil can represent the engineering behavior of by-products, there is an exception to such straightforward extrapolation. For instance engineering behavior of foundry sands, paper mill sludge, and fly ash are similar to soil (Kleven et al., 1998; Kraus et al., 1997), but the engineering behavior of tire chips and sand mixtures are different from soil (Edil and Bosscher, 1994; Bosscher et al., 1997). The shear strength of mixtures of tire chips and sandy silt can be represented by a linear Mohr–Coulomb failure criteria (Tatlisoz et al., 1997), whereas for a mixture of tire chips and dense sand the property is represented by a bilinear failure envelop (Fig. 8.1).

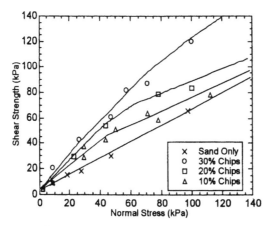

FIG. 8.1   Shear strength envelopes for dense portage sand–tire chip mixtures. Reproduced by permission of the publisher, ASCE, from Edil and Benson (1998).

Because of such differences in engineering behavior, and the absence of standard tests, construction specifications for industrial by-products should be developed carefully. Field trials and demonstration projects are undertaken to develop a proper understanding of the behavior of the by-product before construction specifications are developed for full-scale use. A study of the existing literature regarding reuse projects for the chosen by-product is helpful in developing project specifications. Long-term data, such as the use of fly ash in concrete (Naik et al., 1994), are important in generating confidence regarding reuse projects using industrial by-products.

8.2   REUSE OF INDUSTRIAL BY-PRODUCTS IN CIVIL ENGINEERING PROJECTS

Reuse of industrial by-products in civil engineering projects has a long history. Industrial by-products such as fly ash, bottom ash, and blast furnace slag have been used in highway projects for quite some time (Miller and Collins, 1976). Table 8.1. provides a list, partial at best, of widely used industrial by-products in civil engineering projects. Besides the use of by-products in huge volume in civil engineering construction projects, researchers have suggested many innovative uses such as the use of discarded lime from wastewater treatment process (Fahoum, 1998), paper mill sludge (Quiroz and Zimme,

TABLE 8.1   By-Products and Their Reuse in Civil Engineering Projects

Byproduct	Reuse
Fly ash	Partial replacement for Portland cement
	Mineral filler in asphalt
	Embankments or backfills
Bottom ash	Road base and subbase
	Backfills
	Asphaltic concrete
	Fine aggregate (crushed bottom ash) in concrete
	Masonry
Blast furnace slag	Aggregate in asphalt mix
	Aggregate in concrete mix
	Partial replacement (crushed slag) for cement
Foundry sand	Embankments and backfills
	Final and daily covers in landfills
	Road base and subbase
Tire chips	Embankment fills
	Road construction
	Liner, drainage layer, and final cover in landfills

Based on Ramaswami and Aziz, 1992; Ramaswami et al., 1986; Collins and Ciesielski, 1992; Hall, 1991; Humphrey et al., 1993; Bosscher et al., 1992; Freber, 1996; Abichou et al., 1998; Ramme and Tharanivil, 1999.

1998), glass (Egosi, 1992), fly ash and recycled foamed plastic (Nicholls, 1992), sludge frits (Chesner, 1992), shredded commingled plastics (Benson and Khire, 1994), mixture of flue gas desulfurization sludge and class C or F fly ash (Vipulanandan and Basheer, 1998), sulfur polymer cement (McBee et al., 1992), wood waste (Collins and Ciesielski, 1992), and crushed concrete (Kim et al., 1992). A database on reuse of foundry sand has been reported by Abichou et al., 1998). In addition to the four commonly used by-products included in Table 8.1, concrete from pavement and building demolition are also extensively used in many civil engineering projects (Papp et al., 1998; Maher et al., 1997; Epps et al., 1980). Ramme and Tharaniyil (1999) have reported the utilization of coal combustion by-products in construction. One needs to keep in touch with the current literature to update the reuse list.

8.3 ENVIRONMENTAL ISSUES

An industrial by-product can be used to replace naturally occurring earthen materials only if it does not leach pollutants. The chance of groundwater degradation due to the leaching of pollutants is highest when a by-product is used in such a way that the possibility of water percolation through it is high (e.g., use as backfill). In the absence of a regulatory mandate, concentrations of various parameters included in Table 8.2 may be considered as allowable concentrations for different by-products. If the concentration of various parameters in leachate, obtained by using a water leach test, is equal or less than the concentration indicated in Table 8.2, then detailed waste characterization for the by-product may be skipped. If regulations allow, these by-products can be used in construction projects without any permit from the regulatory agency.

In many instances the regulatory approval process is not well established. In such cases any proposal for reuse of a by-product should be discussed with the regulatory agency first. Necessary physical and chemical tests and design specifications should be developed after such meetings. Significant changes in design specifications may affect the cost of the project. Since industrial by-products are generated continuously, the goal should be to obtain blanket approval for a specific by-product. A long-term action plan should be discussed with the regulatory agency to achieve that goal. After obtaining permission for a specific project, periodic meetings with the agency should be held as the project progresses. The experience gained from the project should be discussed in such meetings. Long-term performance data help in establishing confidence about the performance of the by-product from an environmental standpoint.

When reusing industrial by-products, attention must be paid to both their engineering properties suitable for the project and the environmental risk for the proposed use. It must be assured that such reuses do not pose a nuisance or any risk to human health and welfare as well as the ecology. The reuse should be such that it does not

TABLE 8.2   List of Parameters to Be Tested Using Water Leach Test

Parameter
Aluminum (Al)
Antimony (Sb)
Arsenic (As)
Barium (Ba)
Beryllium (Be)
Cadmium (Cd)
Chloride (Cl)
Chromium, tot. (Cr)
Copper (Cu)
Total cyanide
Fluoride (F)
Iron (Fe)
Lead (Pb)
Manganese (Mn)
Mercury (Hg)
Molybdenum (Mo)
Nickel (Ni)
Nitrite & Nitrate (NO_2+NO_3-N)
Phenol
Selenium (Se)
Silver (Ag)
Sulfate
Thallium (Tl)
Zinc (Zn)

After Wisconsin Department of Natural Resources (1997).

1. Adversely impact the wetlands
2. Adversely impact areas of critical habitat
3. Adversely impact surface water bodies or groundwater
4. Create explosion hazard due to the accumulation of explosive gases
5. Emit any hazardous air contaminants

8.3.1   Waste Characterization

Both bulk chemical analysis and water leach testing should be performed on the waste. In the absence of any regulatory mandate the list of parameters in Tables 8.2 and 8.3 may be used for such tests. Such a parameter list is developed after studying several waste characterization and leach test results of an industrial by-product. This list may be modified based on knowledge about the waste. For instance, nitrate and nitrite need not be included in the param-

TABLE 8.3 List of Parameters to Be Used for Bulk Chemical Analysis

Parameter
Antimony (Sb)
Arsenic (As)
Barium (Ba)
Beryllium (Be)
Boron (B)
Cadmium (Cd)
Chromium, hex. (Cr)
Lead (Pb)
Mercury (Hg)
Molybdenum (Mo)
Nickel (Ni)
Phenol
Selenium (Se)
Silver (Ag)
Strontium (Sr)
Thallium (Tl)
Vanadium (V)
Zinc (Zn)
Acenaphthene
Acenaphthylene
Anthracene
Benz(a)anthracene
Benzo(a)pyrene
Benzo(b)fluoranthene
Benzo(ghi)perylene
Benzo(k)fluoranthene
Chrysene
Dibenz(ah)anthracene
Fluoranthene
Fluorene
Indeno(123-cd)pyrene
1-Methyl naphthalene
2-Methyl naphthalene
Naphthalene
Phenanthrene
Pyrene

After Wisconsin Department of Natural Resources (1997).

eter list for water leach test of ferrous foundry excess system sand. Phenol, selenium (Se), silver (Ag), and strontium (Sr) need not be included in the list for bulk chemical analysis of ferrous foundry excess system sand, ferrous foundry slag, and coal ash. The allowable concentration indicated in the tables may be used in the absence of any regulatory standards. In addition to the chemical characteristics, the physical properties of the by-product must also be assessed. The desirable physical properties of the by-product depend on the intended reuse. For instance, if a by-product is used for incorporation in concrete, then the concrete mix should be tested for construction-related properties (e.g., slump test) as well as for strength-related properties (e.g., compressive strength).

8.3.2 Monitoring

Environmental monitoring of projects in which industrial by-products are used as raw material for the manufacture of a product is not needed if the leaching, emission, and decomposition potential of the by-product are substantially eliminated. Products that would meet these criteria include cement, lightweight aggregate, structural or ornamental concrete or ceramic materials, concrete pavement, roofing materials, plastics, paint, fiberglass, mineral wool, wallboard, plaster, and so forth. Monitoring is also not needed if the by-product is used as a daily cover at a containment-type landfill or used as supplemental fuels for the production of energy. Environmental monitoring is usually needed when more than 30,000 yd^3 (23,000 m^3) of the by-product is used directly on the ground for projects such as highway embankments. A cover consisting of 2-ft-thick (60-cm) recompacted clay with a permeability of 1×10^{-7} cm/sec should be constructed over the area. The clay should be vegetated. For large projects in which the total volume of by-product used is 50,000 yd^3 (38,250 m^3) or more, a liner and lysimeter may be placed underneath the area containing the by-product. A typical design for such lysimeter can be found in Chapter 22. The area of the lysimeter should be 100 ft^2 (9 m^2). The volume of fluid collected in the lysimeter should be monitored twice a year. If the volume of collected fluid exceed 200 gal (757 liters) at the time of monitoring, then the exfiltrated fluid should be tested for all parameters listed in Table 8.2. If the concentration of one or more parameters exceeds the standard identified in the table, then the groundwater surrounding the by-product use area should be monitored for all health and welfare parameters specified by the regulatory agency. In the absence of any regulatory mandate, the parameters indicated in Table 14.2 may be used. The number and location of such monitoring wells will depend on the hydrogeology of the area. The aim of a groundwater monitoring network would be to assess whether the contaminants are expected to adversely impact a groundwater withdrawal point. If the groundwater is adversely impacted, then remedial action should be undertaken to restore water quality. However, if the by-product is tested properly, then it is not expected to adversely impact the groundwater. Thus, if a by-product is reused in a backfill or embankment project, then the leachate data should be studied carefully to ensure that it will not impact the groundwater.

9

CONTAMINATED
SOIL REMEDIATION

Knowledge about contaminated soil remediation is necessary for both redevelopment of contaminated land and landfill remediation projects. Solid waste professionals are involved in both types of projects. Redevelopment of contaminated lands may be considered as recycling because it is essentially use of resources after processing. Although redevelopment of contaminated lands is not considered as part of integrated solid waste management, such redeveloped lands can be used for siting material recovery facilities or refuse-derived fuel plants (see Chapter 7). Therefore, this chapter is included because it considers the bigger picture of integrated solid waste management. The redevelopment topic is divided into two chapters: This chapter summarizes the engineering issues and Chapter 10 considers the nonengineering issues.

Contaminated soil may be treated on-site or off-site. There are two options for on-site treatments: The contaminated soil may either be excavated out of the ground and then treated on-site or be treated *in situ* without excavation. In off-site treatment the soil is excavated out of the ground and transported to another site for treatment. A number of physical, chemical, and biological treatment methods are available for treatment of soil on-site or off-site. Selection of a remediation process depends on the geology and hydrogeology of a site, chemical composition, volume and concentration of the contaminants, ease of excavating the soil out of the ground, treatment cost, allowable time for completion of treatment, and regulatory requirements for allowable residual concentration of the contaminant(s) in the soil after treatment. If both the vadose zone and the aquifer are contaminated at a site, then remediation for each zone may be dealt with separately. Studies providing helpful information for selecting a treatment method are available (Ram et al., 1993; American Petroleum Institute, 1990; Mercer et al., 1997). Suggestions regarding design, operation, and maintenance of remediation systems are also available (Katin, 1995).

An in-depth discussion regarding various physical, chemical, and biological processes involved in treatment is beyond the scope of this book. However, the basics of the processes, design, and operational issues are included to help readers gain a fundamental concept regarding remediation of contaminated soil. Many texts and research articles are available for an in-depth study of the subject (Riser-Roberts, 1998; Norris, 1994; Devine, 1994; Weisman et al., 1994; Palmer and Fish 1997; Nyer et al., 1996).

Although risk analysis should be undertaken prior to implementing a remediation action, it is traditionally not done for small volumes of contaminated soil. Basic concepts of risk analysis are included in Chapter 10.

In many cases a combination of treatment technologies, also known as treatment train, is used to effectively remediate or control subsurface contamination. In some cases a portion of the plume with lower contaminant concentration is treated with a method different from the method used for another portion of the plume with higher contaminant concentration. In many instances chemicals are used to enhance a physical method. Currently, use of a biological method following a physical/chemical method is widely practiced. Usually remediation is divided into two parts: immediate and long term. Immediate action is taken to control the source (i.e., containment or removal), which is followed by plume remediation. Plume remediation is a long-term activity.

9.1   SOIL CONTAMINATION PROCESS

Soil consists of three phases: solid phase, water phase, and the gas phase. A chemical can contaminate all three phases; in other words it can be present in all three phases of a soil. The contaminant can be present in the soil pores as pure substance adsorbed on soil particles, as solute in the water phase, or as vapor in the gas phase. When a liquid chemical spills on ground, it flows downward, due to gravity, through the unsaturated zone of soil commonly known as the vadose zone (The vadose zone is the subsurface above the water table, which also includes the capillary fringe). Whether the entire volume of contaminants will migrate completely into the saturated part of the soil depends primarily on the capillary forces of the vadose zone and also on elapsed time. If the volume of the chemical is more than the holding capacity of the soil pores of the vadose zone, then the excess volume will continue to move downward due to gravity. Capillary forces decrease with the increase in particle size. So in a vadose zone consisting primarily of sandy soil, almost all of the contaminants will migrate downward into the aquifer if sufficient time is allowed. Whereas in a vadose zone consisting primarily of silty or clayey soil, a significant volume of contaminant will be held back. In many instances silty or clayey lenses are present in a sandy vadose zone; these lenses will hold back contaminants. This downward movement of contaminants is influenced primarily by the permeability of each of the soil stratums in the subsurface and the absorptive and adsorptive capacity of the solid and liquid phases of soil. If the density of the contaminant is lower than water (e.g., petroleum, diesel, aromatic compounds), then part of the contaminant volume that reaches the aquifer will float near the top of the water table, and the remaining part will dissolve in the groundwater forming a contaminated plume. The contaminants that are lighter than water and do not get dissolved in it are called light nonaqueous phase liquid (LNAPL). Figure 9.1 shows a typical LNAPL plume. The liquid contaminants that are immiscible and denser than water (e.g., chlorinated solvents, polyaromatic hydrocarbons) are called dense nonaqueous phase liquid (DNAPL). DNAPLs sink deep into the aquifer. It may even reach the bedrock surface depending on the volume of spill, the elapsed time, and the depth to the bedrock. Figure 9.2 shows a typical DNAPL plume. Since the plume geometry for LNAPLs and DNAPLs are different, their treatment strategies are also different.

9.1.1 Phases of Soil

As mentioned earlier, there are three phases of soil: solid, water (or liquid), and gas. Contaminants can be present in one or more of any of these phases.

9.1.1.1 Solid Phase The solid phase consists of mineral particles and organic fractions. Both fractions play important roles in the contamination process. The particle size of the mineral fraction varies from 0.002 mm (2 μm) and less (clay) to several centimeters (boulders). Based on the particle size, mineral fraction is subdivided into clay (0.002 mm and less), silt (0.06–0.002 mm), sand (2–0.06 mm), and gravel (2 mm and more). In general, particles bigger than gravel are called boulders. The bigger the particle size of mineral

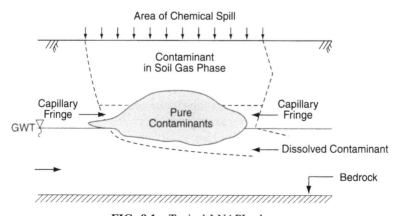

FIG. 9.1 Typical LNAPL plume.

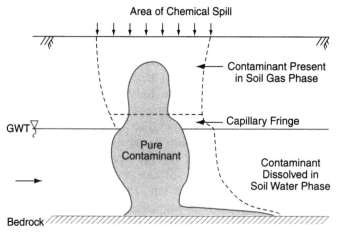

FIG. 9.2   Typical DNAPL plume.

fraction the higher is its permeability. The binding of contaminants (also known as adsorption) on the mineral fraction is directly proportional to the specific surface area of soil (m^2/g). The specific surface area is the sum total of surface areas of all particles of 1 g of soil. The specific surface area increases with decreasing particle size. Thus the adsorptive capacity of sand is lower than that of silt and is even lower compared to clay. Clay has the highest capacity to adsorb contaminants. Absorption refers to the retention of the contaminant within the mass of solid phase. The term sorption refers to both adsorption and absorption. The clay particles are usually negatively charged minerals. The space between two clay particles contains water with cations (positively charged ions) and anions (negatively charged ions). The clay surface, together with the liquid phase adjacent to it is known as the diffused double layer (A discussion on double layer can be found in Section 18.1). Adsorption of contaminant by mineral fraction depends on the chemistry of the contaminants and that of the double layer. The organic fraction consists of dead as well as living organic substances. The dead organic fraction is essentially organic remains of plants and animals, which are yet to be completely decomposed to inorganic compounds and elements. This fraction is commonly known as *humus*. In peat-type soil humus content is high. The living fraction essentially consists of various microorganisms and living plant materials. Like clay particles, organic substances are also negatively charged. So, the organic substances can adsorb positively charged metallic ions of contaminants (e.g., lead, copper, and zinc) easily. Some of the microorganisms are capable of breaking down the contaminants. Generally, the abundance of organic fraction decreases with depth.

9.1.1.2   Water Phase or Liquid Phase   The pores of soil, except in the dry upper part of the vadose zone, are either partially or completely filled with water. At and below the groundwater table the water pressure is equal to atmospheric pressure. Because of capillary pressure, pores of soil above the water table are filled with water. This layer is known as capillary fringe. The thickness of the capillary fringe depends on the particle size of soil above the water table. In clayey soil the capillary fringe can be several feet, whereas in sandy soil the capillary fringe can only be less than an inch. The chemistry of the water phase depends primarily on the mineral and organic fraction of the soil. The downward movement of liquid pollutant in the vadose zone depends primarily on the volume and size of pores and the adsorptive capacity of the soil. Below the water table the transport of liquid contaminants depends on its solubility in water, the chemistry of the soil double layer, the permeability of the liquid, and the relative density of the contaminants.

9.1.1.3 Gas Phase The gas phase is present primarily in the vadose zone. The composition of the gas phase near the surface is almost the same as the ambient air. If vegetation is present at the ground surface, then plant roots consume oxygen from the air in the soil pores. Microorganisms and other living animals (e.g., worms) also consume oxygen and produce carbon dioxide, which causes concentration of carbon dioxide to increase in the soil gas with depth. Transport of contaminants in the gas phase occurs primarily due to diffusion, which depends on the concentration gradient.

9.1.2 Distribution of Contaminants in Soil Phases

The adsorption isotherm of organic compounds, also known as the Freundlich isotherm, is based on a relationship between the amount of contaminants adsorbed onto a solid phase and the concentration of the solute (contaminants) in solution at equilibrium. The Freundlich isotherm is expressed as:

$$A_c = KC^b \tag{9.1}$$

where A_c = amount of constituent adsorbed per unit dry weight of soil
K = soil adsorption constant
C = contaminant concentration
b = constant; K and b depend primarily on the contaminant and soil.

The greater the value of K for a contaminant the stronger is its adsorption to the mineral fraction of soil. Table 9.1 shows the K values of selected hydrophobic organics and Table 9.2 shows K values of selected hydrophilic organics. For the organic fraction the K value depends primarily on the organic carbon content (Riser-Roberts, 1998). Dissolved heavy metals are usually present as cations (e.g., lead, zinc) that bind to negatively charged clay minerals and organic fractions. The binding of heavy metals are influenced by pH and the chemistry of the pore water.

The water phase contains minerals, which react with the contaminants and changes its chemistry. The water phase primarily acts as the transport medium for spreading the contaminants. It also dictates the remediation process. Thus

TABLE 9.1 Soil Adsorption Constants and USEPA Water Quality Criteria for Hydrophobic Organics

Substance	Soil Adsorption Constant K	USEPA Water Quality Criteria (ppm)
Anthracene	700	—
Benz(a)anthracene	60,000	2.8×10^{-6a}
Benzo(a)pyrene	40,000	2.8×10^{-6a}
Pyrene	2,000	2.8×10^{-6a}
Naphthalene	600	—
Oil	$(30,000)^b$	—
Grease	$(5,000,000)^c$	—
Ethyl benzene	50	1.4
Cyclohexane	70	—
Benzo(b)pyrene	40,000	2.8×10^{-6a}

[a]Corresponds to an incremental increase in cancer risk of 10^{-6}.
[b]Estimated based on n-C_{15}.
[c]Estimated based on n-C_{25}.
After Nash (1987).

the chemistry of the water phase, together with physical and chemical properties of the contaminants, influences the choice of treatment method.

Part of the contaminant volatilizes and occupies the gas phase. Equation (9.2) represents the relationship between the concentration of the organic contaminant in the water phase and in the gas phase:

$$C_g = HC_w \qquad (9.2)$$

TABLE 9.2 Soil Adsorption Constants and USEPA Water Quality Criteria for Hydrophilic Organics

Substance	Soil Adsorption Constant K	USEPA Water Quality Criteria (ppm)
Xylene	30	—
Phenol	20	3.5
Toluene	30	14.3
Methylene chloride	5	—
Methyl isobutyl ketone	5	—
Benzene	10	6.6×10^{-4a}
Tetrahydropyran	4	—

[a]Corresponds to an incremental increase in cancer risk of 10^{-6}.
After Nash (1987).

where C_g = concentration of the contaminant in the gas phase
  C_w = concentration of the contaminant in the water phase
  H = Henry's constant

Henry's constant depends on the solubility of the contaminant in pore water, the saturated vapor pressure of the contaminant in the pore gas, molecular weight of the contaminant, and soil temperature. Henry's law is applicable to a single contaminant not for a mixture of contaminants.

9.2   FACTORS AFFECTING REMEDIAL PROCESSES

Remedial processes are influenced by various physicochemical interactions involving the three phases of soil and the contaminants. In addition to the major factors discussed in the following sections, the following factors also influence the remedial processes: composition, physical state, carbon–nitrogen ratio of the contaminant, biological oxygen demand (BOD), chemical oxygen demand (COD), pH, temperature of the subsoil environment, and concentration of the contaminant spill. It may be necessary to seek expert opinion regarding such factors particularly for cases of a spill of uncommon contaminants for which sufficient remedial case histories are not available.

9.2.1   Solubility

Solubility of a chemical in a solvent (e.g., water) is dependent on both the solute and the solvent. Addition of certain chemicals improves solubility. The solubility of organic compounds in water varies widely. Many synthetic chemicals have low water solubility. Microbes have been found to grow on the soluble portion of compounds (Stucki and Alexander, 1987). Soluble hydrocarbons are readily biodegraded and can be partially captured by recovery wells for further treatment. The solubility of petroleum hydrocarbons is low (Norris, 1994). Compounds with low water solubility and high n-octanol/water partition coefficient (termed as K_{ow}, which characterizes the hydrophobic nature of the compound and indicates the tendency of the compound to sorb into soil organic portion) tend to sorb to the solid phase of soil, whereby their movement is retarded. On the other hand, compounds with high solubility and low K_{ow} are quite mobile in aquifers (Bouwer, 1994). Since the solubility of oxygen is relatively low in water, aerobic biodegradation of contaminants is low at a significant depth below the groundwater table.

9.2.2   Volatility

Volatility is an important property related to remediation primarily in the vadose zone. The rate of volatilization is dependent on the contaminant chemistry and the environment surrounding a spill. The effectiveness of *in situ* remediation in a vadose zone depends heavily on the volatility of the contaminant. The following factors affect volatilization of organics in the soil environment: contaminant vapor pressure, contaminant concentration, Henry's constant for the waste, soil/chemical adsorption reactions, contaminant solubility in soil water and soil organic matter, temperature, water content, organic fraction content, porosity, and bulk density (Spencer and Cliath, 1977; Ehrenfeld et al., 1986). Liquid gasoline contains up to 250 constituents, whereas its vapor phase contains 15–70 constituents (Riser-Roberts, 1998). The rate of volatilization is dependent on the properties of the contaminant and its surrounding environment (Dupont and Reineman, 1986). Volatilization is not an important factor in transport of the contaminant within the aquifer (Freeze and Cherry, 1979).

9.2.3 Chemical Structure

Biodegradation of high-molecular-weight hydrocarbons (especially aromatics) takes place at a slower rate, whereby they persist longer in the environment. Generally the biodegradation rate is slower for hydrocarbons with larger and more complex structures (Riser-Roberts, 1998). The chemical structure of a contaminant resists biodegradation in two ways: (a) The molecule contains groups whose chemical bonds cannot be broken, and (b) the structure's physical state (e.g., adsorbed in solid phase) is not conducive for biodegradation (Hutzinger and Veerkamp, 1981). Linear nonbranched compounds are more easily biodegraded compared to branched forms and rings (Pettyjohn and Hainslow, 1983).

9.2.4 Viscosity

Dispersion of a contaminant is dependent in part on its viscosity. The available surface area is higher for more dispersed hydrocarbons; thus biodegradation becomes easier. Viscosity and surface wetting properties affect the transport of a contaminant. Significant volumes of immiscible contaminants can get stored in soil as droplets even after removing a large volume of the contaminants. These droplets may continue to dissolve in water slowly, which will continue to pollute groundwater (Mackay et al., 1985). The remedial methods for highly water-soluble contaminants are different from those that float on the water table, such as gasoline. Biodegradation of highly viscous hydrocarbons is difficult (Norris, 1994).

9.2.5 Toxicity

Biodegradation of toxic contaminants is difficult. The rate and extent of biodegradation of toxic contaminants can be determined through laboratory tests. The toxicity of polycyclic aromatic hydrocarbons (PAHs) is related to their water solubility (Riser-Roberts, 1998). The vapor phase of short-chain alkanes is less toxic than the liquid phase. Water-soluble aromatic hydrocarbons are toxic to aquatic organism. Although low-molecular-weight aromatic hydrocarbons are toxic to microorganisms, they can be metabolized if the concentration is low. Floating oil tends to concentrate hydrophobic pollutant contaminants; thus the toxicity increases, which in turn reduces microbial degradation (Bartha and Atlas, 1977).

9.3   CONTAMINANT TRANSPORT

Contaminants move through the ground essentially in two stages. When a contaminant is released on the ground, it moves downward primarily due to gravity. On reaching the ground water table, it moves primarily due to hydraulic gradient, although diffusion and dispersion also play an important role in the transport of contaminants. (For a discussion on diffusion and dispersion refer to Section 15.2.4.) Apart from the primary process, contaminant transport through unfractured porous media is influenced by size, shape, and continuity of pores, physicochemical reaction with the geologic media, sorption–disorption, double-layer properties, and so on (Evangelou, 1998). The influence of some of these factors is discussed next.

9.3.1   Influence of Various Processes on Contaminant Transport

The transport of contaminants in porous media is influenced by various factors. The concentration of a contaminant may be reduced due to the following processes termed *natural attenuation:* evaporation, adsorption, transformation through microorganisms (Brown et al., 1986), and chemical reactions with soil, dilution, and filtration. For natural-attenuation-based biodegradation of contaminants to be effective the site must have a high natural supply of nutrients and oxygen (Hart, 1996). Some of these factors help reduce the concentration of contaminants away from the source. Knowledge about these factors is helpful in understanding transport modeling. Volatility, which also influences contaminant transport, has been discussed in Section 9.2.

9.3.1.1   Advection   In coarse-grain aquifers, soluble contaminants are transported by flowing groundwater, a process known as advection. The velocity of groundwater in an aquifer is due to the hydraulic gradient. Influence from a well can increase the hydraulic gradient, resulting in contaminant travel faster and further downgradient. The water flowing below the source can result in substantial dilution. In general, the larger the quantity of water flowing the greater the dilution.

9.3.1.2   Dispersion and Diffusion   A contaminant is transported due to the difference in concentration between two points. This can result in contaminant movement contrary to the hydraulic gradient. This method of transport is known as diffusion. Dispersion on the other hand depends primarily on the pore geometry. Dissolved contaminants spread as they move with ground-water. Diffusion and dispersion reduces a contaminant concentration down gradient from the source. Additional discussion of this topic is included in Section 15.2.4.

9.3.1.3 Sorption/Desorption   Contaminant molecules can be held back by the solid phase of soil due to sorption. The extent to which a contaminant is sorbed depends on the contaminant chemistry, soil composition, groundwater chemistry, and the flow velocity. This phenomenon can be important in transport modeling. Sorption influences volatilization, diffusion, leaching, microbial degradation and chemical degradation (Riser-Roberts, 1998). Many compounds, especially hydrophobic, are readily bound to the organic fraction due to sorption (Alexander, 1999).

Desorption of contaminants is an important factor to consider for treatment effectiveness. Although both sorption and desorption generally follow the Freundlich isotherm, the constants for each of the processes are different. Desorption of a contaminant can become more difficult, if it remains in soil for long time (Alexander, 1999).

9.3.2 Influence of Contaminant Properties on Transport

Density and solute mobility greatly influence transport of contaminants in porous media.

9.3.2.1 Density   The density of a contaminant controls much of the plume geometry of the contaminant. Low-density nonaqueous phase liquids (NAPLs) tend to float so that the contaminant is found near the top of the aquifer or in the capillary fringe. Some of the contaminant generally remains in the soil as a thin film when the water table moves downward. NAPLs that are denser than the groundwater sink to the bottom of the aquifer. Such chemicals do not move necessarily with the groundwater under the hydraulic gradient because they have a strong vertical component (Riser-Roberts, 1998).

9.3.2.2 Solute Mobility   The mobility of a nonreactive contaminant in soil is equal to that of water. However, the mobility of a reactive contaminant is less than that of water (Evangelou, 1998). The velocity of a reactive contaminant is given by (McBride, 1994)

$$V_c = \frac{V_w}{(1 + \gamma_s/n)K_d} \tag{9.3}$$

where V_c = velocity of the contaminant
V_w = velocity of water
γ_s = bulk density of soil
n = soil porosity
K_d = distribution coefficient

Thus the velocity of a reactive contaminant differs by a factor of $(1 + \gamma_s/n)K_d$, known as the retardation factor. As the retardation factor increases, the velocity of contaminant decreases relative to the velocity of water. In coarse-grain soil the contaminant transport is influenced heavily by mass flow, whereas in fine-grained soil the diffusion–dispersion influences the transport.

9.3.3   Mathematical Model

In general, advection, diffusion, dispersion, chemical, or biological reaction governs the transport of contaminants in soils. A mathematical model of contaminant transport is based on governing initial and boundary conditions.

In advective transport the average seepage velocity can be expressed as

$$V_s = \frac{V_d}{n} = \frac{Q}{An} \tag{9.4}$$

where V_s = seepage velocity
  V_d = discharge velocity
  n = soil porosity
  Q = discharge volume
  A = area of discharge

According to Darcy's law:

$$V_d = ki \tag{9.5}$$

Thus, $V_s = ki/n$, where k is the hydraulic permeability of soil and i is the hydraulic gradient.

If the water is not of uniform density and viscosity, then for a three-dimensional situation, the Darcy velocity can be written as (Zeng and Bennett, 1995)

$$V_{sx} = -\left(\frac{k_x}{n\mu}\right)\left(\frac{\delta P}{\delta x}\right) \tag{9.6}$$

$$V_{sy} = -\left(\frac{k_y}{n\mu}\right)\left(\frac{\delta P}{\delta y}\right) \tag{9.7}$$

$$V_{sz} = -\left(\frac{k_z}{n\mu}\right)\left(\frac{\delta P}{\delta z} + \gamma_w g\right) \tag{9.8}$$

where k_x, k_y, k_z = intrinsic permeability in x, y, z coordinates, respectively
  μ = dynamic viscosity of water
  γ_w = density of water

g = acceleration due to gravity
P = water pressure
V_{sx}, V_{sy}, V_{sz} = seepage velocity in x, y, z coordinates, respectively

In soil, the effective porosity (n_e) governs the travel time of transport. Effective porosity is the volume of void space that actually conducts flow divided by the total volume of soil. Effective porosity ratio for soil (n/n_e), which can be dependent on flow gradient, is less than one.

To include the effect of dispersion, a flux due to dispersion needs to be added to the advection transport equation. Let us consider a volume element of dimension $\Delta x \, \Delta y \, \Delta z$ within a pore, with the flow direction parallel to x-axis (Fig. 9.3). The area normal to the x axis is $\Delta y \, \Delta z$. The advective transport of the contaminant of concentration C through the $\Delta y \, \Delta z$ area is

$$V_d C \, \Delta y \, \Delta x \tag{9.9}$$

The net difference between inflow and outflow for the element is

$$\frac{\delta}{\delta x} (V_d C) \, \Delta x \, \Delta y \, \Delta z \tag{9.10}$$

Since longitudinal dispersion occurs through area $\Delta y \, \Delta z$, the dispersive mass transport is given by

$$-(D_x) \left(\frac{\delta C}{\delta x} \right) n_e \, \Delta y \, \Delta z \tag{9.11}$$

where D_x is the longitudinal dispersion coefficient.

Considering the possibility of longitudinal velocity variation within the element, the change of dispersive mass transport across the element is change in $(D_x \, \delta C / \delta x)$. Thus the change in mass across the element due to longitudinal dispersion is

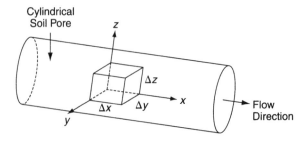

FIG. 9.3   Elemental volume with flow parallel to x axis.

$$\frac{\delta}{\delta x}\left(\frac{D_x \, \delta C}{\delta x}\right) n_e \, \Delta x \, \Delta z \tag{9.12}$$

Similar equations can be developed for change in mass due to dispersion in y and z directions. The rate of mass accumulation within the element is

$$n_e \, \Delta x \, \Delta y \, \Delta z \left(\frac{\delta C}{\delta t}\right) \tag{9.13}$$

Equating the mass accumulation to the change in mass equations, in all three direction, the following equation can be developed (Zeng and Bennett, 1995):

$$\frac{dC}{\delta t} = -\frac{\delta}{\delta x}(V_s C) + \frac{\delta}{\delta x}\left(D_x \frac{\delta c}{\delta x}\right) + \frac{\delta}{\delta y}\left(D_y \frac{\delta C}{\delta y}\right) + \frac{\delta}{\delta z}\left(D_z \frac{\delta c}{\delta z}\right) \tag{9.14}$$

Equation (9.14) represents unidirectional advective-dispersive transport without any sink. In developing Eq. (9.14) the flow direction was assumed to be parallel to the x axis.

The two other processes that should be included in contaminant transport are chemical reaction and retardation. Fully coupled transport equations, which include chemical reaction in addition to advection and dispersion, have been derived by Mangold et al. (1991). Although equations governing various reactions in the transport process are available (Liu and Narashimhan, 1989; Friedly and Rubin, 1992), computational difficulties and uncertainties in natural processes render such attempts unrealistic. Instead equations with reasonable simplifications reflecting realistic assumptions are used for addressing contaminant transport. The following is a commonly used partial differential equation governing the three-dimensional transport of a single contaminant; the equation includes advection, dispersion, sink/source term, sorption, and first-order irreversible rate reaction (Zeng and Bennet, 1995):

$$\frac{R \, \delta C}{\delta t} = \frac{\delta}{\delta x_i}\left(D_{ij} \frac{\delta C}{\delta x_j}\right) - \frac{\delta}{\delta x_i}(V_s C) + \left(\frac{q_s}{n_e}\right)V_s - \lambda\left[C + \left(\frac{\gamma_s}{n_e}\right)C_b\right] \tag{9.15}$$

where $R = (1 + \gamma_b/n_e)\,\delta C_b/\delta c$
  C_b = sorbed concentration (a function of dissolved concentration, C, as defined by a sorption isotherm)
  D_{ij} = dispersion coefficient tensor
  λ = reaction rate constant
  γ_s = bulk density of soil

Equation (9.15) is linked to governing equation for flow:

$$V_s = -\left(\frac{k_{ij}}{n_e}\right)\frac{\delta h}{\delta x_j} \tag{9.16}$$

where h is the hydraulic head and k_{ij} is the hydraulic conductivity tensor.

A general equation applicable for a variable flow direction and a term to reflect addition or reduction of contaminant concentration (called source or sink) is as follows (Zeng and Bennett, 1995):

$$
\begin{aligned}
\frac{dC}{dt} = {} & \frac{\delta}{\delta x}\left(D_{xx}\frac{\delta C}{\delta x} + D_{xy}\frac{\delta C}{\delta y} + D_{xz}\frac{\delta C}{\delta x}\right) - \frac{\delta}{\delta x}(V_{sx}C) \\
& + \frac{\delta}{\delta y}\left(D_{yx}\frac{\delta C}{\delta x} + D_{yy}\frac{\delta C}{\delta y} + D_{yz}\frac{\delta C}{\delta z}\right) - \frac{\delta}{\delta y}(V_{sy}C) \\
& + \frac{\delta}{\delta z}\left(D_{zx}\frac{\delta C}{\delta x} + D_{zy}\frac{\delta C}{\delta y} + D_{zz}\frac{\delta C}{\delta z}\right) - \frac{\delta}{\delta z}(V_{sz}C) + \frac{q_s}{n_e}C_s \quad (9.17)
\end{aligned}
$$

where D_{xx}, D_{xy}, and so on represent the components of the dispersion coefficient along appropriate axes; q_s is the flow of a fluid source/sink per unit volume, and C_s is the concentration associated with source or sink.

In several commonly used transport models such as MOC (Konikow and Bredehoeft, 1978), RANDOMWALK (Prickett et al., 1981), and MT3D (Zeng, 1990) the flow equation and the transport equations are solved independently. The governing equations are solved either analytically or numerically. Analytical solutions are obtained for exact initial and boundary conditions. The relative simplicity of analytical models makes them attractive for initial decision-making processes. Numerical solutions can be used for more complex situations and in general provide a better transport solution. Both finite-difference and finite-element methods are used for obtaining numerical solutions.

In general, the advection plays a dominant role in contaminant transport. So, a purely advective-dispersive transport model can provide a fair estimate of contaminant movement. If it is felt that sorption is important in a site, then use of a retardation factor coupled with advection would provide a simple yet fairly accurate transport model.

9.3.4   Model Selection and Data Needs

It is essential to determine the purpose of developing a transport model. Models can be used to define the dominant transport processes (e.g., advection), estimate the time when contamination began, probable effect of a remedial action, assess a "no-action" scenario, and help select a remedial option (Zeng and Bennett, 1995; Childs, 1985). Models are not replacement for field information but are rather a compliment to observed data. So, knowledgeable professionals must undertake the selection, application, and interpretation of model output.

All relevant local and regional data should be reviewed prior to model development. Primarily soil data from boring logs, chemical and hydrogeologic data from groundwater wells, and information about the source(s) are needed for developing a transport model. The type of data needed and details regarding site characterization for developing a successful transport model have been discussed by many (Domenico and Schwartz, 1990; Fetter, 1993; Mercer and Waddell, 1993).

If the aim of transport modeling is delineation of the approximate extent of contamination for designing a capture zone strategy, then a simple model using advective transport can be sufficient [e.g., Pollock (1988) and USGS MODPATH by Pollock (1989)]. Such models can also provide approximate travel time of contaminants. However, if field data indicate that the effect of dispersion and retardation may be significant, a comprehensive model should be used [e.g., MT3D by Zeng (1990), coupled with MODFLOW by McDonald and Harbaugh (1988)]. There is no one solution for all sites. In critical cases more than one model may be needed to increase confidence in results (Zeng and Bennett, 1995). Due attention must be given to selecting a model with clear documentation and user-friendly instructions. Cost of the model software, additional hardware and software requirements, and cost of user training must be assessed before selecting a model. While selecting a model, reliability and acceptance of the model by all parties involved (e.g., regulatory personnel, project owner) must be considered. The American Society of Civil Engineers (1995) and the National Research Council (1990) have published guidelines for assessing reliability for selecting models.

9.3.5   Conceptual Model Development and Computer Code Selection

Prior to developing the conceptual model, information about the geology, hydrogeology, and geochemistry is compiled from existing sources (e.g., published reports, driller's log). Based on this information, a conceptual model is developed using simplifying assumptions and qualitative interpretation about possible flow and transport processes at the site. Discussion regarding site characterization requirements relevant to contaminant transport can be found elsewhere (Bedient et al., 1994). The observed contaminant plume at the site is then compared with the predictions based on the conceptual model. The comparison will show whether the assumed transport system (e.g., advective, advective-dispersive) is actually operative at the site.

After a conceptual model is finalized, a proper computer code needs to be selected. The code selection depends primarily on the transport process and dimensions of analysis (i.e., two or three dimensions). Code requirements for simulating advective transport is much simpler than simulating advective-dispersive transport. If the contaminant density is expected to influence transport, then appropriate computer code should be selected. Other factors that influence code selection are:

1. Clarity of documentation of the proposed code
2. Cost of the code
3. Compatibility of the existing computer hardware
4. Training needs
5. Reliability of the code
6. Acceptability of the code to the parties involved in the project

9.3.6   Calibration and Sensitivity Analysis of Contaminant Transport Models

After developing a model for a specific site, it is necessary to calibrate the model. Calibration is a process in which model input parameters are adjusted until model output variables (or dependent variables) match field-observed values to a reasonable degree. Even after thorough site investigation, input parameters remain associated with various uncertainties. Error in input data will influence model predictions. As a preliminary step, the model output should be compared with field-observed values. Goodness of fit is assessed using statistical methods such as mean of residual errors, variance of residual errors, mean of absolute residual errors, and so on (Benjamin and Cornell, 1970; Press et al., 1992). The goal of calibration is to minimize residual errors. In some situations changes in input parameters may have significant effect on errors as calculated by a statistical method but have negligible effect on model predictions. A calibration may be considered adequate for a specific purpose if changes in input parameters have little effect on model predictions. If the purpose of a modeling effort is to choose a remediation design, then significant importance must be given to calibration. However, if the goal of modeling is to judge several action scenarios, then large residual errors are allowable. In addition to traditional approaches, a model can be calibrated using trial and error. An in-depth discussion of the topic can be found elsewhere (Zeng and Bennett, 1995).

The sensitivity of a model to parameters selection must also be tested before field application. In a sensitivity analysis, the effect on change of one factor on another factor is studied (McElwee, 1982). A sensitivity analysis provides insight into the model behavior whereby a designer can estimate the accuracy necessary for data collection.

9.4   MAINTENANCE OF WELLS

Many wells are constructed for most *in situ* remediation and monitoring of contaminated groundwater. In addition to proper design, construction, and development of these wells, it is essential to maintain them for long-term use; rehabilitation of monitoring wells established prior to remediation are also necessary. Since well installation is essential for most *in situ* remediation methods, knowledge about design, construction, development, long-term maintenance, and rehabilitation is needed for success in design, operation, and monitoring of any of these methods. Comments regarding design, construction, and development of wells is included in Section 22.4. Only the remediation and rehabilitation of wells will be discussed in this section. Special well-related design or construction issues for any particular remediation method is included in this chapter along with the method.

Deterioration of a monitoring well increases both the direct cost of rehabilitation/replacement and indirect cost of consequences of erroneous data. Erroneous data can have a negative impact in public hearings and in legal proceedings. It may also have environmental or health-related consequences. Thus, great importance must be given to long-term maintenance of monitoring wells and efficiency of extraction wells associated with remediation.

9.4.1 Performance Deterioration

Deterioration in the performance of wells is caused by faulty design and installation, natural causes, and chemical and biological interaction between the screen material and the groundwater. Extraction wells are very likely to be clogged if the filter pack material is not chosen and installed properly. Improper filter in a monitoring well may produce turbid water, which needs to be filtered. Such filtering may alter contaminant concentration because part of the contaminant may be absorbed in the finer particles, which is discarded during filtration to remove turbidity. The main causes of performance deterioration are discussed in the following sections.

9.4.1.1 Faulty Design and Installation   The well screen must be designed for structural and functional integrity. The slots of a screen for a deep well should be such that it can withstand the vertical load of the casing and horizontal pressure during pumping. Failure of screens because of faulty installation and/or design of the filter pack is rather common. In many cases, well installation crews rush to finish the job, leading to faulty installation. Well design and installation should be done by qualified groundwater professionals, experienced in well hydraulics and construction (Powers, 1992). In addition to determining the location, diameter, depth at which well screen should be located, screen slot design for proper intake velocity, selection of casing and screen materials, grouting to prevent infiltration from surface or upper aquifer, well development procedure, and selection of proper drilling method including drilling fluid are all important issues for design and installation of extraction and monitoring wells. Plastic pipes are used for extraction and monitoring wells primarily because they are not susceptible to corrosion. However, a high percentage of plasticizer in polyvinyl chloride (PVC) can cause deterioration of the pipe. In addition, hydrocarbons can soften PVC, which may promote biodegradation. So the interactions between the pipe material and contaminants should be considered while choosing the pipe material. Properly designed and installed wells can provide long-term service without significant deterioration. Pressure relief screens (Fig. 9.4) should be installed in wells in which the draw down outside the well may be significantly above the pumping level inside the casing. The high-pressure differential between the inside and outside of the well may cause water to flow upward through the filter pack. The upward pressure may be high enough to dislodge the filter pack from its position. This uplifting of the filter pack is most likely to occur in tight or highly stratified formations (Driscoll, 1986). A short pressure relief screen should be installed in the riser pipe just above the bottom of the casing. Pressure differentials are relieved through the short screen.

9.4.1.2   Natural Causes   A well may fail due to ground movement during an earthquake or because of ground subsidence. Decline in water level, localized or zonal, may lead to plugging or encrustation of the borehole, screen, or the filter pack. This plugging increases the entrance velocity of water, which increases the risk of entraining fine soil particles. The fine soil particles tend to erode the slot opening causing larger particles (e.g., sand) to pass through the well screen and eventually pass through the pump. Subsequently, the increase in sand loading will cause pump failure.

A well casing may fail due to the collapse of the borehole. This type of collapse may occur because of rock falling in the underground geologic cavern or in rock boreholes.

9.4.1.3   Chemical and Biological Interaction   Due to the pressure and solubility changes at the well screens, carbonate and sulfate salts of Ca and Mg may get precipitated out of the groundwater. This phenomenon is known as chemical encrustation. Chemical encrustation could be a secondary effect of biofouling oxidation or corrosion. Encrustation affects efficiency, specific capacity, and sample quality (Smith, 1995).

The second major cause of well performance deterioration is corrosion, which results from chemical and electrochemical processes. Chemical corrosion occurs due to the removal of materials by carbon dioxide, oxygen, hydrogen sulfide, and so on. In electrochemical corrosion, flow of electric current cause corrosion of the metal. An electrical potential may develop due to the presence of two kinds of metal in close proximity or between two separate areas on the pipe surface in close proximity. For example, electric potential can develop between areas around torch cut slots or in breaks in surface coatings (e.g., paint and millscale). A cathode and anode develop in such instances and metal is removed from the anode (Fig. 9.5). If iron or steel is corroded, the corrosion product is usually ferric hydroxide or oxide

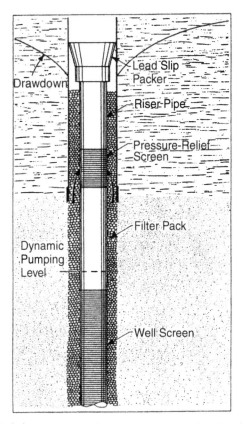

FIG. 9.4   Pressure relief screen. From Driscoll (1986).

(Driscoll, 1986). Corrosion causes either enlargement of slot opening leading to sand pumping or deposition of corrosion products leading to reduction in yield. The water quality is also affected due to corrosion.

Biofouling occurs because of biological formation and deposition of Fe, Mn, and S compounds. Iron bacteria plug wells through oxidation of iron and manganese. The bacteria obtain their energy by oxidizing ferrous iron to ferric iron and in the process create precipitates of hydrated ferric hydroxide. The precipitation, along with a rapid growth of the bacteria, plugs the screen pores (Driscoll, 1986). Biofouling affects both the efficiency of the well and sample quality.

9.4.2   Prevention and Rehabilitation

Prevention starts with the good design, installation, and development of wells. These issues are discussed in Section 12.4. Treatment and rehabilitation meth-

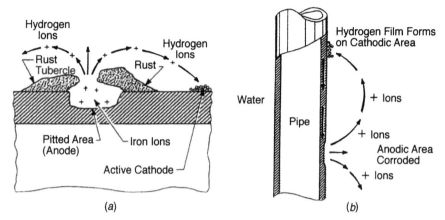

FIG. 9.5   Well corrosion. (*a*) Anodes and cathodes develop in adjacent areas, resulting in corrosion. (*b*) Corrosion of iron at anodic areas leads to the deposition of iron hydroxide and oxide rust at cathodic areas. From Driscoll (1986).

ods following a problem detection will be discussed in this section. Problem detection can only be done through routine monitoring.

9.4.2.1 Monitoring   The first sign of performance deterioration is a reduction in the yield of a well. It is relatively easy to keep track of yield from extraction wells compared to monitoring wells. Motor characteristics (e.g., significant increase in power consumption) also serve as an indicator of deterioration of extraction wells (Borch et al.,1993). Visual inspection using a borehole camera is a relatively new addition for monitoring encrustation. Chemical analyses of water samples may be done for early detection of a problem. The sample analysis should include soluble ferrous iron, total Fe, total Mn, Ca^{2+} total S^{2-}, pH, and temperature. Redox potential should also be monitored using the electrode method (Smith, 1995). Sediment content in bailed or pumped water samples from monitoring wells will provide signs of pack and screen deterioration. Total suspended solids (TSS) in water samples are also a good indicator parameter, which can be measured easily using turbidometers. Particle counters, a recent innovation, provide better information than turbidometers. Particle counters can count as well as determine particle size in water samples (Hargesheimer et al., 1992). Biofouling monitoring involves visual inspection of samples through ordinary or electron microscope. Microorganisms can also be detected through bioculture methods (Smith, 1995). Table 9.3 provides a summary of monitoring parameters for detecting well deterioration.

9.4.2.2 Treatment/Rehabilitation Methods   Total prevention of deterioration of wells is not possible. However, various methods are used to reduce the problem. One of the methods, called the Vyredox system, uses a series

TABLE 9.3   Monitoring Parameters for Detecting Well Deterioration

Fe (total, Fe^{2+}/Fe^{3+}, Fe minerals and complexes):
  Indications of clogging potential, presence of biofouling, Eh shifts.
Mn (total, Mn^{2+}/Mn^{4+}, Mn minerals and complexes):
  Indications of clogging potential, presence of biofouling, Eh shifts.
S (total, $S^{2-}/S^0/SO_4^{2-}$, S minerals and complexes):
  Indications of clogging potential, presence of biofouling, Eh shifts.
Eh (redox potential):
  Direct indication of probable metallic ion states, microbial activity. Usually bulk
    Eh, which is a composite of microenvironments.
pH:
  Indication of acidity/basicity and likelihood of corrosion and/or mineral
    encrustation. Combined with Eh to determine likely metallic mineral states
    present, and with conductivity and alkalinity to assess inorganic salts
    occurrence.
Conductivity:
  Indication of total solids content and a component of corrosivity assessment.
Turbidity:
  Indication of suspended particles content, suitable for assessment of relative
    changes indicating changes in particle pumping or biofouling.
Particle counts:
  Indication of suspended particles content, suitable for assessment of relative
    changes indicating changes in particle pumping or biofouling.
Sand/silt content (v/v, w/v):
  Indication of success of development/redevelopment, potential for abrasion and
    clogging.

Reprinted with permission from Smith (1995). Copyright CRC Press, Boca Raton, Florida.

of injection wells encircling the target wells. Oxygenated water is introduced through the injection wells to oxidize iron, which promotes the growth of iron bacteria. This helps in the reduction of iron concentration in the water around the target well (Hallberg and Martinell, 1976). Chemical encrustation can be removed through acid treatment and mechanical methods. Hydrochloric (HCl), sulfamic (H_3NO_3S), and hydroxyacetic ($C_2H_4O_3$) acids are most commonly used for treatment (Driscoll, 1986). Liquid (or pellets) acid is carefully and gradually introduced into the well through a small-diameter pipe, up to the top of the screen. Then an equal volume of water is poured that forces the acid solution to exit through the slot. A surge block or jetting is used to agitate the solution. The loosened dissolved material along with the solution is pumped out. In a mechanical method the screen is cleaned using a wire brush. The dislodged material is pumped or bailed out. In many instances the mechanical method is used prior to acid treatment.

Sediment plugging can be prevented primarily through proper development after well installation. Dispersing and chelating compounds (e.g., polyphosphates) are used to disperse the silt and clay particles clogging a well. Surfactants are also used for removing oil. The sediment is then pumped or bailed out following agitation (Driscoll, 1986).

Prevention of corrosion is more related to proper selection of screen materials. Corrosion is also dependent on temperature. Hence, where heat is used for remediation, screen should be chosen to avoid corrosion during remediation. Stainless steel screens can be affected by heat. Corrosion can be reduced significantly by cathodic protection. A sacrificial anode (e.g., magnesium block) is connected to the casing with a coated copper wire. The flow of current through the soil provides protection against galvanic corrosion (Driscoll, 1986).

If the iron concentration is high in a aquifer, great care should be taken to introduce iron bacteria during drilling. The drilling fluid may be chlorinated (to 50 g/L) for reducing bacterial contamination. The drill rods and other tools including the filter pack should be chlorinated where needed. Wells affected with biofouling may be treated with chlorine compounds (e.g., calcium hypochlorite). Potassium permanganate is also used for treating biofouling. The well must be pumped after mechanical agitation to remove the dislodged organic slime. Treating the affected well by circulating hot water [176°F (80°C)] has also been reported (Cullimore, 1981). The Vyredox technique can also be used to control iron bacteria.

Since most of the treatment/rehabilitation methods use a chemical, the chemistry of water around the well gets affected. Sufficient time should be allowed to stabilize the water chemistry around a monitoring well after treatment. Monitoring data obtained immediately following treatment do not represent the chemistry of water surrounding the well.

9.5   SITE INVESTIGATION

Remediation involves site investigation and use of appropriate techniques for decontamination. The discussion is focused on sites contaminated by a long-term activity rather than a one-time spill situation. However, many of the issues discussed can be used for investigating one-time spills. Normally risk assessment is a part of site investigation, which is discussed in Chapter 10.

Prior to embarking on soil and groundwater sampling, all existing records relative to the contamination are reviewed. Then soil and groundwater sampling is undertaken. Subsequently, the physical data is analyzed and a decision is made regarding suitable options. Input should be sought from the regulatory agency in every step of the process to ensure ease of obtaining necessary permits for the remediation technique to be selected in the future.

9.5.1   Preliminary Review

All existing information relative to the contamination should be reviewed first. This information gathering may be divided into two steps. In the first step, the following information is obtained: past and present product(s) of the plant (if a plant is involved), geologic site features, information regarding plant process (if a plant is involved), history of disposal practices, review of records, concerns of the regulatory agency, potential health and environmental hazards from the soil and groundwater contamination (which essentially involves risk assessment), reconnaissance survey of the site for visual impact of contamination on trees and other vegetation, and development of a rough sketch of the site. As part of the first step, a report summarizing the information and comments regarding the extent and significance of the contamination should be developed. If the findings in the first step point toward probable remedial action, then a second step involving more detailed review of the site should be undertaken.

The following information should be obtained as part of the review process in the second step: the cause of contamination (i.e., a spill or discharge from a long-term industrial operation), estimated volume of contaminants discharged, geology and hydrogeology of the area based on available existing records, surface water drainage pattern in the area, distance of groundwater withdrawal point(s) (both shallow and deep), proximity of lake(s) and river(s), proximity of dwellings, existence of endangered species in the general area, and accessibility. If the depth to groundwater and its flow direction is unknown for the area, then a minimum of four to five groundwater-monitoring wells should be established. The number of wells will depend on the contaminant volume and the geology of the site. Data from such wells will be useful for preliminary review as well as for detailed fieldwork in future. A site visit for visual assessment of the source of contamination is helpful in the selection of a remediation process in the future. A site visit jointly with regulatory personnel can be helpful in assessing regulatory concerns. A report documenting findings of the preliminary review along with an assessment regarding significance and extent of groundwater contamination and the potential for environmental and health hazards should be developed. This report may need to be submitted to the regulatory agency for review/comment. It is possible that based on the data a decision can be made not to undertake a detailed site investigation or even cleanup. If a "no-action" decision is made, then the justification for such a decision must be clearly documented.

9.5.2 Fieldwork

Fieldwork plays a significant role in remedial actions. Prior to undertaking fieldwork the preliminary review report should be studied to develop a well-organized plan for sampling and data analysis. A detailed reconnaissance survey of the area under investigation needs to be undertaken to establish surface and subsurface soil and groundwater sampling points. A preliminary groundwater sampling using existing groundwater or privately owned wells, if available, can be very helpful in developing sampling points.

Soil samples are collected from the surface, backhoe pits, and boreholes. A detailed bore log must be developed at the time of collecting soil samples. A borehole location plan should be developed carefully so that some of the boreholes can be converted to groundwater monitoring wells if necessary. The boreholes should be deep enough to profile the entire contaminated plume and geologic cross section of the site. Usually, a first set of proposed soil sampling points covers a rather big area while a second (in some cases a third) set of sampling points are used to define the extent of contaminant more precisely. Gas probes may be installed to obtain information regarding the presence of volatile organics in the vadose zone.

Experienced professionals should be involved in both subsoil investigation and groundwater monitoring point installation. Personnel capable of providing visual interpretation of soil samples should be present at the site at all times when soil sampling is done. Proper entry in soil bore logs is extremely important in the correct interpretation of any change in subsoil condition. Soil type should be closely observed to identify each stratum clearly. Extent of the vadose zone, presence of perched layer, soil type in each layer, and so on are important for proper choice and planning of a remedial method. It is a good idea to involve the regulatory agency personnel during the site investigation. Additional soil samples and groundwater sampling points, if required by the regulatory agency, can be undertaken easily while the soil sampling equipment is on site. It is costly to bring the equipment on site a second time.

Laboratory analysis should be carried out for all chemicals suspected of causing contamination at the site. Proper quality assessment/quality control (QA/QC) procedures should be used for sample collection and testing. In many instances regulatory agencies specify the QA/QC procedures to be followed. All tests should be performed in reputable laboratories. In many instances the regulatory agency requires that the tests be performed at laboratories approved by the agency. A list of such certified laboratories are available from the regulatory agency. All regulatory requirements regarding testing must be followed to obtain a permit for a proposed remedial method. It is essential to involve the staff from the laboratory at least from the time samples are collected because use of proper procedures for collection, preservation, and transportation of samples are key to the reliability of the data. In addition, establishing and implementing chain-of-custody rules are important from a legal standpoint. Data obtained without proper chain of custody is not acceptable in a court of law.

Necessary health and safety rules must be followed in all site investigation work. Thus it is imperative to appoint experienced consultants and contractors for carrying out a site investigation.

A report containing all factual information, data evaluation, and hazard assessment should be prepared. The reporting requirements mandated and/or outlined by the regulatory agency must be followed for obtaining permits. If the hazard assessment indicates that the risk of no action is below the limits mandated by environmental regulations, then there is no need to proceed for remediation. However, if the risk is above the acceptable limits, selection of an appropriate technique for remediating the contaminated soil must be undertaken.

9.6 *IN SITU* REMEDIATION METHODS

Currently, many *in situ* remedial methods are available. The methods can be divided into three broad groups, namely physical, chemical, and biological. In practice most remediation methods are a combination of two or all three of these groups. Various issues must be considered before selecting a method. There is no single method that can be useful for all sites and for all situations.

9.6.1 Physical Processes

In a purely physical process no chemical or microorganisms are used for treating the contaminants. However, as mentioned above, a physical method alone is seldom used for completing remediation, except perhaps for shallow vadose zone remediation. It should be mentioned that a physical process is needed for most chemical and biological processes. The physical process essentially works as a delivery system for the other two processes. So, good knowledge of the physical processes is essential to grasp the remedial technology.

9.6.1.1 Horizontal Barriers   If the spill is at or near the surface, the volume of contaminants is rather low, and the vadose zone is deep, then leaching of contaminants can be eliminated by constructing a low permeability layer on top of the contaminated zone. This low permeability layer acts as a horizontal barrier, which is commonly known as cover or cap. The area of the cover should be greater than the contaminated area to ensure that the infiltrating water from precipitation does not contact the contaminated soil (Fig. 9.6). A horizontal barrier is justifiable only in cases where the spill volume is so low that long-term downward movement of the contaminants is ruled out. However, in most sites a horizontal barrier is constructed along with a vertical barrier (see Section 9.6.1.2). In addition a horizontal barrier is also constructed over sites where a soil mixing process (see Section 9.6.2.1) is used, especially if the contaminated zone is shallow.

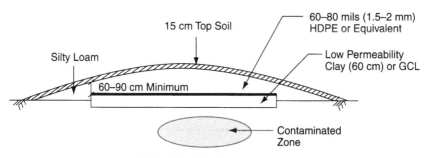

FIG. 9.6   Horizontal barrier.

A horizontal barrier can also be constructed below a contaminated zone to act as a liner. Construction of a series of consecutive horizontal tunnels, filled with low permeability material, to act as a horizontal bottom barrier has been reported in the literature. Although several field trials are reported, construction of horizontal bottom barrier is still in a developmental stage (Daniel, 2000).

9.6.1.2 Vertical Barriers

Vertical barriers, commonly known as cut-off walls, are used either to capture waste contact liquid or redirect groundwater. Of the various types of cut-off walls (e.g., slurry walls, vibrating beam cut-off walls), slurry walls are extensively used for pollution control projects.

Slurry Walls Both vertical and horizontal configurations of slurry walls are important for effective remediation. A slurry wall can be either keyed into an underlying low permeability layer (commonly known as keyed in slurry walls, Fig. 9.7) or constructed deeper than the lower boundary of the plume, but not keyed into a low permeability layer (commonly known as hanging slurry walls, Fig. 9.8) to capture the contaminant plume (Sharma and Lewis, 1994). Both these types represent the vertical configuration. Based on the horizontal configuration, there are three types of slurry walls: upgradient, down gradient, and circumferential (USEPA, 1984). An upgradient slurry wall is constructed at a distance away from the contaminated zone to divert uncontaminated groundwater away from the contaminated zone. A circumferential slurry wall (Fig. 9.9) is constructed all around the contaminated zone with an attempt to minimize contact of groundwater with the plume. Extraction well(s) may be placed within the enclosed area to pump out contaminated water. If the groundwater flow volume is low (e.g., near a groundwater divide), then a

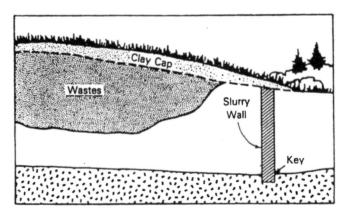

FIG. 9.7 Keyed-in slurry wall. From USEPA (1984).

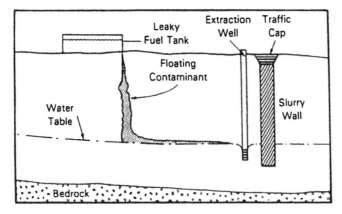

FIG. 9.8 Hanging slurry wall. From USEPA (1984).

down-gradient slurry wall with extraction well(s) is constructed for remediation (Fig. 9.10). A horizontal barrier is also constructed over the contaminated area to minimize infiltration. The horizontal barrier must extend beyond the slurry wall plan location for effective control of infiltration of precipitation into the contaminated zone.

Construction A slurry wall construction is essentially a two-step process: (1) A trench is cut that is filled with slurry to keep it open. (2) The trench is then filled with a backfill material forming a low permeability wall.

The trench is usually cut with a backhoe. In some cases clamshells or draglines are also used. The selection of equipment depends on depth of excavation, depth of groundwater table, soil type, and access to trench site. A backhoe can be used to excavate up to 50-ft deep trenches. Table 9.4

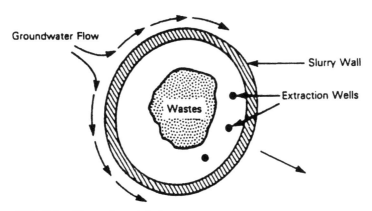

FIG. 9.9 Circumferential slurry wall (plan). From USEPA (1984).

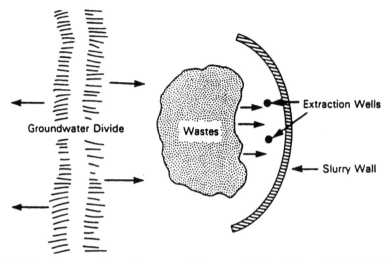

FIG. 9.10 Down-gradient placement of slurry wall (plan). From USEPA (1984).

provides a list of excavation equipment used for slurry trench construction. The soil profile of the site should be studied carefully for choice of proper equipment. Air lifting of loose materials from trench bottom is done where necessary to ensure full contact of the slurry wall with the underlying undisturbed stratum. The slurry is mixed either in a pond or in a mechanical mixing

TABLE 9.4 Excavation Equipment Used for Slurry Trench Construction

Type	Trench Width (ft)	Trench Depth (ft)	Comments
Standard backhoe	1–5	50	Most rapid and least costly excavation method
Modified backhoe	2–5	80	Uses an extended dipper stick, modified engine and counter-weighted frame; is also rapid and relatively low cost
Clamshell	1–5	>150	Attached to a Kelly bar or crane; needs ≥18-ton crane; can be mechanical or hydraulic
Dragline	4–10	>120	Primarily used for wide, deep SB trenches
Rotary drill, percussion drill, or large chisel	—	—	Used to break up boulders and to key into hard rock aquicludes; can slow construction and result in irregular trench walls

After USEPA (1984).

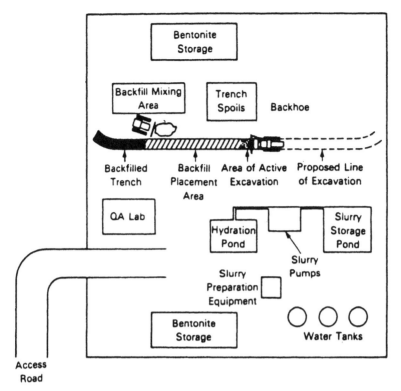

FIG. 9.11 Typical site layout for slurry wall construction. From Spooner et al. (1985).

unit. Sufficient time should be allowed for hydration. Figure 9.11 shows a plan view of a typical slurry wall construction site.

The trench is excavated from one end while the slurry is pumped into the trench. After the trench reaches the required depth, backfill material is pushed into the trench (Fig. 9.12). If the trench is constructed in sections, then each section must be overlapped so that the sections perform as one trench.

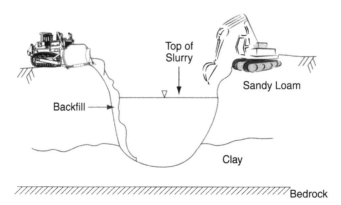

FIG. 9.12 Slurry-trench construction.

The slurry and backfill mix must meet certain specifications for proper construction for achieving design permeability. The slurry consists of dispersed clay minerals in water. Either bentonite or attapulgite is used for slurry. The recommended viscosity for slurry is 38–45 sec-Marsh (about 15 cP). Usually a 5–7% bentonite by weight makes a good slurry (Spooner et al., 1985).

Bentonite is a clay mineral of the smectite group. It has platelike morphology, is highly plastic, and a colloidal expansive clay. Bentonite tends to flocculate when exposed to salt solution. Its permeability may increase in presence of certain electrolytes (Mitchell and Madsen, 1987). So the effect of the contaminated groundwater of the site on the permeability of bentonite should be studied before using it in slurry.

In sites where bentonite is found to be incompatible, attapulgite may be used. Although attapulgite has low swell potential, it is only slightly affected by electrolytes. The morphology of attapulgite is threadlike (Mitchell, 1993). Use of attapulgite may require special construction equipment, which increases the cost of construction (Tobin and Wild, 1986). Tallard (1992) reported the use of synthetic biopolymers as an alternative to bentonite for slurry wall construction (Sharma, 1992).

Water used for slurry should be free from acid, alkali, and organic matters. The water should have a near neutral pH, organic content <50 ppm; hardness <50 ppm, and total dissolved solids <500 ppm (Xanthakos, 1979; Spooner et al., 1985).

Filter Cake   The filter cake is an essentially thin gluelike layer of soil, the pores of which are filled with bentonite (Fig 9.13). The permeability of the filter can be as low as 1×10^{-9} cm/sec (Spooner et al., 1985; Xanthakos, 1979).

Permeability of Slurry Walls   Permeability of both the slurry and the backfill are important for constructing a successful vertical barrier. The permeability of the slurry material should be tested using the contaminated groundwater

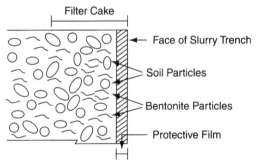

FIG. 9.13   Filter cake.

from the site. The effect of various chemicals on the permeability of soil–bentonite mixture is shown in Table 9.5. If it is found that the permeability of the slurry is affected by the contaminated groundwater, then a test should be done on the proposed backfill material also to find out the effect of the contaminated groundwater on it. If it is found that the contaminated groundwater is expected to increase the permeability above the design value (usually 1×10^{-7} cm/sec), then alternative slurry materials should be tested first. If no suitable slurry material is found, then an alternative type of vertical barrier wall (discussed later) should be used for the site.

It is preferable to use the on site contaminated soil for backfill mix provided the design permeability is achieved (D'Appolonia, 1980). The overall permeability of the slurry wall depends on the permeability of the filter cake and the backfill mix.

The permeability of slurry walls can be calculated using the following equation (D'Appolonia, 1980):

TABLE 9.5   Effects of Various Pollutants on Soil–Bentonite Slurry[a]

Pollutant	Backfill[b]
Ca^{2+} or Mg^{2+} at 1000 ppm	N
Ca^{2+} or Mg^{2+} at 10,000 ppm	M
NH_4NO_3 at 10,000 ppm	M
Acid (pH > 1)	N
Strong acid (pH < 1)	M/H*
Base (pH < 11)	N/M
Strong base (pH > 11)	M/H*
HCl (1%)	N
H_2SO_4 (1%)	N
HCl (5%)	M/H*
NaOH (1%)	M
CaOH (1%)	M
NaOH (5%)	M/H*
Benzene	N
Phenol solution	N
Seawater	N/M
Brine (specific gravity = 1.2)	M
Acid mine drainage ($FeSO_4$ pH ~ 3)	N
Lignin (in Ca^{2+} solution)	N
Organic residues from pesticide manufacture	N
Alcohol	M/H

[a] Abbreviations: N, No significant effect; permeability increase by about a factor of 2 or less at steady state. M, moderate effect; permeability increase by factor of 2–5 at steady state. H, permeability increase by factor of 5 to 10. Asterisk (*), Significant dissolution likely.
[b] Silty or clayey sand, 30–40% fines.

From Spooner et al. (1985).

$$k = \frac{t_b}{(t_b/k_b) + (2t_c/k_c)}$$

(9.18)

where k = permeability of slurry wall
  t_c = thickness of filter cake
  t_b = thickness of soil backfill
  k_c = permeability of filter cake
  k_b = permeability of soil backfill

Figure 9.14 shows the relationship between permeability of an 80-cm-thick slurry wall for various permeabilities of the filter cake and backfill mix. The relationship shows that the upper limit of slurry wall permeability is 1×10^{-6} cm/sec even for a highly permeable backfill (1×10^{-4} cm/sec) and high k_c/t_c ($= 25 \times 10^{-9}$ per sec) value.

Slurry and Backfill Mix Specifications In addition to permeability several other properties of slurry and backfill mix must be specified. In many instances additives are mixed with the slurry materials to aid in performing specific tasks (e.g., sodium tetraphosphate as a thinner and dispersing agent, potassium aluminate to control flocculation) of the material (Xanthakos, 1979).

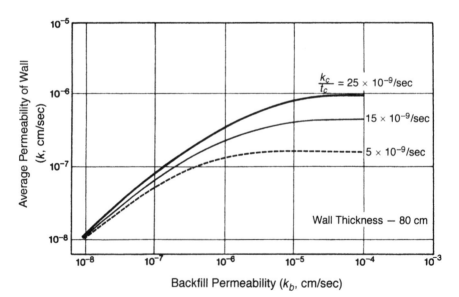

FIG. 9.14 Relationship between slurry wall permeability, backfill permeability, and filter cake permeability. Reproduced by permission of the publisher, ASCE, from D'Appolonia (1980).

Viscosity, thixotropy (or gel strength), and density are important properties of the slurry. Viscosity represents the flow resistance of a fluid. The slurry viscosity should be high enough to stabilize the trench but low enough to allow construction activities. The recommended viscosity of bentonite slurry is 40 sec, as measured in a Marsh cone (D'Appolinia, 1980). Thixotropy is that property of a material that makes it stiff at rest but softens (even liquefies) when remolded or agitated. Thixotropy of a slurry is measured by using a Fann rotational viscometer. For a high-quality bentonite, the 10-min gel strength should be slightly higher than a 10-sec gel strength. A typical value of gel strength of bentonite is around 15 lb/ft^2 (0.72 kPa) (Xanthakos, 1979). The density of a fresh bentonite slurry is about 65 lb/ft^3 (1.04 g/cm^3) or higher (Spooner et al., 1985).

The particle size distribution of the soil, the presence of deleterious materials, and the presence of contaminants and water content affect the properties of the backfill material. In addition, the method of placement also influences the performance of the slurry wall as a barrier. In general, the primary function of a slurry trench as pollution migration control is to provide a low permeability barrier. In order to achieve a low permeability (usually 1 $\times$ 10^{-7} cm/sec), the bentonite requirement increases with decreasing fine content in the backfill soil. Fine soil particle passing a 200 sieve (0.7 mm diameter) exert a significant influence on the permeability of soil. Usually the soil should have 20–40% fines for achieving a low permeability mix. If the strength of the cut-off wall also becomes important, a higher concentration of coarser particles should be used (Spooner et al., 1985). The water content of backfill should be controlled to achieve proper permeability of the mix. If the infield water content of the backfill material is 25%, then the dry bentonite should be spread over thin lifts of soil before mixing. The ideal slump of backfill material is 25–35% (Spooner et al., 1985). The backfill mix must not contain a high volume of soil organic matter, calcium containing material (e.g., gypsum), or a high concentration of soluble salts (e.g., sodium chloride). If the on-site contaminated soil is shown to be affecting permeability, then it should be mixed with uncontaminated soil and tested to see whether the permeability of the mix is expected to achieve the design permeability. The early exposure of the mix to contaminated water is expected to reduce the subsequent change in permeability. The final backfill mix must be a homogenous material with the consistency of cement mortar. It must be stiff enough to stand a 10 : 1 slope, yet flow easily and must be denser than the slurry.

Trench Stability   The gel strength, or the shear strength, of the slurry helps stabilize an open trench appreciably. The critical height (H_{cr}) of slurry trenches, given by the equations below (Xanthakos, 1979), assumes that the trench remains open temporarily; in general, a trench remains open for few days only.

The critical height for a slurry trench in sand is

$$H_{cr} = \Pi \frac{\Gamma_f}{\gamma k_\alpha - \gamma_f - \gamma_f/a} \tag{9.19}$$

where a = trench width
Γ_f = gel strength
γ = density of sand
γ_f = density of slurry
k_α = coefficient of lateral active earth pressure

The critical height (H_{cr}) for a slurry trench in clay is

$$H_{cr} = \frac{4c + \Pi \Gamma_f}{\gamma - \gamma_f - \Gamma_f/a} \tag{9.20}$$

where c is soil cohesion.

The above equations for critical height show that the factor of safety is greater for narrower trenches. A study on the effect of moving load near an open trench showed that although some localized sloughing near the top of the trench may occur, the stability of the trench as a whole is not affected by the moving load (Spooner et al., 1985).

Other Types of Backfill Materials   Although soil bentonite is the most common backfill material, other types of backfields are also used for constructing slurry walls. Each type of backfill material has an advantage over the soil bentonite backfill in specific situations.

SOIL ATTAPULGITE BACKFILL   This backfill mix is used where the permeability of the soil–bentonite mix is adversely affected by on-site-contaminated groundwater. As mentioned earlier, use of attapulgite raises the construction cost due to higher cost of material and construction equipment.

SOIL–CEMENT BACKFILL   This backfill mix is used where the barrier wall is subject to external hydraulic gradient. The soil should have a high percentage of fines.

CEMENT–BENTONITE BACKFILL   This backfill mix consists of cement, bentonite, and soil. The finished slurry wall is quite flexible for accommodating differential settlement and ground movement. The bentonite slurry used in the trench is mixed with cement and soil to form the backfill. Usually 2–4% bentonite and 15–20% cement is used for the mix. The water content is 65–70% (Xanthakos, 1979).

PLASTIC CONCRETE BACKFILL   The advantage of plastic concrete backfill is that it is flexible enough to accommodate settlement and ground movement yet is strong enough to resist overburden pressure. It can develop a low permeability barrier when appropriate mix ratio is used. The mix consists of bentonite (about 2–3%), cement (about 10%), and sand (about 70%). The water content of the mix should be sufficient to produce a slump of 6 in. (15 cm). The material can withstand a 7–8% strain although a permanent deformation is very possible (Xanthakos, 1979). Use of fly ash and bottom ash as aggregate has also been reported in the literature (Evans et al., 1987).

Other Types of Vertical Barriers   Several other types of vertical barriers are used in pollution control projects. Although each of the following types have been used for pollution control projects and have advantages over slurry walls in certain situations, field application of these types of walls for pollution control projects are not as common as the slurry walls. A designer should choose a type of vertical barrier based on project needs and constrains.

COMPOSITE SLURRY WALL   A composite barrier wall essentially consists of a sheet of geomembrane placed within a slurry wall. The backfill can be any of the slurry wall backfills mentioned above, namely soil–bentonite, cement–bentonite, or plastic concrete. The geomembrane is usually a 100-mil high-density polyethylene (HDPE). The HDPE is essentially sandwiched over a geofabric, which provides ease of handling the HDPE during installation. Cavalli (1992) provided detailed steps for constructing a composite barrier wall. The construction essentially involves welding the HDPE sheet to two HDPE pipes as a curtain (Fig. 9.15). Two holes are drilled (slurry is used to stabilize the holes) at a distance to match the width of the HDPE curtain. An HDPE connection along with a steel member is installed in the holes. The ground in between the holes is dug using a clamshell; a bentonite slurry is used to stabilize the trench. The HDPE curtain is installed in the slurry trench. The trench is backfilled with a designed mix using a tremie. Figure 9.16 shows a cross section of the composite barrier wall.

Druback and Artola (1985) proposed the use of sand instead of the geofabric in between the HDPE sheet. However, the system is not widely used and is difficult to install (Ryan, 1987). The specifications for the bentonite slurry and the backfill mix for the composite slurry are the same as the slurry trench materials mentioned earlier when discussing construction.

SOIL MIXED WALLS (SMW)   A multishaft unit is used to drill holes in the ground along the proposed plan configuration. The first shaft is used to redrill the last hole location to provide continuity (Fig. 9.17). A bentonite fluid slurry is injected into the hole simultaneously with the soil drilling. The bentonite fluid mixes with the on-site soil creating a low permeability shaft. The permeability reduction is not as much compared to a soil–bentonite slurry wall.

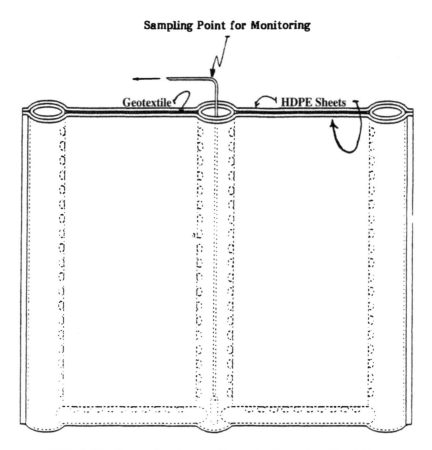

FIG. 9.15 Composite slurry wall curtain. From Cavalli (1992).

A dry jet mixing (DJM) system in which dry bentonite powder is injected is also available. The DJM method is more useful for sensitive clays and peat. Both techniques are more expensive than slurry walls (Ryan, 1987) but are useful where available construction area is rather limited.

PRESSURE GROUTED WALLS   This type of barrier wall is useful for rock formation. A hole is drilled into the ground, which is pressure grouted using a slurry. The grout fills the voids and fissures creating a vertical barrier. The holes should be spaced close enough to achieve a continuous grout curtain. The quality assurance and continuity of the grout curtain is nearly impossible. The cost of construction of pressure-grouted walls is very high (Ryan, 1987). However, in a rocky formation, with all its imperfections, a pressure-grouted wall certainly reduces migration of contaminants further downstream.

JET GROUTING   This type of barrier wall is constructed by drilling a hole in the ground, which is grouted under high pressure. The pressure jets blast and displace soil, which gets mixed with the grout. Although adjacent columns

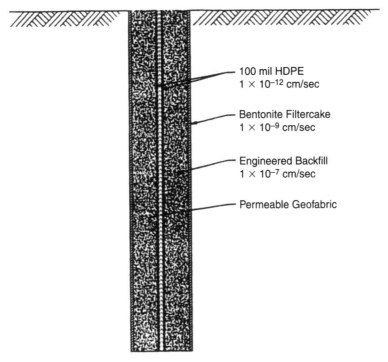

FIG. 9.16   Cross section of a composite slurry wall. From Cavalli (1992).

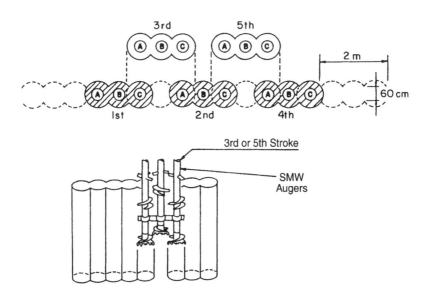

FIG. 9.17   Installation sequence for soil-mixed wall barrier. Reproduced by permission of the publisher, ASCE, from Ryan (1987).

are formed in contact with the previous one, forming a continuous barrier, it is difficult to verify continuity (Andromalos and Pettit, 1986). In double-jet grouting, compressed air (in addition to the high-pressure grout injection) is used at the tip of the drill bit. The introduction of the compressed air increases the penetration depth of the grout, which often doubles the size of the treated soil column (Kauschinger et al., 1992).

VIBRATED BEAM WALL   This type of barrier wall is constructed by grouting, with a cement–bentonite slurry, the hole formed by driving an I-beam in the soil using a pile hammer. The beams are driven with an overlapping pattern to achieve continuity. Although several studies have shown that a low permeability wall can be constructed using this technique (Leonards et al., 1985), it is difficult to construct vibrated beam walls in dense soil with cobbles or larger size particles. Researchers have expressed reservations regarding the use of this type of wall in critical projects and suggest that slurry, with at least 15–20% fines, be designed for each site (Jepsen and Place, 1985). Because of these shortcomings, this type of barrier wall should be used for less dense homogenous soil in which large size particles (cobbles and bigger) are absent. It is preferable to use vertical barriers for shallow depths.

SHEET PILE WALLS   Steel sheet piles can be used for vertical barrier construction. The interlock cavities should be filled with low permeability grout, viz. cement–bentonite. It is suitable for low depth installation specially in the vadose zone (Smyth and Cherry, 1997). Sheet pile is difficult to install through cobbles and boulders. Because of the vibration and noise caused during installation, sheet pile construction should be avoided in populated areas.

9.6.1.3 Soil Vapor Extraction
Soil vapor extraction (SVE), also known as soil venting, generally works well with gasoline and common solvents such as trichloroethene. Extraction of heavier hydrocarbons (e.g., jet fuel, diesel oil) using SVE is possible, but the rate of remediation is slow. Effectiveness of the method depends on the permeability of the soil(s) at the site and volatility of the contaminants. In heterogeneous sites and in fractured clay till the extraction depends on the diffusion rate of the contaminant (Wisconsin Department of Natural Resources, 1993a). The SVE process may consist of only extraction wells or may also include air injection wells in addition to the extraction wells. A schematic diagram of the SVE process is shown in Fig. 9.18. The suction created by the extraction wells may lift the water table. The effectiveness of SVE is reduced if the uplifted water table submerges the contaminated area. Improperly designed well placement or the existence of subsurface blockage may result in stagnation zones. Stagnation zones are areas of no or minimal airflow. Stagnation zones usually occur in between extraction wells whose radius of influence (ROI) do not overlap or due to the presence of a subsurface structure blocking the airflow. To avoid formation of stagnation zones or lifting of the water table, the geologic condition of the

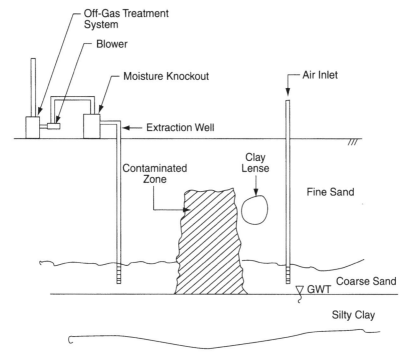

FIG. 9.18   SVE process schematic.

site should be characterized in sufficient detail. The following issues should be taken into consideration before adopting an SVE process for remediation.

Geologic Factors   An experienced subsoil investigator should be on site to ensure proper entry into bore logs. In addition to mentioning soil classification, the logs should include color, geologic origin of soil, moisture content, visual presence of secondary permeability, voids, and layering. It is essential to identify any soil stratification. The presence of a low permeability layer will greatly influence the number of extraction wells, their spacing, and screen location. The seasonal fluctuation of the water table should be measured carefully to ensure proper well screen length and depth. The presence of a surface seal, such as pavement, greatly enhances SVE.

Contaminant Characteristics   Volatility of the contaminants should be assessed to estimate the total mass of contaminants needing extraction. An estimate regarding the total mass of contaminants is necessary to evaluate the overall design of the SVE process, its completion time, and cost. The total extractable mass also dictates the need for air emission control and necessary air permit. During site investigation, the degradation product(s) of the contaminants, if any, should also be assessed for complete remediation.

Pilot Test A pilot test is preferred over a laboratory grain size analysis for estimating airflow rate. Usually one to three extraction wells (Fig. 9.19) are installed for a pilot test. A small blower (100 scfm or less) is sufficient if the permeability of the soil is estimated to be high. For low permeability soil, a high vacuum blower is needed. The vacuum created in the soil due to extraction can be measured in another well or soil gas probes that are installed to the same depth as the extraction well(s). Although frequency of monitoring is not dictated by regulatory agencies during pilot test, at least two gas samples should be collected for complete analysis. Figure 9.20 shows a typical setup for a pilot test.

A detailed pilot test report should be developed that will help in SVE system design. The report should include a site map indicating the location

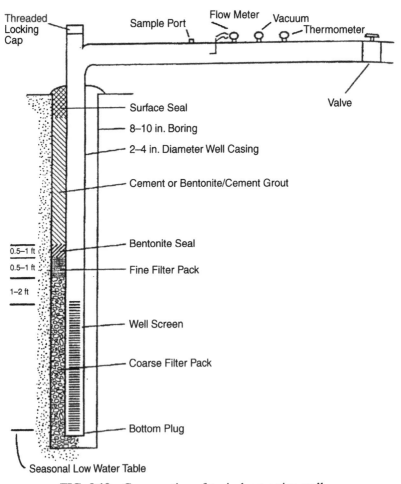

FIG. 9.19 Cross section of typical extraction well.

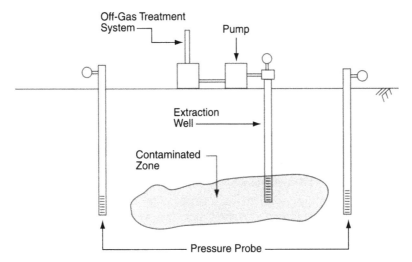

FIG. 9.20   SVE process pilot test setup schematic.

of extraction well(s) and vacuum measuring points, suspected or known source locations, paved areas and building (if any), buried utility trenches (if any), and water table map. A cross section showing the soil stratification, contour lines of vacuum readings, details of extraction well(s), and vacuum measuring points should also be developed. The flow rates, vacuum readings, soil gas temperature, ambient temperature, and pressure should be included in the report. If possible a conceptual design should also be included in the report.

A pilot test is not essential if the unsaturated zone is homogenous, the volume of contaminated soil is small, and total mass of contaminant released is also relatively low. The regulatory agency should be consulted for concurrence regarding a pilot test. One of the primary reasons for undertaking a pilot test is to estimate the ROI for the site. The ROI is the area from which an extraction well effectively draws air. Plots of vacuum (in inches of water column) versus distance from air extraction well for varying ratios of horizontal (K_h) to vertical (K_v) permeability (from 0.67 to 10) using a model developed by Shan et al. (1992) shows that vacuum at any distance from the extraction well decreases with decrease in the K_h/K_v ratio (Wisconsin Department of Natural Resources, 1993a). Closer well spacing in sandy soil speeds up cleanup but increases cost of the project. However, for low permeability soil closer well spacing is needed for effective cleanup. For sites with shallow water tables, closer well spacing is needed to offset the air entry from ground surface. In sites with distinct soil layers well spacing in relatively low permeability layers should be closer. The airflow rate is less important if diffusion is the controlling factor. The design radius should be at a distance where the vacuum is 1.0–0.1% of the vacuum at the extraction well (Buscheck and Peargin, 1991). Generally, the well spacing ranges from 20 to 65 ft (6 to 20 m).

Researchers have expressed reservation regarding the use of ROI (Johnson and Ettinger, 1997). In their opinion, a soil venting system should be designed based on zone of remediation (ZOR), which is the zone remediated to a specified level; it is dependent on the length of time the system is operated. In lieu of modeling, ROI obtained by a pilot test should be reduced by 50–70% to overcome the inherent design-related flaw discussed by Johnson and Ettinger (1997).

Typically a pilot test involves the following steps (Fam, 1996):

1. Measurement of extracted airflow rate and well head vacuum
2. Measurement of vacuum in soil probes or other wells selected for the test
3. Collection of lab samples before beginning of the test, at the middle of the test, and after completion of the test. Proper instruments should be used for the collection of samples in the field.
4. Measurement of oxygen and carbon dioxide content of the withdrawn air. This may become useful if bioventing (see Section 9.6.3.1) is contemplated.

Modeling of SVE Systems Analytical models for homogenous soil (Shan et al., 1992) and numerical models for use in nonhomogeneous soil (Jordon et al., 1995) are available for calculating the rate of extraction. Computer-based models are also available for designing SVE systems. The Dynkin VES design model (cited in Fam, 1996) is used to calculate vacuum distribution at each extraction well and pressure and flow in pipes, extraction wells, and air treatment outlet(s). Flow-type models such as Airflow SVE, HYPER VENTILATE, and AIR 3D are available to locate extraction wells with a predetermined vacuum. These models provide two-dimensional or three-dimensional airflow distribution in an SVE system. The comprehensive multiphase model MORTRAN should be used for bigger and more complicated sites. A multiphase model, although expensive, is capable of optimizing the overall design of an SVE system including prediction of closer time (Fam, 1996).

Design Issues Models should be used for larger projects with complicated geologic conditions. For smaller projects, empirical design based on past experiences and data obtained from the literature may be used. Pilot tests are used to supplement an SVE system design developed empirically or by using a model. When used properly, modeling provides a higher confidence level. However, correct interpretation of geologic factors and selection of proper input are essential for meaningful model output. As a rule a model is costlier compared to an empirical design. So while choosing modeling, the designer needs to assess whether the benefit of higher confidence level at a higher cost is justifiable for a project.

Equipment   Usually one of the following three types of blower is used: centrifugal, regenerative, or rotary lobe. Centrifugal blowers perform best in high-flow but low-vacuum applications, Regenerative blowers develop vacuum higher than centrifugal blowers. Regenerative blowers are typically used for small projects with relatively uniform geology (e.g., sandy soil). Rotary lobe blowers are capable of producing very high vacuum [up to 15 in. (38 cm) of mercury]. Although this type of blower needs frequent maintenance, its use is required for low permeability soils (e.g., silty clay). Liquid ring blowers may also be used for such high-resistance applications. The motor and controls for blowers should be explosion-proof, especially if there is any possibility of igniting the contaminants. Except for centrifugal blowers, all other blower types should be equipped with an air filter. Chlorinated polyvinyl chloride (CPVC) pipe rather than PVC pipe should be used as stack if the discharge temperature is expected to reach 140°F (60°C). A drain at the base of the stack will help drain moisture. Blowers should include filters and moisture knockouts. In sites where a significant amount of water is expected to accumulate in the knockout, a transfer pump is used to discharge the water to a groundwater treatment system.

If the emission of any contaminant is expected to exceed concentration permitted by regulation, then removal or destruction of the contaminant must be undertaken. Primarily three types of air treatment devices are used for such situations: incineration, catalytic destruction, and granular activated carbon. Incineration is used when contaminant emission is high. Catalytic destruction is used when the emission is medium. Granulated activated carbon is used for low concentration of emission. Biofilters and other treatment devices may also be used in some sites.

Process Enhancement   The performance of an SVE process can be improved by injecting air or by heating the soil. These enhancement techniques are discussed in the following sections.

AIR INJECTION   Either air is forced through a vent or the vent is used passively for suction of air from the atmosphere. Air injection helps create better airflow in capillary fringe as well as around buried structures. The air injection rate should not be more than 25% of extraction rate. The injected air should not be colder than the underground air because the colder air will reduce the volatilization of contaminants. The injected air may be heated for better and quicker removal.

HEATING   The performance of SVE can be improved by heating the soil. Heating influences both the contaminant and soil characteristics. The following effects of heating has been reported: increase of vapor pressure and Henry's law constant, removal of soil water due to the decrease in tortuosity of flow path leading to a higher degree of contaminant removal, increase in volatilization rate of contaminants and change of liquid water to vapor phase ratio leading to improved advective flow (Davis, 1997). The soil needs to be heated to at least 100°C (212°F) to get the complete benefit of heating. The energy requirement for heating soil depends on porosity, density of solids, and degree of saturation. The following five techniques are available for heating soil: conductive heating, electrical heating, radio frequency heating, steam heating, and hot air injection. The mechanism of heat transfer to the soil is different in each method. The success of a technique for a site depends on choosing the correct heat transfer mechanism.

Conductive Heating In conductive heating electrical heaters are used to heat the soil. Both electrical heating rods and blankets are used for this purpose. Thermocouples are installed in an array to measure the temperature, which can rise as high as 800°C (1472°F). In the In Situ Thermal Desorption (ISTD) process organic contaminants break down due to high temperature, which accelerates contaminant removal. The results of about 10 projects have shown that the ISTD process can reduce PCB concentration from a few percent to less than detectable concentration (Murdock, 2000).

Electrical Heating Electrical current is passed through electrodes inserted in the ground. The soil behaves as a resistor. The power (P) generated due to a passage of current (I) in a resistor (of resistance R) is given by

$$P = I^2R \tag{9.21}$$

The heating is proportional to the power dissipated. In the field, the voltage is adjusted to produce the desired current required for raising the temperature of the soil. A six-phase electricity is used for connecting to an array of six electrodes with a central neutral electrode for this type of heating (Fig. 9.21). Six-phase heating is an effective method for heating a homogeneous soil volume uniformly. Since the electrical conductivity depends mainly on soil moisture and since low permeability soils such as silt and clay have higher moisture content, they are heated more quickly. If the subsoil has layers of silt and sand (say), then sand will be less heated than the silt. This preferential heating helps in removing contaminants from lower permeability soil whereby overall removal from each layer is enhanced greatly (Murdock, 2000).

Radio Frequency Heating Radio frequency (RF) heats soil by interacting with water molecules in soil. Water being a bipolar molecule (Mitchell, 1993), it rotates due to the reversing of electromagnetic field. In RF heating the water molecules are rotated at extremely high frequencies, which raises the temperature of the water. The rise in temperature of the material being heated depends primarily on the dipole moment of the molecule of the material. The dielectric constant of water is much higher than the solid phase of the soil.

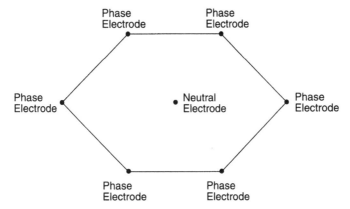

FIG. 9.21 Typical electrode array plan for soil heating.

So the water content of soil is an important factor for heating the soil. Soil can be heated to 100°C (212°F) using RF heating. It is effective in removing high-boiling-point organic contaminants. As heating proceeds, soil moisture is reduced, which changes the dielectric properties of the contaminated soil. So the RF must be adjusted with time. For most remediation sites the RF ranges from 6.8 to 2.5 MHz (Johns and Nyer, et al., 1996). Although the heating process can help in removing high-boiling-point organic compounds, the cost of operation is high and the process may sterilize the soil whereby biodegradation is inhibited. Figure 9.22 shows a typical arrangement for RF heating.

Steam Injection The process involves injection of steam into wells with screens slightly below the level of the contaminated soil zone. This process is very effective in removing NAPL. The heat supplied by the steam volatilizes contaminants whereby its extraction becomes easier. Figure 9.23 shows an SVE process coupled with steam injection. The advantages of steam heating is that it reduces the life cycle of an SVE process and can remove semivolatile organic compounds (SVOC). However, the mobilized contaminants can move downward with the condensed steam into the groundwater. The process can also inhibit bioremediation (Johns and Nyer et al., 1996).

Hot Air Injection This process is similar to steam injection except hot air is injected in place of steam. Hot compressed air is injected into a series of wells. The hot air volatilizes SVOCs and VOCs, which are extracted by an SVE process. The absence of condensed water produced from steam injection eliminates the need for water treatment of the extracted liquid. In addition, use of air eliminates the risk of contaminated condensate entering the aquifer. Hot air injection may promote biodegradation. However, hot air injection is ineffective in saturated soils. Since the heat capacity of hot air is lower than

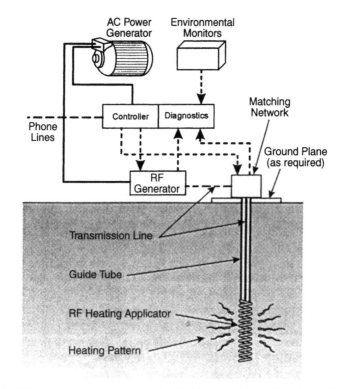

FIG. 9.22 Typical arrangement for RF heating. From Murdoch (2000).

water, the temperature of the injected air needs to be sufficiently high for heating the soil (Johns and Nyer et al., 1996).

9.6.1.4 Passive Soil Vapor Extraction   Natural variation in atmospheric pressure influences air pressure in the vadose zone. This variation can cause gases to flow in or out of wells screened in the vadose zone. This natural phenomenon can be utilized in remediating vadose zone soils. Because of site access difficulties, installation of SVE systems may pose problems. In such sites passive soil vapor extraction (PSVE) may be used (Ellerd et al., 1999; Rossabi et al., 1994).

9.6.1.5 Pump and Treat   This method of remediation has been extensively used over the years. In this process groundwater wells are installed within and around the contaminant plume (Fig. 9.24). The system should be upgraded if needed by installing newer wells or changing pumping rate for effective remediation. The extracted water should be monitored for contaminant concentration to ascertain effectiveness of the system. The extracted water must be treated before discharging into a surface water body or reinjecting it into the aquifer (if permitted).

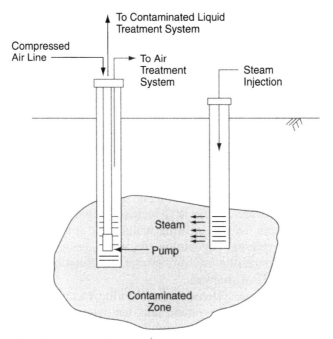

FIG. 9.23 Schematic of SVE system setup coupled with steam injection.

Movement of many contaminants in an aquifer (e.g., NAPL) is different from the movement of water. These contaminants move vertically downward as pure liquid whereas water moves primarily in the horizontal direction. Many of these contaminants adsorb in soil whereby water extraction cannot remediate the contaminated zone. That is why a pump-and-treat method is not successful in many cases (National Research Council, 1994). However, pump and treat being an advection controlled process, it can stop the movement of plumes, which is driven primarily by advection. If designed and operated properly, pump and treat can remediate a site successfully. The main limitation of pump and treat is due to the fact that water is used as the carrier medium. To overcome the limitation of pump and treat, a vacuum is added

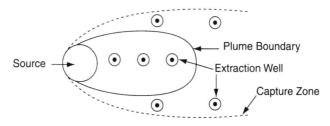

FIG. 9.24 Typical extraction well installation plan for pump-and-treat method.

to the extraction wells to enhance recovery using air as the carrier. The advantages of use of air as the carrier are exchange rate of air is much higher than the exchange rate of water, air can come in direct contact with contaminants that are adsorbed in soil, and air increases volatilization of contaminants whereby their removal becomes easier (Nyer, 1996a).

Vacuum-enhanced pump-and-treat systems are very effective in removing LNAPLs because air becomes the primary carrier in the upper portion of the water table due to a high draw down. However, it is not any better than pump-and-treat methods for dissolved contaminants (Nyer, 1996a). In a vacuum-enhanced pump-and-treat system, vacuum is applied to an extraction well. The well screen is longer and is established in such a way that while in operation, part of it is above the water table. The use of vacuum adds to the effective draw down whereby the ROI increases significantly. This process can be used to increase the rate of water withdrawal or to extract both liquid and vapor from the plume. By putting the well screen within a confined formation, the process can be used to dewater and for subsequent vapor extraction within the ROI of the well.

Two types of extraction systems are used in the vacuum-enhanced pump-and-treat process: single-pump systems and dual-pump systems. In a single-pump system, a withdrawal pipe is installed within the extraction well for withdrawing both air and water. In a dual-pump system, a submersible pump in installed within the extraction well and vacuum is applied to the well using a separate pipe (Fig. 9.25). Single-pump systems are used for shallow aquifers and dual-pump systems are used for deeper aquifers. The dual-pump systems are easier to balance in sites involving five or more extraction wells (Palmer, 1996a).

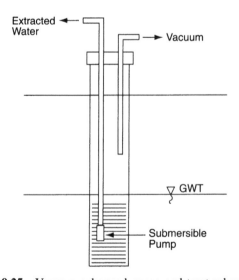

FIG. 9.25 Vacuum-enhanced pump-and-treat schematic.

A pilot test, similar to the SVE process, is conducted for designing this system. The test involves one extraction well surrounded by several monitoring wells. The following items are monitored: fluid level and vacuum in monitoring wells, vacuum applied at the extraction well, and rate of liquid and vapor extracted by the extraction well. Calculations based on draw down due to the combined effect of pumping rate and applied vacuum, together with the pilot test results, are used for estimating the extraction rate for water vapor and well spacing. Selection of vacuum pressure to be applied depends on the desired air and water extraction rates and geologic factors. Increase in vacuum pressure will increase mass removal but will also increase the cost.

The type of blower depends on the applied vacuum. A liquid ring blower is used for high vacuum [15 in. (38 cm) of mercury], a lobe blower is used for midrange vacuum [8 in. (20 cm) mercury] and a regenerative blower is used for low vacuum [less than 8 in. (20 cm) mercury] (Palmer, 1996a).

Estimation of a capture zone is important for designing a pump-and-treat system. A simple two-dimensional model (Todd, 1980; Javanel and Tsang, 1986; Grubb, 1993) may be used for small sites with relatively homogenous sandy aquifer; usually a narrow plume develops in such sites. For relatively complex geologic conditions, a model proposed by Strack (1989) may be used. At low permeability sites, contaminants migrate radially away from the source primarily due to diffusion. In these sites a system, which will induce an inward gradient of 0.01 or more at the plume boundary, should be used.

9.6.1.6 Air Sparging   This process involves the introduction of pressurized air below a contaminated zone within an aquifer. An SVE system needs to be installed along with air sparging for the recovery of dislodged contaminants (Fig. 9.26). So air sparging may be viewed as an enhancer of the SVE

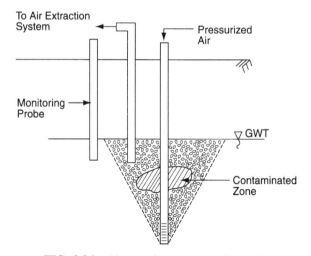

FIG. 9.26 Air sparging process schematic.

process. In addition, air sparging can strip VOCs from soil and reduce iron and manganese concentrations in the saturated zone (Brown, 1997). Air sparging is more cost effective than pump-and-treat methods. The process is capable of removing contaminants from smear as well as the saturated zone, which is not accomplished by SVE alone (Chen et al., 1996). The pressurized air exits through a well screen and moves laterally and vertically through soil forming distorted U- or V-shaped plumes. As the sparge pressure is increased, the degree of horizontal travel increases (Suthersan, 1996). Air flows through continuous air channels if the particle size of the soil is smaller than 2 mm. However, if the particle size of the soil is greater than 2 mm, then the injected air travels upward through the saturated zone in the form of air bubbles (Ji et al., 1993). The air distribution is more uniform in low permeability stratum than high permeability stratum (Rutherford and Johnson, 1996). Both horizontal and vertical permeability's influences the upward plume geometry or the airflow path. In general, a stratum with high silt or clay contents where the permeability is less than 10^{-5} cm/sec is not responsive to air sparging process. There should be no low permeability stratum above the sparge location. Such a barrier will trap the air and will spread the contamination away from the source. To avoid failure, the geology and hydrogeology of the site must be established accurately. In doubtful situations a gas tracer (e.g., helium) may be used to estimate the airflow path geometry (Brown, 1997).

Both airflow rate and air pressure are important for successful air sparging. At low to moderate flow rate partitioning of contaminants into groundwater occurs. At moderate to high flow rates, dissolved contaminants are stripped. At very high flow rates a turbulence is created whereby contaminant starts spreading further away from the original contaminated zone. The ability of the SVE system also dictates the sparging flow rate. In general, the flow rate should be 1–10 cfm (1700–17000 L/min) per sparging point (Suthersan, 1996). However, flow rates of 0.25–20 cfm (425–34000 L/min) have also been recommended (Wisconsin Department of Natural Resources, 1993b).

The minimum injection pressure required for air to flow out must be greater than hydrostatic head at the point of injection. Although capillary entry resistance can impede airflow, the value of the resistive pressure is not significant for the target soil types. The maximum injection pressure must be such that it does not create a turbulent flow, which will spread the contaminants. Two approaches are available for calculating the maximum allowable pressure. One approach suggests that maximum pressure should be three times the hydrostatic head at the point of injection (Brown, 1997). This approach suggests that the maximum allowable pressure should be 80% of the total pressure at the point of injection (Marley and Bruell, 1995). Examples showing both the calculations are included at the end of this section.

Air sparging causes the water table to mound. The mound disappears with time. Field study indicates only 5% of initial upwelling remained after several days of operation (Martinson and Linck, 1993). Upwelling is not expected to cause spreading of contaminants if the SVE system is operated properly.

The temperature of the injected air must be at least the same as that of the aquifer water. Although compression of air will raise its temperature, the heat may be lost in the delivery pipe system, especially in cold weather locations. Similarly, in hot weather the injected air temperature may rise during transport from the compressor. The effect of ambient temperature on the injected air must be taken into consideration when designing the system. The air injection influences decontamination. There are four basic modes of operating an air sparging system: (1) the air may be injected in continuous mode. It provides rapid reduction of contaminants. This pattern of injection is best suited for homogenous, high permeability formation. (2) The second mode is pulsing, that is, the rate of injection is varied with time. The advantages of pulse operation are reduction of long-term channeling effect due to change in pathways, minimization of costs, and better noise control. The disadvantage is that it does not provide a long-term solution and may require pump-and-treat method to finish cleanup. The pulse pattern is best suited for heterogeneous low permeability formation. (3) The third mode is use of multiple wells surrounding and within the plume. The operation begins with the outermost wells and subsequently inner wells are sparged. This provides a barrier against migration whereby further groundwater control is not necessary. (4) The fourth mode involves use of low flow. A low flow maximizes biodegradation but reduces volatilization. It may be used toward the end of the decontamination process to maximize cleaning. It is best suited for contaminants with low volatility and solubility (Brown, 1997).

Although air sparging is quite effective for *in situ* treatment, it is not suitable for all sites and all types of contaminants. A decision tree (Brown, 1997) may be used to assess the applicability of air sparging for a particular site (Fig. 9.27). The system must be monitored to judge the effectiveness. In addition to monitoring the SVE system, two or more down-gradient groundwater monitoring wells should be monitored for the contaminants and dissolved oxygen.

Example 9.1 Calculate the maximum pressure to be used for a site where the depth of groundwater is 10 ft below ground level. The sparge well screen is located at 30 ft below ground. The unit weight of dry soil is 110 lb/ft^3. The soil has 40% porosity.

Method 1 The hydrostatic head at the level of injection is 20 × 62.4/144 = 8.6 psi. The allowable maximum pressure of sparging is 3 × 8.6 = 26 psi.

Method 2 The total pressure at the level of injection is

$$\frac{30 \times 110 + (30 - 10) \times 0.4 \times 62.4}{144} = 26.4 \text{ psi}$$

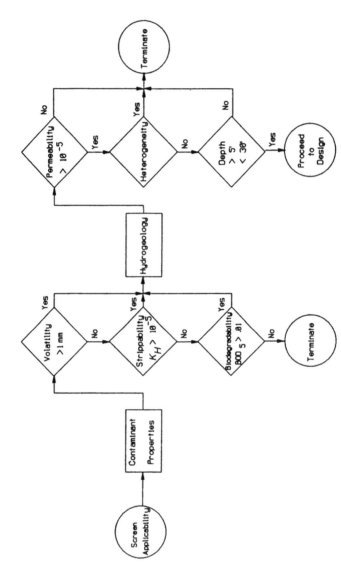

FIG. 9.27 Decision tree to assess the applicability of air sparging. From Brown (1997).

Maximum allowable pressure of sparging is

$$0.8 \times 26.4 = 21 \text{ psi}$$

Note: The sparging may be started with lower of these two values and increased to the higher value, if necessary.

9.6.1.7 *Soil Fracturing*   This process uses pressurized fluid to increase the permeability of soil through fracturing. It is not a remediation process by itself. Soil fracturing has to be combined with some other remediation process for decontaminating a site. Soil fracturing is undertaken at sites with low permeability. This process can increase permeability of the following formations: low permeability formations of sand, silt and clay, sandstone, silt stone, limestone, and shale (Kidd, 1996). Fracturing can be done using either pressurized gel-type liquid (hydraulic fracturing) or air (pneumatic fracturing).

In hydraulic fracturing a lance is installed in a borehole that creates a pointed tip at the end of the borehole. Then a notch is created by using a high-pressure water jet (Fig. 9.28). Subsequently a fracturing fluid (usually a food-quality crosslinked guar mixed with silica sand) is injected at a pressure of about 100 psi (690 kPa). Fractures are created at various levels at 2- to 5-ft (0.6- to 1.5-m) intervals. The guar breaks down (enzyme is added to reduce breakdown time) while the sand remains in the fractures, which helps to keep the fracture open. The silica sand is called proppants. Other types of proppants tested for fracturing are sodium percarbonate, iron fillings, and a graphite-based substance (Kidd, 1996). After advancing the borehole, with fractures at above stated intervals, to the desired depth, an extraction well is established for remediation purposes.

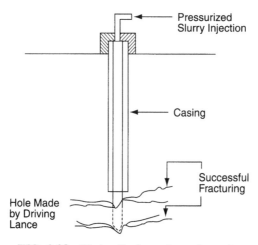

FIG. 9.28 Hydraulic fracturing schematic.

For pneumatic fracturing, the auger is withdrawn after completion of the borehole up to the desired depth. Then a dual packer is inserted and inflated at the first level of desired fracturing. This seals off the air delivery pipe at the desired level (Fig. 9.29). Through the delivery pipe pressurized air is injected at 100–150 psi (690–1035 kPa) with a flow rate of up to 500 cfm (850,000 L/min). It only takes up to a maximum of 1 min to complete fracturing. After completing the fracturing at the desired level, the dual packer is lowered to the next level and the process is repeated. After completion of fracturing of the entire depth of the borehole, a well is installed for remediation purposes.

9.6.1.8 Vitrification   This process is used to immobilize inorganic contaminants through encapsulation. Electrodes are placed next to the contaminated area. Power is applied on the electrodes for heating. At the beginning of the process flaked graphite and glass frits are spread on the surface for increasing soil conductivity. Once melted, soil becomes a conductor of heat. As the soil starts melting, the electrodes are lowered further into the ground at the rate of 1–2 in. (2.5–5 cm) per hour. The process volatilizes some of the contaminants, which are captured by vapor extraction and/or by placing a hood over the treatment area (Johns and Nyer, 1996; USEPA, 1994a). Since the process requires high power (800–1000 kWh per ton of soil), a high-voltage (14 kV) transmission line is needed at the site. Most soil can be treated with this process. However, the process cannot be used if there is a large void or rubble, or if a significant amount of combustible soil is present within the contaminated zone targeted for remediation. Vitrification of saturated soil cannot start until the water is vaporized.

A plasma arc technology can also be used for vitrifying soil. The plasma torch can be lowered into a borehole for converting the adjacent soil to a vitrified material (Riser-Roberts, 1998).

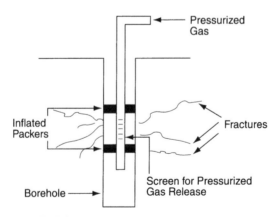

FIG. 9.29   Pneumatic fracturing schematic.

9.6.1.9 Electrokinetic Remediation (EKR) Passage of electric current through soil induces many phenomena such as ion diffusion and exchange, development of osmotic and pH gradients, electroosmosis, and oxidation and reduction reactions. The applied electric field causes the cations and the anions to move to cathode and anodes, respectively. The moving ions carry their water of hydration with them. Since the number of cations are more than anions in clay, there is a net flow of water toward the cathode. This phenomenon is known as electroosmosis (Mitchell, 1993). Electroosmotic water transport is dependent on the coefficient of electroosmotic hydraulic conductivity, K_e. Based on the Helmholtz–Smoluchowski theory and laboratory data, it may be concluded that K_e is relatively independent of pore size. The average value of K_e is about 5×10^{-5} cm^2/sec V for a hydraulic conductivity range of 2×10^{-5} to 1×10^{-9} cm/sec [based on data reported by Mitchell (1993)]. The electrokinetic process is useful for remediation in low permeability soils. Mathematical equations for flow under simultaneously applied hydraulic, electrical, and chemical gradients has been developed by Yeung and Mitchell (1992).

Of the many phenomena induced because of the application of electrical fields two are important for remediation purposes: (1) migration of contaminant anions and cations and (2) increase of water flow rate due to electroosmosis. The migrated contaminants transported to electrodes are removed by excavation or by circulating fluid in the electrode (USEPA, 1994b). A typical scheme for an electrokinetic remediation system setup is shown in Fig. 9.30.

Electrodes are installed on either side of the contaminated zone, at a depth where contamination has been identified. A direct current of 12 A/m of electrode is recommended to avoid heating problems (Lindgren and Brady, 1995). The electrodes may be placed horizontally or vertically. The electrodes are porous ceramic rods that contain an electrolyte solution. The electrodes are connected to a plastic well casing to the top, which is connected to a withdrawal system. If a circulating fluid is used for contaminant removal, the chemical reaction of the contaminants with the circulating fluid must be taken into consideration for proper functioning of the system (Johns and Nyer, 1996).

In a modified process, a highly permeable zone is created by constructing horizontal layers or vertical columns in between the electrodes. The permeable zone will act as a remediation reaction area using biological or chemical processes (e.g., placement of granular activated carbon for adsorption) (Trombley, 1994). Attempts are being made to enhance the electrokinetic remediation process through the use of an acoustic source (USEPA, 1994c).

Electrokinetics improves microbial growth and transport, which can be used to enhance bioremediation. Ionic forms of nutrients can be introduced at electrodes (Riser- Roberts, 1998). The concentration of nutrients should be roughly equivalent to the ionic strength of the pore water. If the electrode

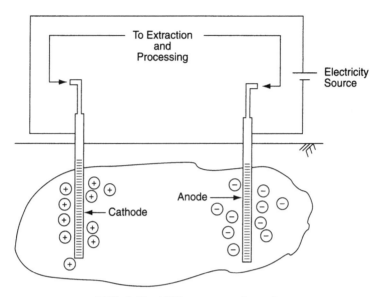

FIG. 9.30   EKR process schematic.

reactions are not buffered for maintaining a neutral pH, then the hydrogen or hydroxyl ions formed will tend to carry the electrical current instead of the nutrients or contaminants (Lindgren and Brady, 1995). Electrokinetics has been used to transport both bacteria and nutrients through a clayey soil for enhancing bioremediation (Nowatzki et al., 1994). Use of electrokinetics for enhancing bioremediation of the soil and groundwater contaminated by gasoline, diesel, and kerosene has been reported (Loo, 1994). Nutrient enhancement achievable through electrokinetics may not be sufficient to support bioremediation totally, but it can provide necessary stimulus in unsaturated soil (Lindgren and Brady, 1995).

9.6.1.10 In-Well Air Stripping   In this process a large-diameter vertical shaft is installed within the contaminated zone of an aquifer. Water is pumped into the well, which is drawn up to a second chamber for removing volatile components from the water. The stripped-off water is recharged into the aquifer. The process will be successful only if the contaminant is volatile, and the aquifer is homogenous and has relatively high permeability. Although it is estimated that the maximum area of influence is 60–80 ft (18–24 m) surrounding the well, additional study needs to be done to establish the influence (Johns and Nyer, 1996).

9.6.2   Chemical Processes

In general, *in situ* chemical remediation processes are implemented along with the physical process described in Section 9.6.1. In the aquifer, transport of dissolved contaminants is controlled by advection and diffusion. The primary aim of chemical processes is to attempt desorption of contaminants from the solid phase and/or influence the chemical reaction whereby the contaminants are either destructed *in situ* or become available for transport by the extraction system. For a chemical process to be successful, the following issues must be addressed: (a) delivery of reactive agent to the contaminated zone; (b) choice of proper reactive agent for interacting with the chemical; (c) estimation of necessary volume for chemical interaction; and (d) proper method for removal of contaminants if *in situ* destruction (through chemical reaction or biodegradation) is not targeted. In general, chemicals are used with either pump-and-treat method or the SVE process. For treatment of contaminants present only in the vadose zone, chemicals are mixed directly with soil for destruction or immobilization of contaminants.

9.6.2.1 Soil Mixing   This process involves mixing a chemical with soil contaminated at or below ground surface. The process is also called stabilization or solidification. If the contaminated zone is shallow, a rototiller may even be used for mixing purposes. For deeper areas in general, mixing is done with large 3- to 12-ft (90- to 365-cm) diameter augers. The stabilizing chemical is added to the soil while the rotating anger moves down. The process can be used for treating soil up to a depth of 100 ft (30 m) (Johns and Nyer, 1996). Mixing with an auger may be repeated to ensure thorough mixing of the stabilizing agent with the soil. The area is left undisturbed for curing. Augers can be used simply to loosen the soil without mixing in any chemical. Turning or churning of the soil can be helpful for enhancing an SVE system. If a thin clay layer overlying a sandy zone exists at a shallow depth, then loosening the soil with an auger will remove the blockage from the clay layer whereby SVE system will function at a higher efficiency.

Proper choice of chemical additive is essential for a successful soil mixing process. The process can be applied to inorganic as well as organic contaminants. For a deeper contaminated zone, the process may come out to be cheaper compared to an *ex situ* process. However, the area may need to be covered with a horizontal barrier because the soil mixing process merely immobilize the contaminants. It does not remove the contaminants. The area may heave because of the addition of the chemical. So, it is recommended to allow the necessary amount of time for the reaction to complete before construction of a horizontal barrier.

9.6.2.2 Soil Flushing   This is essentially a pump-and-treat process in which a chemical or water is injected to increase the mobility of the contaminants. The additive dissolves the contaminant(s) adsorbed in the solid and/or vapor phase; the elutriate is then removed by the pump-and-treat process. The chemical solution is added to either mobilize any free products or to increase the solubility of the contaminants. The chemical can be delivered to the contaminated zone either by gravity (e.g., creating a seepage pond) or through injection. The unused chemical is removed by extraction or allowing it to drain to a porous trench (where geologic settings permits). The elutriate can be recycled back. In some instances heat is used to enhance the extraction process (Riser-Roberts, 1998). To deliver the additive through gravity, the soil above the contaminated area should have sufficient permeability to allow seepage onto the contaminated zone. If the additive is injected, then the injection wells must be upgradient of the contaminated zone and must be screened at appropriate depth so that the additive flows through the contaminated zone.

Additives that have a potential for use include water, acidic aqueous solutions (e.g., sulfuric acid), basic solutions (e.g., sodium hydroxide), surfactants (e.g. alkybenzene), complexing, chelating, and reducing agents (Nash, 1987; Sims and Bass, 1984; USEPA, 1985c). The selection of additive must include safety to humans and environment, purification possibility for removing from the extracted water, and ease of use.

Large interfacial tension exists between DNAPL and water whereby pump-and-treat methods cannot remove it efficiently. The addition of surfactants can lower the interfacial tension by four orders of magnitude, which greatly enhances extraction (Fountain, 1997). Sodium dodecyl sulfate, an anionic surfactant, is frequently used for soil flushing (Liu and Roy, 1995). The biggest problem when adding surfactants for DNAPL removal is that the removal of interfacial tension may make the contaminants move deeper into the aquifer. The downward movement depends on site geology and extraction rate. These issues must be addressed when designing a system for enhanced extraction of DNAPL using surfactants. The efficiency and effectiveness of using surfactants for soil flushing depends on their mobility in soil.

Water can be added to a contaminated zone to enhance the removal of water-soluble contaminants (e.g., phenols). Water is also effective in increasing the extraction of soluble salts such as carbonates of nickel, zinc, and copper (Riser-Roberts, 1998).

The following five types of reactive agents can be used to remove metallic contaminants: acids and bases, oxidants and reductants, competitive additives, complexing agents, and solid-phase substituents. An increase in pH can enhance removal of anionic contaminants, and a decrease in pH can enhance the removal of cationic contaminants (e.g., Zn, Cd). However, the ensuing reaction can result in precipitation of a solid phase, which may retard the remediation rate. Since soil systems have good buffers, it is difficult to adjust the pH of the subsoil. The redox state of most priority pollutants (except Zn, Cd, and Ni) can be altered by using oxidants and reductants. Since redox reactions also alter the pH of a system, use of reductants and oxidants may also require the addition of acids and bases (Palmer and Fish, 1997). Removal of surfactant from the extracted water is very important because reuse of surfactant is essential for cost effectiveness. The following alternatives for removing surfactant and contaminants from the extracted water have been tried: flocculation/coagulation/sedimentation from fractionation, sorbent adsorption, ultrafiltration, surfactant hydrolysis/phase separation, centrifugation, and solvent extraction. The experiments showed that dilute alkaline hydrolysis of a surfactant in the extracted water, followed by neutralization, was effective in separating surfactants (Ellis et al., 1985).

The effectiveness of soil flushing depends primarily on the direct injection contact of the chemical with the contaminants and the hydraulic conductivity of the soil (Sims and Bass, 1984). High silt and clay content reduces movement of the chemicals in subsurface. Soils with 1% or more of organic substances and high clay content would show high sorption characteristics and hence are not suitable for soil flushing. Removal of hydrocarbons may need several rounds of flushing with surfactant (Edwards et al., 1994).

Organic and inorganic contaminants soluble in a low-cost solvent can be removed efficiently using soil flushing. The method can be combined with bioventing (see Section 9.6.3.1) for remediating contaminated groundwater. However, soil flushing is not applicable when the contaminants contain a large number of chemicals. Soil flushing generates a large volume of extracted water, which contains a dilute solution of the contaminants and chemical additive(s). Treatment and/or disposal of this contaminated water is costly (USEPA, 1985c). Soil flushing is not effective with insoluble contaminants.

9.6.2.3 *Treatment Walls*   A permeable wall intercepting the contaminated plume is constructed in the subsurface (Fig. 9.31). Contaminants are removed or degraded because of the reaction with chemical(s) present in the wall. The depth of the treatment wall must be large enough to intercept the entire plume. A treatment wall can be used to treat multiple contaminants. As shown in Fig. 9.32 different chemicals can be placed in series for the removal of various contaminants. Treatment walls can also be used simply to replace the native low permeability soil with a more pervious soil in which an air sparging point can be installed to remediate contaminants (e.g., VOCs) (Palmer, 1996b). The

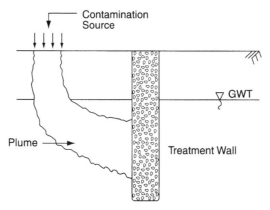

FIG. 9.31 Treatment wall.

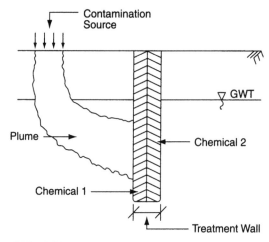

FIG. 9.32 Treatment wall with two chemicals.

treatment wall concept can also be used for biodegradation in a more controlled way (Devlin and Barker, 1993). The treatment wall must be thick enough so that the desired reaction occurs as the contaminants are passing through it. For first-order reactions, the residence time can be calculated by using the following formula (Palmer, 1996b):

$$N_{1/2} = \frac{\ln(C_{eff} \, C_{inf})}{\ln(\frac{1}{2})} \qquad (9.22)$$

where $N_{1/2}$ = required number of half lives
C_{eff} = desired concentration of contaminant in effluent
C_{inf} = concentration of contaminant in plume

Once the retention time is calculated, an estimate regarding the thickness of the treatment wall can be calculated using the permeability of the treatment wall material relative to the contaminant.

9.6.3 Biological Processes

In this process, biochemical reactions are utilized for remediation. Biological processes can be used only if the contaminant is biodegradable. Since microorganisms are found in both the unsaturated zone and in the aquifer, biological processes will occur naturally in most contaminated sites. The aim of biological processes is to create an environment where microorganisms can be more active.

There are two approaches used for bioremediation: (1) microbial approach in which conditioned microorganisms targeted to degrade specific compounds (also known as "superbugs") along with appropriate nutrients are transported to a contaminated zone, and (2) microbial ecology approach in which the environment of the contaminated zone is altered so that the native microorganisms can perform biodegradation more efficiently (Pitrowski, 1991). Bioremediation can also be classified as intrinsic (also known as natural attenuation) or engineered. *In situ* bioremediation is often integrated with a physical process for increasing its effectiveness.

Microorganisms found in soil include bacteria, fungi, algae, protozoa, and metazoa. Biological processes rely primarily on bacteria whose size range from 0.5 to 5 μm in diameter. Microorganisms obtain energy by degrading organic compounds. Biological degradation can occur under aerobic (requiring oxygen for biochemical reaction), anaerobic (does not require oxygen but relies on nutrients for biochemical reaction), and fermentative (relies on electron donors and acceptors for completing the necessary reaction). The largest component of bacteria is carbon. Thus the carbon of contaminants provides a "food source" for the microorganisms.

Although a variety of micronutrients are required for microbial growth, usually only nitrogen and phosphorus are added to enhance bioremediation (Norris et al., 1994). Too many nutrients must not be injected simultaneously. The volume of nutrients must also be controlled so that there is no residue after cleanup. Excess unused nutrients may call for further remedial action. The presence of water also enhances the biodegradation process. Physical (e.g., temperature change) and chemical (e.g., presence of metals, high concentration of a contaminant) conditions may inhibit the biodegradation process. Biodegradation is enhanced by creating an environment conducive to microbial activity through the addition of oxygen and/or nutrients and removing toxic conditions (e.g., reducing contaminant concentration). In many instances, a physical (e.g., air sparging) or chemical (e.g., soil flushing) process automatically creates a favorable environment whereby bioremediation proceeds as a follow-up process. Monitoring of appropriate parameters are key to assess the progress of bioremediation. The monitoring program must include sampling points in up-gradient, down-gradient, and impacted regions of the subsurface.

Multiparameter probes may be used to monitor biogeochemical data such as: dissolved oxygen (DO) redox potential, pH, temperature, and specific conductance (Nyer et al., 1996). These probes are inserted in monitoring wells, and readings are taken both before and after sparging the well. In addition chemical analysis needs to be performed for assessing bioactivity. Microbial counts are also used to monitor bioremediation. Several methods (e.g., direct microscopic counts, most probable number) are available for counting microbial population. Biomolecular methods are also available for assessing organisms in field samples (Riser-Roberts, 1998). Total organic content (TOC) and dissolved organic content (DOC) are also good indicator parameters. Site-specific degraded organic by-product of the contaminant may be monitored for judging the status of bioremediation. Table 9.6 includes a list of suggested parameters for assessing the success of bioremediation.

Several models [such as BIOPLUME II and SLAEM, cited in Nyer et al. (1996)] are available to address bioremediation. Both hydrogeologic factors and biochemical factors need to be included to predict bioremediation.

9.6.3.1 Bioventing Bioventing process setup is the same as the SVE setup. For bioventing to occur, at most sites in addition to extracting air, air is injected to increase the supply of oxygen. By adjusting the airflow rate, biodegradation may be increased to 85% (Riser-Roberts, 1998). Since aerobic bacteria are involved in the biochemical process associated with bioventing, high vacuums inhibit microbial growth. Moisture and nutrients may also need to be added to enhance the biodegradation process.

It is essential to establish viability of bioventing through laboratory testing. The primary objectives of the laboratory studies are to determine whether aerobic bacteria capable of remediating are present and whether sufficient quantities of nutrients are present. The study should also find out whether the pH and moisture content of the site is conducive to microbial action. After the viability of bioventing is established, on-site testing is done to establish optimum airflow for the site. For this purpose air is injected into the subsoil and the oxygen uptake rate is measured at a definite interval (usually 4–5 days) after turning off the air injection.

In general, compounds with low vapor pressure (around 1 mm of mercury) are good candidates for bioventing. Compounds whose vapor pressure is high (760 mm or more of mercury) are gases at ambient temperature. These compounds are not good candidates for bioventing. Many of the commonly spilled petroleum hydrocarbons lie within the low vapor pressure range (e.g., benzene, toluene) and hence can be treated by bioventing (Norris et al., 1994). After remediating the site to the maximum possible extent, natural attenuation is used to where possible. Rifai (1997) has summarized several natural bio-attenuation case histories.

TABLE 9.6 Suggested Parameters for Monitoring Bioremediation

Site-specific contaminants	Obtain through plume monitoring at regular interval
Indicator parameters	DO, Redox, pH, temperature, specific conductance, TOC, DOC, COD, BOD
Electron acceptor parameters	Nitrate, nitrite, amonia, nitrogen, iron (total, ferrous and ferric), sulfate, sulfide
Organic by-products (optional)	By-product list should be developed for each site
Biological parameters	Microbial count (before startup, during cleanup, and after cleanup)

9.6.3.2 Bioslurping   This process is capable of removing LNAPLs floating on the water table. It combines recovery of light free product along with remediation using bioventing. One or more well(s), with the screen partly above the water table (Fig. 9.33) is installed within the contaminated region (Keet, 1995). An aboveground vacuum pump is connected to the wells, which "slurps" the free product along with the water. The free product is separated from the extracted water. After recovery of the free product, continual use of the vacuum pump triggers bioventing (Leeson et al., 1995).

9.6.3.3 Natural Bioattenuation   This is a natural process operating in most spill sites. Bioattenuation can be used if it can be ascertained that proper bacterial populations and environment are present. Laboratory studies, as discussed in bioventing (Section 9.6.3.1), should be undertaken to ensure that natural bioattenuation can be resorted to for the site. In most situations an active remediation using any of the physical and/or chemical processes described in earlier sections is undertaken first.

If natural attenuation is chosen as a process, then the parties need to realize that a portion of the contaminated zone will remain contaminated for a long period of time. In most cases, variance from the regulatory authority is needed to implement natural bioattenuation (Borden, 1994).

9.6.3.4 Phytoremediation   In this process plants are used for remediation purposes. The ability of plants to bioaccumulate metals or to degrade organic contaminants through biochemical process is utilized for remediating a site. Although the process has been used in the field, most of the findings reported in the literature are based on laboratory studies. Pesticides, heavy metals, and a wide range of organic compounds have been removed from soil and groundwater using phytoremediation (Johns and Nyer, 1996). Organic compounds

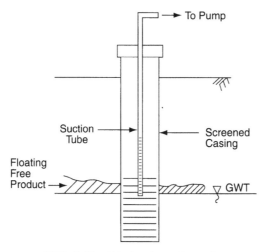

FIG. 9.33   Bioslurping schematic.

are degraded either through metabolization by the plant or through microbial degradation by microbe population surrounding the root system (Alexander, 1999). Trees whose roots go deep (e.g., poplar, willow) draw water from the aquifer and release it to the atmosphere through evapotranspiration. Willow trees have been reported to evapotranspirate 5000 gal of water in a single summer day (Johns and Nyer, 1996). A group of trees can extract several thousand gallons of water, which can have a significant effect on groundwater flow. A field study using 809 hybrid poplar and willow trees has been reported (Quinn et al., 2001). A numerical modeling for the site indicated a large decrease in groundwater contamination despite the seasonal variation of water usage by the trees. The idea of using poplar trees to influence groundwater flow regimes has been supported by Landmeyer (2001) and Ferro et al. (2001). Research is also underway in the use of phytoremediation in dealing with near surface contamination and in wetlands (Ramaswami et al., 2001; Krishna Raj et al., 2000; Krauter, 2002). Phytoremediation involving trees is suitable for tight soil, which can provide stability to the root system. Phyto-remediation for groundwater cleanup can be used for shallow aquifers [up to 20 ft (60 m) below ground level]. The concentration of contaminants also influences feasibility of using phytoremediation. Beyond certain concentrations, even nutrients become toxic to a plant let alone contaminants. Specific plants are capable of degrading or uptaking specific contaminants. So, phytoremediation involves choosing the right type of plant for a specific contaminant. Although extensive research on phytoremediation is underway, currently success stories of field application are rather limited.

Phytoremediation should be considered as part of a remedial train. Phytoremediation alone may not be used for remediating a contaminated site, but it may be used as a final cleanup tool following a cleanup of a majority of the contaminants using a physical or chemical process.

9.7 *EX SITU* REMEDIATION METHODS

In *ex situ* remedial methods, the contaminated soil is excavated out of the ground and treated on site or transported to a different site for treatment. The remediated soil is either put in its original place filling up the hole or is used in other projects. In general, if the contaminated soil is treated at a different geographical location, then it is used in a nearby project. For instance, when petroleum contaminated soil from various underground tank remediation sites are transported to a central landfill for biological treatment, the treated soil is used as daily cover or berm construction within the landfill. *Ex situ* remediation always involves additional cost of excavation and in some cases transportation. Excavation of soil from a contaminated site can release contaminated vapor. So the method can pose an environmental problem in populated areas. Excavation may be restricted due to underground utility lines, sewers, water mains, and so on. In general *ex situ* treatment is costlier compared to *in situ* treatment. However, *ex situ* remediation is quicker compared to an *in situ* process and is easier to reach the reduction target, that is, the allowable residual concentration of the contaminants. As in *in situ* processes, there are three types of *ex situ* processes: physical, chemical, and biological.

9.7.1 Physical Process

Most of the physical processes would require transportation of the contaminated soil to a different geographical location where the treatment plant is located. In many instances, a chemical is added to a physical process that enhances remediation. In addition to the methods described in the following sections, use of volatilization and steam extraction has also been reported in the literature (Ram et al., 1993; Hudel et al., 1995).

9.7.1.1 Incineration   Most incinerators used for remediation are transportable-type systems. In cases where the volume of contaminated soil is high, a semipermanent incinerator is installed on site. However, the life of these on-site semipermanent incinerators is usually low, about 5 years. The parts are usually not salvageable after the useful life of the incinerator. Mobile units are brought on site and parked until completion of the project. These units are mounted on wheels. It is also possible to incinerate contaminated soil in an incinerator used for incinerating other waste types (e.g., municipal waste). Detailed information on incineration is included in Chapter 5. So the topic is not discussed here to avoid duplication.

9.7.1.2 Thermal Treatment   The process is applied primarily for treating organic compounds. Heating causes the contaminants to volatilize which separates them from the solids. Usually soil is heated to 450–600°C (842°F–1112°F) in a heat exchanger (Velazquez et al., 1993; Jensen et al., 1994). Continuous-feed rotary kilns are very effective in removing volatile and semivolatile organic compounds from soil. Desorption rate depends on temperature, residence time, volatility of the contaminant, and velocity of the purge gas (Cheru and Bozzelli, 1994).

Thermal desorption can be combined with a flameless oxidation process, which reduces emission to nearly zero concentration and allows heat recovery at a low cost. After the organic contaminants are separated from the soil, a thermal oxidizer unit is used to treat the gaseous contaminants. The heat produced is used to heat the contaminated soil for desorption (Wilbourn et al., 1994).

9.7.1.3 Soil Washing   In this process the soil is excavated out and washed with water. The process is useful for remediating soil contaminated with oil. Mobile soil-washing units has also been used (Assink and Rulkens, 1984). The sludge produced from a soil-washing unit usually needs to be treated as hazardous waste. Use of soil flushing for remediating soil contaminated with semivolatile and nonvolatile organic compounds (e.g., PAHs, lubricating oil) has been reported (McBean and Anderson, 1996). The contaminated soil is washed to extract hydrocarbons. If necessary, 1% surfactant is added to enhance contaminant removal. Biological treatment is used to decontaminate the extracted water (or leachate). In another process the soil washing is done under elevated temperature and pressure (Amiran and Wilde, 1994). A proprietary chemical is used during soil washing to enhance removal of specific contaminants. The extracted water is treated using biosurfactants. Soil washing can be enhanced using ultrasound in combination with surfactant. A 100% removal of coal tar using this technique has been reported (Meegoda et al., 1995). Soil washing can also be used to treat soil contaminated with cadmium, chromium, cyanide, and zinc (Riser-Roberts, 1998).

9.7.2 Chemical Processes

Although many chemical processes are used for *in situ* remediation, use of chemical(s) alone for *ex situ* remediation is not common. In fact, in most cases chemicals are added during a physical process to enhance remediation. The primary reason is that a physical process is necessary for delivering the chemical within the soil matrix. Spraying of peroxide (Ram et al., 1993) and submersion of contaminated soil in alkaline solution for *ex situ* remediation has been reported. The following is a chemical process that has been used in a large-scale remediation process.

Supercritical Fluid Oxidation   Supercritical water can oxidize many hazardous chemicals readily. The process produces a clean soil with a residual hydrocarbon contamination of <200 ppm and an exhaust steam with high concentrations of CO_2. The process essentially consists of treating the contaminated solid with a mixture of oxygen and water heated and pressurized to a supercritical state (Riser-Roberts, 1998).

9.7.3 Biological Processes

Biological processes to treat petroleum-contaminated soil has been practiced for several decades (Martin et al., 1986). Microorganisms are capable of degrading a wide range of contaminants to carbon dioxide, water, and innocuous by-products. In general, clay or silty soil is more amenable to *ex situ* biotreatment compared to sand or gravel because finer particles provide more surface area for developing microorganism colonies. Since, in general, aerobic degradation is faster than anaerobic degradation, *ex situ* biological treatment may have more promise than *in situ* biological processes. If necessary, anaerobic conditions can easily be established in a vessel for treating recalcitrant compounds. However, the long-term exposure of the open pile of contaminated soil to elements during the treatment may generate leachate and air emissions, which may contaminate groundwater and air, respectively. So, proper care must be taken to protect groundwater and air. In some instances an enclosed space or vessel is used for treatment purposes.

9.7.3.1 Bioreactors   Several types of bioreactors are available. In a bioreactor, the contaminated soil and microorganisms are unified in a vessel or in a lined lagoon. In bioslurry reactors, the vessel size ranges from 15,000 to 30,000 gal (57,000 to 114,000 liters). The process allows aeration and addition of nutrients and surfactants. This type of bioreactor is very effective in treating concentrated residues from soil scrubbing (Bhandari et al., 1994). For faster and more complete degradation, pulverization and pretreatment of soil is recommended. These two processes help to adjust the pH of the soil and increase the surface area for microorganism activities (Black et al., 1991). Biodegradation occurs at a rapid rate in these reactors; treatment time ranges from about 1 to 6 months. The treatment time depends on the concentration of contaminants and target cleanup concentration (Glaser et al., 1995).

There are many types of bioreactors, each with a variation—dual injected turbulant suspension reactors (Geerdiuk et al., 1996), combination reactors (Irvine and Cassidy, 1995), blade mixing reactors (Hupe et al., 1995), and so on.

9.7.3.2 Land Treatment   In this process the contaminated soil is spread on open land and rototilled to incorporate it into the top soil. The contaminants form leachate and penetrate into the ground. Microorganisms in the top 5 ft or so of soil act on the contaminants. Nutrients are added and tilling is done to increase efficiency of biodegradation (Arora et al., 1982). Microorganisms are added at the beginning to enhance biodegradation. Monitoring is done to evaluate the performance as well as downward movement of contaminants. Land treatment is not a passive storage through spreading but a controlled and managed treatment involving tilling, addition of moisture or nutrients, and so on.

Biodegradation rate in land treatment is dependent on temperature, moisture, nutrients, soil pH, and availability of oxygen (Riser-Roberts, 1998). Except for temperature, attempts should be made to optimize all other factors for efficient biodegradation. Sufficient care must be taken to protect groundwater and air from land treatment processes. Not all contaminants can be cleaned up through land treatment; hence careful screening of contaminants is essential prior to choosing this method.

9.7.3.3 Composting   This process, when managed properly, can surpass the treatment capability of incineration. It is also less costly compared to incineration. The treatment time is much shorter compared to land treatment but much longer compared to incineration. Like composting of yard waste, contaminated soil is piled up and mixed with straw or wood chips (Alexander, 1999). Nutrients are added to the pile as needed. The pile, also known as windrow, can be static (static windrow) or turned (turned windrow). Composting can also be done in vessels. Aeration, moisture content, nutrients, and temperature are the key factors for a successful compost pile. The chemistry of contaminants primarily dictates the aeration requirements. Sufficient aeration increases rate and percentage of destruction. Aeration also influences the temperature of the pile. However, anaerobic conditions are useful for the destruction of certain pesticides and most halogenated compounds. In general, biodegradation is greatest at moisture content of 50–70% of field capacity of the soil (Savage et al., 1985). Fungi and actinomycetes are more tolerant of lower moisture content compared to bacteria (Riser-Roberts, 1998). Nutrient requirements must be determined prior to adding any nutrients. Both high and low nutrient concentrations inhibit bacterial growth.

Composting in vessels offers two advantages: emission control and aeration control. If the contaminants are likely to emit hazardous gas, then composting should be done in a vessel. In addition, use of a vessel would allow anaerobic digestion for degrading recalcitrant contaminants. However, if the volume of contaminated soil is too high, composting in a vessel becomes impractical.

9.7.3.4 Vacuum Heap Bioremediation This process is like static compost windrows except aeration is done through slotted pipes placed at different levels in the pile. The contaminated soil is excavated, pulverized when needed, and placed in windrows with layers of embedded slotted pipes. Vacuum is applied in alternate layers of pipe, where air is drawn through the pipes in between two layers of vacuum pipes (Fig. 9.34). Proprietary bacterial nutrient and water is sprinkled through a pipe system at the top of the pile. The pile is covered with plastic to minimize air emission and maintain temperature within the pile. The exhaust air from the vacuum pump is run through an air filter, where necessary, to provide further air emission control. This process is used extensively for treating petroleum-contaminated soil. The piles are located either on a lined area or on an open partially filled landfill surface. If the pile is located on a lined area, then the leachate is collected and usually

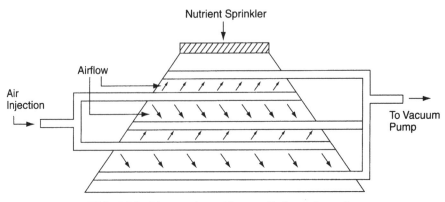

FIG. 9.34 Vacuum heap bioremediation schematic.

recirculated into the pile. A report summarizing results obtained from the treatment of 1 million tons of petroleum-contaminated soil indicates that the process reduced the level of VOCs to less than 10 ppm and PAHs to less than 20 ppm. On an average, 95% reduction of diesel range organics/gasoline range organics (DRO/GRO) was reported. The report further indicates that the treatment time varies between 3 and 6 months (longer time required in cold weather sites, e.g., Wisconsin). The process was found to be effective for treating 3000–60,000 tons of petroleum-contaminated soil in a single pile; the pile size had no effect on treatment time. According to the report VOC and PAH measurements can be used to monitor contaminants in the soil after treatment rather than DRO/GRO measurements (Hamblin and Hater, 1998).

9.8 PROJECT LIFE ESTIMATION

Estimation regarding the life of a project is more important for *in situ* remediation than *ex situ* remediation. However, in both cases the allowable concentration at the end of the treatment process is the key issue. Although toxicological data and risk analysis may help in obtaining an allowable concentration at the end of treatment, technology may not be available to reach that concentration goal. In such cases the allowable concentration for a cleanup project could be higher than the concentration of a contaminant obtained from toxicology or risk analysis. This issue becomes more critical for *in situ* remediation projects. Present experience with *in situ* projects indicates that removal of a significant mass of contaminants can be done, but total cleanup may take a very long time (50–100 years?). So, there is a need to follow a different strategy for remediation projects especially for *in situ* treatments.

The first step in reducing the project life is to remove the source of contamination. For instance, if the source is a leaking underground tank, then the tank should be removed or replaced to minimize time as well as overall cost of the remediation project. The "smear zone" can act as a source of contamination if not cleaned up properly. The smear zone is created because of fluctuations in groundwater level. This allows contaminants, especially LNAPs, to be absorbed in the soil. This issue should be taken into consideration while estimating project life. In an *in situ* treatment the concentration of contaminant decreases as the remedial action progresses. Part of contaminant removal is controlled by advection and part of removal is controlled by diffusion. As shown in Fig. 9.35 reduction of concentration is very rapid at an early stage of remediation because it is mostly controlled by advection. At lower concentrations diffusion controls the remediation. At this stage cleanup target concentration dictates the remediation process capable of reaching the goal.

The object of remediation is to clean up a site so that it no longer poses an environmental risk. The debate that is associated with cleanup is "how clean is clean." One answer to this debate is to clean up to the original state

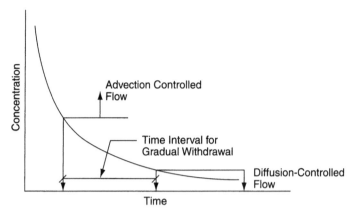

FIG. 9.35   Idealized time–concentration curve for *in situ* remediation projects.

before any contamination occurred. This may be an easy answer but may be almost impossible to address. Since the contaminants were absent prior to contamination of the site, the "original state" would mean a zero concentration of the contaminant. Even a concentration at the detection limit means there is a residual mass because otherwise there would be no detection. So, realistically, there is a need to set up a target concentration that must be achieved before a project can be terminated. Currently, there are three approaches used to set up a target concentration: risk analysis, regulatory directive, and limit of detection. Although risk analysis has developed significantly, a well-accepted standard approach is yet to emerge. One of the issues is whether the risk to humans or the ecology as a whole should be the target. Variation in basic assumptions results in variation of cleanup concentration.

Regulatory directives provide an allowable specific maximum concentration levels (MCLs) for many contaminants. In most cases these numbers are for drinking water. The organics absorbed in soil will release the contaminants to air or water, the concentration of which may or may not be above the MCLs. So, MCLs for soil cannot be the same as that for drinking water MCLs.

The detection limits of contaminants are well established but undergoing change because of better laboratory techniques. It is not uncommon to come up with a target cleanup concentration that is lower than the detection limit. In such cases the detection limit is used to establish a cleanup target. Since the detection limits are subject to lowering with time, a target concentration set up today will not be valid if the detection limit gets lowered due to better analytical techniques available in the future.

All of the above approaches are based on human risk factors. Currently, ecological risk assessment is gaining ground. According to this concept, risk assessment should target the ultimate receptor of a contamination. For example, if a plume discharges into a lake or stream, the allowable concentration should be based on toxicological data from fish or fauna in lakes or streams.

In summary, the cleanup target is a difficult one to establish. However, there needs to be a clear understanding between the regulatory agency and the designer regarding cleanup standards. Without establishing a specific target, remediation will always be without end. Once the target concentration is set, the technical issues should be analyzed to set a plan of action, which will help establish project life. As shown in Fig. 9.35 the diffusion-controlled removal is slow and it is asymptotic to the target concentration. Researchers (Rafai et al., 1988; Nyer, 1991) have shown that active remediation after the contaminant concentration has reached a certain limit is not helpful in remediation. In many situations natural attenuation through biological degradation can realistically halt plume advancement provided the concentration is very low and the proper environment for bacterial activities is present.

Thus a remediation project could probably be divided into two phases. The first part should consist of active removal, and the second part should consist of reliability on natural attenuation or engineered biodegradation (Nyer, 1996b). Depending on the concentration at the end of active remediation using nonbioremediation processes, an active bioremediation may be used. After further reduction in concentration, natural attenuation may be used. This approach, known as treatment train, is gaining ground. Treatment train based project life estimation depends primarily on the experience of the designer. If case histories of all projects undertaken within a nation (better yet for the entire world) could be summarized, then a better estimation regarding the project life for each type of contaminant would have been easier. The problem is to define the time of end of active removal. As shown in Fig, 9.36, the end of the active removal process probably should not be specified; it should be done over a period of time. Data from monitoring of extraction systems could be plotted to see when the concentration versus time curve is somewhat close to becoming asymptotic to the target concentration. At that time the operation of the active removal system may be shut off temporarily. If the concentration of contaminants starts going up again, then the active removal system should be started again. This off-and-on process may go on for some time before a stabilized situation can be reached. At that time the active removal system equipment can be removed. This will be the beginning of the second phase of remediation. In many sites this phase may consist of active bioremediation followed by natural attenuation.

In cases where natural attenuation is not an option, pump-and-treat, barrier wall, or reactive wall methods may need to be established to ensure containment of the plume within a set limit or point of compliance.

9.9   PROJECT COST ESTIMATION

The detail cost analysis associated with each method is unique because equipment and construction items are different. Rather than providing cost estimates for each method, general guidelines for estimating cost for remediation will be discussed. Like discussion of site investigation (Section 9.5), the cost estimation discussed in this section pertains to clean up of sites contaminated over the long term. However, the items identified can be tailored to estimate cost of one-time or short-term spills. Cost for a remediation project can be divided essentially into three parts: investigation, implementation and operation, and maintenance and monitoring.

9.9.1   Investigation Cost

Investigation involves collection of existing file data to identify the source of contamination. The first step for investigation involves time spent in searching records, interviewing employees, and writing a short report including hand sketches and so on. The cost can be calculated by estimating total hours needed for the job and the hourly rate of personnel involved in the investigation. While one person can be sufficient for gathering data for a small job, two or more persons are necessary for big jobs. The team leader for a big job should have prior experience with similar jobs.

The next step is to undertake on-site investigation. There are two major expenses associated with this phase: subsoil investigation and monitoring. Usually subsoil investigation involves collecting soil samples, testing soil samples (e.g., grain size analysis, permeability), and preparing a report to provide a geologic cross section for the site. Experienced professionals should be present at the site for subsoil investigation. So the total cost for subsoil investigation will include cost for sample collection as well as cost for professionals involved in the work. Usually monitoring wells are installed during subsoil investigation. So the cost for well installation should also be included into the cost of the site investigation. The next big item is groundwater monitoring. This will include cost of sample collection and analysis. Usually more than one round of groundwater samples are analyzed. Hence, the cost estimate should reflect the cost of multiple sampling.

9.9.2   Implementation

The first step for the implementation of a remedial action involves data interpretation and design. Cost comparison for various options are also done at this stage. Meeting with clients and regulators and participation in public hearings (e.g., informational hearing, contested case hearing) are also part of the first step. Cost for all of these items should be included. However, since it is very difficult to estimate exact personnel hours for this stage, it is a good idea to develop a range for cost of design (e.g. X–Y). This is a step in which gross underestimate or, in some instances, overestimate of costs can occur. Although reasonable overestimate is good for both owner and contractor, gross underestimate can lead to an incomplete job, which will eventually increase the project cost. If bidding for the job is involved, overestimates can kill a bid. However, seldom is a bid awarded to the lowest bidder; items such as contractor's experience and personnel rosters are also considered before awarding a job. Because of the uncertainty involved, contractors usually submit a set cost estimate for known fixed items and an hourly rate for additional cost beyond the items listed in the bid.

The next step is cost estimate for construction of the remedial process items. It is somewhat easy to estimate *ex situ* remediation cost than *in situ* remediation cost. Cost estimates for equipment (e.g., backhoe for excavation, extraction wells, pumps) and chemicals (if needed) are included in construction costs. Some flexibility for unexpected additional work (e.g., installation of additional wells) should be built into the construction cost estimate. In both these steps, cost for meeting with client and regulatory agency should be included.

9.9.3 Operation, Maintenance, and Monitoring

The cost of operation for *ex situ* remediation projects will usually include equipment mobilization and operation, chemicals (if used), transportation (if used), off-site treatment (e.g., cost of incineration), site supervision, and monitoring costs. These are the major items. Cost for one or more meetings with regulatory agency personnel should be included.

For *in situ* remediation, the major cost estimate items for operation are operating the equipment (e.g., extraction pumps), chemicals (if used), site visits by consultants, and meeting with regulators. For long-term maintenance the estimate should include prevention and maintenance of wells and repair (or replacement) of pumps and other similar equipment. Total cost of maintenance can be relatively high in harsh environments (e.g., chances of corrosion of well screen due to groundwater chemistry is high). The cost will also depend on project life. So, all probable costs per month should be carefully estimated. Monitoring cost includes sample collection and laboratory testing. The total cost of monitoring depends primarily on the length of project life and remediation process.

The total cost of operation and maintenance is difficult to estimate without prior experience with similar project(s). If no experience is available for a similar project, then all probable worst-case scenarios should be discussed and the most probable situation should be used. It is better to overestimate than underestimate. Any cost savings in a month or year should be saved for "bad days." Many times, sudden unexpected expenditures becomes necessary, but the action is postponed due to lack of funds. If project life is expected to be more than 2 years, then starting an escrow-type fund for long-term operation and maintenance should be considered.

9.10 COMMENTS ON REMEDIATION METHOD SELECTION

Many factors need to be considered before selecting a remediation method. It is not possible to provide a definitive approach or decision tree that is applicable to all remediation sites. Some general issues are included that are useful in selecting a method for a site.

One of the primary issues for remediation is proper site characterization. The aim of site characterization is to develop a clear picture of the subsoil stratigraphy and the extent of contamination. Knowledge regarding the distribution of contaminant concentration and chemistry of the contaminants are also important for the selection of a remedial method. For instance, the designer needs to know whether cleanup can be confined to the vadose zone only or cleanup of the aquifer is also needed. Knowledge about the chemistry of the contaminants will help to identify whether the contaminants are essentially LNAPLs or DNAPLs.

Once the contaminated zone is identified then the designer needs to choose between *in situ* and *ex situ* remediation. Although up-front costs for *ex situ* methods are higher compared to *in situ* methods, the long-term costs for *in situ* methods are in general higher than *ex situ* methods. However, cleanup time is shorter for *ex situ* methods compared to *in situ* methods. The costs of *ex situ* remediation could be significantly high if (a) the contaminated soil needs to be transported a long distance for treatment, (b) the contaminated zone is deeper than 15–20 ft (4.5 ft–6 m) and the total volume needing treatment is significant [1000 yd^3 (765 m^3) or more], and (c) the chemistry of the contaminants needs a sophisticated treatment technology. *Ex situ* remediation can be very cost effective if contaminated soil from several sites with the same or similar contaminant chemistry can be treated at a centralized location. One classic example is the treatment of petroleum-contaminated soil from leaking underground tanks. The contaminated soils are transported to a centralized location (usually to an active landfill) where the soils are treated using a vacuum heat bioremediation method (see Section 9.7.3.4).

In situ methods are usually slow, hence cleanup takes several months to several years. However, location of the contaminated zone may be such (e.g., within an aquifer) that *ex situ* remediation is not possible. In some instances, even if the contaminated zone is shallow, an *in situ* method is chosen over *ex situ* methods. Such a choice is made for one or more of the following reasons: (a) the contaminants are such that excavation of the area may release gaseous contaminants that may exceed the allowable concentration limit in the air; (b) the contaminated zone is located in a densely populated area; (c) excavation is restricted due to underground structure(s); (d) cost of *ex situ* treatment is significantly higher compared to *in situ* treatment.

In general, regulatory approval is easier to obtain for physical methods (especially for *in situ* processes) compared to chemical or biological methods. However, since certain types of contaminants in certain geologic formations are susceptible to sorption, chemical or biological methods can be easily justified for these contaminants.

For remediation methods, which rely primarily on convective transport, the contaminants must be volatile or soluble in water. Permeability of the soil is also an important factor for methods that rely on convective transport. Use of electrokinetic remediation may be considered for low permeability soils. For extracting pure contaminants, its sorption in soil should be investigated properly.

Bioremediation can be undertaken only for biodegradable contaminants. In general aerobic degradation is faster than anaerobic degradation. However, it should be noted that anaerobic degradation is more effective for certain recalcitrant contaminants. Active bioremediation is faster than natural attenuation. However, in certain circumstances natural attenuation is a better choice (e.g., if the concentration of the contaminant is low and laboratory data indicate the viability of natural attenuation).

For remediation in a vadose zone, permeability of the soil for air should be assessed properly. Usually permeability of air is 50 times the permeability of water (Otten et al., 1997). Proper investigation of *in situ* permeability of air must be undertaken for nonhomogenous subsoil conditions and for low permeability soil (e.g., silty clay). Heterogeneous conditions may cause preferential flow through relatively permeable pockets whereby the entire contaminated zone will not be remediated properly.

LIST OF SYMBOLS

$2a$ = trench width
A = area of discharge
A_c = amount of constituents adsorbed per unit dry weight of soil
B = constant
c = soil cohesion
C = contaminant concentration
C_b = sorbed concentration
C_{eff} = desired concentration of contaminant in effluent
C_g = concentration of contaminant in gas phase
C_{inf} = concentration of contaminant in plume
C_s = concentration associated with source or sink
C_w = concentration of contaminant in water phase
D_{ij} = dispersion coefficient tensor
D_x = longitudinal dispersion coefficient
$D_{xx}, D_{xy}, \ldots$ = components of disperson coefficient along appropriate axes
h = hydraulic head
H = Henry's constant
H_{cr} = critical height of slurry trench
I = current
k = permeability of the slurry wall
k_b = permeability of backfill
k_c = permeability of filter cake
k_{ij} = permeability tensor
k_x, k_y, k_y = intrinsic permeability of x, y, z coordinates, respectively
k_α = coefficient of lateral earth pressure

K = soil adsorption constant
K_d = distribution coefficient
n = soil porosity
n_e = effective porosity
$N_{1/2}$ = required number of half-lives
P = power
q_s = flow of fluid source/sink per unit volume
Q = discharge volume
R = resistance
t_b = permeability of soil backfill
t_c = thickness of filter cake
V_c = velocity of contaminant
V_d = discharge velocity
V_s = seepage velocity
V_{sx}, V_{sy}, V_{sz} = seepage velocity in x, y, z coordinates, respectively
V_w = velocity of water
γ = density of sand
γ_f = density of slurry
γ_s = bulk density of soil
Γ_f = gel strength
λ = reaction rate constant
μ = dynamic viscosity of water

10

REDEVELOPMENT OF CONTAMINATED LAND

In the past, most contaminated sites were abandoned. The concept of redeveloping these lands was absent. However, with the advent of new soil cleanup technologies and the need for land in urban areas, more and more of these lands are being reused primarily for commercial or industrial purposes. This chapter includes discussion primarily on redevelopment of lands contaminated by industrial activities. In general, lands contaminated because of industrial activities involve hazardous chemicals, whereas lands contaminated because of municipal solid waste disposal involve nonhazardous chemicals and gas generated through to the decomposition of putrescible waste. The following issues need to be addressed before contaminated land can be redeveloped:

1. Proper cleanup of the land
2. The risk of using the land after redevelopment
3. Proper legal protection of seller, buyer, and developer from future liability
4. Adequate funding for cleanup and marketing of the land.

The last three issues are addressed in this chapter. Information regarding contaminated land cleanup is considered in Chapter 9. Discussion regarding construction on cleaned up lands and solid waste landfills are included in Chapter 20.

10.1   RISK ASSESSMENT

Risk assessment is a tool to identify hazard from a situation or a problem. The process is based on science that estimates human health and environmental risks associated with chemicals of concern present in the environment. An approach, known as the risk-based corrective action (RBCA) is undertaken for cleaning up contaminated land. RBCA is a streamlined approach in which risk assessment practices are integrated with corrective action process(es) to ensure selection of appropriate and cost-effective remedies and proper allocation of resources. There are two goals of RBCA: protection of human health and environment and adopting a technically sound corrective action. The first step is to find out the baseline risk. Figure 10.1 shows the various steps necessary to assess baseline risk. Based on RBCA, "no action" can also be an option for a particular site. There are two aspects of risk assessment:

1. *Fate and Transport* Involve the prediction of the concentrations of chemicals at all locations at all times in all three media (i.e., air, water, and soil).

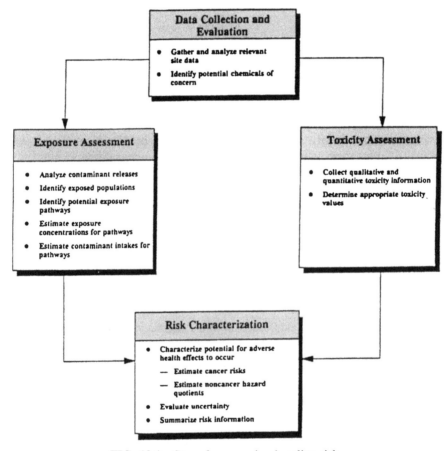

FIG. 10.1   Steps for assessing baseline risk.

2. *Exposure Assessment* Involves the estimation of the extent of exposure of a species through all three media (i.e., air, water, and soil), which relates directly to the proposed use of the property. Risks differ based on use, and so does the remedy.

The total dose and toxicity of the chemicals are estimated to assess the risk to human health or environment. In this book, discussion is aimed primarily at human health risk. Risk is dependent on toxicity and exposure. Exposure is estimated based on the concentration and rate of intake. The intake depends on complete exposure pathways. A complete pathway requires:

1. Source of release
2. Mechanism of release from the source
3. Medium to transport (air, groundwater, soil) the chemical to the receptor
4. Point of contact of the receptor with the medium
5. Route through which the chemical can enter the body (inhalation, ingestion, or dermal contact)

Potential receptors are considered when assessing exposure. The potential receptors or population for exposure are residents residing on or near the contaminated land, workers employed by the industry/commercial establishment(s) constructed on the redeveloped land, and any future human population or animal population exposed through one or more exposure pathways. The redevelopment goal is to reduce or control risks compatible with land use.

Dilution and attenuation of the chemical during transport are considered while assessing exposure pathways. For any receptor the routes of exposure are groundwater ingestion, air inhalation, soil ingestion, or dermal contact. All of these exposure routes are considered when assessing risk for humans. However, for animals and vegetation all of the above exposure routes may not be applicable. Figure 10.2 includes an illustration of exposure pathways for humans.

There are various sources of uncertainty in risk assessment. The reduction of concentration of a chemical from the source to the receptor is dependent on various factors. There is a debate regarding the concentration to be used for risk assessment. For example, the release of a chemical from a waste site to groundwater depends on the infiltration rate, which is dependent in turn on precipitation. Precipitation on the waste pile varies annually and also over the years. So the concentration of the chemical in the leachate is a time-dependent variable. As it enters the groundwater, it may or may not dissolve completely in groundwater. The geology and hydraulic conductivity of the aquifer primarily influence the plume geometry. The concentration of the chemical within the plume is not uniform. Hence, if monitoring wells are not located properly, the predicted concentration of the chemical will be errone-

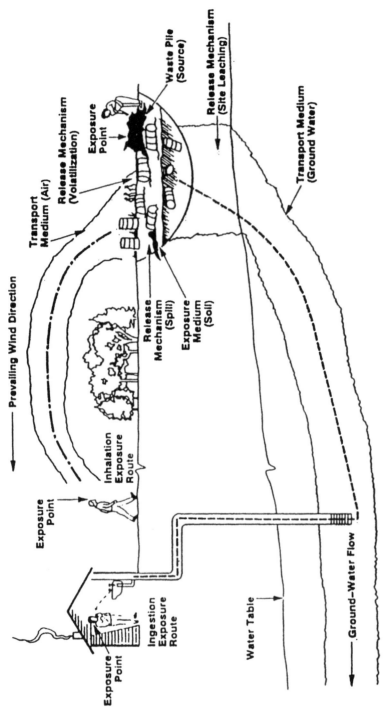

FIG. 10.2 Exposure pathways.

Prevailing Wind Direction

Transport Medium (Air)

Release Mechanism (Volatilization)

Exposure Point

Waste Pile (Source)

Release Mechanism (Site Leaching)

Transport Medium (Ground Water)

Release Mechanism (Spill)

Exposure Medium (Soil)

Inhalation Exposure Route

Exposure Point

Ingestion Exposure Route

Exposure Point

Water Table

Ground-Water Flow

228

ous. Because of these sources of variation, the estimate may reflect maximum, average, or minimum concentrations of the chemical in the groundwater. Because of such variation, the assessment may reflect maximum, average, or minimum risks for the chemical. However, if groundwater is not used, or proposed for use, there may be no risk associated in this situation. In summary, it may be said that there is an inherent uncertainty regarding the predictive model and the parameter concentration used for risk assessment.

In addition to the predictive model uncertainty, there is a toxicology-related uncertainty. Concentration at which a certain chemical will be considered toxic to humans, animals, or vegetation is based primarily on short-term animal studies. Whether these data can be extended to humans is a debatable issue. However, it should be mentioned that in recent years these animal studies have been refined.

The toxicological data are widely accepted and risk assessment, with all its uncertainties, has gained significant ground. Besides, currently the analytical technique is the only tool available for assessing risks. Note that the actual steps and allowable concentrations are not standardized internationally. There are variations in the assumptions used in analysis in different states within the United States and between various countries of the world. The analytical technique recommended by a regulatory agency must be followed in assessing the risk from a contaminated site. Environmental contamination has health, environmental, political, and socioeconomic implications. Risk assessment is one of the best mechanisms available for decision making. It is important to use systematic and technically sound methods of approach for environmental management purposes. Risk assessment should be considered as an integral part of environmental management programs.

10.1.1 Hazard Effect Determination

Toxic characteristics of a substance are categorized according to the organ they affect (e.g., kidney, liver) or the disease they cause (e.g., birth defect). One of the goals of toxicity studies is to establish the no observed effect level (NOEL). NOEL is the dose at which no toxic effect is seen in an organism. If the incidence of effect increases (not necessarily the severity) with the increase in dose, then the response is characterized as probabilistic. It means that the probability of occurrences increases due to an increase in dose. The duration of exposure influences both the NOEL and probability.

Toxic responses vary in the degree of reversibility. For some toxic substance, an effect disappears following cessation of exposure. In some other toxic substances, the exposure causes permanent injury. Seriousness of exposure to toxic substances is also important. For certain types of damage exposure to a toxic substance is clearly adverse and poses a definite threat to health, whereas in other cases the effect may not have obvious or immediate health significance.

In toxicological studies the dose–response curve is developed to study the response (e.g., the effect observed because of exposure to a chemical). The dose–response curve is referred to as a stressor–response profile to characterize ecological effects. Dose–response curve falls into two general categories. For certain chemicals no response is observed until a threshold (i.e., minimum) dose is reached (Fig. 10.3; curve I); these chemicals are known as threshold chemicals. For certain other types of chemicals response is observed for any dose (Fig. 10.3; curve II); these chemicals are known as nonthreshold chemicals. Figure 10.4 shows a typical dose–response curve used for characterizing the ecological effects of chemicals. In threshold chemicals the highest dose that does not produce a no observed adverse effect level (NOAEL) and the lowest dose that produces an observable adverse effect [known as lowest observed adverse effect level (LOAEL)] are important.

10.1.1.1 Risk Assessment for Human Health The potential to human health risk is determined by integrating chemical exposure information with the toxicological data regarding the chemical of concern. The following parameters or factors are to be considered while performing human-health-related toxicity tests (USEPA, 1985a, 1985b):

1. Route of exposure
2. Duration/frequency of exposure
3. Characteristics of test species
4. Individual characteristics
5. Toxicological end points (note: end point denotes the nature of toxic effects such as carcinogenicity, mutagenicity, neurotoxicity, etc.).

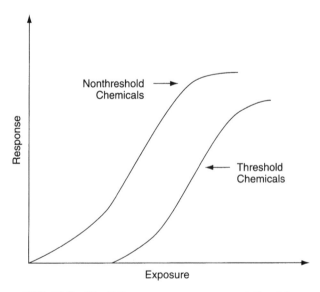

FIG. 10.3 Idealistic exposure–response relationship.

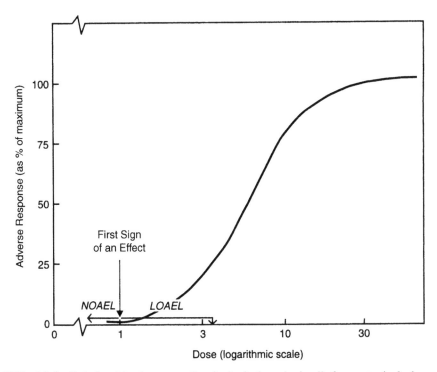

FIG. 10.4 Relationship between "toxicological endpoints" for a typical dose–response curve for threshold chemicals. From Asanten-Duah (1998). Reprinted with permission from John Wiley & Sons.

For human health risk assessment purposes, chemicals are commonly divided into two groups—carcinogenic and noncarcinogenic. In general, it is considered that noncarcinogenic chemicals have a threshold level and carcinogenic chemicals do not have a threshold level. However, some professionals are of the opinion that certain carcinogenic chemicals also have a threshold level (Wilson, 1997). Average daily dose (ADD) and Lifetime average daily dose (LADD) are used for evaluating exposure. ADD is usually used for noncarcinogenic effects and LADD is usually used for carcinogenic effects. Basic relationships for major routes of potential receptors for few exposures are given below. Detail discussion regarding exposure equations can be found elsewhere (McKone and Daniels, 1991; USEPA, 1986b, 1988, 1989a, 1989b, 1991).

$$\text{Inhalation exposure (mg/kg-day)} = (\text{GLC} \times \text{RR} \times \text{CF})/\text{BW} \quad (10.1)$$

where GLC = ground-level concentration (mg/m^3)
    RR = respiration rate (m^3/day)
    CF = conversion factor (1 mg/1000 mg)
    BW = body weight (kg)

$$\text{Water ingestion exposure (mg/kg-day)} = \frac{CW \times WIR \times GI}{BW} \quad (10.2)$$

$$\text{Soil ingestion exposure (mg/kg-day)} = \frac{CS \times SIR \times GI}{BW} \quad (10.3)$$

$$\text{Crop ingestion exposure (mg/kg-day)} = \frac{CS \times RUF \times CIR \times GI}{BW} \quad (10.4)$$

where   CW, CS = chemical concentration in water and soil, respectively
         (mg/L)
WIR, SIR, CIR = consumption rate of water, soil, and crop, respectively
         (L/day or kg/day)
         RUF = root uptake factor
          GI = gastrointestinal absorption factor
          BW = body weight (kg)

Chemical ingestion depends on the absorption across the gastrointestinal lining. Estimates of GI for various chemicals are available in the literature. If the GI value is not available, then absorption may be conservatively assumed to be 100% (Asanten-Duah, 1998).

10.1.1.2 Risk Assessment for Ecological Impact   The risk assessment procedure used to determine probable ecological impact is similar to the procedure used for assessing human health risks. Steps for assessing ecological risks are shown in Fig 10.5. Standardized procedures for ecological risk assessment is difficult to develop because of complex relationships between the various parts of the ecosystem (Kolluru et al., 1996). Like human health risk assessment, the chemical exposure information and toxicological data are necessary to assess the risk.

Each ecosystem is a unique combination of physical, chemical, and biological characteristics; their response to a contamination is unique. An ecosystem is classified into two broad categories: terrestrial and aquatic. Wetlands are considered as a transition between these two broad classifications. A screening of chemicals at the target ecosystem is done to assess the risk. The following issues are taken into consideration while evaluating an ecological hazard:

1. Nature of the ecosystem (e.g., terrestrial, wetland)
2. Habitat evaluation
3. Selection of target species
4. Selection of end points
5. Identification of ecological effects

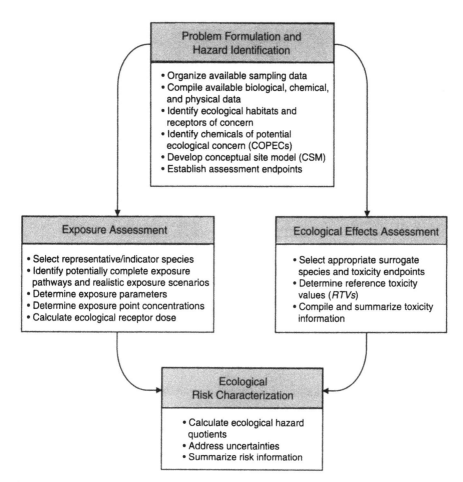

FIG. 10.5   Ecological risk assessment process. From Asanten-Duah (1998). Reprinted with permission from John Wiley & Sons.

Figure 10.6 shows a typical flowchart for assessing ecological risk for a contaminated site. Detail discussion regarding ecological risk assessment can be found elsewhere (Sutter et al., 2000; Calabrese and Baldwin, 1993; Asanten-Duah, 1998).

10.1.2   Error of Analysis

As mentioned earlier, there are many sources of uncertainty in risk assessment. Because of the uncertainties involved, an analysis may overestimate or underestimate the risk. The following are the three main issues that cause uncertainty.

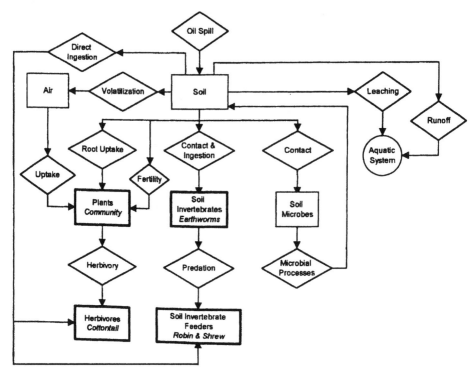

FIG. 10.6   Typical flowchart for assessing ecological risk. Reprinted with permission from Sutter et al. (2000). Copyright Lewis Publishers, an Imprint of CRC Press, Boca Raton, Florida.

10.1.2.1   Toxicological Data   The toxicity of many chemicals are not completely investigated. A chemical may be more toxic or less toxic for humans. In many cases toxicological data are obtained by extrapolating data obtained from animal studies. Since most data are obtained from a short-term exposure, the real effect of a long-term low-level exposure cannot be analyzed using the available data. Then there are uncertainties related to exposure to more than one chemical. There are three possible types of effects from a mixture of two or more chemicals. If the effect of the combination is more than the sum of the inputs, then it is called *synergetic*. If the effect of the combination is less than the inputs, then it is called *antagonistic*. If one of the chemicals has no toxic effect but when combined with another chemical the effect is more, then it is called *underpotentiation*. Chemicals present in a mixture can react with one another creating a new chemical with a different toxic effect.

10.1.2.2   Model Assumptions   Uncertainties are introduced by the mathematical assumption made during development of a mode. Boundary conditions, physical processes, other parameters used during analysis, and data interpretation also introduce error into the risk assessment. In addition, the groundwater monitoring network design and the average values of various chemicals used as the model input also influence the model output, which in turn causes error in risk assessment.

10.1.2.3 Incomplete Source Assessment   All possible hazard sources and exposure possibilities must be analyzed while assessing the risk. If potential risk scenarios are not included in the analysis and if in-depth analyses of individual scenarios are not undertaken, then the resulting assessment will be erroneous.

Because of the inherent problems associated with the elements of the risk assessment procedure, the resulting conclusions may vary. To minimize such variations in outcome, it is essential to use a standardized approach for risk assessment and add safety factors in the model. Risk assessment should not be viewed as a tool for the prediction of expected impact quantity (e.g., actual cancer rate in exposed population) but as a decision-making tool for prioritizing problems and selecting solutions. Sensitivity analysis (Iman and Helton, 1988) and probabilistic analysis (Macintosh et al., 1994; Power and McCarty, 1996) should be undertaken if the cost of necessary mitigation is estimated to be high. Probabilistic analysis can be used to assess the extent of uncertainty. Detailed discussions on use of these techniques can be found elsewhere (Burmaster, 1996; Finley and Paustenbach, 1994; Finley et al., 1994a, 1994b; Smith, 1994).

10.2 RISK-BASED CORRECTIVE ACTION FOR REDEVELOPMENT OF CONTAMINATED LAND

When redeveloping contaminated land, the proposed use of the land needs to be considered to determine the cleanup standards using risk assessment. For example, if the proposed use of a cleaned up site is the establishment of a residential area, then the risk assessment must include direct contact of children with the soil (pica behavior). Whereas consideration of pica behavior is not needed if the proposed use for site is industrial. The risk-based cleanup standard will be different in these two situations. A tiered approach is usually used for cleaning up a site. A site-specific risk-based cleanup standard is developed for the site based on proposed use. Good knowledge regarding mandates of the regulatory agency is essential for completing a project. If redevelopment of a contaminated land is contemplated, then proposed remedial action should be judged against the cleanup cost. The cleanup standards should be such that they address the proposed use. A target cancer risk of 10^{-6} is used for all carcinogenic chemicals. For noncarcinogenic chemicals a hazard index of 1 or less is usually used.

10.3 LEGAL ISSUES

If contaminated land is used after cleanup, then the directives from the regulatory agency should be followed for proper legal protection. Both the user and developer of the land should be protected from legal action. When cleanup is done in accordance with the regulatory standards, it ensures that the site does not pose health risks for the proposed use. A legal document, which will protect the developer from the user's health-related legal actions (if any), should be prepared and signed by all parties. In many situations the city, the developer, and the new owner of the site undertake the corrective action jointly. State or federal law may provide civil immunity to certain economic development corporations if the property is used for economic development of the community. Such civil immunity protection is contingent upon proper cleanup to ensure substantial protection to public health and safety.

10.4 FUNDING

Funding for redevelopment of contaminated lands may be available from state, federal, or local governments. Such funding may be given as loan guarantee, grants, or tax relief. Government funding helps to encourage other investments aimed at redevelopment and economic growth. After proper cleanup the local government may sell the redeveloped land to a developer or to an industry.

If the regulatory agency mandates long-term monitoring as part of the cleanup action, then the local government must ensure continuation of the monitoring. The deed must reflect the monitoring requirements. If contaminated land is developed jointly by a local government, a developer, and the future owner(s), then all may agree to provide partial funding. A document indicating the willingness to provide necessary funding and responsibilities of all parties must be negotiated prior to undertaking a joint cleanup project. In many situations there are cost overruns. The funding agreement must reflect the additional funding responsibilities of each party in case of a cost overrun.

PART II

LANDFILL DESIGN

11

INTRODUCTION

Land disposal of waste has been practiced for centuries. In the past it was generally believed that leachings from waste are completely attenuated (purified) by soil and groundwater and hence contamination of groundwater was not an issue. Thus, disposal of waste on all landforms (e.g., gravel pits and ravines) was an acceptable practice. However, with increasing concern for the environment in the late 1950s landfills came under scrutiny. Within a short period of time several studies [California Water Pollution Control Board (CWPCB), 1954, 1961; Apgar and Langmuir, 1971; Garland and Mosher, 1975; Kimmel and Braids, 1974; Walker, 1969] showed that landfills do contaminate groundwater. Although percolation of leachings from chemical industry waste to groundwater aquifers was considered unsafe, leachings, from nonchemical industrial waste and municipal waste was considered less harmful. As a result waste was divided into two categories: hazardous and nonhazardous. In many countries separate regulations were developed for these two types of wastes. Although collection of leachate from nonhazardous waste was not mandated, leachings from hazardous waste were required to be collected. For nonhazardous waste the emphasis was on transforming waste dumps into "sanitary landfills."

As a result of these different attitudes toward the effect of leachate on groundwater pollution, two separate design concepts evolved: natural attenuation (NA) type of landfills (Fig. 11.1) and containment-type landfills (Fig. 11.2). In NA-type landfills leachate is allowed to percolate into the groundwater aquifer. The design guidelines for NA-type landfills included a minimum allowable thickness of the unsaturated zone, a depth to bedrock, a distance to the nearest home with a private well, and so on. In containment-type landfills the design concept consisted of constructing a low permeability

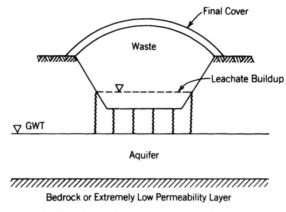

FIG. 11.1   Natural attenuation landfill.

liner to restrict leachate from percolating through the base of the landfill and a pipe system to collect leachate generated within landfills. The second category of landfills, also known as engineered or secured landfills, was mandated for disposal of hazardous waste. In addition to the trench-type landfill configuration in which waste is disposed below grade (Figs. 11.1 and 11.2), there are two other configurations in use. The second configuration may be termed as canyon fill landfills in which waste is disposed in sloping canyons and swales (Fig. 11.3). The third configuration may be termed as at grade landfills in which the waste is disposed at or slightly below grade within the confines of berms (Fig. 11.4). Although most of the design issues are the same for all types of landfills, there are special issues for each landfill type (e.g., slope stability of canyon fill landfills) that should receive due consideration.

Further research indicated that soil cannot attenuate all the contaminants leached even from nonhazardous waste (Bagchi, 1983), no matter how thick the underlying unsaturated zone is or how high the cation-exchange capacity of the soil is (refer to Sections 15.2 and 15.3 for a detailed discussion). Thus, subsequently, restrictions were imposed on NA-type landfills. Currently NA-type landfills are totally banned in some places. Certain changes also took place in the design of hazardous waste landfills. Initially these landfills were required to have a single liner. However, because of the possibility of leakage through a single liner, especially if the liner material is clay, a second liner (Fig. 11.5) was mandated. In many instances a synthetic membrane was required as the first liner. Thus, at present, land disposal of waste is no longer a simple practice. Discussion in this book focuses mainly on nonhazardous solid waste issues.

11.1   COMMENTS ON REGULATORY REQUIREMENTS

Landfill siting is a fairly involved process. Statutes governing standards for waste management have been adopted in many countries. These statutes allow a regulatory agency to adopt minimum standards for landfills.

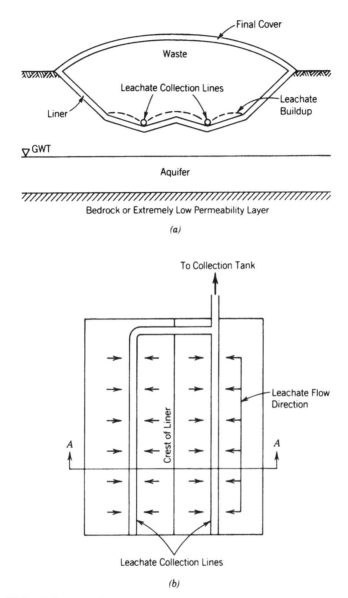

FIG. 11.2 Containment landfill: (*a*) Cross-section A–A; (*b*) plan.

Prior to planning a landfill, it is necessary to obtain all pertinent information about the units of government that may exercise control over different operational issues of a landfill. For instance, for containment-type landfills leachate is collected and treated either on-site or in a treatment plant off-site. In many instances the landfill management is handled by one section and the treatment facility management is handled by a separate section of the regulatory body. Therefore, the planner must become familiar with the require-

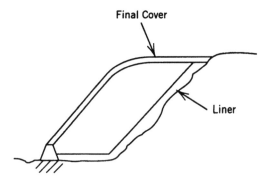

FIG. 11.3   Canyon fill landfill.

ments of both these bureaus or sections. A landfill also has to meet several locational criteria (see Section 12.1.2), separate sections of the regulatory agency may exercise control over each of these locational criteria. Construction of a flow diagram indicating regulatory requirements and the name of the bureau involved in each operation is helpful in obtaining all the necessary approvals from different bureaus. Local governments sometimes exercise control over refuse collection, storage, hours of operation, and so on. In some regions local or regional authorities exercise control over "waste flow" (or how and where the waste generated within the region is to be handled). This type of waste flow control becomes an important issue if an incinerator is located within that region and the waste generated within that region is not high. It may become essential to direct a minimum waste volume to the incinerator so that it can be operated economically. Because landfill technology is evolving and the regulations are changing, it is important to remain up to date about both.

11.2   SCOPE AND ORGANIZATION OF PART II

Rather than discussing the regulatory requirements in detail, Part II concentrates on current theory and practice regarding landfills. Attempts have been

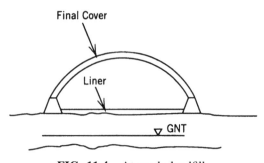

FIG. 11.4   At grade landfill.

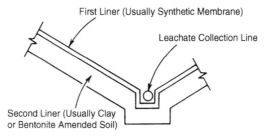

FIG. 11.5 Double liner.

made to discuss the pros and cons of different concepts so the designer can make the correct decision in each case. Landfill siting is usually a three-step process. The first step includes choosing several sites using locational criteria. The second step is to conduct a feasibility study of potential sites developed in step 1 and to determine which site is best for the landfill. A detailed discussion on feasibility study is beyond the scope of this book. However, some guidelines regarding the site selection process is included in Chapter 12. The third step is to develop a detailed design for the site identified in a feasibility study. Part II primarily deals with the issues confronted by the engineers involved in the process. (Note: usually hydrogeologists and engineers work as a team on landfill projects. This part of the book is primarily aimed at engineers, although hydrogeologists will find it informative.) Part II written primarily for practicing engineers and engineering students; however, it will be very helpful to any person interested in learning the fundamentals of landfill technology. Attempts have been made to provide both theory and current practice mainly regarding landfill design and construction. Fundamental concepts of landfill monitoring are included so that engineers can interact with hydrogeologists as a team. In addition to information about landfill technology, information about wetland development and health and safety are included in this part. The contents of each chapter are briefly discussed below.

Chapter 11 includes a brief history of changes in attitudes toward landfilling of waste. Because in most countries landfills are regulated by government, comments are included regarding approval by local regulatory agencies. The scope and organization of the book are also included.

Chapter 12 discusses how to select a site. Both preliminary site selection and the final selection process are discussed briefly.

Chapter 13 discusses how leachate and gas are generated in a landfill. Approaches for assessing leachate quality and quantity and the typical leachate quality of various wastes are included.

Chapter 14 discusses how to characterize waste. Characterization of both hazardous and nonhazardous waste is discussed separately.

Chapter 15 discusses natural attenuation processes and the design approach used for natural attenuation landfills.

Chapter 16 discusses the design of containment-type landfills for both hazardous and nonhazardous waste. The advantages and disadvantages of natural attenuation and containment-type landfills are discussed.

Chapter 17 discusses bioreactor landfill design. The issues that need further attention are also included.

Chapter 18 discusses the various materials used for landfill construction.

Chapter 19 discusses the design of several landfill elements. The design of gas venting systems and how to retrofit an existing natural attenuation landfill to function as a containment-type landfill are also included.

Chapter 20 discusses construction-related issues and the tests usually performed for quality control purposes.

Chapter 21 discusses landfill remediation.

Chapter 22 discusses fundamental concepts regarding performance monitoring. A detailed listing of items usually monitored in a landfill is included.

Chapter 23 discusses landfill operation and long-term care needed for maintaining a sanitary landfill.

Chapter 24 discusses compensatory wetland development. Wetland identification and mitigation issues are also included.

Chapter 25 discusses health and safety issues. Although it is included in this part, the information is useful for everyone involved in solid waste management programs.

Chapter 26 discusses how to estimate the costs of construction, operation, final closure, and long-term monitoring.

Landfill technology is multidisciplinary. It requires application of principles from hydrogeology, civil engineering, chemistry, mathematics, and material science. Although each concept is discussed in simple language, some knowledge of the above disciplines is assumed for readers interested in an in-depth study (i.e., students and professionals). Landfill technology is developing and is fairly complex. Correct, unambiguous answers to all issues are not yet available. Some points made may be in dispute and some discussions may be incomplete. Hopefully these uncertainties can be resolved in the future. Since publication of the first edition research efforts have been reported regarding new ways of disposal and landfill space use. In Europe most countries now require that municipal waste be treated prior to disposal. Currently treatment includes incinerations and composting. However, concepts of "bioreactor landfills" are also being studied. Landfill mining, which essentially consists of digging up old landfills and recovering usable materials, is no longer a dream; several landfills have already been mined. Although, these items are outside the preview of landfill design, brief discussion and references on the subjects are included to help readers in their search for alternatives to landfilling.

Landfill design and construction-related issues are discussed in reasonable detail. In addition, references are cited for readers interested in follow-up reading. Attempts have been made to include detailed design steps whenever possible, but justice could not be done in some cases. Some comments regarding research needs and some worked out examples are included for classroom use.

A list of references is included at the end of the book.

12

SITE SELECTION

A landfill site has to meet several locational and geotechnical design criteria and be acceptable to the public. A preliminary list of potential sites is developed satisfying the first two criteria. For this purpose, usually a circle indicating a "search radius" (the maximum distance the waste generator is willing to haul the waste) is drawn on a road map of the region, keeping the waste generator (a city or industry) at its center (Fig. 12.1). The search radius depends on economics of waste hauling. Hauling of waste is one of the high-cost items in landfill operation. So it is essential to keep the cost as low as possible. One should start with a small search radius and enlarge it if needed. A methodology for estimating the optimal service region for a landfill has been proposed by Wenger and Rhyner (1984). Once a landfill site is identified, this methodology may be used to identify the waste generation points that can use the landfill at a reasonable cost. Both the landfill operating cost and hauling cost are taken into consideration while developing the optimal service region using this approach. This approach is very helpful in identifying the communities that should participate in a municipal waste landfill located in a region. Search radius may have to be increased if potential sites cannot be located within the search area. If more than one waste generator is involved (e.g., several cities within a county) then a compromise location acceptable to the waste generators is used as the center. A discussion with the waste generator(s) regarding the search radius and center of the search area (especially if more than one generator is involved) must be undertaken before site selection begins. A landfill site may be owned by an individual, a company, or a public body (e.g., municipality). Since acceptability to the public is crucial to the landfill siting process, the citizens to be affected should be informed regarding the site selection process as early as possible. Locational criteria dictated by the regulatory body must also be studied. A close look at the

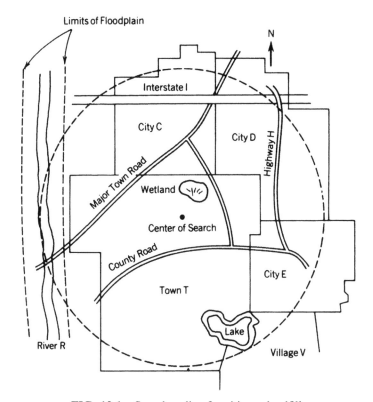

FIG. 12.1 Search radius for siting a landfill.

flexibility of these locational criteria is helpful. For instance, in the United States a landfill can never be sited within a critical habitat area; however, some compromise may be available for landfills located near a highway. The process of selecting a landfill site is complex and it involves three major issues: data collection, locational criteria, and preliminary assessment of public reactions. In the following sections each issue is discussed separately.

12.1 DATA COLLECTION

Several maps and other information need to be studied to collect data within the search radius. The following information is needed: topographic map(s), soil map(s), land-use plan(s), transportation plan(s) (discussed in Sections 12.1.1–12.1.8, and waste type and waste volume (discussed in Sections 12.1.9–12.1.15). Brief discussions on each of these items follows.

12.1.1 Topographic Maps

The topography of the area indicates low and high areas, natural surface water drainage pattern, streams, and wetlands. A topographic map will help find sites that are not on natural surface water drains or within a wetland. Debate exists regarding locating a landfill on a groundwater recharge or discharge zone, an issue to be settled with the local regulatory agency.

12.1.2 Soil Maps

These maps, primarily meant for agricultural use, will show the types of soil near the surface. Although these maps are partly useful for natural attenuation type of landfill siting, they have very little use for containment-type landfills.

12.1.3 Land-Use Plans

These plans are useful in delineating areas with definite zoning restrictions. There may be restrictions on the use of agricultural land or on the use of forest land for landfill purposes. These maps are used to delineate possible sites that are sufficiently away from localities and to satisfy zoning criteria.

12.1.4 Transportation Maps

These maps, which indicate roads and railways and locations of airports, are used to determine the transportation needs in developing a site. If clay needs to be hauled from a distant source for constructing a liner for a site, then these maps are used to estimate the hauling distance. Allowable axle loads on roads leading to a potential site must be studied to find out whether any road improvement will be necessary. Restriction may exist regarding locating a municipal waste landfill within a certain distance of an airport to minimize "bird hazard" for airplanes.

12.1.5 Water-Use Plan

These maps are usually not readily available. However, once potential areas are delineated, the water use in those areas must be investigated. A plan indicating the following items should be developed: private and public wells indicating the capacity of each well, major and minor drinking water supply line(s), water intake jetty located on surface water bodies, and open wells. A safe distance (365 m or more) should be maintained from all drinking water sources. A minimum distance between an existing well and a proposed landfill site may be specified by the regulatory agency.

12.1.6 Floodplain Maps

These maps are used to delineate areas that are within a 100-year floodplain. For hazardous waste landfills a 500-year floodplain may be used. Landfill siting must be avoided within the floodplains of major rivers. A landfill may be constructed near an intermittent stream if additional protection measures (e.g., levee) are implemented.

12.1.7 Geologic Maps

These maps will indicate geologic features and are very important for glaciated regions. A general idea about soil type can be developed from a glacial geologic map. They are also very helpful in identifying clay borrow sources. In nonglaciated regions they may be used to identify predominantly sandy or clayey areas.

12.1.8 Aerial Photographs

Aerial photographs may not exist for the entire search area. Once a list of potential sites is developed, aerial photographs or preferably a photogrammetric survey of each of the potential sites may prove to be extremely helpful. Surface features such as small lakes, intermittent stream beds, and current land use, which may not have been identified in earlier map searches, can be easily identified using aerial photographs.

12.1.9 Waste Type

The first thing to identify is whether the waste is hazardous or nonhazardous. Regulations are significantly different for these two types of waste. If the waste is nonhazardous, then the designer should know whether it is municipal or industrial waste. Municipal wastes are highly mixed types of wastes whereas industrial wastes are usually either monotypic or a mixture of two or three different waste streams with identifiable characteristics. Characterization of the waste (refer to Chapter 14 for details) should be performed to develop proper landfill design.

12.1.10 Waste Volume

The volume of industrial waste (hazardous or nonhazardous) can be easily estimated by studying the previous disposal records. For a new plant the estimate of waste volume should be made from the waste generation rate of similar types of industries elsewhere. Although municipal waste generation rates vary widely, 0.9–1.8 kg (2–4 lb) per person per day is a reasonable estimate; 650–815 kg/m^3 (40–50 lb/ft^3) is a reasonable range of bulk unit weight for municipal garbage. An estimation of population during active life of the landfill should be done; the estimated population in each year is then multiplied by the waste generation rate to obtain the waste volume in each year. For industrial waste the bulk unit weight, if not readily available, should be determined by laboratory testing.

12.1.11 Landfill Volume

Landfill volume is estimated by adding the daily, intermediate (if used) and final cover volume to the waste volume. Daily cover is mandatory for most municipal garbage landfills. If soil is used as daily cover, then a waste to daily cover ratio of $4:1$ to $5:1$ by volume is a reasonable estimate. The intermediate and final cover volume can be estimated from the thicknesses of these covers.

12.1.12 Availability of Landfill Equipment

Although this item is not involved in developing the landfill plan, these data should be gathered at the planning stage. Special hauling trucks are sometimes required for hauling sludges, especially if they have a high liquid content.

12.1.13 Recycling and Incineration Options

A study regarding possible recycling or incineration of all or part of the waste should be undertaken. Recycling of a waste, if technology is available, is sometimes mandated by regulatory authorities. Recycling and incineration, although technically feasible in some cases, may not be an economically acceptable option. Therefore both the technical and economic feasibility of recycling and incineration should be checked. Incineration will reduce the landfill volume (and may change the design if the ash is found to be hazardous) needed for the disposal of ash. From an environmental standpoint, every effort should be made to recycle waste as much as possible and then arrangements should be made to incinerate the rest if it is technically and economically feasible.

12.1.14 Existing Disposal Option

The available landfill volume within a reasonable haul distance should be studied. The cost of disposing of waste in existing landfill could be less than developing and operating a new landfill. There are certain hidden costs of operating a landfill (e.g., monitoring of groundwater wells, payment for bond purchase if required by regulation) that are sometimes overlooked. A list of landfills around the proposed site will also be helpful in emergency situations (e.g., the landfill could not be built by the target date due to litigation).

12.1.15 Funding

The cost of developing a landfill is quite high. Funds for the initial investigation, preparation of the report, landfill construction, and so on must be obtained for planning purposes. A proper estimate for each stage of the proposal preparation and the flow of necessary funds must be studied. Discussion regarding the availability of funds with the person(s) dealing with the budget is essential. Funding is a very critical issue for facilities owned publicly or by a giant corporation.

12.2 LOCATIONAL CRITERIA

Usually a landfill cannot be sited within a certain distance of the following: lakes, ponds, rivers, wetlands, floodplain, highway, critical habitat areas, water supply well, active fault areas, and airports. In addition, landfill siting is not allowed in areas in which a potential for contamination of groundwater or surface water bodies exists. Landfill siting should be avoided in unstable areas and seismic impact zones. No information is available as to how the distances mentioned in the following sections were arrived at, however, these distances are widely accepted. Usually permission from a regulatory agency is required if a proposed landfill site does not meet the locational criteria. If it is absolutely essential to site a landfill within the restricted zone(s), then permission from the regulatory agency should be sought at an early date. In the absence of any regulatory requirements regarding safe distances, the following distances may be used.

12.2.1 Lake or Pond

No landfill should be constructed within 300 m (1000 ft) of any navigable lake, pond, or flowage. This distance may be reduced for a containment-type landfill. However, because of concerns regarding runoff of waste contact water, a surface water monitoring program should be established if a landfill is sited less than 300 m from a lake, pond, or flowage.

12.2.2 River

No landfill should be constructed within 90 m (300 ft) of a navigable river or stream. The distance may be reduced in some instances for nonmeandering rivers, but a minimum of 30 m (100 ft) should be maintained in all cases.

12.2.3 Floodplain

No landfill should be constructed within a 100-year floodplain. Regulations may require a more restrictive floodplain (e.g., 500-year floodplain) siting criteria. A landfill may be built within the floodplains of secondary streams if an embankment is built along the stream side to avoid flooding of the area. However, landfills must not be built within the floodplains of major rivers.

12.2.4 Highway

No landfill should be constructed within 300 m (1000 ft) of the right of way of any state or federal highway. This restriction is mainly for aesthetic reasons. A landfill may be built within the restricted distance if trees or berms are used to screen the landfill site.

12.2.5 Public Parks

No landfill should be constructed within 300 m (1000 ft) of a public park. A landfill may be constructed within the restricted distance if some kind of screening is used. A high fence around the landfill and a secured gate should be constructed to restrict easy entry of unauthorized persons in the landfill.

12.2.6 Critical Habitat Area

No landfill should be constructed within critical habitat areas. A critical habitat area is defined as the area in which one or more endangered species live. It is sometimes difficult to define a critical habitat area. If there is any doubt, then the regulatory agency should be contacted. Siting a landfill within critical habitat areas is not suggested.

12.2.7 Wetlands

No landfill should be constructed within wetlands. It is often difficult to define a wetland area. Maps may be available for some wetlands, but in many cases such maps are absent or are incorrect. If there is any doubt, then the regulatory agency should be contacted. Disturbance of wetlands should be avoided. However, there are situations where impact to wetlands is unavoidable. Detailed discussion regarding compensatory wetland development is included in Chapter 24.

12.2.8 Unstable Areas

Landfills should not be constructed over unstable areas. An area can be unstable due to natural conditions or human activities. Naturally occurring unstable areas include regions with low bearing capacity soil, expansive soil, and karst terrain. Areas considered unstable because of human activities include areas where groundwater withdrawal is high, near a cut or fill, or areas where significant quantities of gas have been extracted. An area underlain by a thick soft clay layer may settle due to the overburden pressure exerted by waste. Significant total or differential settlement can damage the integrity of the liner and the leachate collection system. Areas with expansive soil such as smectite (a clay mineral of the montmorillonite group), vermiculite, and soil rich in sodium sulfate, calcium sulfate or (anhydrite), or iron sulfide (pyrite) may swell with increase in water content (USEPA, 1999). Such expansive soils may also damage the integrity of liners and leachate collection systems. Sinkholes, as big as 30 m (100 ft) deep and 91 m (300 ft) wide, may develop suddenly in karst terrain leading to catastrophic failure of a landfill. Removal of significant quantity of groundwater, oil, or natural gas can cause settlement of the soil. The additional overburden pressure of the waste disposed in a landfill may increase the settlement. A bearing capacity failure can occur if a landfill is sited next to a cut or fill.

Detail subsoil investigation is essential to identify a human-made or naturally occurring unstable area. Karst terrain can be identified by a geophysical technique such as electromagnetic conductivity or ground penetrating radar. Remote-sensing techniques such as aerial photography are also used to gather information regarding karst terrain.

12.2.9 Airports

No landfill should be constructed within 3048 m (10,000 ft) of any airport. This restriction is imposed to reduce bird hazard. Birds are attracted to landfills where food is available (in general, municipal landfills fall in this category). Permission from the proper authority (in most cases the airports are under the administration of a central agency) should be sought if the proposed landfill site is within the restricted zone.

12.2.10 Water Supply Well

No landfill should be constructed within 365 m (1200 ft) of any water supply well. It is strongly suggested that this locational restriction be abided by at least for down-gradient wells. Permission from the regulatory agency may be needed if a landfill is to be sited within the restricted area.

12.2.11 Active Fault Areas

Landfills should not be sited within 60 m (200 ft) of an active fault. Active fault areas are subject to earthquakes, which may cause landslides or soil liquefaction. The vibrating motions caused by an earthquake may turn a saturated granular material into a viscous fluid, a process known as liquefaction. Liquefaction reduces the bearing capacity of soil leading to the failure of the base liner and the berms surrounding a landfill.

Known faults are already mapped and available from the agency involved in earthquake monitoring (note: usually such agencies are associated with the geological survey in a country). For suspected areas, a fault characterization is done.

12.2.12 Seismic Impact Zone

A landfill should not be sited in a seismic impact zone; an area having a 10% or greater probability that the maximum horizontal acceleration caused by an earthquake at the site will exceed $0.1g$ in 250 years. In addition to liquefaction, earthquake-induced ground vibrations can also compact loose granular soils; such compaction could result in large uniform or differential settlement of the landfill base (USEPA, 1999).

Seismic impact zones are already mapped. For areas suspected of seismic activity, the agency responsible for maintaining records of seismic activities should be contacted.

12.3   PRELIMINARY ASSESSMENT OF PUBLIC REACTIONS

The public should be informed regarding the possibility of siting of a landfill in their area as soon as a list of potential sites is developed. The public is less suspicious and more open to discussion if they are informed by the owner rather than getting the news from other sources. Public education regarding the dangers and benefits of a landfill should be undertaken. A preliminary assessment of public opinion regarding all the sites in the list is essential.

A site may be technically and economically feasible yet may be opposed heavily by the public. The "not in my back yard" (NIMBY) sentiment is high initially, however, with proper discussion it can be overcome in some cases. Early assessment regarding how strong the NIMBY sentiment is can significantly reduce the time and money spent on obtaining the final permit for a landfill site. In many instances residents around a proposed site cooperate if the owner's representative listens to concerns of area residents and considers those concerns in designing and monitoring a site. Noise, dust, odor, increases in traffic volume, and reduction in property value concern the area residents more than the fear of groundwater contamination.

To identify the major concerns of the public affected by the siting of a landfill, the owner may conduct interviews with representatives of major community groups such as civic groups, religious organizations, and business associations. Residents living within a quarter mile (0.4 km) of the proposed site should also be interviewed to gauge the opinion. Table 12.1 includes a list of effective methods for notifying the public.

In many cases, existing landfills are expanded. To avoid public opposition to such expansions, it is essential to maintain good public relations both with the immediate neighbors and with the community in which the landfill is located. A community can be informed about the landfill by:

1. Conducting a tour of the landfill
2. Easily accessible documents summarizing the activities in the landfill
3. Short informational news articles published in local newspapers or magazines published by civic or environmental groups

12.4   DEVELOPMENT OF A LIST OF POTENTIAL SITES

After studying the information discussed in Sections 12.1–12.3, areas having potential for site development should be delineated. A road map may be used

TABLE 12.1   Effective Methods for Notifying the Public

Methods	Features	Advantages	Disadvantages
Briefings	Personal visit or phone call to key officials or group leaders to announce a decision, provide background information, or answer questions.	Provides background information. Determines reactions before an issue "goes public." Alerts key people to issues that may affect them.	Requires time.
Mailing of key technical reports or environmental documents	Mailing technical studies or environmental reports to other agencies, leaders of organized groups, or other interested parties.	Provides full and detailed information to people who are most interested. Often increases the credibility of studies because they are fully visible.	Costs money to print and mail. Some people might not read the reports.
News conference	Brief presentation to reporters, followed by a question-and-answer period, often accompanied by handouts of presenter's comments.	Stimulates media interest in a story. Direct quotes often appear in television and radio. Might draw attention to an announcement or generate interest in public meetings.	Reporters will only come if the announcement or presentation is newsworthy. Cannot control how the story is presented, although some direct quotes are likely.
Newsletters	Brief description of what is going on, usually issued at key intervals for all people who have shown interest.	Provides more information than can be presented through the media to those who are most interested. Often used to provide information prior to public meetings or key decision points. Helps to maintain visibility during extended technical studies.	Requires staff time. Costs money to prepare, print and mail. Stories must be objective and credible, or people will react to the newsletters as if they were propaganda.
Newspaper inserts	Much like a newsletter, but distributed as an insert in a newspaper.	Reaches the entire community with important information. Is one of the few mechanisms for reaching everyone in the community through which you can tell the story your way.	Requires staff time to prepare the insert, and distribution costs money. Must be prepared to newspaper's layout specification.
Paid advertisements	Advertising space purchased in newspapers or on the radio or television.	Effective for announcing meetings or key decisions or as background material for future media stories.	Advertising space can be costly. Radio and television may entail expensive production costs to prepare the ad.
News releases	A short announcement or news story issued to the media to get interest in media coverage of the story.	Might stimulate interest from the media. Useful for announcing meetings or major decisions or as background material for future media stories.	Might be ignored or not read. Cannot control how the information is used.

Source: USEPA, 1999.

255

to show the potential sites that are technically acceptable and satisfy the regulatory locational criteria. The sites should be ranked based on public reactions. This first list of potential sites should then be discussed with the owner(s) of the site. It is a good idea to develop stronger public relations in the areas that are identified as prime candidates for landfill development. Preliminary geotechnical investigation should be undertaken at each of the potential sites where public opposition is low. If possible three or more sites should be studied initially. After studying the geotechnical information a potential list of two or more sites should be developed. A report indicating geotechnical information, conceptual design, and discussion regarding locational criteria should be developed for each of the potential sites. These reports should then be discussed with the regulatory authority, the area residents of the potential sites, and the municipality in which the site is to be located.

12.5   FINAL SITE SELECTION

Based on the discussion of the reports mentioned in Section 12.4, one or two sites are selected finally. A detailed investigation for each of these sites needs to be undertaken to develop a feasibility report. In some states/countries, requirements of a feasibility report are clearly spelled out in administrative code(s) issued by the regulatory agency. These codes should be obtained and followed when writing a feasibility report. Two major items of this report with which engineers get involved are on-site geotechnical investigation and borrow source investigation. These two items are discussed below.

12.5.1   On-Site Geotechnical Investigation

The purpose of geotechnical investigation is primarily to obtain data to study the different soil stratum present at the site and to prepare a groundwater map for the site.

12.5.1.1 Subsoil Investigation   Continuous soil samples need to be collected to determine soil stratification. Mechanical properties (strength and consolidation characteristics) are not a prime concern in subsoil investigation for landfills, although consolidation characteristics of highly organic (e.g., peat) clay layer(s) and strength characteristics of suspected collapsible soil should be studied carefully. Permeability of the soil layers (both laboratory permeability of undisturbed soil samples and field permeability) should be studied carefully. For major projects both horizontal and vertical permeability should be investigated. Ratios of horizontal to vertical permeability of undisturbed soil samples between less than one to seven were reported for several clay types (Mitchell, 1956). Olsen (1962) predicted a ratio of up to 20 for kaolinite and 100 for other clay types for totally horizontal particle orientation. He used a tortuous flow path model for this analysis.

In addition to the items mentioned in the above paragraph, the following items are of interest in soil investigation for a landfill project: Atterberg limits of fine-grained soils, grain size distribution of soil samples, existence of fracture in the clayey layer, thickness of each stratum, depth to bedrock, identification of bedrock (this can be done by studying a geologic map of the area), natural moisture content, and degree of saturation of the clayey strata. The last two items should be studied to discover the probable location of the groundwater table (GWT) in the clayey stratum and the existence of perched GWT. This will also help in installing groundwater monitoring wells to define GWT. An experienced soil engineer or hydrogeologist should be present during the entire soil investigation program. A detailed borelog should be prepared, which should include comments and observations.

Differences of opinion may exist regarding the total number of borings required to define the soil stratigraphy and groundwater condition at a site. A clearer concept of the soil formation process (geology) at the site is helpful in developing a strategy. In the absence of specific requirements regarding the number of borings the following guidelines may be followed:

1. The borings should be distributed in such a way that it covers an area at least 25% larger than the proposed waste limits.
2. Five borings should be done for the first 2 hectares (ha) (5 acres) or less and two additional borings for each additional hectare. The boring should be well distributed over the entire area. Refer to Section 12.6.2 for suggested number of groundwater wells.
3. The borings should extend at least 7.5 m (25 ft) below the proposed base of the landfill.

As far as landfill design goes, soil borings help to identify the soil type(s) and bedrock depth and the depth and thickness of usable groundwater acquifers. In some sites the bedrock or the aquifer may be too deep. In those sites regional geologic data may be used to assess the bedrock and water table depth. One or two borings should extend at least 2 m (6 ft) into the aquifer/bedrock for verification purposes.

Knowledge of local geology and hydrogeology is essential for proper planning of a soil boring program. It must be borne in mind that the above-mentioned numbers of borings provide a guideline only. The exact number of boreholes and groundwater table wells may be far in excess of the minimum suggested number. The reverse may also be true in some cases, that is, far fewer borings may be sufficient to define the soil stratigraphy of a site completely. Usually there is a lump-sum cost for bringing the drill rig on site. This cost can be saved if all the necessary borings can be done when the rig is on site.

12.5.1.2 *Seismic Hazard Investigation*

In sites located within the seismic impact zone, additional investigations to find seismic design parameters should be undertaken. The two mechanisms that can cause damage to landfills are (1) strong ground motion and (2) displacement of the ground below or adjacent to the landfill base due to movement along a fault. The strong ground motion may cause liquefaction of subsoil leading to substantial settlement of the landfill base.

The magnitude of an earthquake is measured by the Richter scale, which was developed by Charles Richter in the early 1900s. In general, earthquakes of 5.0 and above on the Richter scale can cause significant damage to structures. It should be noted that humans cannot sense earthquakes below 2.0 on the Richter scale. The following parameters need to be assessed for use in seismic design: maximum horizontal acceleration (MHA) and maximum horizontal velocity (MHV). Although vertical motions are also observed during earthquakes leading to a momentary reduction in weight, such vertical motions are not important for bulky soil structures such as landfills and dams.

Design earthquake ground motions can be found from existing seismic hazard maps (note: these maps are available from appropriate government organizations) or by conducting site-specific seismic hazard analysis. Note, however, that published maps usually provide high values of MHA and MHV. On the other hand, site-specific seismic data collection and analysis are time consuming and costly. It is suggested that values obtained from existing maps be used for sites with low or moderate seismic activity. For sites located in high seismic activity regions, a site-specific seismic study should be undertaken.

Certain soil types (e.g., saturated loose sand) are prone to liquefaction. So the liquefaction potential of soils from various strata below the landfill base should be studied carefully during subsoil investigation. It is also recommended that investigations regarding the existence of faults below a proposed site should be undertaken for sites located within a seismic impact zone (note: see Section 19.6). Such investigations of the existence of faults are costly. A stepped approach may be used, which will help avoid unnecessary cost and time involved in the investigation. The suggested steps are (Weiler et al., 1993):

1. Review of the published seismic data.
2. Review of subsurface exploration data to determine if a fault exists.
3. A geologic reconnaissance survey of the area.
4. Review of the regional seismological and geological history.
5. Geophysical investigation utilizing one or more of the following: seismic refraction/reflection, gravimetric survey, and magnetic survey.
6. Angular boring.
7. Test trenching to search for evidence of recent faulting.

Usually the first four steps will provide needed information regarding the possibility of the existence of a fault. Steps 5–7 are undertaken only when strong evidence regarding the existence of a fault is found in the first four steps.

12.5.2   Borrow Source Investigation

A conceptual design of the site should be developed based on the subsoil investigation report and the waste type. This conceptual design should then be discussed with the regulatory agency to find out whether it agrees with the design. Material requirements for natural attenuation type of landfills are minimal whereas those for containment-type sites are significant. In a containment-type landfill, liner design is a major issue. The allowable type of material for both liner and cap will dictate borrow source investigation. If the liner material is clay and the drainage blanket is clean sand, then identification of borrow sources for these two materials will become critical to the construction of the landfill. If the recommendations given in Section 12.1 are followed, then initial data regarding the availability of these materials may already be known. The next step is to embark on a detailed investigation regarding the properties of these materials. The following sections discuss guidelines for investigation that are to be followed for each type of geologic materials used in landfill construction.

12.5.2.1   *Clay*   Clay may be used as primary or secondary (if a double liner is proposed) liner material. Test pits and borings are used to find the vertical and horizontal extent of the potential clay borrow source. The total number of test pits and borings required needs to be discussed with the regulatory agency. Five test pits and 5 borings (a total of 10) for the first 2 ha (5 acres) or less and 2 test pits or borings for each additional hectare are reasonable. The locations of the test pits and borings should be well distributed on a uniform grid pattern. Logs should identify the geologic origin, testing results, soil classification (using a system acceptable to the technical community of the region), and visual description of each major soil layer. The layer or layers of soil should be identified based on test pit logs and boring logs. It is a good idea to avoid the use of a clay layer for liner construction that is less than 1.5 m (5 ft) in thickness. It may become extremely difficult to procure clay from such a thin layer. Grain size distribution curves and Atterberg limits should be developed for at least two to three samples obtained from each potential clay layer(s).

Differences of opinion may exist regarding specifications for clay, which when compacted in the field will provide a low permeability (1×10^{-7} cm/sec or less) layer; in the absence of guidelines the following specifications may be used to identify a clay borrow source that can provide a low permeability liner: liquid limit between 20 and 30%; plasticity index between 10 and 20%; 0.074 mm or less fraction (P200 content: 50% or more; clay fraction (0.002 mm or less size): 25% or more. Grain size distribution significantly influences the permeability of soil. Thus, a soil that has a classical "inverted S-type" particle size distribution can be easily compacted to achieve low permeability whereas a soil with a very high clay content but with poor particle size distribution may not be easily compacted to achieve low permeability. Five representative samples for the first 4 ha (10 acres) or less and one additional sample for each additional 2 ha (5 acres) or less should be used to develop modified proctor curves. Each proctor curve should be developed based on a minimum of five points. Thus, there will be at least five modified proctor curves for the first 4 ha (10 acres) and one additional modified proctor curve for each additional 2 ha (5 acres) of clay borrow area. It is essential to study the relationship between compaction, moisture content during compaction, and recompacted permeability of the clay layer(s). For this purpose permeability of some of the modified proctor samples should be studied to find out the range of permeability at different moisture content and compaction. Literature indicates (Mitchell, 1976) that soils compacted wet of optimum moisture will develop lower permeability than the ones compacted low of optimum. Compaction at wet of optimum moisture provides a better kneading effect in the field. Refer to Sections 18.1.8 and 20.2.1 for additional information.

12.5.2.2 Sand   Sand is used as the drainage blanket and sometimes as a protective layer over a clay cap. The primary property needed for the drainage layer is high permeability (1×10^{-2} to 1×10^{-3} cm/sec). Usually clean sand containing a maximum of 5% below the 0.074 mm fraction (P200) is capable of providing high permeability. Coarse sand can be easily washed to satisfy the permeability criteria.

The number of test pits and borings used to identify a clay borrow source is also applicable for sand borrow source identification. Borrowing of sand from a sand layer less than 1.5 m (5 ft) thick should be avoided for the same reason described in the above section. As far as the test goes, only two types of tests need to be done on representative samples. Two or three samples from each test pit/boring should be collected and tested individually for grain size distribution and permeability at 80–90% relative density (*Note:* Permeability of clean sand does not depend heavily on compaction.)

12.5.2.3 Silty Soil   A 60- to 90-cm (2–3 ft) or more silty soil layer is preferred as the protective layer over the barrier (clay) layer in the landfill final cover to protect the barrier layer from freeze–thaw and desiccation effects (*Note:* Freeze–thaw cycles and desiccation cracks increase permeability of compacted clay.) The number of test pits and boring necessary to identify a silty soil borrow source can be the same as suggested for clay. No specification is generally assigned to the soil in that layer, but it is preferred that the soil be a silty loam.

12.5.2.4 Topsoil   Topsoil is used on the final cover of a landfill. Usually the area stripped for the construction of a landfill has topsoil that can be stockpiled for future use. However, if there is not enough topsoil at the site, additional borrow sources should be identified. Local agricultural experts/ horticulturists may be contacted for this purpose. Sometimes an additional nutritional requirement and pH adjustment are necessary before a soil can be used for planting. Although detailed nutritional testing is not necessary at the feasibility stage, it is essential that a topsoil source be identified if needed.

12.5.2.5 Leachate Collection Pipes and Synthetic Membrane for Liner
Supplier(s) for leachate collection pipes and synthetic membrane should also be identified if synthetic membrane is to be used for liner and/or final cover construction. A list of available products should be obtained for future use.

12.6 PREPARATION OF FEASIBILITY REPORT

The purpose of a feasibility report is to determine whether a particular site has the potential for use as a landfill for the disposal of a particular waste type. The submittal requirements for this type of report are clearly indicated by regulatory agencies in some state/country. A familiarity with the technical information requirements and other necessary legal steps (e.g., notifying the town board in which the landfill is proposed) for obtaining a permit is essential. Detailed discussion of the items is beyond the scope of this book. However, the following sections will summarize some of the major issues generally associated with a feasibility report, some of which have already been discussed in earlier sections of this chapter.

12.6.1 Geotechnical Information

A detailed description of the geology of the area including a bedrock map or depth, distribution of unconsolidated units, and surface topography must be included. The topography of the area should be mapped using 30- to 60-cm (1–2 ft) contour interval.

12.6.2 Hydrogeology

The depth to groundwater, direction of groundwater flow, vertical and horizontal hydraulic gradients, existence of any groundwater divide, and the depth to usable aquifer should be investigated. To obtain this information, groundwater wells must be installed. The total number of wells necessary to define the hydrogeology of a site must be resolved with the regulatory agency. In the absence of a specific requirement the following guidelines may be used: two water table wells/hectare (2.5 acre) for the first 2 ha (5 acres). At least at 25% of the groundwater table locations a second well should be installed to form a piezometer nest to find the vertical gradient. All water table wells and piezometers must be developed properly. The purpose of developing a well is to remove fine particles, drill cuttings, and so on, and clean the well screen to ensure a proper inflow of water. Water quality in a properly developed well remains stable. The well development process usually involves pumping two to three "well volumes" of water from the well. Because of this water removal, the water level remains unstable for a period of time; recovery time depends on the fineness of the medium in which the well is placed. Recovery time is maximum for clayey soil and minimum for sandy soil. It may be noted that the water level in a well fluctuates within the year (seasonal fluctuation) and also fluctuates on a long-term basis. Therefore the water level should be measured for at least 1 year to find the seasonally high water table.

The existing water quality should also be established prior to disposing of waste in a landfill. Therefore water samples from all the wells and piezometers should be collected and tested for a list of parameters appropriate for the waste type. It is recommended that at least four (preferably eight) water samples collected at least one month apart from each well be tested to form some idea about the background (existing) water quality at the site.

12.6.3   Environmental Impact

This may become an important issue for some landfills. The landfill may impact flora, fauna, groundwater, surface water, ambiant air quality, and so on. A detailed discussion of each is beyond the scope of this book. Thorough discussions should be held with the regulatory agency to find out the exact requirements in each case.

12.6.4   Conceptual Design

A conceptual design of the landfill should be presented in the report. Decisions must be made regarding subbase level, volume of landfill, material and thickness of liner, leachate collection system design, final contour, final cover design, routing of surface water, leachate treatment (on-site or off-site), and final use of the landfill. For major landfills some engineering analysis may need to be performed at this stage.

12.6.5   Contingency Plan

A discussion should be included in the report regarding actions to be taken in case the groundwater at the site is impacted. This may not be a serious issue for smaller landfills but for a large hazardous waste landfill a contingency remedial action plan is worth investigating at an early stage. The analysis may show that remedial action at the proposed site may be very expensive or technically not feasible. Such findings may require a more stringent liner and leachate collection system design or may lead to discarding the proposed site.

13

LEACHATE AND GAS GENERATION

The subjects of Chapters 13 and 14 are interrelated. Both chapters should be read to understand the importance of leachate characteristics in landfill design. The main aim of this chapter is to discuss leachate and gas generation as they relate to landfill design. The quality and quantity of leachate generated during the active life and after closure of a landfill are important in managing a landfill. In addition, the quality of leachate is an important issue for leachate treatment.

Leachate is generated as a result of the percolation of water or other liquid through any waste and the squeezing of the waste due to self-weight. Thus, leachate can be defined as a liquid that is produced when water or another liquid comes in contact with waste. Leachate is a contaminated liquid that contains a number of dissolved and suspended materials. Part of the precipitation (snow or rain) that falls on a landfill reacts (both physically and chemically) with the waste while percolating downward (Fig. 13.1). During this percolation process it dissolves some of the chemicals produced in the waste through chemical reaction. The percolating water may also dissolve the liquid that is squeezed out due to weight of the waste (e.g., squeezing out of pore liquid of papermill sludge). Many studies were conducted to determine the role of microbial activities in decomposing municipal waste and subsequent leachate formation (Rovers and Farquhar, 1973; Caffrey and Ham, 1974; CWPCB, 1961). In a municipal waste, methane, carbon dioxide, ammonia, and hydrogen sulfide gases are generated due to anaerobic decomposition of the waste. (Additional information regarding microbial activities in landfills is included in Chapter 17.) These gases may dissolve in water and react with the waste or dissolved constituents of the percolating water. For instance, carbon dioxide combines with water to form carbonic acid, which then dis-

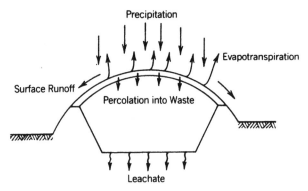

FIG. 13.1 How leachate is generated.

solves minerals from the waste [American Public Works Association (APWA), 1966]. Several other chemical reactions also take place releasing a wide range of chemicals, depending on the waste type. The percolating water plays a significant role in leachate generation. It should be noted that even if no water is allowed to percolate through the waste, a small volume of contaminated liquid is always expected to form due to biological and chemical reactions. The concentration of chemical compounds in such liquid is expected to be very high. The percolating water dilutes the contaminants in addition to aiding its formation. The quantity of leachate increases due to the percolation of water, but at the same time the percolating water dilutes the concentration of contaminants. Both quality and quantity of leachate are important issues for landfill design.

In general, to minimize leachate generation, the present concept is to construct final cover on a landfill as soon as the waste reaches the designed final grade. However, increasingly it has been suggested that the final cover construction be delayed. The advantage of prompt construction of final cover is a significant reduction in leachate quantity within a short period of time after the construction of final cover. The disadvantage of prompt construction of final cover is that the leachate that will need treatment is likely to be produced for several years after closure. Although the current thought is that leachate that needs treatment will not be generated 40 years after landfill closure, there are no definite data to support the theory. In a recent study on an old landfill, sufficient amount of biodegradable matter was identified 40 years after disposal (Suflita et al., 1992). This uncertainty regarding the length of time for which leachate treatment will be needed has economic and legal ramifications. The advantage of not constructing a final cover promptly after reaching final grade is that the contaminants from the waste are expected to flush out sooner. Although data from actual landfills are not available, results from three large test cells (25,000–35,000 tons of municipal waste) reported by Lechner et al. (1993) show that biodegradable matter reduced significantly within 21 months after disposal. The disadvantage of not constructing a final cover promptly is that the gas produced from a landfill cannot be collected efficiently; odor from such an "open landfill" is also expected to pose problems. In many countries in Europe only treated waste (i.e., after incineration or some form of biodegradation) is allowed to be land disposed. Chapter 17 includes discussion on bioreactor landfills, which utilize enhanced aerobic and/or anaerobic biodegradation.

13.1 FACTORS THAT INFLUENCE LEACHATE QUALITY

Various factors influence leachate quality. In general, leachate quality of the same waste type may be different in landfills located in different climatic regions; landfill operational practices also influence leachate quality. The following sections discuss the basic reasons why such variations are observed.

13.1.1 Refuse Composition

Variation in refuse composition is probably at a maximum in municipal waste and at a minimum in industrial waste. Because of this variation in refuse composition, the quality of municipal leachate varies widely (Garland and Mosher, 1975; Lu et al., 1985). In general, quality variation is higher for putrescible wastes than for nonputrescible waste.

13.1.2 Elapsed Time

Leachate quality varies with time. In general the overall quality of leachate generated in year 1 will be less strong than that generated in subsequent years. Leachate quality reaches a peak value after a few years and then gradually declines. Figure 13.2 shows an idealized relationship of leachate quality with time (Ham, 1980; Ham and Anderson, 1974; Pohland, 1975). For an actual landfill leachate the quality variation is not as smooth, although distinctive zones of upward and downward trends can be observed if quality variation is plotted with time. All the contaminants do not peak at the same time, and the

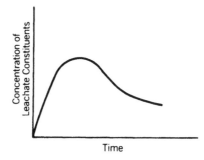

FIG. 13.2 Idealized variation of leachate quality with time.

time versus concentration variation plots of all contaminants from the same landfill may not be similar in shape.

13.1.3   Ambient Temperature

The atmospheric temperature at the landfill site influences leachate quality. The temperature affects both bacterial growth and chemical reactions. Subzero temperatures freeze some waste mass, which reduces the leachable waste mass and may cause inhibition of some chemical reactions. There are no reports of freezing an entire landfill during winter in cold regions, although chunks of frozen mass may be found in landfill. There are several studies on landfill leachate temperature (Chian and Dewalle, 1977; Fungaroli and Steiner, 1979; Wigh, 1979; Leckie et al., 1979), but no study is reported relating ambient temperature with leachate quality.

13.1.4   Available Moisture

Water plays a significant role in biodegradation and subsequent leaching of chemicals out of a waste. Leachate quality from waste disposed in a wet climate is expected to be different from the leachate quality of the same waste disposed in a dry climate.

13.1.5   Available Oxygen

The effect of available oxygen is notable for putrescible waste. Chemicals released due to aerobic decomposition are significantly different from those released due to anaerobic decomposition. The anaerobic condition in a landfill develops due to frequent covering of waste with soil (daily/weekly cover) or with fresh waste. The supply of oxygen starts to become depleted as soon as the waste is covered (either with soil or with more waste). A predominantly anaerobic condition develops in thicker refuse beds.

13.2   FACTORS THAT INFLUENCE LEACHATE QUANTITY

Like quality, leachate quantity is also dependent on weather and operational practices. The following sections discuss the basic reasons why such variations are observed.

13.2.1   Precipitation

The amount of rain and snow falling on a landfill influences leachate quantity significantly. Precipitation depends on geographical location.

13.2.2 Groundwater Intrusion

Sometimes landfill base is constructed below the groundwater table. In these landfills groundwater intrusion may increase leachate quantity.

13.2.3 Moisture Content of Waste

Leachate quantity will increase if, because of is own weight, the waste releases pore water when squeezed (e.g., sludges). Unsaturated waste continues to absorb water until it reaches field capacity (a water saturation state). So dry waste will reduce leachate formation. However, it must be noted that in actuality, channeling causes water to flow through the waste without being absorbed by the waste; thus, the water absorption is much less than that predicted by laboratory or small-scale field studies. Co-disposal of sludge or liquid waste in municipal landfills will increase the leachate quantity in a landfill.

13.2.4 Final Cover Design

Leachate volume is reduced significantly after a landfill is closed and finally covered because of two reasons: vegetation grown in the topsoil of a final cover reduces infiltratable moisture significantly by evapotranspiration and the low permeability layer reduces percolation. A properly designed final cover will reduce postclosure leachate quantity significantly.

13.3 ASSESSING PROBABLE QUALITY OF LEACHATE

Assessing leachate quality is difficult and is more so for putrescible waste. Both laboratory and field-scale studies have been done mostly on municipal waste. It should be mentioned that leachate quality predicted by laboratory tests may vary widely from the actual leachate obtained from a matured landfill. Before expanding any further on the test details, it is essential to understand why assessment of leachate quality is important for landfill design purposes. There are basically four reasons for assessing leachate quality at an early stage: (1) to identify whether the waste is hazardous, (2) to choose a landfill design, (3) to design or gain access to a suitable leachate treatment plant, and (4) to develop a list of chemicals for the groundwater monitoring program. To assess the leachate quality of a waste, the normal practice is to perform laboratory leachate tests wherever possible and to compare the data with the quality of actual landfill leachate, if available. Difficulty arises when field data are not available for a particular waste type. In such cases it is better to take a conservative approach while performing/designing laboratory leachate tests on the waste.

Three approaches are available for assessing leachate quality: (1) laboratory test, (2) field study using lysimeters, and (3) predictive modeling. The laboratory tests are not applicable to biodegradable wastes in which bacteria play an important role.

13.3.1   Laboratory Tests

Several laboratory procedures are available: the water leach test, the standard leachate tests, toxicity characteristic leaching procedure (TCLP), synthetic precipitation leachate procedure (SPLP), and multiple extraction procedure (MEP).

The leaching test to be performed depends on chemical, biological, and physical characteristics of the waste and the environment in which the waste is placed. The physical state of the waste should be considered while choosing an analytical procedure. Leachate characteristics of an industrial waste, co-disposed with another industrial or municipal waste, may be influenced by the other waste. An appropriate leachate test should be chosen after consulting the regulatory agency.

13.3.1.1   Water Leach Test   ASTM water leach test (D 3987-85) is used to predict leachate quality under laboratory conditions. The test data may not provide a representative quality of a "field leachate." In this method usually 70 g of waste is mixed with 1400 ml (or a ratio of 1 : 20) of water (meeting the specification of D1193 of ASTM) in a 2-liter watertight container (note: a container with a venting mechanism should be used for samples in which gases may be released; it should be noted that such venting may affect the concentration of the volatile components in the extract) and agitated for approximately 18 hr at 18–27°C. The agitation is done using a motor with an axial rotation of 29 rpm (Fig. 13.3). The extract is analyzed for specific constituents using an available standard method. The solid liquid ratio and the rpm of the motor may significantly influence the leachate quality. The method is not suitable for use with organic, monolithic, and solidified wastes.

13.3.1.2   Standard Leach Test   This test (Ham et al., 1979) consists of two alternative mixing series. Procedure R is intended to indicate the maximum quantity of contaminants likely to release. Procedure C is intended to indicate the probable maximum concentration of contaminants in the leachate. In both tests either distilled water or a "synthetic leachate" can be used as a leaching medium. The distilled water medium is intended to model a monofill scenario, and the synthetic leachate medium is intended to model a co-disposal (with municipal waste) scenario. The synthetic leachate consists of a complex mixture of organic and inorganic chemicals to model a municipal solid waste leachate. This leachate must be used in an anaerobic test environment. The leaching is similar to the ASTM procedure. However, a different agitation method and solid-to-liquid ratio are used. In method C new waste is added

FIG. 13.3 Shaking apparatus. (Courtesy of RMT Inc., Madison, WI.)

during the test to maximize contaminant levels in the extract. Currently this test is not used widely.

13.3.1.3 TCLP Test The toxicity characteristic leaching procedure (TCLP) was developed by USEPA. Table 13.1 lists the 40 chemicals for which the TLCP test can be used and their toxic concentration level. The TCPL test uses two leaching media: pH 5 acetate buffer and 0.5 *N* acetic acid. The choice of leaching medium depends on whether the waste is classified as highly alkaline. The 0.5 *N* acetic acid is to be used for highly alkaline waste and the pH 5 acetate buffer is recommended for all other waste types. A comparison of the extraction procedure (EP) toxicity test (a test procedure used by USEPA that was subsequently discontinued by the agency) with the TCLP test is included in Table 13.2 (Duranceau, 1987). The toxic concentration levels for EP toxicity tests were chosen arbitrarily, whereas the toxic concentration levels for the TCLP test are based on toxicological health data and groundwater models. A device known as a zero-headspace extractor (ZHE) is used to capture volatile compounds released during the test (Fig. 13.4).

13.3.1.4 Synthetic Precipitation Leachate Procedure The extraction fluid used in the method is dictated by the site location and whether leachability of cyanide and volatiles needs to be determined. The following three types of fluids are used in this method: (1) Fluid 1: a 60/40 wt% mixture of sulfuric

TABLE 13.1   Toxicity Characteristic Contaminants and Levels

Contaminant	Regulatory Level (mg/liter)
Arsenic	5.0
Barium	100.0
Benzene	0.5
Cadmium	1.0
Carbon tetrachloride	0.5
Chlordane	0.03
Chlorobenzene	100.0
Chloroform	6.0
Chromium	5.0
m-Cresol	200.0
o-Cresol	200.0
p-Cresol	200.0
Cresol	200.0
2,4-D	10.0
1,4-Dichlorobenzene	7.5
1,2-Dichloroethane	0.5
1,1-Dichloroethylene	0.7
2,4-Dinitrotoluene	0.13
Endrin	0.02
Heptachlor (and its epoxide)	0.008
Hexachlorobenzene	0.13
Hexachlorobutadiene	0.5
Hexachloroethane	3.0
Lead	5.0
Lindane	0.4
Mercury	0.2
Methoxychlor	10.0
Methyl ethyl ketone	200.0
Nitrobenzene	2.0
Pentachlorophenol	100.0
Pyridine	5.0
Selenium	1.0
Silver	5.0
Tetrachloroethylene	0.7
Toxaphene	0.5
Trichloroethylene	0.5
2,4,5-Trichlorophenol	400.0
2,4,6-Trichlorophenol	2.0
2,4,5-TP (Silvex)	1.0
Vinyl chloride	0.2

TABLE 13.2 Comparison of EP Toxicity and TCLP Leaching Tests

Topic	EP	TCLP	Comments
1. Extraction device	Rotary mixer or tumbler	Tumbler at 30 ± 2 rpm	—
2. Structural integrity	Test with falling weight for monolithic waste	*All* samples, to be crushed, ground, or broken to pass through ³/₈-in. sieve	Could change leachability of waste (surface area to liquid ratio)
3. Filtering	Use of 0.45-μm filter	Use of 0.6- to 0.8-μm filter	More particulates will pass and may change results from EP test
4. Leaching period	24–28 hr	18 hr	—
5. Leaching method	Manual pH adjustment	All leachant added at start of test	—
6. Leachant media	0.5 *N* acetic acid	Two media 0.5 *N* acetic acid Acetate buffer at pH 5	Acetic acid for highly alkaline waste only Acetate buffer may change leachability of some constituents
7. Temperature	Room temperature	22 ± 3°C	Need to control temperature
8. Highly alkaline waste	Procedure same for all wastes	Screening procedure to be used to identify highly alkaline wastes and leachant to be used	Adds time to test
9. Quality control requirements	All analyses by standard additions	Standard additions only required where accuracy is <50% or >150% or when test results are within 20% of regulatory level	More reasonable approach but no time saver
10. Leaching bottles	Plastic or glass	Teflon, glass, or ZHE	—
11. ZHE (zero-headspace extractor)	Not used	Used for volatiles	Separate device and leaching procedure
12. Laboratory analysis of leachates	Smaller number of chemicals	Many more chemicals to test	Analytical problems anticipated due to wide range of matrices

After Duranceau (1987); courtesy of Madison waste conf. committee.

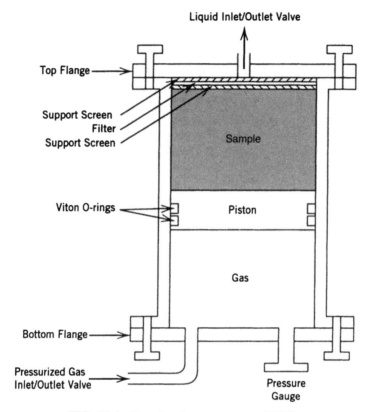

FIG. 13.4   Zero-headspace extractor (ZHE).

and nitric acids is mixed with reagent water (ASTM type 2 or equivalent) until the pH is 4.2 ± 0.5. The fluid is used to determine the leachability of waste disposed in a site east of the Mississippi River in the United States. (2) Fluid 2: the composition of the fluid is the same as fluid 1 except that the pH is adjusted to 5.00 ± 0.05. The fluid is used to determine leachability of waste in a site west of the Mississippi River in the United States. (3) Fluid 3: this is reagent water (ASTM type II or equivalent) that is used to determine cyanide and volatiles leachability from a waste. Figure 13.5 shows a flowchart used for the test. Details about the test can be found elsewhere (USEPA, 1986).

13.3.1.5   Multiple Extraction Procedure (MEP)   The MEP is intended to simulate 1000 years of freeze and thaw cycles and prolonged exposure to a leaching medium. The MEP can be used to evaluate liquid, solid, and multiphase samples. A liquid-to-solid ratio of 16 : 1 by weight is used for an extraction period of 24 hr. The entire extraction procedure is repeated until the concentration of target chemicals decreases. Since the procedure removes

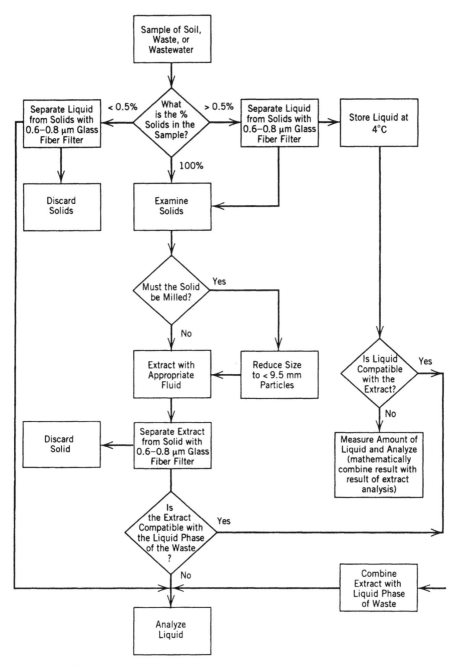

FIG. 13.5 Synthetic precipitation leachate procedure flowchart.

excess alkalinity, the behavior of metal contaminants due to decreasing pH can be evaluated; a decrease in pH increases solubility of most metals. Details of this test (number 1320) can be found elsewhere (USEPA, 1986).

13.3.2 Leachate Study Using Lysimeters

Rectangular or cylindrical lysimeters are used to study refuse leaching. Lysimeter studies can be conducted indoors or outdoors. Several researchers used lysimeters to study leaching of municipal refuse (Quasim and Burchinal, 1970b; Rovers and Farquhar, 1973; Pohland, 1975; EMCON Associates, 1975; Wigh, 1979; Walsh and Kinman, 1979; Fungaroli and Steiner, 1979; Ham, 1980). Both leachate quality and quantity can be monitored by using lysimeter study. Large lysimeters (15–60 m long × 6–9 m wide × 2.5–3 m deep) constructed outdoors and exposed to natural climatic conditions will provide valuable data if studied for a long period of time (5–6 years minimum). Refuse is not stabilized in a short period of time and hence data from a short-term lysimeter study, even if it is constructed outdoors, will not provide reliable results.

13.3.3 Predictive Modeling

Two approaches to model leachate composition have been reported in the literature. In one approach an attempt is made to quantify the physical, chemical, and biological processes. In the other approach an attempt is made to develop empirical equations to predict leachate concentration with time. Quasim and Burchinal (1970a) attempted physical process modeling to predict the concentration of leachate constituents. The experimental and theoretical concentrations of leachate constituents were fairly close. Straub and Lynch (1982a,b) proposed a model to predict both leachate quantity and concentration of both inorganic and organic constituents. It is essentially a one-dimensional kinetic model. Demetracopoulas et al. (1986) proposed a mathematical model that is solved numerically. It predicts a hydrograph-like contaminant concentration history at the bottom of the landfill that showed good qualitative agreement with the measured concentration history. Revah and Avnimeleik (1979) developed an empirical relationship to model the variations of concentration with time for the following leachate constituents: organic carbon, TKN, NH_3, NO_3, Fe, Mn, and volatile acids.

Leachate modeling has not developed enough to predict field leachate quality. However, the usefulness of leachate modeling must not be underestimated.

13.4 ESTIMATION OF LEACHATE QUANTITY

Leachate quantity depends heavily on precipitation, which is difficult to predict. The preclosure and postclosure leachate generation rates in a landfill vary significantly and the methods used to calculate them are also different.

An estimation of the preclosure leachate generation rate is needed to determine the spacing of the leachate collection pipe at the base of the landfill, the size of the leachate collection tank, and the design of an on site/off site plant for treating the leachate. Based on a field study of 13 municipal landfills in north West Germany Ehrig (1983) reported the following leachate generation rate, which depends on type of compactor used for compacting the waste: 15–25% of annual precipitation if steel wheel compactors are used; 25–50% of annual precipitation if crawler tractors are used. An estimation of the postclosure leachate generation is needed primarily to determine the long-term care cost (refer to Section 26.1.3 for details). The leachate generation rate is higher during the active life of the landfill and is reduced gradually after construction of the final cover. The following sections discuss how to estimate the preclosure and postclosure leachate quantity.

13.4.1 Preclosure Generation Rate

Leachate is generated primarily as a result of the precipitation and squeezing out of pore liquid in waste disposed in the landfill. Decomposition of putrescible waste mass can also release water/liquid. In a study conducted in a California landfill, the leachate generated due to decomposition from water was reported to be 0.5 in./ft of waste (CWPCB, 1961). For practical design purposes the volume of leachate generated due to decomposition from water is negligible. Surface run-on water may also cause an increase in leachate quantity (Lu et al., 1985); however, in a properly designed landfill surface water should not be allowed to run on into the waste. So this issue is also not addressed here. However, if for an existing landfill surface run-on water is unavoidable, then the volume of runon water must be estimated using principles of hydrology to calculate the volume of leachate. The preclosure leachate generation rate is guided by Eq. (13.1). However, in reality a model needs to be used to predict the preclosure leachate generation rate (refer to Section 16.5).

$$L_v = P + S - E - AW \qquad (13.1)$$

where L_v = preclosure leachate volume
$\quad\quad S$ = volume of pore squeeze liquid
$\quad\quad P$ = volume of precipitation
$\quad\quad E$ = volume lost through evaporation
$\quad AW$ = volume lost through absorption in waste

It is difficult to estimate S, E, and AW in a real landfill. Discussion of all the variables except precipitation follows.

13.4.1.1 Leachate Volume due to Pore Squeeze   When a layer of sludge is disposed in a landfill, the liquid within the pores of the sludge layer is released due to the self-weight of the sludge and the weight of the layers above it. The pore water is released essentially because of the consolidation of sludge. Both primary and secondary consolidation can take place. Although secondary consolidation may be high for putrescible waste, partly due to the creep of fibers and partly due to the microbial decomposition of organic matter present in the waste, the total volume of liquid drained due to pore squeeze is not expected to be high. Usually primary consolidation accounts for the majority of the pore squeeze liquid, which can be predicted reasonably well using laboratory values (Charlie and Wardwell, 1979). Charlie and Wardwell (1979) developed a mathematical relationship between the leachate generation rate and the primary consolidation properties of sludge.

Terzaghi's one-dimensional consolidation theory was used in developing the relationship. Although doubt was expressed regarding the use of Terzaghi's consolidation theory in predicting the settlement (Bagchi, 1980), studies conducted by Mar (1980) on municipal digested sludge supported Charlie and Wardwell's (1979) approach of predicting the leachate generation rate. Usually the following laboratory testing is used to predict leachate generation from sludge: The sludge is placed in a mold (usually a proctor's mold) and pressure is applied on the sludge that is equal to the anticipated maximum weight of the sludge in the field. The pressure is applied for several days and the settlement at the end of the period is recorded. It is assumed that the settlement is solely due to release of pore liquid. Based on this assumption the field leachate volume is estimated for the entire sludge volume to be disposed in the landfill. Solseng (1978) proposed a different approach for estimating pore squeeze liquid using standard consolidation data. Pore squeeze liquid from mechanically pressed sludges will be negligible because the pressure applied to the sludge mechanically is much higher than the pressure on the sludge after disposal. So the landfill designer needs to know whether the sludge is mechanically pressed prior to disposal. Estimation of leachate quantity in a co-disposal situation (e.g., disposal of sludge and municipal waste in the same landfill) is difficult. The absorbing capacity of the municipal waste will influence the volume of leachate generated.

13.4.1.2 Loss of Leachate due to Evaporation Precipitation moisture or the moisture already present in a landfill may evaporate under favorable conditions. Evaporation depends on factors such as ambient temperature, wind velocity, difference of vapor pressure between the evaporating surface and air, atmospheric pressure, and the specific gravity of the evaporating liquid. A 1% decrease in evaporation rate due to each 1% rise in specific gravity of the evaporating liquid has been reported (Keen et al., 1926; Fisher, 1927; Penman, 1948; Veihmeyer and Henderickson, 1955; Chow, 1964). Soil tends to bind water molecules by an attractive force that depends on the moisture content of the soil and its characteristics. The evaporation rate of unsaturated soils is almost constant over a range of moisture content of the soil. A shallow surface layer of soil (approximately 10 cm for clays and 20 cm for sand) will continue to evaporate until the layer reaches a permanent wilting point (the point at which the moisture content of the soil prevents the soil from supplying water at a sufficient rate essentially due to intermolecular surface tension). Evaporation from deeper soil is negligible (Chow, 1964). The water budget method, energy budget method, and mass transfer techniques have been used to predict evaporation from open water bodies (Viessman et al., 1977). As discussed above, evaporation opportunity depends on the availability of water. It is 100% from saturated soil but nearly zero from dry soil. Evaporation from open water bodies can be measured directly by pan evaporation. In the United States usually an unpainted galvanized iron pan 4 ft in diameter and 10 in. in height is used in determining pan evaporation. The pan is mounted 12 in. above the ground, on a wooden frame. The evaporation observed from the pan is multiplied by a factor of 0.67–0.81, known as the pan coefficient, to determine evaporation from large open water bodies such as lakes (Linsley and Franzini, 1972). The estimate must be based on long-term observations to avoid significant error. The average evaporation from an active landfill surface will be much lower than pan evaporation because of unsaturated conditions.

13.4.1.3   *Loss of Leachate due to Absorption in Waste* Waste may absorb some moisture before allowing it to percolate through. Theoretically, once the field capacity of the refuse is reached, all precipitation that falls on the waste will show up as leachate. The field capacity of the waste is defined as the maximum moisture content that waste can retain against gravitational forces without producing a downward flow of liquid. However, moisture absorption by waste is not uniform. A high heterogeneity of waste mass exists in a landfill; as a result channeling of precipitated water occurs in a landfill. The absorptive capacity of the waste depends on the composition of the waste. A detailed study of water absorption capacity of waste components was conducted by Stone (1974). The study indicated that field capacity of any refuse can be estimated with reasonable accuracy if the relative percentage of each waste component is known. The initial moisture content and field capacity of municipal solid waste as reported in several studies are summarized in Table 13.3 (Rovers and Farquhar, 1973; Walsh and Kinman, 1979, 1981; Wigh and Brunner, 1981; Fungaroli and Steiner, 1979). The data in Table 13.3 indicate that, on average, a field capacity of 33 cm/m (4 in./ft) is reasonable for municipal solid waste. From Table 13.3 the average initial moisture content of municipal solid waste can be assumed to be 12 cm/m (1.5 in./ft). Thus, on average, a municipal waste can absorb an additional 21 cm/m (2.5 in./ft) of moisture. However, in actual field situations absorption of moisture to the full field capacity is reduced due to channeling. Sludges are mostly saturated, hence reduction in leachate volume due to absorption may be neglected for sludges. The field capacity would be very low for a sandy nonputrescible waste (e.g., foundry sand). Thus, loss of moisture due to absorption depends

TABLE 13.3   Summary of Field Capacity of Waste

S1 Number	Data Source	Wet Density (lb/yd^3)	Dry Density (lb/yd^3)	Initial Moisture Content (in./ft)	Field Capacity (in./ft)
1	Rovers and Farquhar (1973)	530	Not available	1.92	3.62
2	Walsh and Kinman (1979)	808	526	2.0	3.82
3	Walsh and Kinman (1981)	798	520	1.98	4.85
4	Wigh (1979)	658	510	1.0	4.4
5	Fungaroli (1979)	563	476	0.62	4.1

on waste type—a point that should be borne in mind while estimating pre-closure leachate generation rate. Sequencing of waste placement to allow maximum moisture absorption can reduce leachate quantity.

13.4.1.4 Computer Model   Two computer-based models are available for predicting preclosure leachate generation rate (Schroeder et al., 1984; Bagchi and Ganguly, 1990). A study reported by Mbela et al. (1991) for four Wisconsin landfills indicates that error range in predicting preclosure leachate generation rate using the Help model (Schroeder et al., 1984) is between 84.1 and 196.7%. The error range using a model reported by Bagchi and Ganguly (1990) for two of the four landfills studied by Mbela et al. (1991) is between −65.2 and −7.7%. Reasonable accuracy in prediction of preclosure leachate generation on a daily basis is needed in the following two situations: (1) if the leachate is to be treated in relatively small municipal or industrial waste water treatment plants and (2) if a pretreatment or on-site treatment plant is needed for treatment of the leachate.

13.4.2 Postclosure Generation Rate

After the construction of the final cover, only the water that can infiltrate through the final cover percolates through the waste and generates leachate. Five approaches are available to predict the long-term leachate generation rate: the water balance method, computer modeling in conjunction with water balance method, empirical equation, mathematical modeling, and direct infiltration measurements. Descriptions of each of these methods and summary comments are included in the following sections.

13.4.2.1 Water Balance Method   Up to the early 1980s the water balance method was used to predict the long-term leachate generation rate. In simple terms the water balance equation can be written as

$$L'_v = P - \mathrm{ET} - R - \Delta S \tag{13.2}$$

where L'_v = postclosure leachate volume
$\quad\;\; P$ = volume of precipitation
$\quad$ ET = volume lost through evapotranspiration
$\quad\;\; R$ = volume of surface runoff
$\quad\; \Delta S$ = volume of soil and waste moisture storage

When precipitation falls on a covered landfill, part of it runs off the surface (R) and part of it is used up by vegetation (ET). The remaining part infiltrates the cover (Fig. 13.6), but part of it is held up by soil and waste (ΔS). The water balance method is applicable only for landfills in which a relatively high permeable layer of soil is used as final cover. A significantly lesser amount of water will infiltrate into a landfill if it is covered with a low permeability clay layer or synthetic membrane.

Evapotranspiration   Evapotranspiration is a term that combines evaporation and transpiration. Evaporation, discussed in detail in Section 13.4.1, is the loss of water that occurs from the soil surface. Transpiration on the other hand is the loss of water from the soil due to uptake by plants and its subsequent partial release to the atmosphere. Because of the difficulties in measuring the two items separately, they are measured as one item and termed evapotranspiration. Since the goal in a water budget is to "predict" the future leachate generation rate, potential evapotranspiration rather than actual evapotranspiration is of interest to the designer. Essentially two methods are available for predicting potential evapotranspiration.

USE OF AN EMPIRICAL RELATIONSHIP   The rate of transpiration is approximately equal to the pan evaporation rate from a free water surface reduced by the pan evaporation coefficient, provided plants continue to get sufficient water (Chow, 1964; Linsley and Franzini, 1972). Because the type of vegetation greatly influences evapotranspiration (Foth and Turk, 1943), the approach may over- or underestimate potential evapotranspiration.

EMPIRICAL/THEORETICAL APPROACHES   Several empirical/theoretical equations are available for estimating the potential evapotranspiration rate (Veihmeyer, 1964). A brief description of the equations used to predict monthly/daily evapotranspiration rates is given below.

Blaney–Morin Equation   This equation, proposed in 1942, essentially predicts evapotranspiration empirically using percentage daytime hours, mean monthly temperature, and mean monthly relative humidity. The equation takes into account the seasonal consumptive use of several irrigated crops.

Thornthwaite Equation   This equation, originally proposed in 1944, uses an exponential relationship between mean monthly temperature and mean monthly heat index. This method for predicting evapotranspiration was further developed by providing additional tables necessary for calculation (Thornthwaite and Mather, 1957). The relationship is based on studies conducted mostly in the central and eastern United States. The method is widely used to predict evapotranspiration from landfill cover. A detailed discussion on how to use the method for estimating leachate production in landfills is provided by Fenn et al. (1975).

Penman Equation   This is a theoretical equation based on absorption of radiation energy by ground surface. The values of variables used in the equation can be obtained from graphs and tables found elsewhere (Veihmeyer, 1964). Daily evapotranspiration can be calculated using this equation. This method is also widely used to predict evapotranspiration from landfill covers.

Blaney–Criddle Equation   This is a revised form of the Blaney–Morin equation that does not consider the annual mean relative humidity used in the Blaney–Morin equation.

As mentioned earlier, of all these equations Thornthwaite's and Penman's equations are most widely used. Thornthwaite's equation requires extensive use of tables that were developed mostly from observations in the central and eastern United States. However, because the studies were performed at different locations, the effect of latitude on evapotranspiration can be accounted for by using this approach. Thornthwaite and Mather (1955) recognized that evapotranspiration is dependent on root zone and vegetation type and may vary 400-fold from one location to another. They therefore issued a caveat regarding the use of the tables for precise estimates. Since the tables were developed mostly from observation stations in the United States, its worldwide application may not be useful, a fact recognized by the authors of the method.

Surface Runoff   Approaches for estimating surface runoff are different for water and snow. Surface runoff for water is discussed under the headings Field Measurement and Empirical Relationship. For snow, the infiltration rather than runoff from snow melt is estimated, which is discussed under the heading Snow Melt.

FIELD MEASUREMENT   For field measurement of surface runoff a test plot needs to be fenced to collect the runoff from the enclosed area. A precipitation gauge must be located next to the fenced area to measure precipitation at definite intervals of time (but not more than an hour apart). Several areas, with different slopes but each with the same type of topsoil and vegetation as proposed for the landfill, must be studied. The need for this type of study cannot be justified because runoff from different soils and slopes can be predicted fairly accurately using empirical relationships described in the next section. However, if the designer is certain that at the location where the landfill is to be sited use of empirical equations will not provide reasonable runoff estimates, then use of the field measurement technique to estimate surface runoff is justified.

EMPIRICAL RELATIONSHIP   These relationships were essentially developed from extensive field measurements. Several methods are available for surface runoff measurements (Chow, 1964; Varshney, 1979). However, only the two methods widely used in the United States are discussed. Foot-pound system (F.P.S.) units are used for both methods.

Rational Method   The following equation is used to calculate peak surface runoff (R) in ft^3/sec:

$$R = CIA_s \qquad (13.3)$$

where I = uniform precipitation rate in inches
A_s = area of the landfill surface in acres
C = runoff coefficient

The surface runoff can be predicted fairly accurately if a proper value of C is chosen. Different sets of values for C for different surface conditions are available [American Society of Civil Engineers (ASCE), 1960; Chow, 1964; Perry, 1976; Salvato et al., 1971]. Of these sets of values the ones by Salvato et al. (1971) were part of a landfill study. The rational method cannot account for the relationship between duration of precipitation and runoff, antecedent soil moisture content, frequency of precipitation, and permeability of the cover material.

Example 13.1 (in F.P.S. Units)   Calculate the surface runoff for a 10.5-acre landfill. Based on precipitation data, the 10-year 24-hr storm intensity is found to be 2.7 in./hr; the landfill has a cover that consists of the following layers: 1 ft of sand over the waste, 2 ft of recompacted clay, 2.5 ft of silty sand, and 6 in. of topsoil. The landfill has good vegetative cover and the top slope varies between 2 and 5%.
    The surface runoff is over a sandy loam with grass cover; the surface slope is 2–5%. From Table 13.4 the value of C is between 0.15 (sandy soil with a 2–7% slope) and 0.22 (heavy soil with a 2–7% slope). Assume an average value of 0.18. [*Note:* For this case, values of C obtained from other sources (Chow, 1964; Perry, 1976; Salvato et al., 1971) vary between 0.3 and 0.45.] It is a good idea to minimize surface runoff while predicting leachate volume and to maximize surface runoff when designing storm water drainage systems.
    For the example landfill, C = 0.18, I = 2.7 in./hr, A_s = 10.5 acres; R = $0.18 \times 2.7 \times 10.5 = 5.1$ ft^3/sec.

Curve Number Method   The curve number method proposed by the Soil Conservation Service of the United States is used to predict surface runoff

TABLE 13.4   Runoff Coefficients for Storms of 5- to 10-Year Frequency

S1 Number	Description of Area	Runoff Coefficients
1	Unimproved areas	0.10–0.30
2	Lawns: sandy soil	
	Flat, 2%	0.05–0.1
	Average, 2–7%	0.1–0.15
	Steep, 7%	0.15–0.2
3	Lawns: heavy soil	
	Flat, 2%	0.13–0.17
	Average, 2–7%	0.18–0.22
	Steep, 7%	0.25–0.35

After American Society of Civil Engineers (ASCE) (1960).

from agricultural land [Soil Conservation Service (SCS), 1975]. In addition to rainfall volume, soil type, and land cover, the method accounts for land use and antecedent moisture conditions. The antecedent moisture condition is first divided into three groups based on season (dormant and growing) and a 5-day total antecedent rainfall in inches. Soil is grouped into four different types based on ability to cause runoff (e.g., clayey soil has high runoff potential and sand or gravel has low runoff potential; all other soil types are classified between these two extremes). The land use and land cover are then determined. The weighted curve number is then established using tables. The direct runoff can then be estimated for different rainfall using Eq. (13.4):

$$R_i = \frac{\{W_p - 0.2[(1000/CN) - 10]\}^2}{W_p + 0.8[(1000/CN) - 10]} \tag{13.4}$$

where R_i = surface runoff in inches, W_p = rainfall in inches, and CN = curve number.

Example 13.2   A landfill has a surface area of 4.3 acres and has a 2-ft sandy silt final cover with 6 in. of topsoil. The landfill has poor pasture cover. Assume that the 10-yr 24-hr storm intensity for the area is 2.65 in./hr, which occurred in a dormant season. The total 5-day antecedent rainfall was 0.45 in. Estimate surface runoff using the curve number method.

1. From Table 13.5 the antecedent moisture condition = AMC I.
2. From Table 13.6 the hydrologic soil group is *C*.
3. From Table 13.7 CN = 86 for AMC II.
4. From Table 13.8 obtain CN for AMC II to CN for AMC I = 72.
5. From Eq. (13.4) for CN = 72 and W_p = 2.65 in. the direct runoff = 0.8 in.

TABLE 13.5 Antecedent Moisture Class for 5-Day Rainfall

	5-Day Antecedent Rainfall (in.)		
S1 Number	Dormant Season	Growing Season	Moisture Condition Class
1	<1.1	>2.1	III
2	0.5–1.1	1.4–2.1	II
3	<0.5	<1.4	I

After Soil Conservation Service (SCS) (1972).

TABLE 13.6 Soil Groups Relevant to Landfill Cover Design

S1 Number	Description of Soil	Soil Group
1	Soils having moderate infiltration rates when thoroughly wetted; moderately coarse to moderately fine textured soil	B
2	Soils having slow infiltration rates when thoroughly wetted; moderately fine to fine textured soil	C
3	Soils having very slow infiltration rates when thoroughly wetted; chiefly clayey soils with low permeability	D

After Soil Conservation Service (SCS) (1972).

TABLE 13.7 Runoff Curve Number for Different Soil Groups and Land-Use Conditions Relevant to Landfill Cover Design

			Hydrologic	Soil Groups		
S1 Number	Land Use	Agricultural Practice	Condition[a]	B	C	D
1	Fallow	Contoured	Poor	79	84	88
		Contoured	Good	75	82	86
		Contoured, terraced	Poor	74	80	82
		Contoured, terraced	Good	71	78	81
2	Pasture or range	Contoured	Poor	57	81	88
		Contoured	Fair	59	75	83
		Contoured	Good	35	70	79

After Soil Conservation Service (SCS) (1972).

[a] Poor hydrologic conditions mean heavily grazed, with no mulching a surface or less than 50% of the area covered with plants. Fair hydrologic conditions mean moderately grazed with plant cover or 50–75% of the area. Good hydrologic conditions mean lightly grazed with plant cover on more than 75% of the area.

TABLE 13.8 Runoff Curve Numbers (CN) Relevant to Landfill Cover Design

CN for Condition II	CN for Antecedent Moisture Condition	
	I	II
90	78	96
88	75	95
86	72	94
84	68	93
82	66	92
80	63	91
78	60	90
76	58	89
74	56	88
72	53	86
70	51	85
68	48	84
66	46	82
64	44	81
62	42	79
60	40	78
58	38	76
56	36	75
54	34	73
52	32	71
70	31	70
48	29	68
46	27	66
44	25	64
42	24	62
40	22	60
38	21	58
36	19	56
34	18	54

After Soil Conservation Service (SCS) (1972).

Snowmelt Infiltration In many areas leachate generated as a result of infiltration during snowmelt is significant. The majority of snowmelt usually occurs in early spring. Infiltration from snow depends on the condition of the ground (frozen or unfrozen), ambient temperature and its duration (snowmelt will depend on whether a temperature of 32°F and above prevails for 1 day or several days), radiation energy received (more snowmelt occurs on sunny days than on a cloudy day), rainfall during snow melting (rainfall accelerates the snow melting process), and so on. Because of the variables involved, it is difficult to predict snowmelt runoff or infiltration. Two methods are usually used: the degree-day method and the U.S. Army Corps of Engineers equation. Because it is simpler, only the degree-day method is discussed here. A detailed discussion of the U.S. Army method can be found elsewhere (Lu et al., 1985; Chow, 1964).

DEGREE-DAY METHOD   The following equation is used to estimate snowmelt infiltration (SCS, 1975):

$$SM = K(T - 32°F) \qquad (13.5)$$

where SM = potential daily snowmelt infiltration in inches of water
       K = constant that depends on the watershed condition
       T = ambient temperature above 32°F

$T - 32°$ is the number of degrees per day. The total snowmelt infiltration predicted must not exceed the total water equivalent of precipitated snow (note: 1 in. of water = 10 in. of snow).

Example 13.3 (in F.P.S. Units)   Estimate the snowmelt infiltration from an 18-in. snowpack during spring. The average daily air temperatures for the next 5 days were 33, 34, 29, 31, and 36°F.
    From Table 13.9, the value of $K = 0.02$. (*Note:* Assume low runoff to maximize leachate production.)

$$18 \text{ in. of snow} = 18 \times (\tfrac{1}{10}) = 1.8 \text{ in. of water}$$

$$\text{Total expected infiltration} = 0.2 \times [(33 - 32) + (34 - 32)$$
$$+ (36 - 32)]$$
$$= 0.14 \text{ in.}$$

(*Note:* All temperatures below 32°F are to be neglected because snow will not melt if the temperature is 32°F or below.)

TABLE 13.9 Values of K for the Degree-Day Equation Relevant to Landfill Cover Design

S1 Number	Watershed Condition	K
1	Average heavily forested area	
	North facing slopes	0.04–0.06
2	South facing slopes	0.06–0.08
3	High runoff potential	0.03

After Soil Conservation Service (SCS) (1972).

Soil Moisture Storage   Part of the infiltrating water will be stored by the soil. Only part of this stored water is available for use by vegetation. Soil moisture storage capacity is expressed as

$$\Delta S = \text{field capacity} - \text{wilting point} \tag{13.6}$$

Soil moisture storage depends on soil type, state of compaction, and thickness of the soil cover. Lutton et al. (1977) published data for soil moisture storage for different types of soil. The effect of state of compaction is not accounted for in this study.

Comments on Water Balance Method   The preceding paragraphs included discussion on how to estimate evapotranspiration, surface runoff and soil storage. Postclosure leachate volume production is calculated by subtracting evapotranspiration, surface runoff, and soil and waste moisture storage from the precipitation for each day. Use of a computer facilitates calculation. Test cell data and other field verification of the water balance method indicate that the margin of error is very high. Wigh (1979) and Wigh and Brunner (1981) found a difference of 43% between the leachate volume predicted using average climatic data and the actual leachate collection. In general, test cell data for the verification of the water balance method are in good agreement (Fungaroli and Steiner, 1979; Walsh and Kinman, 1981). SCS engineers (1976) collected leachate generation data from 5 existing sites located in 5 different geographic areas. Twenty-five different methods were used to predict leachate generation; different combinations of methods were used to estimate surface runoff, infiltration, and evapotranspiration. Thus, in all, the study reports on 125 individual cases (25 methods × 5 sites). The data indicate that in 54 cases (42.4%) the leachate generation was underestimated and in 71 cases (56.8%) the leachate generation was overestimated. The average error for the 25 methods varied between 83 and 1543%. So the study shows that use of the water balance method for predicting the leachate generation rate can be highly erroneous. The error range for predicting the leachate generation rate using the HELP model (Schroeder et al., 1984) is between −96 and +449% (Peyton and Schroeder, 1988). In the past the water balance method was used widely for estimating the leachate generation rate and more than 100 approaches are available for its calculations. A study is available that compares all the approaches and identifies a method for estimating the long-term leachate generation rate (Lu et al., 1985). Dass et al. (1977) and Perrier and Gibson (1980) conducted a parametric study to determine the sensitivity of the predicted leachate volume. These studies show that the sensitivity is moderate to high for all of the parameters involved in the water balance equation. Therefore, caution should be exercised in using the water balance method for predicting the long-term leachate generation rate.

The following are the drawbacks of using the water balance method alone for estimating postclosure leachate generation:

1. The method does not take into account the permeability of the barrier layer, which is an essential design feature for minimizing leachate generation.

2. In an attempt to maximize evapotranspiration, a designer may use a very thick vegetative layer without realizing that the root system of the chosen vegetation may not penetrate the entire thickness of the layer. In addition, one may note that approximately 80% of the moisture uptake from soil by a root system is performed by only the top few centimeters of roots of grass/shrub-type vegetation (Foth and Turk, 1943) commonly used in landfill cover. So the total length of root should not be used for estimating evapotranspiration. Thus increasing the vegetative layer beyond a certain depth will not increase evapotranspiration and thus will not reduce infiltration by increasing evapotranspiration.

13.4.2.2 *Computer Modeling in Conjunction with the Water Balance Method*

Several computer models available for predicting the postclosure leachate generation rate. The HELP model developed by Schroeder et al. (1984) uses the water balance equation but also considers permeability of the barrier layer to predict infiltration into the landfill. In the model, apportionment of infiltrating water after evapotranspiration, into vertical percolation through the barrier layer, and horizontal runoff through the barrier layer are estimated (Fig. 13.6). A simulation study by Peyton and Schroeder (1988) indicated that the HELP model can predict leachate generation fairly accurately. However, it is interesting to note that in the study the input of model variables was based on judgment, making the validity of the prediction questionable. In addition, the model has to share the inaccuracies associated with the water balance method. Miller and Mishra (1989) also raised doubts about several aspects of the verification process. Of most importance were the inability of the model to simulate infiltration under unsaturated conditions and the effect of macropores (due to desiccation cracks) on the infiltration rate. In their response Peyton and Schroeder (1989) indicated that the HELP model computes unsaturated hydraulic conductivity using an equation proposed by Campbell (1974) and that the most appropriate use of the model is for comparison of designs rather than prediction of quantities. It should be mentioned, however, that the HELP model is used widely in the United States.

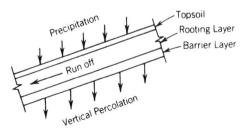

FIG. 13.6 Apportionment of precipitation through landfill cover.

In addition, three computer-based models are also available for predicting postclosure leachate generation rate (Bagchi and Ganguly, 1990; Demetracopoulas et al., 1984; Scharch et al., 1985). The models by Demetracopoulos et al. (1984) and Scharch et al. (1985) are modifications of Wong's model (Wong, 1977); in these models infiltration through the cover soil is an input that is estimated using a water balance method. Both models can predict the total postclosure leachate generation rate in a year but the error is high for an estimate of the daily leachate generation. The model reported by Bagchi and Ganguly (1990) needs field data regarding the postclosure leachate generation rate from a closed landfill located within the same climatic region for calibrating the model (refer to Section 16.6 for a detail discussion of these models).

13.4.2.3 Empirical Equation   Gee (1986) proposed an empirical relationship for predicting percolation through the top cover:

$$\text{Percolation} = \exp[K + \beta_1 \ln W_p + \beta_2 \ln(W_i/W_f) + \beta_3 \ln \gamma_d + \beta_4 \ln \alpha + \beta_5 C_c] \tag{13.7}$$

where W_p = rainfall in inches
  α = slope of the landfill surface (%)
  W_i = initial soil moisture content of the cover (%)
  W_f = field capacity of the soil (%)
  γ_d = soil density in lb/ft^3
  C_c = coefficient of curvature of the soil
β_1 to β_5 = constants

The accuracy of the equation was judged against field observations. From the comparison of the equation with Thornthwaite's water balance method, the HELP model, and field data, it was concluded that the equation can predict percolation through the landfill cover fairly accurately.

13.4.2.4 Mathematical Model   Korfiatis et al. (1984) proposed a mathematical model using the theory of unsaturated flow. The model was tested in the laboratory. Comparisons of predicted and measured leachate quantity showed reasonable agreement.

13.4.2.5 Direct Infiltration Measurements   Published data are available regarding the average long-term percolation rate in soil in the northern United States. A 16-year average yearly percolation rate (in Ohio) in a silt loam was observed to be 17% (range 13–27%) (cited in Foth and Turk, 1943). Percolation into solid waste test cells was reported by Ham (1980). The average yearly percolation was about 18% for a 7-year period. Of this 18%, approximately 80% took place during the spring. Thus, an average yearly percolation rate of 20% may be used for rough estimation of the long-term leachate generation rate.

13.4.2.6   Summary Comments on Long-Term Leachate Generation   Prediction of the postclosure leachate generation rate is a difficult task. So far as landfill management goes, the long-term rate is needed to establish a fund for long-term maintenance of the landfill (refer to Section 26.2 for more details). The full advantage of evapotranspiration is not available until a stable vegetative growth is established over the entire landfill, which takes at least 1 to 2 years. The leachate volume remains high at least for those 1 or 2 years after closure. So the money set aside for long-term leachate treatment will be insufficient if evapotranspiration is taken into consideration to predict the postclosure leachate generation rate in the first few years. Thus, prediction based on models that use evapotranspiration becomes more of a theoretical exercise than of any practical use, at least for the first few years after closure. The infiltration rate may be estimated roughly by assuming 20–30% of precipitation as infiltration from landfill cover to calculate the fund requirements for long-term care. The fund can be adjusted in subsequent years based on actual field data. This approach will provide a better safeguard against a low long-term care fund. As an alternative the author's model (refer to Section 15.6), which uses a field calibration technique, may be used for better prediction of postclosure leachate generation rate. However, if regulatory requirements dictate the use of a water balance method for predicting the postclosure leachate generation rate, then long-term climatic records (20–30 years) should be used for the estimation. The computer models mentioned in Section 13.4.2.2 are useful for parametric studies in determining a better final cover design option and for comparing landfill designs.

13.5   TYPICAL LEACHATE QUALITY OF VARIOUS NONHAZARDOUS WASTES

The ranges of concentrations of different parameters in leachate of various nonhazardous wastes are included in Tables 13.10–13.15. The aim is to provide some ideas about leachate quality for different waste types. As discussed earlier, leachate quality for any waste type is not unique. Hence the constituents and their concentrations are not recommended for automatic use for landfill design but are provided here essentially to form a database. The leachate quality for incinerator ash needs particular attention because the concentration of lead and cadmium may exceed the permissible level in some cases and therefore the ash may become hazardous (Sopcich and Bagchi, 1988; U.S. Congress, 1989).

TABLE 13.10   Typical Leachate Quality of Iron Foundry Waste (Composite Sample)

S1 Number	Parameter	Concentration (mg/liter except as indicated)
1	Arsenic	0.113
2	Barium	<0.2
3	Cadmium	<0.01
4	Chloride	139
5	Total chromium	<0.05
6	COD	150
7	Cyanide	<0.05
8	Fluoride	1.60
9	Iron	<0.03
10	Lead	<0.005
11	Manganese	<0.01
12	Mercury	<0.0002
13	Nickel	<0.04
14	Phenols	0.118
15	Selenium	0.030
16	TDS	1990
17	Sulfate	5.1
18	Zinc	<0.01
19	pH	12.3 units

13.6   LEACHATE TREATMENT

As discussed above, landfill leachate is highly contaminated liquid that cannot be discharged directly into surface water bodies. Studies have shown that discharge of raw municipal leachate into streams impacts aquatic life and causes degradation of water quality (Hansen, 1980; Nutall, 1973; Cameron and McDonald, 1982).

The following traditional techniques used for wastewater systems are also used for treating landfill leachate: biological treatment (aerobic and anaerobic biological stabilization) and physical/chemical treatment (precipitation, adsorption, coagulation, chemical oxidation, and reverse osmosis). As indicated earlier, both short-term and long-term variability of leachate characteristics are expected. The variation of leachate characteristics makes the design of a treatment system difficult. Leachate from municipal waste landfills can have quite high BOD concentration and significant concentrations of metals and trace organics. The concentrations of chemicals in leachates of other types of waste (both hazardous and nonhazardous) are usually significantly high. Leachate from each landfill is unique; however, some generalization can be made for leachate based on waste type and landfill location. Thus, a general approach of treatment for a particular waste type from all landfills located in a region appears to be reasonable. Although bench-scale studies may be

TABLE 13.11   Range of Concentration of Different Parameters in Leachate of Municipal Incinerator Ash

S1 Number	Parameters	Range of Concentration (mg/liter except as indicated)
1	Aluminum	2.3–88.8
2	Arsenic	0.005–0.218
3	Barium	0.055–2.48
4	Boron	0.42–3.2
5	Cadmium	<0.001–0.3
6	Calcium	21–3200
7	Chromium	<0.002–1.53
8	Cobalt	0.007–0.04
9	Copper	<0.005–24
10	Iron	<0.01–121
11	Lead	<0.0005–2.92
12	Magnesium	0.006–41
13	Manganese	0.103–22.4
14	Mercury	<0.00005–0.008
15	Molybdenum	<0.03
16	Nickel	<0.005–0.412
17	Potassium	3.66–4300
18	Selenium	0.0025–0.037
19	Silver	<0.001–0.07
20	Sodium	11.5–7300
21	Strontium	0.07–1.03
22	Tin	0.005–0.013
23	Zinc	0.002–0.32
24	Chloride	32.6–305
25	Fluoride	0.1–3.39
26	Hardness	49–742
27	Nitrate–nitrogen	0.011–0.59
28	Phosphate	0.16–0.43
29	Specific conductivity	253–1,874 μmho/cm
30	Sulfate	105–4900
31	Alkalinity	60.9–243
32	pH	8.47–9.94 units
33	Benzaldehyde	ND–0.008
34	Biphenyl	ND–0.051
35	Dimethyl propane diol	ND–0.120
36	Dioxins (ng/liter)	
	total	0.06–543
	2,3,7,8-TCDD	0.025–1.6
37	Ethyl hexyl phthalate	ND–0.08
38	Furans, total (mg/liter)	0.04–280
39	Hexa tiepane	ND–0.082
40	PCBs (ng/μl)	<1
41	Sulfonylbis sulfur	ND–0.011
42	Thiolane	ND–0.400

Based on Bagchi and Sopcich (1989) and U.S. Congress (1989).

TABLE 13.12 Range of Concentration of Different Parameters in Leachate of Papermill Sludge

S1 Number	Parameter	Range of Concentration (mg/liter except as indicated)
1	pH	5.4–9.0 units
2	TDS	289–9,810
3	TSS	80–320
4	Conductivity	70–14,370 μmho/cm
5	Alkalinity	174–5,500
6	Hardness	682–6,600
7	BOD	36–10,000
8	COD	4–43,000
9	Sulfate	0.9–550
10	Sodium	9–4,500
11	Calcium	5.5–2,400
12	Aluminum	0.008–18
13	Chloride	1–1,200
14	Iron	<0.1–950
15	Zinc	<0.018–0.03
16	Color	1,315–38,300 color units
17	Turbidity	NR[a] turb. units
18	Phenols	0.0011–4.5
19	Tannin-lig	13–90
20	Kjeldahl-nitrogen	34.5–385
21	Ammonia-nitrogen	<0.1
22	Nitrate	<0.1–15
23	Nitrite	<0.01–0.018
24	Sulfite	4–64
25	Sulfide	ND[a]
26	Phosphate	0.11–0.58
27	Total volatile solids	211–483
28	Total fixed solids	144–266
29	Barium	0.011–1.1
30	Bromide	ND[a]
31	Cadmium	0.006–0.02
32	Chromium	0–0.15
33	Cobalt	0.005–0.014
34	Copper	<0.01–0.21
35	Lead	0.037–0.1
36	Magnesium	3.8–6,000
37	Manganese	0.1–200
38	Mercury	<0.01–7 μg/liter
39	Nickel	<0.005–0.024
40	Potassium	140
41	Selenium	75
42	Tin	<0.1
43	Titanium	0.04
44	Vanadium	<0.01
45	TOC	1,350
46	Silicon	<3
47	Phosphorous	0.65
48	Arsenic	0.029
49	Cyanide	0.017

After Benson (1980).

[a]NR, not reported; ND not detected.

TABLE 13.13 Range of Concentration of Different Parameters in Leachate of Coal Burner Fly Ash

S1 Number	Parameter	Range of Concentration (mg/liter except as indicated)
1	Aluminum	0.85–1.7
2	Antimony	<0.02
3	Arsenic	0.135–0.41
4	Barium	<0.1
5	Boron	1.8–2.3
6	Cadmium	<0.01
7	Calcium	60–22
8	Chromium	0.03–0.29
9	Cobalt	<0.01–0.02
10	Copper	<0.01–0.04
11	Germanium	<3.0
12	Iron	0.07–0.24
13	Lead	<0.01–0.04
14	Magnesium	1.1–4.3
15	Manganese	0.01–0.05
16	Mercury	<0.009
17	Molybdenum	0.29–3.8
18	Nickel	0.03–0.06
19	Potassium	20–29
20	Rubidium	<0.09
21	Selenium	0.05–0.18
22	Silica	5.1–51.0
23	Sodium	9.0–50.0
24	Strontium	<0.04–0.99
25	Sulfur	53.3–222
26	Tin	<1.0
27	Titanium	<0.5
28	Uranium	<0.005
29	Vanadium	0.26–0.92
30	Zinc	0.02–0.04
31	Alkalinity ($CaCO_3$)	37–50
32	COD	6–16
33	Chloride	<1.0–1.7
34	Conductivity	409–1213 μmho/cm
35	Dissolved solids	390–1240
36	Hardness ($CaCO_3$)	216–596
37	pH	7.83–9.05 units
38	Phosphorus	0.04–0.08

TABLE 13.14   Range of Concentration of Different Parameters in Leachate of Municipal Waste[a]

S1 Number	Parameter	Range of Concentration (mg/liter except as indicated)
1	TDS	584–55,000
2	Specific conductance	480–72,500 μmho/cm
3	Total suspended solids	2–140,900
4	BOD	ND–195,000
5	COD	6.6–99,000
6	TOC	ND–40,000
7	pH	3.7–8.9 units
8	Total alkalinity	ND–15,050
9	Hardness	0.1–225,000
10	Chloride	2–11,375
11	Calcium	3.0–2,500
12	Sodium	12–6,010
13	Total Kjeldahl nitrogen	2–3,320
14	Iron	ND–4,000
15	Potassium	ND–3,200
16	Magnesium	4.0–780
17	Ammonia-nitrogen	ND–1,200
18	Sulfate	ND–1,850
19	Aluminum	ND–85
20	Zinc	ND–731
21	Manganese	ND–400
22	Total phosphorus	ND–234
23	Boron	0.87–13
24	Barium	ND–12.5
25	Nickel	ND–7.5
26	Nitrate–nitrogen	ND–250
27	Lead	ND–14.2
28	Chromium	ND–5.6
29	Antimony	ND–3.19
30	Copper	ND–9.0
31	Thallium	ND–0.78
32	Cyanide	ND–6
33	Arsenic	ND–70.2
34	Molybdenum	0.01–1.43
35	Tin	ND–0.16
36	Nitrite–nitrogen	ND–1.46
37	Selenium	ND–1.85
38	Cadmium	ND–0.4
39	Silver	ND–1.96
40	Beryllium	ND–0.36
41	Mercury	ND–3.0
42	Turbidity	40–500 Jackson units

Based on McGinley and Kmet (1984), Lu et al. (1981), and Tharp (1991).

[a] Several bacteria and fungi species and several priority pollutants are found in the leachate.

TABLE 13.15   Range of Concentration of Different Parameters in Leachate of Construction/Demolition Waste[a]

S1 Number	Parameter	Range of Concentration (mg/liter except as indicated)
1	pH	6.5–7.3
2	Specific conductance	2,920–6,850 (μmho/cm)
3	BOD, 5 day	100–320
4	COD	3,080–11,200
5	Total dissolved solids	2,412–4,270
6	Total suspended solids	1,000–43,000
7	Total organic carbon	76–1,080
8	Chloride	125–240
9	Calcium	148–578
10	Magnesium	92–192
11	Sodium	256–1,290
12	Potassium	118–618
13	Carbonate	0
14	Bicarbonate	2,090–7,950
15	Sulfate	<40
16	Fluoride	<0.1–0.4
17	Nitrate	4–13
18	Nitrite	—
19	Ammonia (N)	30–184
20	Phenolphthalein alkalinity ($CaCo_3$)	0
21	Alkalinity ($CaCo_3$)	1,710–6,520
22	Hardness ($CaCo_3$)	597–1,516
23	Phosphorus	2.5–3.89
24	Oil and grease	18–47
25	Iron	29–172
26	Filtered iron	0.24–11
27	Manganese	1–4.9
28	Boron	1.4–3.9
29	Cyanide	<0.10
30	Phenol	0.7–2.99
31	Nickel	0
32	Arsenic	0.017–0.075
33	Barium	1.5–8.0
34	Cadmium	0.02–0.03
35	Chromium	0.1–0.25
36	Hex chromium	0.18–4.92
37	Copper	0.14–0.49
38	Lead	0.22–2.13
39	Mercury	<0.002–0.009
40	Selenium	<0.001
41	Silver	<0.01–0.03
42	Zinc	1.7–8.63

Based on Norstrom et al. (1991).

[a]Leachate quality depends on type of demolition debris disposed in the landfill.

needed for leachate from major landfills (especially from hazardous waste), certain general trends are being observed regarding leachate treatment. The following comments on the subject are based on the current trend in leachate treatment technology [Boyle and Ham, 1974; Carlson and Johansen, 1975; Cook and Foree, 1974; Uloth and Mavinic, 1977; Steiner et al., 1979; Stegmann, 1979; Rebhun and Galil, 1987; Chian and Dewalle, 1977; American Society of Civil Engineers/Water Pollution Control Federation (ASCE/ WPCF), 1977; Cadena and Jeffers, 1987; Kremer et al., 1987; Metcalf & Eddy, Inc., 1979; Yong, 1986; Lange et al., 1987; Meidl and Peterson, 1987; Hoffman and Oettinger, 1987; McShane et al., 1986; Ying et al., 1987; Osantowski et al., 1989; Ehrig, 1984]:

1. Generally landfill leachate is treated either in an on-site leachate treatment plant (common for extremely large municipal waste landfills and most hazardous waste landfills) or in an off-site existing wastewater treatment plant (common for most nonhazardous waste, which also includes municipal waste landfills). In some instances pretreatment of the leachate (in an on-site or off-site treatment plant) is done and the effluent is discharged in an existing wastewater treatment plant.

2. In many instances a combination of biological and physical/chemical processes is utilized for treating leachate.

3. Leachate with a high organic content is best treated with a biological process, whereas leachate with a low organic content is best treated with a physical/chemical process.

4. To avoid shock to a treatment plant, leachate should be slowly introduced into the treatment stream. Necessary leachate storage at the treatment plant should be arranged if a slow introduction is envisioned. This type of storage is important where the available capacity of the treatment plant is low and the leachate is hauled to the treatment plant using trucks.

5. The projected variation of leachate quality and quantity with time (daily, seasonal, and long-term) needs to be communicated to the wastewater treatment plant designer/operator who is responsible for designing/maintaining the effluent quality of the treatment plant.

6. Investigation regarding treatability of leachate from a proposed landfill should be undertaken at an early stage. Current practice is to treat municipal waste leachate and most other types of nonhazardous waste leachate in a municipal wastewater treatment plant. In many instances leachate from an industrial waste landfill is treated in the wastewater treatment plant of the same industry. A detailed bench-scale study is generally undertaken for hazardous waste landfills; in many instances an on-site pretreatment is used and the effluent is discharged in an existing wastewater treatment plant.

7. Both granular and powdered activated carbon are found to be useful in removing organic compound from leachate.

8. A reverse osmosis technique is also used for leachate treatment (Longman, 1990).

Figure 13.7 shows a flow diagram of a leachate treatment system that combines physical, chemical, and biological processes. The flow diagram also includes air stripping, an option useful for volatile organic compound (VOC) removal.

Recirculation of municipal leachate is promoted as a method for treating leachate. Although leachate recirculation reduces BOD and COD concentration, the concentration of metals and chloride increases (EMCON Associates, 1975; Robinson and Maris, 1985; Stegmann, 1979). It is argued that recirculation will reduce leachate volume due to increased evaporation and absorption in the waste. Problems such as reduction in permeability of the cover, perching of leachate, and odor have been reported by researchers (McGinley and Kmet, 1984; Robinson and Maris, 1985; Lechner et al., 1993). It appears that recirculation of municipal waste leachate may be successful initially but not in the long run. No data are available on recirculation of leachate from other waste types. Although research may be undertaken to study leachate recirculation for other waste types, recirculation does not appear to be a viable option for leachate treatment.

Land disposal for municipal leachate, as a means for treatment, has also been studied by some researchers (Chan et al., 1978; Bramble, 1973). Symptoms of toxicity were observed on plants grown in fields irrigated with municipal waste leachate (Menser and Winant, 1980). The concentrations of chemicals in leachate of most waste types are expected to be high. Leachate from those waste types, which needs to be disposed of in containment-type landfills, will impact the groundwater if land disposed. Thus, in general, land disposing of leachate as a means of treating it is not a logical approach. If leachate from a landfill in which only one type of waste has been disposed is observed to be of such quality that it can be landspread or directly discharged into surface water bodies, then one may reconsider the need for disposal of the waste in a landfill.

13.7 GAS GENERATION

Although gas generated within a few waste-type landfills may be negligible (e.g., foundry waste), most waste type is expected to generate a significant quantity of gas. It should not be assumed that only putrescible waste can generate gas. Gas generation from nonputrescible waste should be studied carefully prior to designing a landfill for the waste. The quality of gas depends mainly on the waste type. As with leachate, the quality and quantity of landfill gas vary with time. The discussion on quality and quantity of gas that follows pertains mainly to municipal waste landfills.

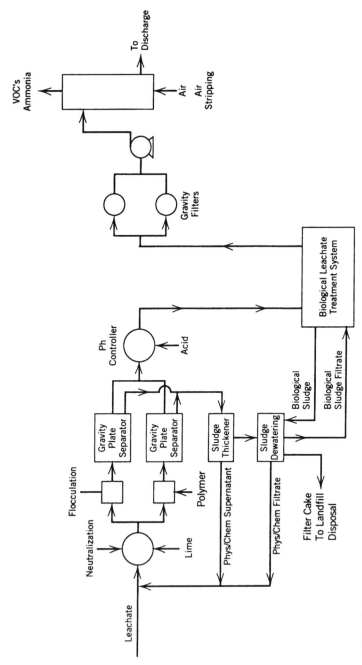

FIG. 13.7 Integrated physical/chemical and biological treatment system. After Osantowski et al. (1989).

There are usually three distinct but closely related stages of anaerobic digestion of biodegradable waste (e.g., municipal waste): (1) hydrolysis and fermentation due to bacterial activities, (2) acetogenesis and dehydrogenation, and (3) methanogenesis (Kayhanian et al., 1991). The first stage is dominated by the formation of acidic and propionic acid due to bacterial fermentation. In this stage the pH of the landfill leachate drops significantly along with a simultaneous rise of COD and BOD. In the second stage the soluble materials are oxidized to low-molecular-weight organic acids. Hydrogen gas is produced in this stage. In the third stage methane fermentation takes place, which primarily leads to the formation of methane and carbon dioxide; however, small amounts of other gases such as hydrogen sulfide, hydrogen, and nitrogen also form in this stage.

Within a relatively short period of time a stable methane phase is reached whereby gas composition (consisting mainly of methane, carbon dioxide, and a low percentage of nitrogen) and gas production rate remain constant over a longer period of time. Finally, the biological activity decreases, leading to a gradual decrease in gas production, which may be looked at as a fourth stage of anaerobic decomposition (Lechner et al., 1993). A detailed discussion on refuse decomposition can be found elsewhere (Barlaz and Ham, 1993). The time dependency of the percentage of methane is critical for landfill gas recovery and reuse projects. The typical quality of gas generated in municipal waste landfills is included in Table 13.16 (EMCON Associates, 1980; Kester and Van Slyke, 1987; Wood and Porter, 1987; Kalfka, 1986). It may be noted that percentage/concentration of various parameters depends on age and composition of waste in a landfill.

TABLE 13.16   Typical Composition of Stabilized Municipal Landfill Gas (Waste Volume <0.35 million m³)

S1 Number	Parameter	Percentage or Concentration
1	Methane	30–53%
2	Carbon dioxide	34–51%
3	Nitrogen	1–21%
4	Oxygen	1–2%
5	Benzene	ND–32[a] ppm
6	Vinyl chloride	ND–44[a] ppm
7	Toluene	150[a] ppm
8	t-1,2-Dichloroethane	59[a] ppm
9	$CHCl_3$	0.69[a] ppm
10	1,2-Dichloroethane	19[a] ppm
11	1,1,1-Trichlorethane	3.6[a] ppm
12	CCl_4	0.011[a] ppm
13	Trichloroethane	13[a] ppm
14	Perchloroethane	19[a] ppm

[a] Maximum concentraiton obtained from a survey of 20 landfills. ND, not detected.

The quantity of gas generated depends on waste volume and time since deposition in landfill. Gas production may be increased by adding sewage sludge or agricultural waste, removal of bulky metallic goods, and use of less daily and intermediate cover soil. The methane production rate ranges from 1.2 to 7.5 liters/kg/year (0.04–0.24 ft^3/lb/year) (EMCON Associates, 1980).

If gas is expected to be generated from a landfill, then proper arrangements should be made for venting/extraction and subsequent treatment (where necessary). Whether gas should be vented/extracted from a landfill is sometimes argued by design professionals. The following issues should be considered before deciding not to vent gas from the landfill:

1. *Gas pressure:* Some estimate regarding gas pressure should be made. The estimated gas pressure should be low enough so that it will not cause any rupture of the landfill cover. If the waste is expected to generate gas due to biodegradability and/or other physical/chemical processes, then venting of the gas should be recommended.

2. *Stress on vegetation:* The effect of the gas diffused through the cover on the vegetation should be studied. Stress may cause vegetation to die, which in turn will lead to increased erosion of the final cover.

3. *Toxicity of the gas:* The toxicity of the landfill gas should be studied. Release of the gas, by diffusion, through the final cover is unavoidable. The rate, concentration of release, and toxicity of the gas will determine whether such diffusional release will violate any air quality criteria.

4. *Location of the landfill:* The diffused gas may pose a health risk to the population residing in the immediate vicinity of the landfill.

LIST OF SYMBOLS

A = effective area of the landfill
A_s = area of the landfill surface in acres
C = runoff coefficient
C_c = coefficient of curvature of the soil
CN = curve number
E = volume lost through evaporation
ET = volume lost through evapotranspiration
I = uniform precipitation rate in inches
K = constant that depends on the watershed condition
L_v = preclosure leachate volume
L_v' = postclosure leachate volume
P = precipitation volume
R = surface runoff volume
R_i = surface runoff in inches
S = volume of pore squeeze liquid
T = ambient temperature above 32°F
W_f = field capacity of the soil (%)

W_i = initial soil moisture content of the cover
W_p = rainfall in inches
α = slope of the landfill surface (%)
β = constant
ΔS = soil and waste moisture storage volume
γ_d = soil density in lb/ft^3

14

WASTE CHARACTERIZATION

Waste characterization must be undertaken prior to designing a landfill. In general the characteristics of wastes may vary within a type of industry (e.g., waste from different papermills may not have the same characteristics) and may vary over time as a result of change in the industrial process. Therefore, it is good practice to characterize a new waste and repeat the characterization if a process change occurs. Characterization of municipal garbage is usually not performed because it is extremely difficult to perform tests on the waste and many studies have already been done to characterize the waste. However, since the composition of municipal waste may vary widely across a country (e.g., metropolitan areas versus small towns, industrial versus nonindustrial cities) and in different parts of the world, waste characterization studies of municipal garbage should be undertaken wherever possible. A typical range of major components of municipal garbage in the United States is indicated in Table 14.1. The range of the components is compiled from several studies conducted in the United States (Glaub et al., 1983; Wigh, 1979; Walsh and Kinman, 1981; EMCON Associates, 1975; Fungaroli and Steiner, 1979). The range of components has changed slightly due to recycling efforts. In general, waste characterization is done to address the following issues:

1. Whether the waste is hazardous
2. Whether the waste can be landfilled
3. Probable leachate constituents (necessary for judging liner compatibility, treatment plant design, and groundwater monitoring program design)
4. Volume rate of waste generation

TABLE 14.1   Typical Range of Major Components of Municipal Garbage in the United States

S1 Number	Major Components	Range (% of wet weight)
1	Food waste	4.4–15.3
2	Garden waste	12.5–24.2
3	Glass	6.5–10.9
4	Metals (iron and aluminum)	4.0–9.0
5	Moisture	27.1–35.0
6	Other combustibles	1.6–12.1
7	Other noncombustibles	1.8–11.1
8	Paper	41.6–53.5
9	Plastics	0.76–5.7

5. Physical properties of the waste necessary for the design of a landfill
6. Physical properties of the waste necessary for the operation of a landfill
7. Identification of safety precautions to be observed by landfill operators and inspectors
8. Identification of waste reduction alternatives

Therefore both the physical and chemical properties of the waste must be determined to address these eight issues. The detailed steps for waste characterization for each source can vary widely; therefore a general guideline is provided that must be further developed on a case-by-case basis.

14.1   GENERAL GUIDELINES FOR WASTE CHARACTERIZATION

The guidelines provided in this section are primarily for industrial waste. For characterizing municipal garbage, study can be conducted either by collecting samples from curbside or from garbage trucks arriving at a landfill. The first step in characterizing a waste is to study the material flow. A correct material flow diagram must be developed. Each final waste stream that is disposed in a landfill should be identified with a number. Most industries practice inhouse recycling of waste; these waste recyclings should be noted carefully. Many times these recycled wastes are wrongly identified as "waste stream" and characterized. Only those waste streams that are leaving the building need to be characterized. A typical (partial) flow diagram along with the waste identification number is shown in Fig. 14.1.

The raw material safety data sheet that identifies the chemicals used in the process needs to be studied. This will provide information on the probable constituents of the waste and thereby help narrow the list of parameters for future chemical analysis.

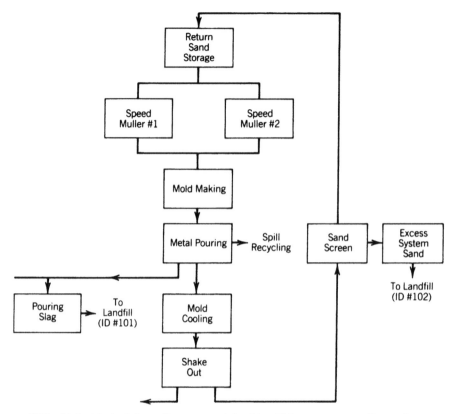

FIG. 14.1   Typical flow diagram used for identifying waste sampling points.

14.1.1   Sampling

Once the sampling points are identified in the flow diagram, the next step is to decide sampling frequency and volume of sample. A defective sampling plan will not provide the true characteristics of the waste. The sampling should be done in such a way that the test data are representative and indicate variability of the waste (USEPA, 1986a).

The methods and equipment used for sampling waste materials will vary with the form and consistency of the waste. Different sampling protocols are used for sampling different waste types. The following is a list of standards tests that may be used in the absence of a regulatory directive on sampling protocol:

1. ASTM standard D140-70 for extremely viscous liquid (ASTM)
2. ASTM standard D346-75 for crushed or powdered waste (ASTM)
3. ASTM standard D420-69 for soil or rock-like waste (ASTM)
4. ASTM standard D1452-65 for soil type waste (ASTM)
5. ASTM standard D2234-76 for fly ash type waste (ASTM)

Sludges may be treated as viscous liquids for the purpose of sampling.

Waste characteristics may vary due to the variability in raw material input or the production process. For instance, the municipal garbage fed into incinerators may vary daily or seasonally. A papermill may vary raw chemicals input depending on the end product in each production period (e.g., different colored paper at different times of the day or days of the week). Such variation must be taken into consideration when deciding sampling frequency. Random sampling provides a better assessment of waste characteristics. If samples are composited, care must be taken to ensure that the composite is representative of the original (e.g., for an incinerator a composite of samples collected in the same season is allowable but a composite of samples collected over several seasons is not recommended). If two or more waste types are disposed of simultaneously in a landfill, then a composite of each waste in the same ratio (volume or weight) in which it is disposed in the landfill is admissible. Individual samples are preferred over composite samples. However, sometimes to obtain a more representative sample, a composite of several samples is practiced (e.g., hourly samples of fly ash from an incinerator may be collected and composited to obtain one sample for the whole day).

14.1.2   Chemical Tests

Two types of chemical tests are done on each waste stream: bulk chemical analysis and the leach test.

14.1.2.1   Bulk Chemical Analysis   The aim of bulk chemical analysis (also termed total analysis) is to determine the chemical makeup of the waste. In general, bulk analysis involves solubilization of waste constituents to the greatest possible extent and then identification of them so that the sum total of all constituents is 99.99% of the original total bulk weight of the sample. No universal procedure for all waste or chemicals is available, but there are several test guidelines [Water Pollution Control Federation (WPCF, 1981; ASTM D-2795-84 and E-886-82; American Society for Testing and Materials (ASTM), 1986; USEPA, 1986a]. Study of the raw material input to the process would help in finalizing the list of chemicals for which tests are to be run. Tables 14.2 and 14.3 include parameters of concern and Table 14.4 includes recommended lists of parameters for different waste types.

The lists are by no means final and should be modified whenever necessary. The lowest detection limits, as permissible by current technology, should be used in both bulk chemical analysis and the leach test. (Limit of detection is the lowest concentration at which a chemical species can be detected and limit of quantification is the lowest concentration at which a chemical species can be quantified.)

The advantages and disadvantages of bulk chemical analysis are discussed below. The advantages are as follows:

TABLE 14.2 Public-Health-Related, Public-Welfare-Related, and Indicator Parameters

Public-health-related parameters	*Public-welfare-related parameters*
Aldicarb	Chloride
Arsenic	Color
Bacteria, total coliform	Copper
Barium	Foaming agents MBAS
Benzene	(methylene-blue active substances)
Cadmium	Iron
Carbofuren	Manganese
Chromium	Odor
Cyanide	Sulfate
1,2-Dibromoethane	Total dissolved solids (TDS)
1,2-Dibromo-3-chloropropane (DBCP)	Zinc
p-Dichlorobenzene	
1,2-Dichloroethane	*Indicator parameters*
1,1-Dichloroethylene	
2,4-Dichlorophenoxyacetic acid	Alkalinity
Dinoseb	Biochemical oxygen demand (BOD_5)
Endrin	Boron
Fluoride	Calcium
Lead	Chemical oxygen demand (COD)
Lindane	Magnesium
Mercury	Nitrogen series
Methoxychlor	Ammonia nitrogen
Methylene chloride	Organic nitrogen
Nitrate + nitrite (as N)	Total nitrogen
Selenium	Potassium
Silver	Sodium
Simazine	Specific conductance
Tetrachloroethylene	Total hardness
Toluene	Total organic carbon (TOC)
Toxaphene	Total organic halogen (TOX)
1,1,1-Trichloroethane	
1,1,2-Trichloroethane	
Trichloroethylene	
2,4,5-Trichlorophenoxypropionic acid	
Vinyl chloride	
Xylene	

1. Sources of error are few because the number of variables to control the test are minimal.
2. Bulk chemical analysis may be the only way of determining the contamination potential from an unstable waste.
3. Bulk chemical analysis will provide the total contaminant content.

TABLE 14.3 Priority Pollutants Suggested for Bulk Chemical Analysis and Leach Test

Metals, cyanide, and total phenols

Antimony
Arsenic
Beryllium
Cadmium
Chromium
Copper
Lead
Mercury
Nickel
Selenium
Silver
Thallium
Zinc
Cyanide
Phenols, total

Dioxin

2,3,7,8-Tetrachlorodibenzo-*p*-doxin

Gas chromatography/mass spectrometry (GC/MS) fraction—volatile compounds (purgeable)

Acrolein
Acrylonitrile
Benzene
bis(Chloromethyl)ether
Bromoform
Carbon tetrachloride
Chlorobenzene
Chlorodibromomethane
Chloroethane
2-Chloroethylvinyl ether
Chloroform
Dichlorobromomethane
Dichlorodifluoromethane
1,1-Dichloroethane
1,2-Dichloroethane
1,1-Dichloroethylene
1,2-Dichloropropane
1,2-Dichloropropylene
Ethylbenzene
Methylbromide
Methylchloride
Methylene chloride
1,1,2,2-Tetrachloroethane

TABLE 14.3   (*Continued*)

Tetrachloroethylene
Toluene
1,2-*trans*-Dichloroethylene
1,1,1-Trichloroethane
1,1,2-Trichloroethane
Trichloroethylene
Trichlorofluoromethane
Vinyl chloride

GC/MS fraction—acid compounds (acid extractable)

2-Chlorophenol
2,4-Dichlorophenol
2,4-Dimethylphenol
4,6-Dinitro-o-cresol
2,4-Dinitrophenol
2-Nitrophenol
4-Nitrophenol
p-Chloro-*m*-cresol
Pentachlorophenol
Phenol
2,4,6-Trichlorophenol

GC/MS fraction—base/neutral compounds
(base/neutral extractable)

Acenaphthene
Acenaphthylene
Anthracene
Benzidine
Benz[*a*]anthracene
Benzo[*a*]pyrene
Benzo[*b*]fluoranthene
Benzo[*g,h,i*]perylene
Benzo[*k*]fluoranthene
Bis(2-chloroethoxy)methane
Bis(2-chloroethyl)ether
Bis(2-chloro-*iso*-propyl)ether
Bis(2-ethylhexyl)phthalate
4-Bromophenyl phenyl ether
Butyl benzyl phthalate
2-Chloronaphthalene
4-Chlorophenyl phenyl ether
Chrysene
Dibenz[*a,h*]anthracene
1,2-Dichlorobenzene
1,3-Dichlorobenzene
1,4-Dichlorobenzene

TABLE 14.3 *(Continued)*

3,3-Dichlorobenzidine
Diethyl phthalate
Dimethyl phthalate
Di-*n*-butyl phthalate
2,4-Dinitrotoluene
2,6-Dinitrotoluene
Di-*n*-octyl phthalate
1,2-Diphenylhydrazine
Fluoranthene
Fluorene
Hexachlorobenzene
Hexachlorobutadiene
Hexachlorocyclopentadiene
Hexachlorethane
Indeno[1,2,3-*c,d*]pyrene
Isophorone
Naphthalene
Nitrobenzene
n-Nitrosodimethylamine
n-Nitrosodi-*n*-propylamine
n-Nitrosodiphenylamine
Phenanthrene
Pyrene
1,2,4-Trichlorobenzene

*GC/MS fraction—pesticides and polychlorinated
biphenyls (PCBs)*[a]

Aldrin
α-BHC
β-BHC
γ-BHC
δ-BHC
Chlordane
4,4'-DDT
4,4'-DDE
4,4'-DDD
Dieldrin
α-Endosulfan
β-Endosulfan
Endosulfan sulfate
Endrin
Endrin aldehyde
Heptachlor
Heptachlor epoxide
PCB-1242
PCB-1254

TABLE 14.3   (*Continued*)

PCB-1221
PCB-1232
PCB-1248
PCB-1260
PCB-1016
Toxaphene

After Weston (1984).

[a]BHC: benzene hexachloride; DDT: dichlorodiphenyltrichloroethane; DDE: dichlorodiphenyldichloroethylene; DDD: dichlorodiphenyldichloroethane.

TABLE 14.4   Parameters Recommended for Bulk Chemical Analysis and the Leach Test for Different Waste Types

Waste Type	Chemical Substances/Parameters
Alum mud	Parameters listed in Table 14.2 plus aluminum, beryllium, hydrogen sulfide, pH, and sulfide
Coal fly ash	Parameters listed in Table 14.2 plus cobalt, molybdenum, nickel, pH, strontium, thallium, and vanadium
Foundry waste	Parameters listed in Table 14.2 plus aluminum, formaldehyde, molybdenum, nickel, pH, phosphorus, phenol, and tin (Note: If other chemicals are identified in the raw materials then those should be included in this list)
Hazardous waste (all sources)	Parameters listed in Table 14.2 plus priority pollutants listed in Table 14.3 plus other chemicals identified in the raw materials used in the process
Municipal waste	Parameters listed in Tables 14.2 plus priority pollutants listed in Table 14.3 plus aluminum, antimony, beryllium, molybdenum, nickel, pH, total phosphorus, total suspended solids, thallium, and tin
Municipal solid waste incinerator ash (both bottom ash and fly ash)	Parameters listed in Table 14.2 plus aluminum, cobalt, strontium, tin, phosphate, and pH
Papermill sludge	Parameters listed in Table 14.2 plus priority pollutants listed in Table 14.3 plus aluminum, bromide, cobalt, phosphate, phosphorus, phenols, pH, tin, titanium, total suspended solids, and vanadium
Other nonhazardous waste (minimum recommendation)	Parameters listed in Table 14.2 plus other chemicals identified in the raw materials used in the process

4. The test data provide a baseline for comparing the waste with other waste or natural material.

5. It provides a basis for studying and ranking the leachate constituents obtained.

The disadvantages of bulk chemical analysis are as follows:

1. Although bulk chemical analysis is aimed at a complete analyses of the waste, in many instances it does not identify 99.99% of the waste components.

2. The highly aggressive nature of the test is not representative of the field situation. The leachable concentration of a component in a waste will seldom equal the total concentration of the component in the waste.

14.1.2.2 Leach Test   Prior to deciding on a leaching media, the situation under which the waste is to be landfilled should be investigated. For monofills water leaching is acceptable; however, for co-disposal with municipal waste, acid leaching should be performed. A synthetic leachate may also be used as a leaching medium for the co-disposal scenario. The list of chemicals to be tested in the elutriate should be the same as for bulk chemical analysis. However, if bulk chemical analysis indicates a very low percentage of certain chemicals that are not expected to be dissolved in the leaching medium, then those chemicals may be deleted from the list of chemicals to be used for the leaching test. A discussion on available leach tests is included in Section 13.3.1.

14.1.3 Physical Tests

The physical tests to be performed depend on the landfill design and the available knowledge about the physical behavior of the waste. The following list includes the physical properties necessary for landfill design; the reader has to use judgment to choose the most appropriate tests: compacted bulk density or unit weight, specific gravity, grain size distribution, permeability, consolidation characteristics, Atterberg limits, and static and dynamic strength characteristics (i.e., angle of internal friction and cohesion). Of the above items specific gravity may not be necessary in most cases. In addition to its design-related use, compacted bulk density is necessary to estimate the tipping fee for disposing of waste in a landfill; determination of Atterberg limits is not necessary for most waste types. Interface friction angle with construction materials (e.g., synthetic membrane) may be needed for stability design.

Standard soil testing methods can be used for testing soil type waste; however, difficulty arises with putrescible waste and sludges (especially those that release gas under pressure). The problem is further multiplied if the waste is heterogeneous (e.g., municipal waste). No standard test methods are available for such waste at present. The physical properties of waste are found by using standard geotechnical equipment and testing procedures with minor modification (Wardwell and Charlie, 1981; Lowe and Andersland, 1981; Zimmerman et al., 1977; Rao et al., 1977; Andersland and Mathew, 1973; Hagerty et al., 1977; Kulhawy et al., 1977; Somogyi and Gray, 1977; Pervaiz and Lewis, 1987). Doubts have been expressed regarding the use of standard geotechnical procedures for testing sludge (Bagchi, 1980; Wardwell and Charlie, 1981). This is an area in which further research may be done.

14.2   IDENTIFICATION OF HAZARDOUS WASTE

Prior to developing any detailed physical and chemical testing program, the first step should be to determine whether the waste is hazardous. Dawson and Mercer (1986) have provided a good discussion on how to define hazardous wastes and how they are defined in different countries. A discussion with the regulatory agency regarding its definition of hazardous waste is suggested. In general, two approaches are taken in the United States to define hazardous waste: by listing and by identification of characteristics. The list is subject to change. A solid waste containing any of the hazardous constituents is considered hazardous unless proven otherwise by the disposer.

In summary the steps involved in identifying a hazardous waste are as follows:

1. Determine whether it is already listed as a hazardous waste.
2. If not, then characteristic identification tests are to be performed to determine whether the waste is hazardous due to any characteristics detailed in Section 14.2.2.

It may be noted that radioactive waste and infectious waste (often termed biohazardous waste) are not included in this discussion and, therefore, any reference to hazardous waste in this book would always mean nonradioactive, noninfectious hazardous waste.

14.2.1   Listing

A list of chemicals whose health hazard is already known is developed. Wastes that contain any of these compounds are characterized as hazardous. A list of such hazardous waste compounds may be obtained from the regulatory agency.

14.2.2   Characteristics Identification

The following four characteristics are used to identify hazardous waste.

Corrosivity   If the waste is aqueous, has a pH less than 2 and greater than 12.5, and corrodes plain carbon steel (carbon content of 0.2%) at the rate of 6.35 mm or greater per year at 55°C, then the waste is characterized as corrosive waste.

TCLP Toxicity   The TCLP test is performed on the waste. Refer to Section 13.3.1 for more details.) If the extract contains any of the substances listed in Table 13.1 at a concentration equal to or greater than the respective value given in that table, then the waste is considered as TCLP toxic waste.

Ignitability   If the waste is liquid other than an aqueous solution containing less than 24% alcohol by volume, and has a flash point of less than 60°C as determined by ASTM test D-93-79 or D-93-80 (ASTM), it is characterized as ignitable waste. A nonliquid waste (other than gaseous) is also characterized as ignitable if it causes fire at 0°C and at a pressure of 1 atm through friction, absorption of moisture, or spontaneous chemical changes or burns vigorously creating a fire hazard. For gaseous waste the ignitability is determined by ASTM test D-323 (ASTM).

Reactivity   A waste is characterized as reactive if it is normally unstable and undergoes violent change without detonating or reacts violently with water, or forms a potentially explosive mixture with water or generates significant quantities of toxic gas when mixed with water endangering human health or the environment.

14.3   RESTRICTION ON LAND DISPOSAL OF HAZARDOUS WASTE

In general hazardous wastes should not be disposed of in a landfill without pretreatment. Although compatibility tests may demonstrate the suitability of the proposed liner material for the waste type, it is good practice to "stabilize" the waste so the mobility of the hazardous constituents is reduced. Waste that produces toxic fumes due to contact with water or other waste in the landfill, or waste with flash points >140°F, should not be disposed of in landfills. In addition, waste having the following characteristics should not be disposed of in landfills (Stanczyk, 1987):

1. Waste with high percentages of volatile organic content
2. Waste with high percentages of aromatic, halogenated, and nonhalogenated compounds
3. Waste with high percentages of metallics, especially arsenic, cadmium, lead, mercury, and selenium
4. Waste with high percentages of cyanide and sulfide

TABLE 14.5 Selected Physical Property Values for MSW

Parameter	Range	Average
Volumetric field capacity[a]	30–53%	44%
Field density[b]	1080–2565 lb/yd^3	1802 lb/yd^3
	(6.3–14.9 kN/m^3)	(10.5 kN/m^3)
Porosity[c]	0.4–0.62	0.51
Hydraulic conductivity[d]	4×10^{-2}–1.5×10^{-4} cm/sec	6×10^{-3} cm/sec
Shear strength parameters[e]	$\phi = 18°$–$43°$	$\phi = 31°$
	$c = 210$–460 lb/ft^2	$c = 356$ lb/ft^2
	(10–22 kPa)	(17 kPa)

[a] Based on Rovers and Farquhar (1973), Wigh (1979), Zornberg et al. (1999), and Blight et al. (1992).
[b] Based on Wigh (1979), Rovers and Farquhar (1973), Zornberg et al. (1990), and Oweis et al. (1990).
[c] Based on Oweis et al. (1990) and Zornberg et al. (1999).
[d] Based on Fungaroli and Steiner (1979), Schroeder et al. (1984a and b), Oweis et al. (1990), Qian et al. (2002), and Landva and Clark (1990).
[e] Based on Landva and Clark (1990) and Richardson and Reynolds (1991).

5. Powdery hazardous waste that may cause dust problems in and around the landfill

6. Waste with very low shear strength that may preclude construction of a final cover for the landfill

7. Waste with high percentages of liquid that may generate too much leachate in the landfill

The regulatory agency may impose additional criteria for restricting land disposal of waste. A discussion with the regulatory agency on this issue is recommended. Currently in the United States, there is a ban on disposal of hazardous waste in landfills. Hazardous waste must be stabilized (i.e., rendered nonhazardous) prior to disposal.

TABLE 14.6 Selected Physical Property values for Paper Mill Sludge

Parameter	Range	Average
Field density[a]	1770–2466 lb/yd^2	2158 lb/yd^3
	(9.83–13.7 kN/m^3)	(11.99 kN/m^3)
Hydraulic conductivity[b]	1.8×10^{-6}–4×10^{-10} cm/sec	6.39×10^{-7} cm/sec
Shear strength parameters[c]	$\phi' = 25°$–$40°$	$\phi' = 34.2°$
	$c' = 59$–189 lb/ft^2	$c' = 134$ lb/ft^2
	(2.8–9 kPa)	(6.36 kPa)

[a] Based on Moo-Young and Zimmie (1996).
[b] Based on Moo-Young and Zimmie (1996), Maltby and Eppstein (1994), NCASI (1990).
[c] Moo-Young and Zimmie (1996).

14.4   IDENTIFICATION OF NONHAZARDOUS WASTE

As previously indicated, complete waste characterization will include determination of both the physical and chemical properties of the waste. For known nonhazardous waste usually only the leach test and some physical tests are performed. However, it is good practice to perform at least the TCLP test to determine whether the waste is TCLP toxic. The regulatory agency should be contacted to determine the test requirements for each waste type. Scattered information on the physical and chemical properties of different waste types is available in the literature. Since waste characterization is rather waste specific, a summary of all available data is not included.

Standard test procedures are applicable for the determination of physical properties of most nonputrescible wastes (e.g., foundry sand). However, it is difficult to determine the physical properties of putrescible wastes. Range and average values of various physical properties of municipal solid waste (MSW) and paper mill sludge are included in Table 14.5 and 14.6 respectively. Discussion regarding the test protocols and their limitations can be found in the literature cited in the tables. While it is best to perform appropriate tests for the determination of each physical property, the values included in the tables can be helpful for preliminary design purposes. It may be noted that researchers have expressed doubt regarding use of Mohr-Coulomb theory while interpreting laboratory data related to waste strength (Singh and Murphy, 1990,

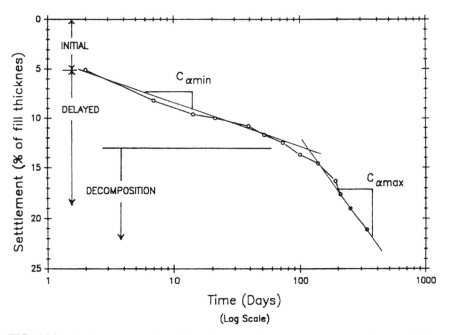

FIG. 14.2   Idealized plot of landfill settlement. (After Bjarngard and Edgers, 1990.)

Quian et al., 2002). Percentage of liquid within the waste mass greatly influences stability of MSW slopes (Koerner and Soong, 2000). The settlement of putrescible waste is influenced heavily by the state of decomposition of the waste mass. As shown in Figure 14.2, MSW landfill settlement is high after the first year (Bjarngard and Edgers, 1990). A 30–50% settlement within 2 to 5 years is possible in bioreactor landfills (USEPA, 2000).

15

NATURAL ATTENUATION LANDFILLS

The design concept for natural attenuation (NA) type of landfills consists of allowing the leachate to percolate through the landfill base with the expectation that the leachate will be attenuated (purified) by the unsaturated soil zone beneath the landfill and by the groundwater aquifer. In the past only NA-type landfills were used for disposal of all types of waste. At that time it was thought that the soil in the unsaturated zone is capable of completely attenuating the leachate. This concept of attenuation by soil has changed significantly. Presently only nonhazardous wastes are disposed of in NA-type landfills. Recent studies indicate that even small NA-type landfills (waste volume up to 50,000 yd³) may impact groundwater (Friedman, 1988). Whether such an impact on groundwater is to be considered as severe depends on the prevailing groundwater rules in the area in which the landfill is to be located. Currently in some countries of the world (e.g., Germany) and in some states of the United States (e.g., Wisconsin), NA-type landfills are not allowed regardless of volume or waste type. Therefore, before designing an NA-type landfill, the designer should discuss the issue with the local regulatory agency.

Two types of filling methods are used in operating NA-type landfills: the area method and trench method. In the area method an entire area is excavated to the subbase grade and filled up from one end (Fig. 15.1). In the trench method, individual trenches are excavated, filled, and covered progressively (Fig. 15.2). The area method would require less land for disposing the same volume of waste compared to the trench method. However, in the area method of filling, because precipitation comes in contact with the entire waste area, the quantity of leachate generated is higher and leachate quality is worse compared to the trench method of filling. Therefore, in choosing between the two types of filling, a designer has to strike a balance between land availability

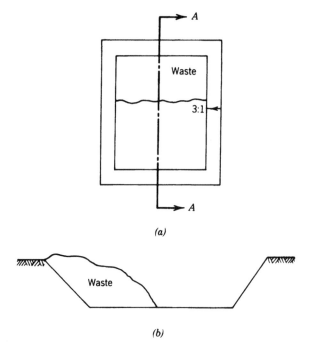

(a)

(b)

FIG. 15.1   Area method of landfilling: (*a*) plan; (*b*) section *A–A*.

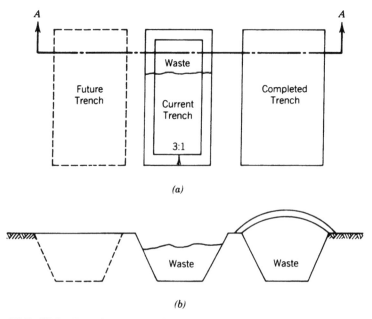

(a)

(b)

FIG. 15.2   Trench method of landfilling: (*a*) plan; (*b*) section *A–A*.

and extent of allowable groundwater impact. A third option of filling has evolved that reduces the leachate generation rate and at the same time needs less disposal area (Fig. 15.3). In this filling method, the landfill area is progressively excavated on one side while the area that has reached final grade on the other side is capped. Usually the subbase has a downward slope in the direction in which excavation proceeds. A small [0.6–0.9 m (2–3 ft)] berm is constructed at a suitable distance so that noncontact water collected between the berm and the end of excavation is pumped out using a small pump. However, if a pump is used for pumping noncontact water, the operator must be made aware of the fact that the waste limit must be inside the small berm to avoid discharging contaminated water into surface water bodies. The designer should contact the regulatory agency prior to using a pump to discharge noncontact water into surface water bodies.

The trench filling method works best when a relatively small volume of waste is disposed of in a short time period (several days) and the interval between such disposals in high (6 months or more). The area method works best when a daily flow of waste volume is expected; the third option of landfill mentioned in the previous paragraph may also be used in this situation.

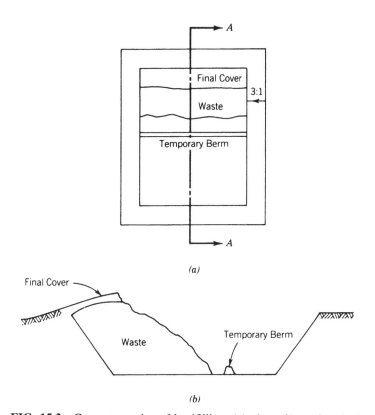

(a)

(b)

FIG. 15.3   Current practice of landfilling: (a) plan; (b) section A–A.

15.1 NATURAL ATTENUATION PROCESS

Figure 15.4 shows a cross section of an ideal NA-type landfill. Generalized soil stratigraphy of an ideal NA-type landfill includes a clayey stratum [which should consist either of ML or ML-CL-type soil per the Unified Soil Classification System (USCS) followed in the United States] directly below the landfill base, which overlies a sandy aquifer. The bedrock underlying the sandy aquifer should be quite deep [ideally 15–18 m (50–60 ft) or greater below the water table].

Attenuation of leachate occurs in two stages. In the first stage, soil in the (initially) unsaturated zone reacts with the leachate constituents and attenuates the leachate in part. The second stage of attenuation occurs in the groundwater equifer (Fig. 15.4).

15.2 MECHANISMS OF ATTENUATION

Natural attenuation may be defined as a process by which the concentration of leachate parameters is reduced to an acceptable level by natural processes. Based on this definition, the following mechanisms of attenuation are identified: (1) adsorption, (2) biological uptake, (3) cation- and anion-exchange reactions, (4) dilution, (5) filtration, and (6) precipitation reactions (Bagchi, 1983). Except for dilution, the other mechanisms can be operative in the unsaturated zone. All the mechanisms except for biological uptake can be operative in the aquifer.

15.2.1 Adsorption

Adsorption is the process by which molecules adhere to the surface of individual clay particles. Because of the difficulty in distinguishing adsorption

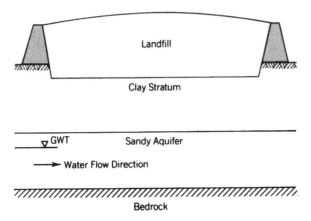

FIG. 15.4 Soil stratification for an ideal natural attenuation landfill.

from exchange reactions (explained in Section 15.2.3) experimentally, it is sometimes referred to as adsorption-exchange reactions (Philips and Nathwani, 1976; Minnesota Pollution Control Agency, 1978; Roberts and Sangrey, 1977). However, there is a basic difference between adsorption and exchange reactions. Adsorption will cause a decrease in the total dissolved solid (TDS) in the leachate, whereas exchange reactions will not. Therefore adsorption may be considered to really attenuate leachate, whereas exchange reactions will simply change the type of ions present in the exfiltrate (the liquid generated after leachate percolation through the unsaturated zone). The adsorption reaction is pH dependent and to determine the adsorption capacity of a particular clay, adsorption isotherms must be established experimentally. An adsorption isotherm is a plot of the amount of ion adsorbed versus the concentration of the ion in the solution. In addition to the theoretical difficulties of adsorption analysis, there is a general lack of data concerning the adsorption by clay minerals of individual ions from a solution of various ions. Much of the available information pertains to adsorption for specific ions, although some studies have included analyses of adsorption isotherms for the leachate–soil system (Griffin and Shimp, 1976; Griffin, 1977). Although some generalization of leachate quality for municipal waste can be made for a region, in most cases, leachate quality is significantly site specific. Thus, in the absence of experimental results using the on-site soil and leachate obtained from the waste to be disposed of at a particular site, site-specific quantitative analysis cannot be performed.

The opposite of adsorption—desorption—occurs in many systems (Gebhard, 1978). The isotherm based on a young system may not be true for an aged adsorbate–solid complex. Hence, even the adsorption isotherm obtained experimentally should be used carefully for site-specific design.

15.2.2 Biological Uptake

Biological uptake is a mechanism by which microorganisms either break down or absorb leachate constituents and thereby attenuate leachate. Microbial growth in soil systems can have a tremendous impact on leachate attenuation initially. Processes that soil microorganisms either perform or mediate include the following (Fuller, 1977; Wood et al., 1975):

1. Breaking down carbonaceous wastes
2. Production of carbon dioxide and subsequent formation of carbonic acids
3. Production of various organic acids
4. Using up available oxygen supplies and creation of an anaerobic environment
5. Participation in metal ion reactions
6. Oxidation or reduction of inorganic compounds

7. Transformation of cyanide to mineral nitrogen compounds and then denitrification of the compound to nitrogen gases
8. Methylation of metals and metalloids
9. Production of complex organic compounds that react with leachate constituents
10. Production of large and small organic molecular species on which leachate constituents can be absorbed
11. Production of small-sized organic debris that can infiltrate pore spaces and thereby reduce soil permeability

Mineralization is the process by which elements or organic matter, microbial tissues, and organic complexes are converted into an inorganic state. Biological immobilization is considered to be the reverse of mineralization. Trace and heavy metals are incorporated into microbial tissues and the mobility is controlled by cells or cell tissues. For elements that are relatively immobile in soils such as inorganic complexes, incorporation into cell materials may be thought of as a mechanism by which they can migrate as minute particles and cell materials when the tissues die and decay. Movement of phosphorous in an organic form is an example of this phenomenon (Hannapel et al., 1964). Thus, the presence of microbes beneath a landfill may be beneficial for attenuating some leachate constituents but could be detrimental for mobilizing others. Organic complexes in typical municipal landfill leachates probably could immobilize many of the trace metal parameters by precipitation under aerobic conditions (Fuller, 1977). However, under anaerobic and acid conditions, attenuation of metals by microbes will be less effective, especially if the pH drops to 3 or less.

Biological uptake in biodegradable (putrescible) waste landfills is basically through anaerobic bacteria. Methane gas generation is an indicator of the anaerobic biological activity. It is known that methane gas generation attains a peak volume and then decreases. The rate of depletion of the generation of gas is related to the rate of decay of the microorganisms. The microorganisms in a landfill can be only heterotropic (surviving on a food source) not autotrophic (capable of food synthesis through the use of sunlight). Since the strength of the landfill leachate decreases with time, causing a decrease in the availability of food, the activity of the microorganisms is bound to decrease. Therefore, it is obvious that the chemical fixation of pollutants is not permanent. As the biological population dies, pollutants once fixed in microbial cells could be released through a mineralization process. It is possible to study the decay rate and subsequent pollutant release of landfill microorganisms using principles of biokinetics; however, this has not been reported in the literature. Because of the uncertainties involved in the decay rate and the fact that the biological uptake is not permanent, it is prudent not to use biological uptake in landfill design, at least at the present time. It is an area in which further research should be undertaken so as to obtain more valid data on the subject. As for now, biological uptake may be considered a safety factor, the value of which is not yet known.

15.2.3 Cation and Anion Exchange

The exchange reactions mainly involve the clay minerals and may be defined as exchange of ions of one type by ions of another type without disturbing the mineral structure (known as isomorphous substitution) (Grim, 1968). The solid phase of a given soil may contain various amounts of crystalline clay and nonclay minerals, noncrystalline clay minerals, organic matter, and salts. Although the amount of nonclay minerals in a given soil is usually considerably greater than the proportion of clay minerals present, cation exchange performed by the clay mineral fraction is quite significant. By and large, the nonclay particles are relatively inert.

Anion exchange increases as the pH of the soil system decreases. since organics are negatively charged ions, attenuation of organic ions in a clayey soil environment will be mostly through anionic exchange. Therefore, a low pH system should attenuate organics well (Mitchell, 1976). Organic anions may be adsorbed by clays and inorganic ions. However, note that the pH of the leachate–soil system converges to a near neutral value (refer to Section 15.2.6). Thus, significant attenuation of organic ions through anion exchange is not expected. Adsorption on particle surfaces in place of previously adsorbed water molecules is a possibility (Van Olphen, 1963).

Time taken to complete anion- and cation-exchange reactions is not well documented. However, it is reported that the exchange reaction in kaolinite is almost instantaneous because of easily accessible sites located at broken bond edges. A longer reaction time is expected in smectites because a majority of the exchange sites are located within layers (Mitchell, 1976). Further research is needed to determine the time required for completion of exchange reactions in clayey soil.

15.2.4 Dilution

This is not a mechanism by which leachate constituents are chemically altered or attenuated by the soil. It reduces the concentration of leachate constituents. To what extent dilution should be used for the design of NA-type landfills depends on the policy of the regulatory agency. If a policy of nondegradation of groundwater is to be pursued, then dilution cannot be taken into account because all natural attenuation type of landfills will degrade groundwater. Chloride, nitrate, hardness, and sulfate found in municipal landfill leachate are not attenuated by soil; the only mechanism by which these parameters are attenuated is dilution. The concentration of these and other parameters may be diluted in a groundwater aquifer to such a level that the quality of down-gradient water at a certain distance is degraded only slightly compared to the background water quality, and therefore the water quality remains acceptable for specific uses. When locating an NA-type landfill, the designer should ensure that groundwater quality at a distance [the distance is sometimes fixed by local regulators; in the absence of a regulation a distance of 300–360 m (1000–1200 ft) or the distance to the nearest down-gradient drinking water well, whichever is less, should be used] is safe for drinking purposes.

The major factors that influence dilution are the density difference of leachate and ambient groundwater, leachate entry velocity, groundwater velocity, diffusion–dispersion coefficients of leachate constituents in the aquifer, soil stratigraphy beneath the landfill base, and the area of the landfill base. These factors are all site specific, although for some a range of value may be estimated. Contaminant transport models can be used to estimate concentration of leachate constituents at a down-gradient point from a landfill; however, these models must be field calibrated, which appears to be a fairly costly proposal. Because most potential users of natural attenuation type of landfills (e.g., small townships and cities) are incapable of funding such costly studies, realistically contaminant transport modeling is of very little help in designing NA-type landfills. An alternative method of estimating the approximate average concentration of leachate constituents at a down-gradient distance from the landfill is available (Bagchi, 1983).

Diffusion and Dispersion   Diffusion and dispersion are two mechanism by which leachate is diluted by the aquifer. Because the leachate has a chemical concentration that is different from the background water, it tries to equilibrate with the ambient water quality through diffusion.

Diffusion is essentially a physicochemical phenomenon. Dispersion, on the other hand, is more of a mechanical phenomenon. Dispersion can occur in a longitudinal or in a transverse direction. Longitudinal dispersion occurs in the direction of flow and is caused by different macroscopic velocities, as some parts of the invading fluid move through wider and less tortuous pores. For example, leachate entering at point 2 or 3 in Fig. 15.5 will advance more slowly than leachate entering at point 1.

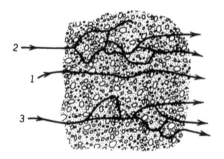

FIG. 15.5   Dispersion of contaminants in soil.

Transverse dispersion occurs normal to the direction of flow and results from the repeated splitting and deflection of the flow by the solid particles in the aquifer. Transverse dispersion is effective only at the edges of a contamination source (Bouwer, 1979). It should be noted that the dispersion theories are applicable to sand and gravel-type deposits. (Harleman et al., 1963). For nonfractured clayey soil deposits with a highly oriented bedding plane, transverse dispersion may not be pronounced.

The relative importance of diffusion and dispersion has been extensively studied (Blackwell, 1959; Bruch and Street, 1967; Raudkivi and Callandar, 1976). It was found experimentally that diffusion was important when Reynold's number is less than 1×10^{-3} (Raimondi et al., 1959). Perkins and Johnston (1963) have provided a comprehensive literature survey on the subject of diffusion and dispersion in porous media. Dispersion in fractured rock or fractured clay deposits will seldom produce a uniform predictable concentration distribution within a plume.

15.2.5 Filtration

Filtration is a mechanism by which leachate constituents are physically trapped. The random pore structure of the soil system serves to physically entrap suspended and settleable solids in a leachate, just as gravity filtration through sand and other media will remove solids in water treatment. Filtration efficiency depends on the pore size and hydraulic gradient of the leachate. Finer soil materials and lower leachate hydraulic gradients will improve filtration. It is difficult to estimate the percentage of attenuation through filtration for any parameter, however, this will be an operative mechanism whenever chemical precipitation, biological growth, and other processes produce undissolved solid particles (Farquhar and Rogers, 1976; Fuller, 1977).

Filtration is a "one-time" phenomenon and is not highly significant so far as leachate attenuation is concerned. Therefore, it is suggested that this mechanism not be taken into account in landfill design.

15.2.6 Precipitation

Chemical precipitation involves a phase change in which dissolved chemical species are crystallized and deposited from a solution because their total concentration exceeds their solubility limit. Adsorption is prominent in microconcentration, but precipitation is the dominant mechanism in macroconcentrations (Fuller, 1977). The solubility limits depend on factors such as ionic species and their concentration, temperature, pH, redox potential (EH), and concentration of dissolved substances (Gebhard 1978). Of these, pH and EH are the most important and can be regarded as "master variables."

The pH of a system controls all acid–base reactions and profoundly influences the equilibrium reactions that determine the relative abundance of hydroxide, carbonate, sulfide, and other ions in a system (Stumm and Morgan,

1970; Ponnamperuma, 1973). The pH of a leachate-saturated soil converges to a near neutral value regardless of the initial pH of the soil; this is because the leachates are typically anaerobic solutions making the soil a reduced media. In acidic soils the reducing reactions consume the available hydrogen ions. In calcereous, alkaline soil reducing reactions increase carbon dioxide pressure. So in both cases the system tends to converge to a near neutral value (Ponnamperuma, 1972, 1973; Roberts and Sangrey, 1977). Therefore, shifting of the pH of groundwater down-gradient from a landfill to a near neutral value may serve as an indicator of plume arrival.

The redox potential also influences precipitation of chemicals. Although solubility plots for many metals under various pH and EH values are available, the main difficulty is in measuring redox potential for a soil–leachate system. Representative values of EH can be obtained only by *in situ* testing of soil. Any determination of redox potential even in Shelby tube samples may not give the actual EH of the soil (Stumm and Morgan, 1970). Thus, although the redox potential influences the solubility of a pollutant, no reliable data regarding the range of EH values of leachate-saturated soil are available for use in a solubility plot.

Table 15.1 provides a summary of the net effect of different attenuation mechanisms.

15.3   EFFECTS OF VARIOUS FACTORS ON ATTENUATION MECHANISMS

The effects of factors such as leachate velocity and degree of saturation of soil are discussed in the following sections. The discussions are qualitative in nature, which will help in designing an NA-type landfill (Bagchi, 1987b).

15.3.1   Leachate Velocity

All attenuation mechanisms are somewhat dependent on leachate velocity. Reaction kinetics will govern four of the five mechanisms: adsorption, biological uptake, cation exchange, anion exchange, and precipitation. Because significant data are not available regarding reaction kinetics of leachate in soil, this is a field worth researching. Presently it is assumed that the flow of leachate is slow enough to complete all applicable attenuation mechanisms.

15.3.2   Degree of Saturation of the Unsaturated Zones

The degree of saturation will influence biological uptake significantly and precipitation reactions to some extent. A higher saturation would mean less available oxygen for biological activity, leading to predominantly anaerobic bacterial growth. Since at worst, soil in the unsaturated zone could be fully

TABLE 15.1 Net Effect of Attenuation Mechanisms

S1 Number	Mechanism	Net Effect
1	Adsorption	Immobilization if change of pH is not high
2	Biological uptake	Temporary immobilization; long-term effect needs additional study
3	Cation and anion exchange	Immobilization of the constituents but elution of some other element causing increase in its concentration
4	Dilution	Reduction of concentration
5	Filtration	Immobilization
6	Precipitation	Immobilization in most cases; change of pH and EH has significant effect

After Bagchi (1987b); courtesy of Waste Management and Research, Denmark.

or nearly saturated, attenuation should be focused on biological uptake under similar conditions.

15.3.3 Organic and Inorganic Matter in Soil, Clay Type, and Soil Fabric

Organic and inorganic matter in clayey soil, usually considerably greater in proportion than clay minerals, influences adsorption, precipitation, and biological uptake. The surfaces of organic matter provide some adsorption sites; in addition, they may serve as an energy source for microorganisms. Inorganic matter such as oxides and hydroxides of iron, aluminum, and manganese participates in the precipitation reaction and influences the pH and EH of the leachate–soil system. Clay types influence cation and anion exchange and precipitation reactions. Soil fabric and texture influence filtration. Note that there are mechanisms other than the cation-exchange capacity of soil that control attenuation of leachate in soil. Under appropriate conditions one or more of these factors could become dominant and exert a controlling influence (Fuller, 1977; Griffin et al., 1976a; Roberts and Sangrey, 1977).

15.3.4 Soil Stratigraphy

Soil stratigraphy or the sequence in which different soil type (i.e., sand, silt, and clay) layers exist at a particular site can significantly influence natural attenuation of leachate. The soil stratigraphy best suited for natural attenuation is (Fig. 15.4) (1) an unsaturated zone consisting mainly of silty clay with a fairly high cation exchange capacity (30–40 mEq/100 g) and a permeability of 1×10^{-4} to 1×10^{-5} cm/sec and (2) a groundwater table that is below the base of the unsaturated silty clay and that is within a sandy layer with fairly high permeability (i.e., 1×10^{-3} cm/sec or more) layer. Because it is difficult to find the ideal soil stratigraphy and because in reality numerous soil stratigraphies exist, it is impossible to discuss each situation. Instead, the following comments regarding soil stratigraphy are included to help in decision making:

1. If a clayey stratum is not available beneath the landfill, construction of a retarder base similar in characteristics to the unsaturated zone mentioned above must be undertaken. Guidelines for retarder base construction are included in Chapter 20.

2. Macrofabric features such as fissuring would cause channeling of leachate without proper contact with the soil in the unsaturated zone. This geologic feature must be properly investigated at the time of subsoil investigation (Rowe, 1972).

3. Uneven weathering of bedrock may create a buried mountain-type bedrock profile in residual soils. Thus, while siting a landfill in residual soils, a bedrock profile below a proposed NA-type landfill site should be properly investigated. Landfill base should not be constructed directly on bedrock.

4. Low permeability (1×10^{-7} cm/sec or less) of the unsaturated zone will not allow proper percolation of leachate, leading to ponding and subsequent leachate seep; a natural attenuation type of landfill should not be sited in such areas.

5. Soil stratigraphy will influence dilution. If the underlying aquifer is not thick enough, then plume development will be restricted, and sufficient dilution may not take place.

6. Soil stratigraphy also influences plume geometry. Paschke and Hoppes (1984) have shown that among other variables the plume shapes depend on the downward velocity of the leachate entering the groundwater and on the velocity of the groundwater. Thus, the permeability of the unsaturated zone, the leachate head within the landfill, and the permeability of the aquifer will influence dilution of leachate parameters in the aquifer.

15.4   ATTENUATION MECHANISMS OF SPECIFIC POLLUTANTS

In the United States 66,000 chemicals are used commercially and worldwide 45,000 substances are traded. It is estimated that 1000 new chemicals are added to the list each year (Richards and Shieh, 1986). No study of the attenuation mechanism(s) for all chemicals in soil is available. The purpose of this section, therefore, is to provide a general overview of the attenuation mechanisms offered by soil and groundwater for the constituents usually found in nonhazardous landfill leachate. Discussions of attenuation mechanisms of pollutants usually found in hazardous waste leachate are not included because hazardous wastes cannot be disposed of in NA-type landfills. However, many of the pollutants mentioned in the following sections may not be present in a particular leachate. At the end of the discussion on each pollutant, a qualitative estimate of the attenuation potential for each pollutant in a clayey environment is included, which will be useful for NA-type landfill design and land-spreading. Attempts have been made to include as many pollutants as possible to form a broad database.

15.4.1   Aluminum

The major attenuation mechanism of aluminum is precipitation. In a high (alkaline) or near neutral pH environment, aluminum readily forms insoluble oxides, hydroxides, and silicates. However, in a low pH (acidic) environment, aluminum is quite soluble. The mobility of aluminum in a clayey environment is low (Griffin, 1977).

15.4.2   Ammonium

The major attenuation mechanisms of ammonium are cation exchange and biological uptake. Biological uptake converts ammonium to nitrate and/or organic nitrogen. These transformations may be reconverted at a later time to ammonia or other products. In the long run, however, denitrification and immobilization as organic nitrogen can combine with sorption and fixation to yield a net reduction of both ammonium and total nitrogen in a leachate (Gebhard, 1978).

Nitrification is a two-step process, carried out by autotropic bacteria. Conditions under which this process proceeds at a maximum rate are adequate supply of ammonium, presence of a moderate pH (optimum around 8.5), adequate oxygen, moderate moisture content, and optimum temperature (about 30°C) (Tisdale and Nelson, 1975). These conditions, particularly the presence of adequate oxygen, may not be fulfilled in a landfill. The ammonium ion may also transform to gaseous ammonia under alkaline conditions in the presence of free lime (Carter and Allison, 1961). More transformation apparently occurs from coarser soils (Gebhard, 1978). The mobility of ammonium in clayey soil is moderate.

15.4.3   Arsenic

The major mechanisms of attenuation of arsenic are precipitation and adsorption. In an aerobic environment, arsenic reacts with iron, aluminum, calcium, and many other metals to form arsenate, which is only slightly soluble. If saturation occurs, arsenic appears in the reduced arsenic form, which is more soluble and mobile than the oxidized form (Swaine and Mitchell, 1960).

When saturated soils return to oxidic conditions, arsenite is oxidized to arsenate (Quastel and Scholefield, 1953).

Adsorption of arsenite at pH 7 is directly proportional to the amount of lime applied and small amounts of arsenite are irreversibly bound to the soil (Fuller, 1977). Maximum removal of arsenate from leachate occurs in the range of about pH 4–6 and generally removal of arsenite increase from pH 3 to 9. Adsorption of arsenite by montmorillonite clay consistently showed an unexplained discontinuous peak at pH 6–7. Montmorillonite clay was found to adsorb approximately twice as much as kaolinite clay. The mobility of arsenic is moderate in clayey soil (Griffin et al., 1976a).

15.4.4   Barium

The major attenuation mechanisms of barium in soils are adsorption, ion exchange, and precipitation. Experiments have shown that adsorption of barium increases with the increasing cation-exchange capacity (CEC) of the soil and decreases with increasing interference from other leachate constituents (Griffin et al., 1976a; Farquhar and Rovers, 1976). Studies with soil organic matter showed that differences in relative stability and retention value of barium, calcium, magnesium, and copper complexes are small. Retention values were ranked in the order copper > barium = calcium > magnesium (Broadbent and Ott, 1957). If free lime is present in soil, barium will be precipitated as barium carbonate. Barium carbonate is only slightly soluble in water and barium is, therefore, effectively attenuated. Soil factors favoring attenuation of barium will include a high clay percentage and the presence of other colloidal material. An alkaline condition and free lime will also tend to favor attenuation by ion exchange and chemical precipitation (Gebhard, 1978). The mobility of barium in clayey soil is low.

15.4.5   Beryllium

The major attenuation mechanisms of beryllium are precipitation and cation exchange. The chemistry of beryllium is similar to aluminum. Beryllium can be mobile in very low or very high pH soil due to hydrolysis (Griffin, 1977). It is highly attenuated in soils, particularly those containing montmorillonite and illite-type clay. Beryllium may displace divalent cations already on common adsorption sites in the exchange complex (Fuller, 1977). In general, the mobility of beryllim is low in clayey soil.

15.4.6   Boron

The major attenuation mechanisms of boron include adsorption as borate on various inorganic surfaces and precipitation or coprecipitation with hydrous iron or aluminum oxides (Gebhard, 1978). Therefore, the activity of boron or borate in soil systems is related to that of aluminum and ferric iron. Leaching experiments using simulated or real landfill leachate have shown a net borate addition to the leachate (Streng, 1976; Griffin et al., 1976a). Its mobility in clayey soil is high (Griffin, 1977).

15.4.7   Cadmium

The major attenuation mechanisms of cadmium are precipitation and adsorption. Adsorption on colloidal surfaces due to coulomb-type forces is said to be primarily responsible for the immobility of cadmium in soils (Fuller, 1977). Cadmium, like zinc, mercury, and lead, undergoes hydrolysis at pH values normally encountered in soil environments. Several studies have shown that pH is the most important factor in controlling the attenuation. Chemical precipitation of cadmium with anions such as phosphate, sulfide, and carbonate can also effectively attenuate cadmium (Santillan-Medrano and Jarinak, 1975, Huang et al., 1977; Griffin et al., 1976b).

Data obtained in leaching experiments using specially prepared cadmium solutions and cadmium-spiked municipal leachate indicate that removal of cadmium by various hydroxides in clayey soils increases from a relatively negligible level to a very significant level as the pH rises from about 6 to 8 and that attenuation is quite stable at pHs greater than 8. Chemical precipitation of cadmium is highly dependent on that available anions, pH, and redox potential (Huang et al., 1977). Precipitation and possible coprecipitation with phosphate, sulfide, and carbonate is likely to occur in a near neutral pH environment (Santillan-Medrano and Jarinak, 1975; Huang et al., 1977). A limestone (calcium carbonate) barrier between landfill materials and soils can provide an additional attenuation capacity for cadmium (Fuller, 1977). However, such use of limestone in an NA-type landfill may decrease permeability of the base, which may cause leachate ponding; an NA landfill will not be successful if leachate ponding occurs.

Houle (1976) found that changes in leachate salts and available complexing ions could alter the degree to which cadmium is attenuated. Thus, it appears that cadmium adsorption in leachate–soil systems is not irreversible and remobilization can easily occur. The mobility of cadmium in clayey soil in moderate.

15.4.8   Calcium

The major attenuation mechanisms for calcium are precipitation and cation exchange. It readily forms carbonate precipitation under alkaline pH. Since calcium is the dominant ion in the soil exchange complex, it is not adsorbed but in most cases is eluted (Griffin, 1977). Montmorillonite was found to elute calcium to a significantly greater degree than illite and kaolinite. It has been postulated that calcium elution from soil is due to exchange of calcium by sodium, potassium, ammonium, and magnesium from leachate. This elution adds to the hardness of the groundwater beneath municipal sanitary landfills. Thus, calcium is not only highly mobile in clays, there is a good possibility of this pollutant being eluted.

15.4.9   Chemical Oxygen Demand (COD)

Rather than being an individual ionic species that can be studied in its various states, COD represents a group of compounds that can be oxidized by a boiling solution of potassium dichromate and sulfuric acid. The molecular species falling into this category include many (but not all) organic materials. For example, straight chain aliphatic compounds, aromatic hydrocarbons, and pyridine are not oxidized by a chromic acid–sulfuric acid solution (Gebhard, 1978).

Dissolved organic compounds occur in waters in such low concentrations that they usually have to be concentrated and separated from the inorganic salts before they can be identified chemically. Bioassay methods are frequently of great sensitivity and permit direct determination of very low concentrations of organic nutrients in water (Stumm and Morgan, 1970).

The most important mechanism of COD attenuation is biological uptake, which produces sulfur dioxide and methane. Filtration is a minor mechanism of attenuation for COD In a municipal landfill, decomposition will usually take place under anaerobic conditions, which results in slower decomposition rates, chemically reducing conditions, and production of acid by-products (Patrick and Mahapatra, 1968).

Vigorous microbial activity (controlled by near neutral pH, adequate supplies of nutrients, and dissolved oxygen where aerobic growth is possible), absorption, and ion exchange (controlled mainly by high soil organic matter, percentage clay, cation-exchange capacity, iron oxide content, etc.) favor attenuation of organic constituents of leachate. It is suggested that for some organic compounds the best overall correlation is with soil organic matter (Greenland, 1970). In a sanitary landfill, maintenance of an aerobic condition is not usual even though experiments were performed to study the feasibility of maintaining such a condition (Stone, 1975). A fine grain soil will probably favor COD attenuation due to increased surface area and improved mixing between solution and solids. COD is relatively mobile in a clayey environment (Gebhard, 1978).

15.4.10   Chloride

Chloride is not attenuated by any soil type and is highly mobile under all conditions (Apgar and Langmuir, 1971; Polkowski and Boyle, 1970; Gerhardt, 1977). Dilution is the only attenuation mechanism for this leachate constituent.

15.4.11   Chromium

Precipitation and cation exchange and/or adsorption are the principal mechanisms of attenuation of chromium in soil (Griffin, 1977; Gebhard, 1978). The redox potential has a marked effect on attenuation. The importance of each mechanism is dependent on the form of chromium. Chromium is found in two valent states: hexavalent and trivalent. Hexavalent chromium is anionic in form and trivalent chromium is cationic in form. The dominant species appears to be trivalent (Fuller, 1977). In a municipal landfill, hexavalent chromium could be of concern. The concentration of chromium in the soil is reduced by adsorption on organic matter, clay minerals and hydrous oxides of iron, manganese, and aluminum and precipitates as an oxide (Basu et al., 1964). Data obtained in leaching experiments indicate that trivalent chromium is attenuated effectively in soil systems (Griffin, 1977). At a pH exceeding 6, migration will be controlled by precipitation as an oxide, carbonate, or sulfide (Griffin et al., 1976b; Fuller, 1977). Below pH 4, trivalent chromium species are effectively attenuated by adsorption on both kaolinite and montmorillonite (Griffin, 1977). Between this pH range, a combination of the two mechanisms is effective with a radical increase in attenuation to a maximum at pH 6 and above. Attenuation of hexavalent chromium was found to be a function of both concentration and soil pH. Results of leaching experiments indicate that montmorillonite is more effective than kaolinite in attenuating chromium (Griffin, 1977). Retention of hexavalent chromium appears to correlate best with amount of iron oxide, manganese, and clay in the soil (Korte et al., 1976). In clayey soil trivalent chromium is immobile whereas hexavalent chromium is highly mobile. Soil materials contributing to attenuation of chromium will include organic matter, clay minerals, and hydrous metal oxides. The impact of pH will depend on the valence state of the chromium. Less attenuation is expected in coarse textured soils than in fine textured soils because of the larger pores, greater permeability, and smaller amounts of clay minerals (Fuller, 1977).

15.4.12   Copper

Most important attenuation mechanisms for copper include adsorption, ion exchange, and chemical precipitation. The majority of the available information concerning the activity of copper in soil–water environments concerns divalent copper (Gebhard, 1978). Attenuation studies indicate that the removal of copper varies somewhat with clay type. The soil pH is the most important factor controlling removal of copper with a given absorbant (Griffin et al., 1976b; Huang et al., 1977). Montmorillonite appears to be more effective in removing copper than kaolinite; however, the amount of copper removed from solutions was not equal to the cation-exchange capacity of the clay mineral. It was suspected that desorbing calcium ions effectively competed with the heavy metals present in solutions (Griffin et al., 1976a). There are some copper compounds that become soluble under acidic conditions. In the pH range of 5–6 precipitation of copper compounds can occur when copper concentrations are high (Griffin et al., 1976b).

Copper attenuation by organic matter is extensive and indicates strong complexing with organic matter. Soil column testing indicates that copper is of very low mobility for a wide range of soils (Korte et al., 1976).

Soil materials favoring attenuation of copper include colloidal matter, free lime, hydrous oxides of manganese and iron, a high clay content, and organic matter content. Use of ground limestone resulted in a significant improvement in attenuation of copper (Fuller and Korte, 1976). It appears that a near neutral pH may be the most effective in copper attenuation by clays (Griffin et al., 1976b), although solubility of copper in soil systems may continue to decrease as the pH rises (Lindsay, 1972). The mobility of copper in a clayey environment is low.

15.4.13   Cyanide

The only attenuation mechanism for cyanide is adsorption. Cyanide is an anion and as such is not strongly retained in soils. The adsorption is dependent on the pH of the soil (Griffin, 1977). Based on the limited information available it appears that cyanide is highly mobile in clayey environments.

15.4.14   Fluoride

The major attenuation mechanism of fluoride in soil appears to be anion exchange. Although Bower and Hatcher (1967) found that acidic soils adsorb fluoride more readily than alkaline soils, Larsen and Widowsen (1971) found that the solubility of fluoride increases in both acidic and alkaline soils. In general, the mobility of fluoride in clayey soil is high.

15.4.15   Iron

Precipitation, cation exchange, adsorption, and biological uptake are the important attenuation mechanisms of iron. Divalent and trivalent iron are present in almost all leachate–soil systems (Gebhard, 1978). Below approximately neutral pH conditions, the solubility of divalent iron increases about 100-fold for each unit decrease in pH (Lindsay, 1972). Iron oxides have been found to be among the most significant factors influencing attenuation processes (Fuller and Korte, 1976). In general, iron compounds appeared to be moderately attenuated in soil (Fuller and Korte, 1976). Although Griffin et al. (1976a) found no significant correlation with cation exchange, Farquhar (1977, cited in Gebhard, 1978) found that iron attenuation did increase with the cation-exchange capacity of soil.

Biological activity can influence iron activity in two ways. Anaerobic growth creates reducing conditions and acid by-products that will convert ferric iron to ferrous iron and increase iron mobility.

A zone may exist in a soil system in which the solubility of iron is considerably greater than the drinking water standard. This zone has near neutral or acidic pH and moderately reducing conditions. Such a zone may form the bulk of migrating leachate plumes. High iron levels in groundwater near landfills may only be a consequence of migration of a moderately reduced zone rather than of migration of iron from the landfill (Roberts and Sangrey, 1977).

The mobility of trivalent iron in soil is low and divalent iron is high (Griffin, 1977). Since it is difficult to assess which species is present, it is better to assume that the mobility of iron is high to moderate in clayey soil.

15.4.16 Lead

The major attenuation mechanisms for lead are adsorption, cation exchange, and precipitation. Although lead may be present in two valence states, practically all the common lead compounds correspond to the divalent state (Gebhard, 1978). Lead attenuation in clays increases as the pH rises above 5 (Griffin, 1977). If was found that the lead removal capacity of montmorillonite is higher than that of kaolinite (Griffin et al., 1976a). In a separate study of soil conducted in New York, the major attenuation mechanism for lead has been reported to be precipitation, which depends on the EH–pH state of the soil (Roberts and Sangrey, 1977). In soil systems, lead can be expected to form poorly soluble precipitates with sulfate, carbonate, phosphate, and sulfide anions. Evidence of a relatively insoluble organic matter complex has also been found (Fuller, 1977). Experimental data indicate that lead hydroxide probably regulates lead activity in soils at pH less than 6.6 (Santillan-Medrano and Jarinak, 1975).

In a municipal landfill, where anaerobic conditions are likely to occur, lead should become more mobile (Fuller, 1977). This is consistent with findings of Griffin et al. (1976a), which indicate that competitive effects of other constituents in municipal landfill leachate can also lower the removal of lead (Gebhard, 1978). Soil materials favoring attenuation of lead in leachate will include organic matter, clays, and free lime. Most effective removal will require a pH greater than 5 or 6. It was also reported that there is a greater affinity between lead and organic matter that results in immobilization of lead. Lead is generally more strongly attenuated than many other divalent heavy metals (Fuller, 1977). The mobility of lead in clayey soil is low.

15.4.17 Magnesium

Cation exchange and precipitation are the major attenuation mechanisms for magnesium. Under neutral to alkaline pH, magnesium may form a carbonate precipitate under favorable conditions (Griffin, 1977). Magnesium attenuates moderately in clayey soil.

15.4.18   Manganese

Precipitation and cation exchange are the major attenuation mechanisms for manganese. Manganese is a very common element in soils and can have valence states of $2+$, $3+$, $4+$, $6+$, and $7+$. It has been suggested that under normal conditions manganese exists in soils primarily as insoluble oxides, although under reducing conditions Mn^{2+} is formed and can increase in solubility to the point at which it can participate in ion exchange (Ellis and Knezek, 1972). A divalent manganese ion is formed from a tetravalent ion when the redox potential is in the range of $+200$ to $+400$ mV (Lucas and Knezek, 1972). Investigations of the activity of manganese in soil systems under the influence of synthetic or real landfill leachate are contradictory. Using leachate collected from an Illinois landfill, Griffin et al. (1976a) found that at near neutral pH, manganese was actually removed from kaolinite and montmorillonite. In contrast, Farquhar (1977, cited in Gebhard, 1978) found that manganese was moderately attenuated in several soils. He also found very limited desorption of the retained manganese with the addition of water. Therefore, he concluded that manganese has a high selectivity in the exchange sequence. Farquhar also found that manganese removal in soils appeared to increase with the cation-exchange capacity.

Under either alternate wetting and drying conditions or saturated (anaerobic) conditions, the absorption of manganese is highest on bentonite (montmorillonite), intermediate on illite, and lowest on kaolinite.

Soil materials favoring manganese attenuation include clays, organic matter, hydrous metal oxides, and free lime. Alkaline conditions and an abundance of anions such as sulfide and carbonate will improve retention (Gebhard, 1978). The mobility of manganese in clayey soil is high.

15.4.19   Mercury

The dominant attenuation mechanisms of mercury are adsorption, precipitation, and redox reactions resulting in volatilization of this heavy metal (Gebhard, 1978). In aerated water with a neutral pH, the inorganic mercury species distribution is dominated by mercurous hydroxide, but under reducing conditions, elemental mercury is formed readily and can be lost by volatilization. The methyl mercury ion is common in nature above pH 6, and decomposes slowly to methane at a temperature of around 20°C. In aqueous systems such as a landfill leachate, the predominant species is often mercurous chloride (Fuller, 1977, cited in Gebhard, 1978).

The conversion of inorganic mercury compounds to toxic mono- or dimethyl mercury has been shown to be the result of activity by various bacteria. Variation of the anion initially associated with divalent mercury will affect the rate at which methylation takes place (Ridley et al., 1977). Enzymes produced by enteric bacteria have been found to be effective in demethylating methyl mercury. The products of those reactions include dimethylselenide, methane, and reduced mercury (Wood et al., 1975).

In clayey soils, mercury compounds are likely to be attenuated by adsorption with iron oxide, organic matter, and clays. Experimental results indicate that adsorption of mercury, in either specially prepared mercurous chloride solutions or in mercury-spiked municipal landfill leachate, is pH dependent and increases steadily as the pH rises from about 2 to 8 (Griffin, 1977). However, large amounts of mercury were also removed from solution in the absence of clay, suggesting that precipitation and/or volatilization accounted for the removal of 70–80% of the mercury from leachate solutions (Griffin, 1977).

Using extraction by 0.1 N HCl, Korte et al. (1976) found that sorbed mercury was more easily remobilized by acid solution, especially with several clay soils. Extraction with water yielded negligible to low remobilization. In another test, use of a crushed limestone liner had a low to moderate effect on mercury attenuation and improvement was better with longer contact times. Mercury applied to the environment will not remain in the same state and dimethyl mercury or reduced mercury is expected to be expelled as gas (Fuller, 1977, cited in Gebhard, 1978).

Results of investigations indicate that maximum removal of mercury from leachate should be expected under alkaline conditions. It appears that attenuation can be improved by soil colloidal matter, especially clays and iron oxides. In general, it appears that mercury is highly mobile in soils (Gebhard, 1978).

15.4.20 Nickel

The major attenuation mechanism of nickel includes sorption and precipitation. Jenne (1968) found that nickel is removed from solution by hydrous metal oxide precipitates. Nickel appears to have much greater affinity for manganese oxide than ferric oxide.

Korte et al. (1976) found that nickel retention in soil from a nickel-spiked municipal landfill corelates will with surface areas, cation-exchange capacity, and clay content. Other soil factors favoring retention of nickel include alkaline conditions, high concentrations of hydrous metal oxides, and free lime. The mobility of nickel is moderate in clayey soil (Griffin, 1977).

15.4.21 Nitrate

The major attenuation mechanism for nitrate is biological uptake. The biological denitrification or reduction to gaseous nitrogen or nitrous oxide requires anaerobic conditions and a carbon source. At a pH of 7 and redox potential of about +225 mV denitrification will begin to take place (Patrick and Mahapatra, 1968). The pH of the system (in the range of 5–8.4) does not seem to have a pronounced effect on the rate of denitrification, but does affect the final product. Limited biological loss of nitrate has also been found to occur by way of direct volatilization of nitrate as nitric acid under conditions of low soil pH or with the presence of high exchangeable aluminum in soil (Gebhard, 1978). Immobilization of nitrate nitrogen by adsorption in the bodies of microorganisms has been suggested (Allison and Klein, 1962). As the carbon source is exhausted and the microorganism population begins to decline, nitrogen with the potential for reconversion to nitrate is again released. Nitrate is considered to be highly mobile in soil (Tisdale and Nelson, 1975; Preul, 1964; Griffin, 1977).

15.4.22   Polychlorinated Biphenyls (PCBs)

The major attenuation mechanisms of PCBs in soil are adsorption and biodegradation. Volatilization and plant uptake are important attenuation mechanisms in the context of landspreading. Attenuation of Aroclor 1254 (one variety of PCB) was found to be dependent on soil type (Iwata et al., 1973). The quantities or sorbent required to sorb 50% of applied PCBs increased in the following order: a peaty muck < montmorillonite < sand < peroxide-treated sand. The presence of activated carbon enhanced the attenuation of PCB (Strek, 1980, cited in Girvin and Sklarew, 1986). Less chlorinated PCBs were observed to have a higher degree of biodegradation (Griffin and Chian, 1980). The position of chlorine atoms also affects attenuation of PCBs in soil. Biodegradation is more significant for monochloro-, dichloro-, and trichloro-PCBs than for pentachloro- and highly chlorinated PCBs (Pal et al., 1980). There is an active debate regarding rates and reversibility of PCB sorption/desorption in soil. In general, the mobility of PCBs in clayey soil is considered to be high to moderate.

15.4.23   Potassium

The major attenuation mechanisms affecting potassium are precipitation and cation exchange. Griffin et al. (1976a) conclude from experiments that potassium is well attenuated in clayey soil. In micalike minerals such as illite, potassium is readily fixed in a nonexchangeable position. The attenuation is maximum under neutral to alkaline conditions. The mobility of potassium in clayey soil is moderate (Griffin, 1977).

15.4.24   Selenium

The major attenuation mechanisms for selenium are adsorption and anion exchange. Selenium is typically present in soils as an inorganic anion associated with iron, calcium, or sodium. Under the action of soil microorganisms and atmospheric agents, selenium can be oxidized and reduced repeatedly (Fuller, 1977).

Griffin et al. (1976a) studied removal of selenium by kaolinite and mont-morillonite using specially prepared selenium-deionized water solutions and selenium-spiked municipal landfill leachate. Selenium removal by montmo-rillonite was found to be two to three times greater than kaolinite. A distinct pH dependency was observed, indicating that selenium removal improved as the pH dropped to values in the range of 2–4, below with selenium removals decreased. The mobility of selenium in clayey soils is moderate.

15.4.25 Silica

The major attenuation mechanism is precipitation. Not much information is available in the literature regarding the attenuation of silica. Silica readily precipitates in silicate mineral phases. It is moderately mobile in soil, how-ever, mobility increases under alkaline condition (Griffin et al., 1976b).

15.4.26 Sodium

Cation exchange is the major attenuation mechanism for sodium. It may be totally attenuated, but since it is a monovalent ion, a low concentration of sodium could pass through soil without being attenuated at all (Griffin et al., 1976a).

15.4.27 Sulfate

The major attenuation mechanism of sulfate is anion exchange. Adsorption on a clay surface, on organic matter, and on hydrous oxides of aluminum and iron has also been reported (Chao et al., 1962). Because sulfate is relatively weakly held, leaching losses are proportional to the amount of water passing through the soil (Tisdale and Nelson, 1975).

Sulfate is also a product and a reactant of soil microorganisms. It is pro-duced from sulfides and free sulfur by the action of aerobic bacteria; optimal conditions for their growth include abundant oxygen, temperatures near 30°C, and low pH (Tisdale and Nelson, 1975). At a pH at 7 sulfate reduction will occur at a redox potential of about 150 mV, which is much lower than the redox potentials necessary for reduction of either nitrate or ferric iron. If iron is not present sulfide loss as hydrogen sulfide gas can occur (Patrick and Mahapatra, 1968). Since sulfate is an anion its mobility is high in soil (Griffin, 1977).

15.4.28 Viruses

Virus survival within soil depends on soil moisture content, temperature, pH, and nutrient availability. Inactivation of viruses near the soil surface is much more rapid than when it penetrates into the soil (Keswich and Gerba, 1980). Transport of viruses for a long distance through sandy soil has been reported (Wellings et al., 1974). Factors that favor virus removal from leachate include clayey soil, low pH, and the presence of cations (Lu et al., 1985). The mobility of viruses in clayey soil is low.

15.4.29   Volatile Organic Compounds (VOC)

These are organic compounds that volatilize at normal temperature and pressure (NTP), with some exceptions. Biological uptake and dilution are the major attenuation mechanisms for these compounds. The existence of several VOCs in landfill leachate has been reported (Sridharan and Didier, 1988; McGinley and Kmet, 1984). Although biodegradation of organics has been demonstrated by several researchers (Callahan et al., 1979; Tabak et al., 1981; Petrasek et al., 1983; Alexander, 1981; Barker et al., 1986), the rate of metabolism is the main factor (Richards and Shieh, 1986). In a soil–leachate system attenuation of VOC is not expected to be high (Edil et al., 1992). VOC concentrations of groundwater around landfills with a volume of more than 38,000 m^3 (50,000 yd^3) were found to be high (Friedman, 1988; Battista and Connelly, 1988).

15.4.30   Zinc

The major attenuation mechanisms for zinc are adsorption, cation exchange, and precipitation. Zinc is a common cation in soil systems. As is true with other cations, the pH of the leachate–soil system is a crucial factor in zinc removal, reflecting the influence of dominant hydrolysis species on both the affinity for soil colloids and the sollubility of zinc (Gebhard, 1978).

The attenuation of zinc was found to increase rapidly for a pH change from 2 to 8 with a significant rise around 6 to 8 (Griffin et al., 1976b). Precipitation of zinc with a variety of anions—including sulfide, phosphate, carbonate, and silicate—has also been found to be important in zinc immobilization (Stumm and Morgan, 1970; Fuller, 1977). Experimental results suggest that the removal of zinc is also dependent on clay type and cation-exchange capacity (Griffin, 1977; Farquhar, 1977). Organic matter improves zinc immobilization (Folett and Lindsay, 1971; Huang et al., 1977; Norcell, 1972), and zinc chelates are most stable at pHs between about 5 and 7.5 (Folett and Lindsay, 1971; Huang et al., 1977; Norcell, 1972). Soil material favoring attenuation of zinc includes clays, organic material, hydrous metal oxides, and free lime. Zinc attenuation will be most favored by an alkaline condition. In general, mobility of zinc in a clayey environment is low (Griffin, 1977).

15.4.31   Summary

Attenuation mechanisms for 30 common landfill leachate constituents are discussed. A summary of the attenuation mechanisms of these constituents is included in Table 15.2. From the above discussion, the following general trends can be observed: (1) Most metals attenuate well in clayey soil, (2) nonmetals are not attenuated well in clayey soil, (3) very low attenuation of nitrate, sulfate, VOC, and COD is expected to occur in clayey soil, and (4) chloride is not attenuated in soil at all: dilution is the only mechanism of attenuation for chloride.

TABLE 15.2 Major Attenuation Mechanism(s) of Landfill Leachate Constituents

Leachate Constituent	Major Attenuation Mechanism	Mobility in Clayey Environment
1. Aluminum	Precipitation	Low
2. Ammonium	Exchange, biological uptake	Moderate
3. Arsenic	Precipitation, adsorption	Moderate
4. Barium	Adsorption, exchange, precipitation	Low
5. Beryllium	Precipitation, exchange	Low
6. Boron	Adsorption, precipitation	High
7. Cadmium	Precipitation, adsorption	Moderate
8. Calcium	Precipitation, exchange	High
9. Chemical oxygen demand	Biological uptake, filtration	Moderate
10. Chloride	Dilution	High
11. Chromium	Precipitation, exchange, adsorption	Low (Cr^{3+}); high (Cr^{6+})
12. Copper	Adsorption, exchange, precipitation	Low
13. Cyanide	Adsorption	High
14. Fluoride	Exchange	High
15. Iron	Precipitation, exchange adsorption	Moderate to high
16. Lead	Adsorption, exchange precipitation	Low
17. Magnesium	Exchange, precipitation	Moderate
18. Manganese	Precipitation, exchange	High
19. Mercury	Adsorption, precipitation	High
20. Nickel	Adsorption, precipitation	Moderate
21. Nitrate	Biological uptake, dilution	High
22. PCBs	Biological uptake, adsorption	Moderate to high
23. Potassium	Adsorption, exchange	Moderate
24. Selenium	Adsorption, exchange	Moderate
25. Silica	Precipitation	Moderate
26. Sodium	Exchange	Low to high
27. Sulfate	Exchange, dilution	High
28. Zinc	Exchange, adsorption, precipitation	Low
29. Virus	Unknown	Low
30. Volatile organic compound	Biological uptake, dilution	Moderate

After Bagchi (1987b); courtesy of Waste Management and Research, Denmark.

15.5   DESIGN APPROACH

Although the methodology discussed in this section is applicable to the design of municipal landfills, the approach may be adapted to design nonhazardous, nonmunicipal landfills. Because of the microorganisms present in municipal landfill leachate, it is difficult to duplicate leachate in a laboratory that can represent a field leachate. The exact composition of leachate and its variation with time cannot be established until the landfill has been operated for some time. Because of this problem, exact quantitative analysis of the chemical reaction between a landfill leachate and soil, before a landfill siting, is not possible. As an alternative, a semiquantitative design method is used utilizing the information included in Sections 15.2, 15.3, and 15.4. The following issues need to be addressed to develop a design approach for NA-type landfills (Bagchi, 1983);

1. A list of leachate constituents with significant concentrations in municipal leachate, and the expected total mass of some of these constituents during the design life of the landfill
2. Attenuation mechanism(s) of the leachate constituents with significant concentrations and the overall impact from the waste-specific leachate on groundwater (if available from field study)
3. The volume of soil in the unsaturated zone that is involved in the attenuation reaction
4. Dilution of leachate constituents in the groundwater aquifer

Soil stratification shown in Fig. 15.4 is assumed in developing this design approach.

15.5.1   Major Leachate Parameters

The pollutants of concern in municipal landfill leachate are copper, lead, zinc, iron, ammonium, potassium, sodium, magnesium, BOD, COD, nitrate, chloride, and sulfate (ASCE, 1976; Chian and Dewalle, 1975; Garland and Mosher, 1975; Ham and Anderson, 1974, 1975a,b). Chemical constituents and their concentrations in leachate vary over a wide range. Refuse composition, landfill age, and climate are the main factors causing variability. The generalized concentration variation plot, shown in Fig. 15.6, is based on field observations and laboratory studies (Griffin et al., 1976a; Meyer, 197This plot is used to calculate the total mass leached for each of the following: ammonium, potassium, sodium, and magnesium.

In Fig. 15.6, "P.O." represents the period from the start of the landfill to closure. For P.O. < 5 years point C coincides with point B; tz is the time (since closure of the landfill) at which the leachate concentration is low

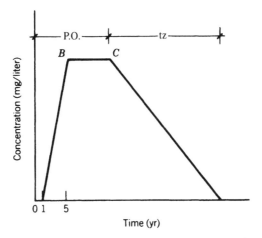

FIG. 15.6 Generalized concentration variation plot for leachate constituents.

enough so as not to cause any environmental problems. Use of this plot to calculate the total mass of ammonium, potassium, sodium, and magnesium is shown in an example, worked out at the end of Section 15.5.5.

15.5.2 Attenuation Mechanisms of Pollutants and Major Impact of Leachate

Five mechanisms, adsorption, biological uptake, cation-exchange reaction, filtration, and precipitation, are operative in the clayey stratum (first stage). Although all six attenuation mechanisms could be operative in the aquifer (second stage), only one mechanism, dilution, is assumed to be operative in the groundwater aquifer. This is because the aquifer sand is assumed to have negligible amounts of fines. This assumption leads to a conservative design.

The first stage of attenuation, which takes place in the initially unsaturated zone, will cause the pH of the leachate–soil system to merge to near neutral value (Roberts and Sangrey, 1977). Attenuation of copper, lead, zinc, iron (in part), ammonium, magnesium, potassium, and sodium will occur in the first stage under the near neutral pH environment. Ammonium, magnesium, and potassium will be exchanged to elute calcium and thus will increase the hardness of the groundwater (Griffin et al., 1976a). BOD, COD, iron (in part), nitrate, chloride, and sulfate are attenuated through dilution.

Field observations on the impact of leachate on groundwater are supportive of the laboratory observations indicated in Section 15.4. Many have observed a predominant increase in hardness of the groundwater in the vicinity of municipal landfills (Andersen and Bornbush, 1967; Walker, 1969; Ziezel et al., 1962).

15.5.3   Soil Volume Involved in Attenuation

Seepage of water from a channel to the groundwater table, the channel bottom being covered with a thin layer of sediment, is discussed by Bear (1969). This model fits the municipal landfill situation. Neglecting the length effect and leachate head buildup in a municipal landfill (which, in a wall-designed municipal landfill, is normally low) the volume of soil involved in attenuation reactions could be approximated by

$$V_s = R \times A \times H \tag{15.1}$$

in which V_s = the soil volume available for attenuation reaction in cubic meters and R = the reduction factor to account for the soil fabric effect (varies between $1/1.2$ and $1/1.3$). The flow of pore fluid through the soil is mostly via large pores, therefore the total cation-exchange capacity of the attenuating soil mass must be reduced to account for the soil that does not come into contact with the leachate. A = the base area of the landfill in square meters and H = the average depth of the unsaturated zone beneath the landfill in meters. The value of V_s is somewhat conservative but is reasonable for practical purposes (Mundell, 1984; Bagchi, 1984).

15.5.4   Dillution of Pollutants

Dillution does not reduce the amount of contaminant in the flow system but does reduce the concentration of contaminant. When the exfiltrate (the leachate after flowing through the unsaturated zone) reaches the groundwater table, the concentration of each pollutant is reduced further either due to density differences or diffusion and dispersion (Bouwer, 1979; Bruch and Street, 1967) or a combination of both. The plume geometry suggested in Fig. 15.7 is developed from laboratory and field data (Bouwer, 1979; Bruch and Street, 1967; Fattah, 1974; Freeze and Cherry, 1979; Kimmel and Braids, 1974; Nicholson et al., 1980; Sykes et al., 1969). The plume geometry is applicable for diluting only nonreactive pollutants in a nonreactive aquifer. The dotted line in Fig. 15.7a represents the more probable plume configuration; however, so far as steady-state dilution is concerned, the discrepancy between the probable plume and suggested plume geometry will not significantly affect results.

The area that will be intercepted by the proposed plume at some horizontal distance is given by

$$A_i = [(L_1 \times X_1)\tan \theta_1 + X_2 \tan \theta_2]$$
$$\times [L_2 + 2(X_1 + L_1)\tan \theta_3 + 2X_2 \tan \theta_4] \tag{15.2}$$

in which A_i = the area intercepted by groundwater for effecting pollutant dilution in square meters. X_1 and X_2 = the horizontal distance as shown in

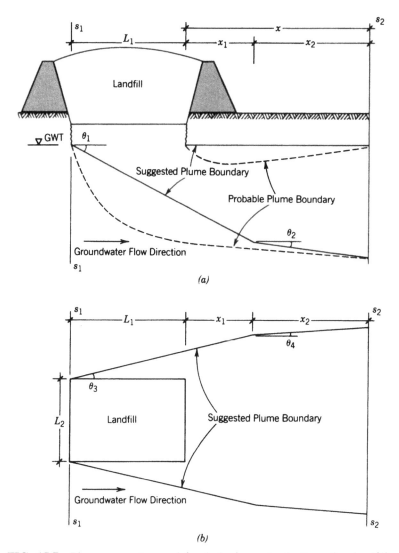

FIG. 15.7 Plume geometry used for designing natural attenuation landfills.

Fig. 15.7 in meters. L_1 = the dimension of the landfill parallel to groundwater flow in meters and L_2 = the dimension of the landfill perpendicular to groundwater flow in meters; θ_1 through θ_4 = divergence angles in degrees.

Equation (15.3) is obtained by equating the mass per unit time of a parameter passing through area A_i at section s_2–s_2 and the sum of pass per unit time of the parameter in the ambient water passing through the same area A_i at section s_1–s_1 and leachate exfiltrate, which enters the proposed plume configuration through the landfill bottom.

$$C_b A_i U + C_l V_e = C_{dx}(A_i U + V_e) \tag{15.3}$$

in which C_b = the average concentration of a parameter in the background water in mg/m³, C_l = the peak concentration of a parameter in the leachate in mg/m³, U = the average groundwater velocity in m/day, V_e = the average volume of leachate exfiltration per unit time in m³/day, C_{dx} = the average concentration of a parameter in the down-gradient water at distance X ($= X_1 + X_2$) in mg/m³. Rearranging the terms

$$\frac{C_b}{1 + (V_e/A_i U)} + \frac{C_l}{1 + (A_i U/V_e)} = C_{dx} \tag{15.4}$$

The plume geometry is applicable close to the landfill, which would mean that X_1 is less than or equal to 90 m (300 ft) and ($X_1 + X_2$) is less than or equal to 300 m (1000 ft). The suggested values of the divergence angles are $\theta_1 = 4–7°$, $\theta_2 = 2–5°$, $\theta_3 = 2–3°$, and $\theta_4 = 1–2°$. These values are based on limited field data; if available; additional field data should be taken into consideration.

15.5.5 Design Procedure

The following procedure is applicable to a typical NA-type landfill site described earlier (Fig. 15.4).

1. Copper, lead, and zinc: Because the concentrations of these heavy metals are low in leachate, they will precipitate out in a clayey environment.

2. Iron: Attenuation of iron is not high in clayey soil. Assume that 50% of peak concentration will be diluted in the sandy aquifer.

3. Ammonium, magnesium, potassium, and sodium: These will undergo cation exchange and in turn will increase the hardness of the exfiltrate from the first stage. First, soil volume should be checked to see whether enough CEC is available for performing the exchange reaction with these four pollutants. If adequate CEC is available, then the hardness of the exfiltrate (the hardness of the leachate plus the increase in hardness due to the cation-exchange reaction of these pollutants) should be checked for dilution.

4. BOD, COD, nitrate, chloride, and sulfate: Dilution is the only long-term attenuation mechanism for these constituents. Therefore their peak concentration should be checked for dilution.

The landfill plan dimensions should be increased, if necessary, until acceptable concentrations for all leachate parameters are met within 300 m (1000 ft) of the landfill or the property boundary, whichever is less. The following example will enumerate the design steps.

Example 15.1   The proposed disposal site has a dimension of 100 m (parallel to the groundwater flow $= L_1$) by 150 m (perpendicular to the groundwater flow $= L_2$). The unsaturated zone is 6 m thick and has clayey soil (CEC $=$ 30 mEq/100 g of soil). The proposed waste volume is 60,000 m³ with a site life of 10 years. The expected peak concentration of parameters in the leachate and in the background water is given in Table 15.3. Groundwater velocity in the sandy aquifer is $1 = 10^{-1}$ m/day. Given $R = 1/1.2$, tz $= 15$ years, unit weight of soil $= 1.9$ g/cm³, $\theta_1 = 5°$, $\theta_2 = 3°$, $\theta_3 = 2.5°$, $\theta_4 = 1°$, $X_1 = 60$ m, $X_2 = 100$ m, and rainfall $= 75$ cm/year.

Calculation for Exchange Reaction   The available cation-exchange capacity (CEC) of soil: substituting relevant values in Eq. (15.1):

$$V_s = (1/1.2) \times 100 \times 150 \times 6 = 7.5 \times 10^4 \text{ m}^3$$

$$\text{Weight of soil} = (7.5 \times 10^4) \times (1.9 \times 10^6) = 1.425 \times 10^{11}$$

$$\text{Available CEC in the soil} = (1.425 \times 10^{11}) \times (30/100)$$

$$= 42.75 \times 10^9 \text{ mEq}$$

Assuming a uniform filling over the 10-year period, the following formula can be obtained from Fig. 15.6 for the total mass for ammonium, potassium, magnesium, and sodium leached from the refuse.

TABLE 15.3   Peak Concentration of Selected Parameters in Leachate

S1 Number	Constituent	Leachate (mg/liter)	Background Water (mg/liter)
1	Ammonium	200.00	0.00
2	BOD (5 day)	5,000.00	0.00
3	Chloride	500.00	100.00
4	COD	12,000.00	0.00
5	Copper	0.5	0.01
6	Hardness (CaCO₃)	4,000.00	50.00
7	Iron	200.00	0.03
8	Lead	0.2	0.005
9	Magnesium	150.00	0.00
10	Potassium	200.00	0.00
11	Sodium	300.00	0.00
12	Sulfate	300.00	50.00

$$\text{Total mass leached} = \frac{V_f C_1}{4} (1 \times 2 + 2 \times 3 + 3 \times 4 + 4 \times 5)$$

$$+ V_f C_1 (6 + 7 + \cdots + 10)$$

$$+ \frac{10 V_f C_1}{15} (1 + 2 + 3 + \cdots + 14) \qquad (15.5)$$

$$\text{where } V_f = \text{landfill filling rate (m}^3/\text{year)}$$

$$= 60{,}000/10 = 6000 \text{ m}^3/\text{year}$$

Sample Calculation

1. First term (year 3): Mass leached in third year $= C_3 \times 3V_f \times 1$. From Fig. 15.6. $C_3/2 = C_1/4$. Therefore mass leached in the third year $= (V_f \times 2C_1 \times 3 \times 1)/4$.
2. Second term (year 7): Mass leached in the seventh year $= 7V_f \times C_1 \times 1 = V_f \times C_1 \times 7$.
3. Third term (year 23): Mass leached in the 23rd year, $= 10V_f \times C_{14} \times 1$. From Fig. 15.6. $C_{23}/2 = C_1/15$. Mass leached in the 23rd year $= (10V_f \times C_1 \times 2 \times 1)/15$.

Note that the first term of Eq. (15.5) will remain the same so far as P.O. > 5 but the second and third terms will depend on the period of operation (for P.O. > 5).

From Table 15.4 the total milliequivalents available in soil is higher than the total leached. [Note: mEq/liter = concentration (mg/liter × valence/atomic weight).] Hence, hardness causing constituents in the leachate will be exchange totally by the clayey stratum. To calculate the increase of hardness in the groundwater, the peak concentration of each of these pollutants shall be used to study the worst case. Peak increase in hardness due to exchange reaction $= 41.6 \times 40.08/2$ (assuming exchange takes place with the calcium ion only) $= 833$ mg/liter. Total hardness of the exfiltrate $= 4833$ mg/liter.

TABLE 15.4   Total and Peak Milliequivalents of Hardness Causing Pollutants Leached

S1 Number	Pollutants	Peak Concentration (mg/liter)	Peak (mEq/liter)	Total Leached (mEq × 10⁹)
1	Ammonium	200	11.11	8
2	Magnesium	150	12.34	8.9
3	Potassium	200	5.11	3.7
4	Sodium	300	13.04	9.4
		Total	41.60	30

After Bagchi (1983).

Dilution Calculation Substituting proper values in Eq. (15.2): $A_i = 3222$ m^2. Assuming 20% infiltration of precipitation, the average exfiltration of leachate into the groundwater system per day will be

$$V_e = (0.72 \times 0.2 \times 100 \times 150)/365 = 6.16 \text{ m}^3/\text{day}$$

Groundwater quality change due to the landfill siting is summarized in Table 15.5.

Sample Calculation (S1 number 3, Table 15.5):

$$C_{dx} = \frac{100.0}{1 + (6.16/3222 \times 10^{-1})} + \frac{500.0}{1 + (3222 \times 10^{-1}/6.16)}$$
$$= 107.5 \text{ mg/liter}$$

15.6 SUMMARY AND COMMENTS

As is apparent from the previous sections, a total quantitative analysis of an NA-type landfill is not possible. All of the following items are site specific: quantity and quality of leachate, chemical composition of the soil and therefore the attenuative capacity of the soil in the unsaturated zone, and plume geometry and therefore the dilutive capacity of the groundwater aquifer. So it is difficult to say how much waste can be deposited safely in an NA-type landfill at a particular site. In general, approximately 35,000–40,000 m^3 of

TABLE 15.5 Change of Groundwater Quality due to the Landfill Siting at 160 m Down-Gradient

S1 Number	Pollutants	Background Concentration (mg/liter)	Average Concentration 160 m Down-Gradient (mg/liter)
1	Ammonium	0.00	0.00
2	BOD (5 day)	0.00	93.8
3	Chloride	100.00	107.5
4	COD	0.00	225.1
5	Copper	0.01	0.01
6	Hardness	50.00	139.7
7	Iron	0.03	1.9
8	Lead	0.005	0.005
9	Magnesium	0.00	0.00
10	Potassium	0.00	0.00
11	Sodium	0.00	0.00
12	Sulfate	5.00	54.7

After Bagchi (1983).

municipal garbage is the maximum volume that can be allowed in a natural attenuation landfill. A higher volume may be allowed for other nonputrescible waste type. The problem is compounded by applicable ground water rules at each site. Some areas follow a qualitative approach (e.g., a policy of no degradation of groundwater or allowing degradation up to a certain percentage above background water quality) and others follow a quantitative approach (i.e., set concentration limits at a set distance from the landfill). Theoretically an NA-type landfill will always alter groundwater quality down-gradient of the landfill irrespective of the volume of waste deposited at the site. So strictly speaking, an NA-type landfill will invariably degrade groundwater. As pre-

TABLE 15.6   Chemicals Usually Found in Municipal Landfill Leachate and Their Potential Contibutors

Parameters	Products
Organics	
Acetone	Carburetor and fuel injection cleaners, paint thinners, paint strippers and removers, adhesives, fingernail polish removers
Xylene	Oil and fuel additives, carburetor and fuel injection cleaners, adhesives, paints, transmission additives
Methylene chloride	Oven cleaners, tar removers, wax, degreasers, spray deodorants, brush cleaners
Toluene	Contact cement, degreasers, paint brush cleaners, perfume, dandruff shampoo, carburetor and fuel injection cleaners, paint thinners, paint strippers and removers, adhesives, paints
1,1,1-Trichloroethane	Drain and pipe cleaners, oven cleaners, shoe polish, household degreasers, deodorizers, leather dyes, photographic supplies
cis/trans-1,2-Dichloroethylene	Contact cement, perfumes, make-up (perfume), upholstery and rug cleaners
Benzene	Adhesives, antiperspirants, deodorants, oven cleaners, tar removers, medicines, solvents and thinners
1,1-Dichloroethane	Degreasers, adhesives
Metals	
Lead	Batteries, electrical solder, paints
Cadmium	Paint, pigment, plastics
Chromium	Cleaners, paint pigments, linoleum, batteries
Nickel	Batteries, spark plugs, electrodes
Zinc	Batteries, solder, TV screens

Based on a study conducted by Minnesota Pollution Control Agency, Minneapolis, Minnesota, 1978.

viously discussed, the thickness of the unsaturated zone does not dictate the overall impact on groundwater. The hardness of the exfiltrate will always be higher than that of leachate; some of the leachate constituents (e.g., chloride, nitrate, and organics) are not attenuated by the unsaturated zone. So, these constituents will degrade the down-gradient water quality compared to background water quality. For some remote sites a minimal impact may be tolerated. However, if the design shows that a severe impact is probable, than an NA-type landfill should not be designed for the site.

Because of the above problems, a parametric study may be used to develop the range of allowable waste volume for a site. The designer may then compare the site against an ideal site and arrive at a conclusion after giving due consideration to the local groundwater rules.

A list of organic chemicals usually found in landfill leachate and their potential contributors is included in Table 15.6. Exclusion of discarded containers of these products from the solid waste steam disposed in a landfill will minimize the risk of groundwater contamination from the chemicals listed in Table 15.6.

LIST OF SYMBOLS

A = base area of landfill

A_i = area intercepted by groundwater for affecting pollutant dilution

C_b = average concentration of a parameter in the background water

C_{dx} = average concentration of a parameter in the down-gradient water at distance X

C_1 = peak concentration of a parameter in the leachate

H = average depth of unsaturated zone beneath the landfill

L_1 = dimension of landfill parallel to groundwater flow

L_2 = dimension of landfill perpendicular to groundwater flow

R = reduction factor to account for soil fabric effect (varies between $1/1.2$ and $1/1.3$)

U = average groundwater velocity

V_e = average volume of leachate exfiltration per unit time

V_s = soil volume available for attenuation reaction

X_1, X_2 = horizontal distances

θ_1 to θ_4 = divergence angles

16

CONTAINMENT LANDFILLS

The design concept for a containment-type landfill consists of restricting leachate seepage into the aquifer so as to minimize groundwater degradation. To satisfy these design criteria, landfills are lined with clay or synthetic membrane or both and a leachate collection system is installed (Fig. 16.1a and 16.1b). For a few waste types and insensitive environments, total containment may be needed. How the regulatory agency defines groundwater must be considered before designing a containment site. In general, groundwater is loosely defined as any water in the ground that can be withdrawn for use; however, it may also be defined as any water that occurs in a saturated subsurface geological formation of rock or soil. Thus, based on the first definition only groundwater aquifers need to be protected whereas the second definition includes saturated clay deposits and perched groundwater (which are usually not "tapped") in addition to groundwater aquifers. Detailed design and groundwater monitoring will depend on how groundwater is defined, which should be clarified by consulting the regulatory agency.

Theoretically leakage through the base of a containment landfill is unavoidable, however, it can be reduced to practically zero. In general, leakage through the liner of a single lined landfill is higher than through the liner of a properly designed and constructed double-lined landfill. It should be mentioned that if the primary mode of transport of the leachate constituent(s) is diffusion, low leakage cannot be achieved using the conventional containment landfill design discussed in this book.

In regions with a shallow groundwater table (GWT), the landfill base may have to be constructed below the ground water table. Therefore, landfills can also be divided into two types, based on whether the GWT is below or above the landfill base.

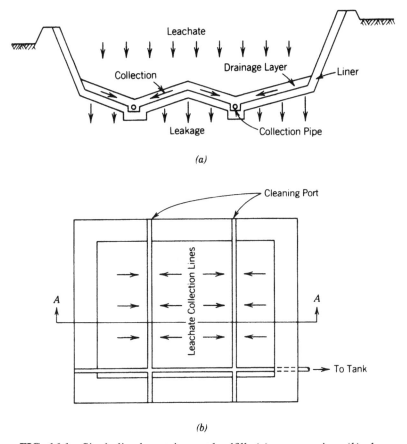

FIG. 16.1 Single-lined containment landfill: (*a*) cross section; (*b*) plan.

Usually clay or synthetic material is used in lining a landfill. A detailed discussion regarding the suitability of liner material can be found in Chapter 18. Discussions on how to determine the spacing for the leachate collection pipe and the thickness of the drainage blanket are included in Section 16.6 and a discussion on liner thickness is included in Section 16.4. There are several other elements (e.g., collection pipe) in a containment landfill that need to be designed. Detailed designs of each of these elements are included in Chapter 19.

16.1 SINGLE-LINED LANDFILLS

Figure 16.1*a* and 16.1*b* shows cross section and plan of a typical single-lined landfill. As mentioned, either clay or a synthetic membrane may be used for lining a site. Synthetic materials allow less leakage but are difficult to protect from damage, whereas clay liners are not easily damaged. The chances of

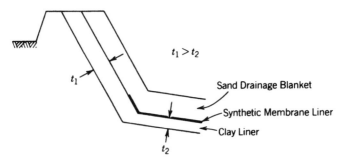

FIG. 16.2   Scheme showing combined clay and synthetic membrane lining.

damaging the liner in nonsludge landfills are higher. Hence clay is preferred as a liner in such landfills. For sludge landfills a synthetic membrane may be used provided care in taken to avoid running the compaction equipment directly on the liner. Sometimes drag lines are used to dispose of sludge in landfills, which may damage the top portion of the liner. In such cases a combination of clay and synthetic lining may be used; one such scheme is shown in Fig. 16.2.

Leakage through a properly constructed clay liner is not high. Gordon et al. (1989) published some data on clay liner leakage that indicate that leakage through a clay liner reduces over time (probably due to a decrease in permeability of the linear). Leakage through a synthetic membrane liner could be higher than anticipated if damage (during installation and landfill operation) goes undetected. Thus, if a synthetic membrane is used as a liner, extreme caution must be exercised to protect the liner during both the construction and operation of a landfill.

16.2   DOUBLE- OR MULTIPLE-LINED LANDFILLS

A double- or multiple-lined landfill may have one or two leachate collection systems. A landfill constructed with two liners, a synthetic membrane liner placed over a clay liner (Fig. 16.3), is called a composite lined landfill. Figures 16.3–16.8 show conceptual drawings for different options for double- or

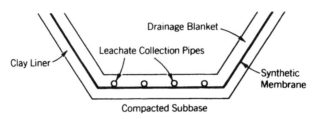

FIG. 16.3   Double-lined landfill with a single collection system: scheme 1.

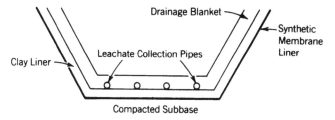

FIG. 16.4 Double-lined landfill with a single collection system: scheme 2.

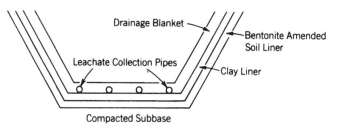

FIG. 16.5 Double-lined landfill with a single collection system: scheme 3.

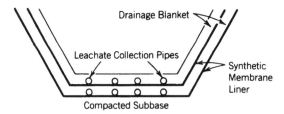

FIG. 16.6 Multiple-lined landfill with two collection systems: scheme 1.

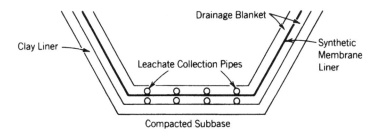

FIG. 16.7 Multiple-lined landfill with two collection systems: scheme 2.

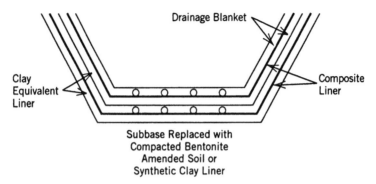

FIG. 16.8   Multiple-lined landfill with two collection systems: scheme 3.

multiple-lined landfill designs. A discussion of each option is included in the following section.

16.2.1   Comments on Design Issues

To design a landfill with two collection systems it is necessary to understand the mechanics of leachate apportionment in a lined landfill. Current leachate apportionment models (see Section 16.6) show that leachate flow toward a collection pipe is due to the gradient of the phreatic surface; the slope of the liner (usually 2–4%) has very little effect, if any, in driving the leachate toward the collection pipe. For a properly designed and constructed liner, leakage through the first linear will be low. If there is a second collection system below the first liner, then the permeability of the second liner installed for the second collection system is critical in effectively collecting leachate. The rate of input on the second liner is the rate of leakage through the first liner. Leachate apportionment models show that leakage through a liner with permeability of 1×10^{-7} cm/sec) is extremely high (80–100% of input rate) if the input rate is low (2–3 cm per year). However, if the second liner has a permeability of 1×10^{-9} cm/sec or less, then the leakage through the second liner will be practically negligible provided advection is the primary mode of transport of the leachate constituents through a porous media.

The next issue that needs to be understood is the thickness of the drainage blanket. All the models discussed in Section 16.6 use a single permeability of the drainage blanket, which means that the thickness of the drainage blanket is more than the maximum height of leachate mound that occurs near the crest of the liner at all times during the active site life of the landfill. The maximum height of the leachate mound fluctuates depending on the amount of leachate volume generated each day. If the permeability of the waste is lower than the permeability of the drainage blanket (e.g., municipal waste, papermill sludge) and if the drainage blanket is not thick enough to accommodate the leachate mound, then part of the leachate mound will penetrate the waste, which will increase the depth of the leachate mound from the point of intersection of the mound with the waste. This increase in depth of leachate mound will increase leakage through the liner (see Section 16.6 for a detailed discussion of this topic).

Leakage through the composite liner shown in Fig. 16.3 will be very low. Chances of damaging the synthetic liner during placement of the sand drainage blanket can be minimized by adopting a proper construction technique (see Section 20.4). Because of construction-related issues the option shown in Fig. 16.4 may not provide as much protection against leakage as the option shown in Fig. 16.3. Theoretically, a synthetic liner will allow significantly lower leakage compared to a clay liner. So the synthetic liner should be able to hold leakage coming from the clay liner. However, damage to the synthetic liner during the construction of the clay liner above it is unavoidable. If the subbase is sandy, then even a small damaged area will allow the majority of the leachate to escape. The change of damaging the first synthetic liner shown in Fig. 16.3 during construction is low; even if there is a damaged area, the clay liner below will allow far less leakage compared to a compacted sandy or clayey subbase.

The option shown in Fig. 16.5 is comparable to the option shown in Fig. 16.3. In general, the use of a thin layer (30–45 cm) of bentonite-amended soil liner alone is not recommended, mainly because of the relatively high probability of deterioration due to desiccation cracks and chemical incompatibility with the leachate. Construction of the clay liner above the bentonite liner can provide protection against both of these. However, the clay liner should be constructed immediately after laying the bentonite-amended soil layer to minimize desiccation cracks.

The option shown in Fig. 16.6 will not provide reliable protection against leakage, although there are two layers of synthetic membrane liners and two collection systems. Damage to any of the two synthetic liners during construction will cause significant leaking because the highly permeable drainage blanket will not sufficiently impede the flow through the liners. In the option shown in Fig. 16.7 the leakage through a damaged synthetic linear will be high because of the sand drainage blanket below it. Leakage through the clay liner will also be high because the relatively low volume of leachate leaked through the first liner will not create a significant leachate mound over the second liner; hence most of the leachate leaked through the first liner will escape through this clay liner.

The option shown in Fig. 16.8 has a multiple liner and two collection systems. Of the options shown in Figs. 16.3–16.8 this will provide maximum protection against leakage. While a clay liner is recommended for the first composite liner, either bentonite-amended soil or a synthetic clay liner may be substituted for the clay linear in the second composite liner. The subbase need not be replaced with bentonite-amended soil or synthetic clay liner if the subbase soil is clayey. Scarification and recompaction of the subbase will provide a high degree of protection.

The following points should be taken into consideration in designing a double- or multiple-lined landfill:

1. A clay, synthetic clay liner, or bentonite liner alone should not be used below the second collection system because, even when constructed properly, a clay or bentonite liner will collect only a small percentage of leachate leaked from the first liner.

2. A synthetic liner should not be used alone; it should always be used in combination with a clay, synthetic clay liner, or bentonite-amended soil. If such composite liner is used, then placement of the synthetic liner over the clay or bentonite-amended soil is recommended.

3. The drainage blanket over the first liner should be thick enough to accommodate the leachate mound at all times during the active site life. Use of a thin synthetic blanket (e.g., geogrid) is not recommended over the first liner.

4. The drainage blanket over the second liner need not be as thick as the first drainage blanket. A thin synthetic blanket may be used as a drainage blanket above the second liner.

5. Leachate apportionment models included in Section 16.6 are not applicable if a thin synthetic blanket is used as a drainage blanket. The leachate apportionment models are applicable only when a drainage blanket of sufficient thickness to accommodate the entire depth of the leachate mound is used.

16.3   LINER MATERIAL SELECTION CRITERIA

Selection of liner material depends on the waste type and landfill operation. The liner material must be compatible with leachate. In other words, the leachate generated from the waste must not degrade the liner material. A detailed discussion of this issue can be found in Chapter 18. For a double- or multiple-lined landfill compatibility of the primary liner with the leachate is important; compatibility of the secondary liner with the leachate may not be as important an issue except for some acutely hazardous wastes. Landfill operation also influences design. The thickness of the drainage blanket should be increased if a synthetic membrane is used as the primary liner where heavy compaction equipment is expected to be used in the landfill (e.g., municipal refuse). If the primary liner is a synthetic membrane, then extreme care should be taken during the construction of the drainage blanket, and during disposal of the first 1.2 m (4 ft) of waste so that the liner is not damaged. A thicker drainage blanket will provide better protection for a synthetic membrane liner.

16.4   COMMENTS ON LINER THICKNESS

The liner should be thick enough to provide a low permeability layer. Although theoretically, a thin low permeability layer is sufficient to reduce leakage, several other factors need to be considered in determining liner thickness.

For synthetic liners, degradation due to ultraviolet rays and puncture resistance are determining factors in choosing linear thickness. In general, synthetic membranes 1.5–2 mm (60–80 mils) thick are used for lining landfills. Additional testing regarding loss of strength due to exposure to leachate must also be considered in choosing the thickness of a synthetic membrane liner. Refer to Section 18.2 for a more detailed discussion of these issues.

Clay liner thickness is more dependent on construction-related issues and degradation due to freeze–thaw and desiccation than on theoretical design. Desiccation crack and freeze–thaw degradation increase the permeability of a compacted clay liner (refer to Section 18.1). Therefore although a 15- to 30-cm (6-in. to 1-ft)-thick clay liner is theoretically acceptable for providing a low permeability layer, a thicker liner should be constructed for the reasons mentioned above.

Benson (1990) reported the use of a stochastic model to evaluate the minimum thickness of a clay liner considering only the hydraulic behavior of soil. The hydraulic conductivity of a soil liner can vary significantly within a few meters (Rogowski et al., 1985). Benson (1990) indicated that a 1500-cm^2 area of soil liner is sufficient to incorporate the variability in hydraulic property. The following factors influence the overall hydraulic conductivity of clay liners: flow through macropores (both vertical and horizontal), variation in soil properties, variation in leachate production rate, degradation by chemicals present in leachate, cracks due to desiccation, and freeze–thaw cycles. Benson's stochastic model incorporated the variability and uncertainty associated with hydraulic properties of soil due to improper construction practices. The schematic of macropores used in the model is shown in Fig. 16.9. The variation of equivalent hydraulic conductivity with liner thickness predicted by

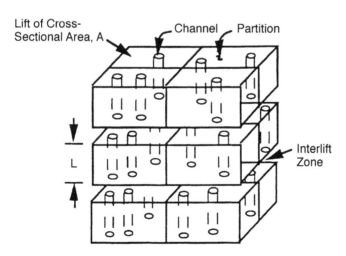

FIG. 16.9 Schematic of soil liner described by the stochastic model. After Benson (1990).

using the models is included in Fig. 16.10, which shows that the sensitivity of uncertainty associated with *in situ* hydraulic conductivity, for certain typical hydraulic conductivities, is negligible for liner thicknesses of 90 cm or more. In other words, an overall hydraulic conductivity of 1×10^{-7} cm/sec can be easily achieved in a 90-cm or more clay liner even when quality control during liner construction is poor. Based on *in situ* hydraulic conductivity measurement of several landfill liners, Benson (1990) concluded that *in situ* hydraulic conductivity of liners is less sensitive to thickness if excellent construction methods are used and that clay liners as thin as 60 cm may perform adequately when excellent construction methods are used.

A liner may be exposed to natural elements for 1 year whereby the permeability of the top 30–60 cm (1–2 ft) may increase by an order of magnitude. A clay liner 1.2–1.5 m (4–5 ft) thick will provide at least a 60- to 90-cm low permeability layer even after degradation of the top 30- to 60-cm layer. In addition, construction of a thick liner (1.2–1.5 m) will ensure an overall low permeability barrier because chances of developing macrofabric features such as stratification, fissuring, and large-scale heterogeneity are low if proper construction and quality control measures are undertaken.

16.5   COMMENTS ON FINAL COVER DESIGN

The design and construction of the final cover on a landfill are as important as the liner at the base of the landfill. The final cover should be such that it will allow infiltration that is less than or equal to the leakage through the base of the landfill. If the infiltration is more than the leakage, then the collection

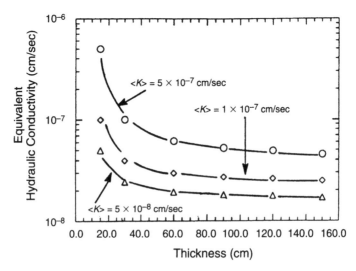

FIG. 16.10   Equivalent hydraulic conductivity as a function of thickness. After Benson (1990).

system has to be maintained in perpetuity to avoid continuous leachate head buildup within the landfill. Such head buildup will cause more leakage and will increase the possibility of seep(s) through the berm. The barrier layer (see Section 19.8) in the final cover should be the same as or better than the primary liner at the base of the landfill. Refer to Section 19.8 for a detailed discussion on final cover design.

There is a debate as to whether a final cover should be constructed over a landfill as soon as a reasonable area (from a construction standpoint) reaches the designed final grade (Note: This debate is applicable to a containment-type landfill only. Final cover on an NA-type landfill should be constructed as soon as a reasonable area reaches the designed final grade) or be deferred to a later date. This debate pertains to municipal waste landfills only; however, the concept is applicable to any putrescible waste that can be composted. The current approach for disposing of municipal waste is to encapsulate it to minimize the impact on groundwater. In the following sections the opponents' and supporters' views on the issue of whether landfill final cover should be constructed as soon as a reasonable portion reaches final grade are summarized.

Opponents' View In theory the encapsulation concept can keep the waste dry, which will not allow waste to decompose, leading to prevention of leachate formation. It is postulated that such a design will minimize pollution of groundwater. This approach has significant deficiencies that will ultimately threaten public heath and the environment. First, leakage through the liner is practically unavoidable (for both synthetic membrane and clay liners) because of manufacturing and construction-related defects. For a synthetic membrane liner the typical warranty is for about 20 years. It is reasonable to assume that leachate formation and subsequent leakage from the landfill are unavoidable in the long run: Encapsulation merely delays the process. Thus the encapsulated waste represents a perpetual threat to groundwater (Lee and Jones, 1990; Harper and Pohland, 1988).

To avoid this threat municipal waste should be composted (or fermented) first in a leach pad whereby all leachable chemicals can be extracted and the waste stabilized. This stabilized waste should then be disposed and encapsulated. Threats to groundwater and public health are negligible from such stabilized waste. It is claimed that recirculation of leachate will enhance fermentation and removal of leachate chemical constituents. It is conceded though that little is known about conditions under which removal of leachables will be maximized.

Supporters' View The concept of "stabilizing" municipal waste appears to be good in theory but not so in practice because (1) the possibility of leachate generation is high even after stabilizing the waste; (2) for a large volume of waste the cost of handling is estimated to be high, the area of leaching pad needed is quite large, the cost of construction and maintenance of such leaching pad is also high, the odor and bird hazard around the landfill will increase significantly; and (3) in a municipal landfill, precipitation water percolates through preferred channels and not through the entire waste mass. The total amount of chemicals leached out is expected to be significantly low compared to the fermentation/compost operation proposed for stabilizing the waste.

Summary Comments From an overall management standpoint it appears that "stabilizing" municipal waste prior to permanently landfilling is not realistic for very large landfills. However, the concept of stabilizing waste may be considered for landfills of small communities where the yearly waste generation rate is rather low. The following additional issues, either in favor of or against the timing of landfill final cover construction, should be considered prior to making a decision for a particular landfill:

1. An early release of gas and leachable constituents for a waste mass will certainly minimize future risks of groundwater contamination. However, the leachate generated during "stablization" is expected to be strong during the early phase. A pretreatment or a dedicated treatment system for treating the leachate should be considered if a stabilizing operation is proposed.

2. An estimate regarding the time to "flush out" all or the majority of contaminants is needed for estimating the leaching pad area required and for planning purposes.

3. Monitoring requirements and remedial action possibility may be significantly less for landfilling a "stabilized" waste but should not be completely neglected. Necessary funds should be set aside for these purposes.

4. For both stabilized and unstabilized waste landfills sufficient funds for monitoring the remedial action should be escrowed during the period when revenue is being generated.

5. Leachate generation from an encapsulated landfill may continue for many years after the final cover construction. Ten years ago it was thought that leachate should be pumped for 20–30 years after closure. However, currently the estimate for such long-term maintenance is a minimum of 40 years. This uncertainty in long-term maintenance requirement poses financial and legal problems, especially for privately owned landfills. In view of the existing laws, unless a clear provision is made in the laws governing landfills, "inheritance of unknown future liability" cannot be forced on an individual or a company. For publicly owned landfills, appropriation of necessary funds may be difficult politically. Setting up "long-term maintenance funds" for all landfills within the jurisdiction of a regulatory agency is an option that should be investigated. Such funds may be accumulated by requiring a disposal fee from all landfills at the time of active disposal. However, disbursing of such funds for long-term maintenance of a landfill may be difficult administratively.

6. From a regulatory standpoint the encapsulation concept is easily enforceable, especially on privately owned landfills because the action is taken

relatively quickly. If the final cover construction is delayed by several years after revenue generation, the owner's action is difficult to control unless the owner has continued interest in the matter (e.g., owns another landfill controlled by the same regulatory agency). A requirement to escrow the funds necessary for cover construction during active site life may help resolve this problem.

16.6   LEACHATE APPORTIONMENT MODELS

Leachate percolates vertically through the waste until it is influenced by the low permeability liner that retards vertical movement. Leachate then starts ponding on the liner. A portion of the ponded liquid flows laterally to the leachate collection pipes and the other portion flows vertically through the liner. Thus, the leachate produced in a landfill is apportioned into a drainage fraction (flowing laterally) and a leakage fraction (flowing vertically). Several models have been developed to predict the apportionment of leachate. Both steady-state and non-steady-state models are available (Wong, 1977; McBean et al., 1982; Korfiatis and Demetracopoulos, 1986; Demetracopoulos et al., 1984; Demetracopoulos and Korfiatis, 1984; Lentz, 1981; Schroeder et al., 1984; Bagchi and Ganguly, 1990). All these models assume a saturated liner. Bureau (1981) proposed a model in which the liner is assumed to be unsaturated. This model, although applicable to short-term waste impoundments, is not discussed here because landfill liners are permanent and are expected to be saturated.

The first apportionment model was proposed by Wong (1977). He assumed that leachate ponded on a liner develops a phreatic surface parallel to the liner and that the phreatic surface is formed instantaneously in response to leachate input. The rate of collection depends on the velocity with which the rectilinear slab of leachate moves to the collection point (Fig. 16.11). At any time (t), the saturated volume (v) above the linear is given by

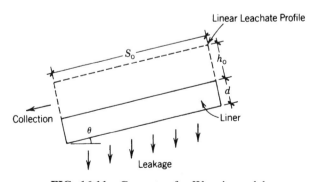

FIG. 16.11   Geometry for Wong's model.

$$v = sh \cos \theta \tag{16.1}$$

for low values of θ, $\cos \theta \approx 1.0$, in which s = length of the rectilinear slab at time t, h = thickness of the rectilinear slab at time t, and θ = slope of the liner. The rate of volume change will be

$$\frac{dv}{dt} = h\left(\frac{ds}{dt}\right) + s\left(\frac{dh}{dt}\right) \tag{16.2}$$

The first term represents the collection rate and the second term represents the leakage rate. Each term is integrated separately to find drainage and leakage volumes.

The following equations were derived to estimate drainage and leakage volumes:

$$\frac{v_1}{v_0} = \left[\frac{1}{K}\right][e^{-Kt/t_1} - 1]\left[1 + \left(\frac{d}{h_0}\right)\right] + \left(\frac{d}{h_0}\right)\left(\frac{t}{t_1}\right) \tag{16.3}$$

$$\frac{v_2}{v_0} = \left[\frac{1}{K}\right][e^{-Kt/t_1} - 1 + K]\left[1 + \frac{d}{h_0}\right] \tag{16.4}$$

$$K = \frac{P}{2d}\frac{K_2}{K_1}\cot \theta \tag{16.5}$$

in which v_1 = the total volume of leachate collection, v_2 = the total volume of leachate leakage, v_0 = the total leachate volume, t_1 = the time for the entire length of the leachate slab to drain into the collection line, d = the thickness of the liner, h_0 = the initial thickness of the saturated volume, K_1 = the permeability of the drainage layer, and K_2 = the permeability of the liner. Wong's model was revised by Kmet et al. (1981); Scharch et al. (1985) and Demetracopoulos et al. (1984) extended the model to provide drainage and leakage estimates for quasi-steady-state conditions.

McBean et al. (1982) proposed a steady-state model that used a nonlinear phreatic surface and a nonleaking liner. An implicit nonlinear equation for leachate mound was derived that transforms to the following simple form for horizontal liners:

$$y = \sqrt{y_0^2 + 2\left(Lx - \frac{x^2}{2}\right)} \tag{16.6}$$

in which $y = y_0$ at $x = 0$. The procedure is helpful in obtaining the leachate mound where leachate collection lines are placed on a liner sloped in one direction, but the approach cannot be used for estimating leakage through the liner.

A non-steady-state model was proposed by Korfiatis and Demetracopoulos (1986). Principles of mass conservation are used to develop the following equation summing all inflow and outflow to an element:

$$\frac{\partial Q_c}{\partial_x} + I - Q_1 = n \frac{\partial H}{\partial t} \tag{16.7}$$

in which Q_c = the leachate collection rate, I = the leachate input rate, Q_1 = the leakage rate, H = the leachate head over the liner at a distance x from the crest of liner, t = time, and n = porosity of the drainage layer.

The following dimensionless nonlinear partial differential equation is developed, which was solved by explicit finite difference approximation using central differences:

$$\frac{\partial}{\partial x^*}\left(h^*\frac{\partial h^*}{\partial x^*} - h^* \sin\theta\right) + I^* - K^*(h^* + 1) = n\frac{\partial h^*}{\partial t^*} \tag{16.8}$$

in which $x^* = x/d$, d = the thickness of the liner, $h^* = h/h$, $I^* = I/Kd$, K_1 = the saturated permeability of the drainage layer, $K^* = K_2/K_1$, K_2 = the saturated permeability of the liner, and $t^* = tK_2/d$.

Lentz (1981) proposed a leachate apportionment model for landfill cover that can predict leakage for water surplus on a daily basis. The model can be used for the bottom liner. Principles of mass balance are used in developing a partial differential equation that was linearized to obtain a solution. Schroeder et al. (1984) proposed a computer-based model known as HELP (Hydrologic Evaluation of Landfill Performance) for estimating leachate apportionment on a daily basis. The model couples both water infiltration into a landfill through cover (also known as water balance) and leachate apportionment at the bottom liner level.

The following linearized semiempirical equation was developed for a steady-state condition for estimating the leachate collection rate:

$$Q_c = \frac{2C_1 K_1 \bar{Y}(Y_0 + \theta p)}{p^2} \tag{16.9}$$

$$C_1 = 0.51 + 0.00205\theta p \tag{16.10}$$

in which $\bar{Y}$ = the average thickness of the water profile above the liner between the leachate collection pipe and the crest of a module, θ = the liner slope, and $2p$ = pipe spacing. Y_0 is estimated as

$$Y_0 = \bar{Y}\left(\frac{\bar{Y}}{\theta p}\right)^{0.16} \tag{16.11}$$

The vertical leakage is essentially estimated by the following equation:

$$Q_1 = K_2 \frac{\overline{Y} + d}{d} \tag{16.12}$$

An iterative procedure with four equal time steps per day is used to estimate the collection and leakage rate for a degree of accuracy of $\pm 5\%$. The model was subsequently updated, which includes a few modifications. Required data for many U.S. cities are included in the program files. Files containing daily precipitation, temperature, and solar radiation data need to be added for use in other countries.

The theory for leachate apportionment developed by the author is discussed below. Error of prediction using this model was found to be low (Bagchi and Ganguly, 1990). A general equation of the leachate mound is developed first, which is then used to predict the daily collection and leakage rate. The liner and collection pipe configuration used in developing the general steady-state equation are shown in Fig. 16.12. Equating the leachate generation for collection and leakage,

$$(Q_x)_c + (Q_x)_1 = (p - x)I \tag{16.13}$$

in which $(Q_x)_c$ = the leachate flow rate toward the collection point, through a section located at a distance of x (i.e., through GH as shown in Fig. 16.12), $(Q_x)_1$ = the leachate leakage rate through the liner between the apex and a point x (i.e., through the liner GF area as shown in Fig. 16.12), p = the distance of the liner apex from the origin, and I = the leachate generation rate (i.e., recharge rate on liner) per unit length of the liner.

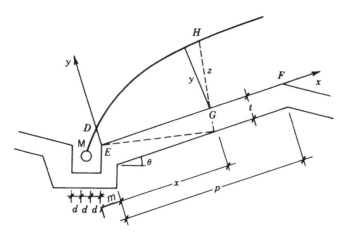

FIG. 16.12   Geometry for the unsteady-state leachate apportionment model.

$$Z = y \sec \theta + x \sin \theta \qquad (16.14)$$

or

$$\frac{dZ}{dx} = \frac{dy}{dx} \sec \theta + \sin \theta \qquad (16.15)$$

$\sin \theta$ can be neglected for small values of θ. Hence,

$$\frac{dZ}{dx} = \left(\frac{dy}{dx}\right) \sec \theta \qquad (16.16)$$

From Darcy's law,

$$(Q_x)_c = K_1 y \left(\frac{dZ}{dx}\right) \qquad (16.17)$$

or

$$(Q_x)_c = K_1 y \sec \theta \left(\frac{dy}{dx}\right) \qquad (16.18)$$

in which K_1 = the permeability of the drainage layer. From Eqs. (16.13) and (16.18)

$$(1 + R)K_1 y \sec \theta \left(\frac{dy}{dx}\right) = (p - x)I \qquad (16.19)$$

in which $R = (Q_x)_1/(Q_x)_c$.

From symmetry, the phreatic surface is assumed to be passing through the centerline of the collection trench. From Fig. 16.12 the coordinates of M are $x_c = -1.5d \sec \theta$ and $y_c = 0$, in which d = diameter of the leachate collection pipe. The general equation of the phreatic surface [Eq. (16.20)] is obtained by integrating Eq. (16.19) and using the coordinates of M. A detailed derivation of Eq. (16.20) can be found elsewhere (Bagchi and Ganguly, 1990). Leachate mound curves (phreatic surface) predicted by this model were found to be parallel to the experimental curves obtained by Schroeder (1985) for a base slope of up to 4%.

$$y^2 = -Ax^2 + Bx + C \qquad (16.20)$$

The general equation of the phreatic surface is used to estimate the daily leakage and collection rate. The following assumptions are used to estimate the daily events:

1. The phreatic surface passes through point M (Fig. 16.12).
2. The gradient of the phreatic surface is zero at $x = p$; i.e. $(dy/dx)_{x=p} = 0$.
3. The slope of the liner is not greater than 4%.

Using the formula for the area of an ellipse and assumption 1,

$$V_d = \frac{n\pi(p + m)h_{max}}{4} \qquad (16.21)$$

or

$$h_{max} = \frac{4V}{n\pi(p + m)} \qquad (16.22)$$

in which V_d = the remaining volume of leachate in a day, n = the porosity of the drainage blanket, p = the width of the module (= $\frac{1}{2}$ pipe spacing approximately), m = the distance shown in Fig. 16.12 (= $d \sec \theta$), θ = the slope of the liner, and h_{max} = the height of the leachate mound at $x = p$. Put $y = 0$ and $x = -m$ in Eq. (16.20):

$$0 = -Am^2 - Bm + C \qquad (16.23)$$

Using assumption 2 and Eq. (6.20)

$$-2Ap + B = 0 \qquad (16.24)$$

Put $y = h_{max}$ and $x = p$ in Eq. (6.20)

$$h_{max}^2 = -Ap^2 + Bp + C \qquad (16.25)$$

From Eqs. (6.22)–(6.25)

$$A = \frac{h_{max}^2}{m^2 + 2pm + p^2}$$

$$B = \frac{2h_{max}^2\, p}{m^2 + 2pm + p^2}$$

$$C = h_{max}^2 \left(\frac{m^2 + 2pm}{m^2 + 2pm + p^2} \right)$$

For a precipitation event V_i is given by

$$V_i = I(p + m) \tag{16.26}$$

in which I = the depth of precipitation and V_i = the volume of precipitation. For any day V_d is the remaining volume, which is the carryover volume from the previous day and the leachate generated due to any precipitation that falls in that day (i.e., V_d = the remaining volume + V_i). The leachate volume generated by the precipitation is added to the remaining volume before apportionment is started for the day. Figure 16.13 shows the routing scheme. The volume of collection V_c for a day is calculated first using the remaining volume, assuming no leakage during that process. The collected volume is subtracted from the remaining volume V_d, which is then used to calculate the leakage for that day. The process is reversed to calculate leakage first and collection next. The averages of the two collection and leakage volumes are taken, which are the respective volumes for the day.

The volume of collection (V_c) and leakage (V_l) per day are calculated as follows:

$$V_c = K_1 \times \text{area } DE \times \left(\frac{dy}{dx}\right)_{x=0} \tag{16.27}$$

$$\text{Area } DE = \sqrt{(-Ax^2 + Bx + C)}_{x=0} = \sqrt{C} \tag{16.28}$$

$$\left(\frac{dy}{dx}\right)_{x=0} = \frac{B}{2\sqrt{C}} \tag{16.29}$$

so

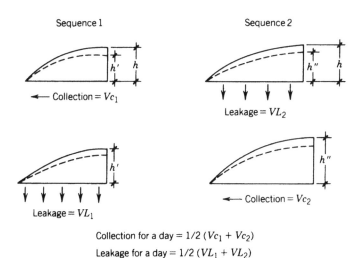

Sequence 1

Collection = Vc_1

Leakage = VL_1

Sequence 2

Leakage = VL_2

Collection = Vc_2

Collection for a day = $1/2\,(Vc_1 + Vc_2)$

Leakage for a day = $1/2\,(VL_1 + VL_2)$

FIG. 16.13 Apportionment schemes.

$$V_c = \frac{K_1 B}{2} \qquad (16.30)$$

$$V_1 = \int_0^p K_2 \frac{y \sec \theta + t}{t} \, dx \qquad (16.31)$$

or

$$V_1 = K_2 p + \int_0^p \frac{K_2 Y \sec \theta}{t} \, dx \qquad (16.32)$$

Y can be expressed as

$$Y = \sqrt{A}(\sqrt{V^2 - U^2}) \qquad (16.33)$$

in which

$$U = x - \frac{B}{2A} \qquad V = \sqrt{\frac{C}{A} + \frac{B^2}{4A^2}}$$

Since $(dy/dx)_{x=p} = 0$, $B = 2AP$ when $x = p$, $U = 0$, and when $x = 0$, $U = -p$, and $dU = dx$. Equation (16.32) can be written as

$$V_1 = K_2 P + \frac{K_2 \sec \theta \sqrt{A}}{t} \int_{-p}^0 \sqrt{V^2 - U^2} \, dU \qquad (16.34)$$

Integrating Eq. (16.34)

$$V_1 = K_2 p + \frac{K_2 \sec \theta \sqrt{A}}{t} \left[\frac{1}{2} U(\sqrt{V^2 - U^2}) + V^2 \sin^{-1} \frac{U}{V} \right]_{-p}^0 \qquad (16.35)$$

Inserting limits and simplifying Eq. (16.35)

$$V_1 = K_2 p + \frac{K_2 \sec \theta \sqrt{A}}{t} \left[\frac{p}{2} \sqrt{\frac{C}{A}} + \frac{1}{2} \left(\frac{C}{A} + p^2 \right) \sin^{-1} \frac{p}{\sqrt{C/A + p^2}} \right] \qquad (16.36)$$

The leachate volume in a landfill is less than the precipitation that falls on a landfill. Part of the precipitation is lost by evaporation from the surface of the landfill and part is absorbed by the waste (nonsludge-type waste only). In landfills located in cold regions, snow accumulated during the winter months melts in spring. Thus, the precipitation needs to be adjusted to reflect the effects of these factors. Leachate data obtained from an operating landfill need to be used to determine the daily values of the three factors (namely evaporation factor, absorption factor, and snow accumulation factor). Once these factors are obtained using field data for at least 1 year, they can be used to predict future apportionment. The thickness of the drainage blanket is dictated by the maximum depth of the leachate mound at the crest (h) in a year. The drainage blanket should be thick enough to accommodate the leachate mound throughout the year or most of the year. The model can be used to predict the daily leachate collection and leakage rate, the variation of h over a year for a set of design parameters (namely, pipe spacing, permeabilities of the drainage blanket and liner, and thickness and slope of the liner). The observed leachate head at the midpoint of a module for a landfill in Wisconsin was reported to be in good agreement with the value predicted by the model (Bagchi and Ganguly, 1990). Variation of h in a year for a typical landfill in midwest United States is expected to be between 30 and 58 cm. Variation of h is dependent primarily on precipitation and waste type. If the drainage blanket is not thick enough to accommodate the leachate mound in a day, then the mound will rise (Fig. 16.14), leading to an increase in leakage; this rise in leachate mound will occur only if the permeability of the waste is lower than the permeability of the drainage blanket. Leakage from part of the module between point a and point b will increase because of the increased height of the phreatic surface between these two points. The increase in leakage rate will depend on the permeability of the waste (which will dictate the increase in the height of the phreatic surface) and the distance between point a and b (which is dependent on the thickness of the drainage blanket). A designer should attempt to minimize the occurrence of the situation (e.g., not

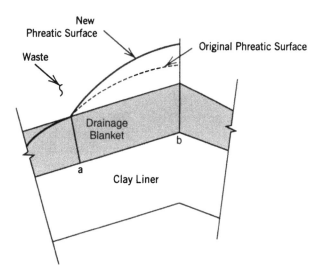

FIG. 16.14 Change in phreatic surface due to inadequate drainage blanket thickness.

more than 30 days) depicted in Fig. 16.14 at a predetermined point on the module (e.g., $ab = \frac{1}{4}$ of module width). In other words the drainage blanket should be thick enough such that the leachate mound does not cross the drainage blanket line at point a for (say) more than 30 days a year.

The leachate apportionment models indicate that leakage through the landfill base depends heavily on the liner permeability and pipe spacing. Leakage can be reduced by lowering the liner permeability and pipe spacing. However, one should note that the required drainage blanket thickness increases with a decrease in pipe spacing. A leachate apportionment model should be used to estimate collection and leakage. Both peak and average collection volumes are important for treatment plant design and may become critical for a treatment plant with marginal available capacity. In the absence of access to an apportionment model, the following guidelines may be used for landfill liner and collection system design:

1. *Liner*
    Thickness: 1–1.5 m (3–5 ft)
    Permeability: 1×10^{-7} cm/sec or less
2. *Pipe spacing*
    15–30 m (50–100 ft)
3. *Diameter of collection pipes*
    15 cm (6 in) minimum
4. *Slope of liner*
    2–4%
5. *Thickness of the drainage blanket* (Note: The drainage blanket thickness increases with decrease in pipe spacing)
    30–120 cm (1–4 ft)

16.7   COMMENTS ON DESIGNING THE LANDFILL BASE BELOW THE GROUNDWATER TABLE

The landfill base should be above the seasonally high GWT. However, it is difficult to follow this guideline at all times. The design approach is different if the subbase is entirely clayey or sandy. A liner should be constructed even in a clayey environment because in most cases clays have sand seams, vertical and horizontal fractures that are difficult to ascertain during subsoil investigation. Even field permeability tests may not detect these problems if testing is not performed carefully. Excessive leakage may take place through the cracks and so on. Some regulatory agencies may require a minimum thickness of the saturated clay stratum, which should be verified with the agency before designing a landfill in a saturated clayey environment. The changes of leakage through the liner of such landfills is not higher than a landfill sited above the groundwater table.

In a sandy environment the groundwater table may be lowered temporarily by dewatering during construction or permanently by installing a groundwater dewatering system (Fig. 16.15). A groundwater dewatering system, if installed, must be maintained properly by cleaning the pipes at least once a year. Raising of the base grade by constructing a retarder subbase may be considered in some cases to avoid the construction of a dewatering system. Since maintenance of the dewatering system is likely to become a perpetual activity, its installation should be avoided as far as possible. The following approach may be used to estimate pipe spacing.

The Donnan formula [Eq. (16.37)] is used to calculate pipe spacing in draining agricultural fields; a steady-state condition is assumed:

$$L^2 = \frac{4K(b^2 - a^2)}{Q_d} \qquad (16.37)$$

in which L = drain spacing (ft), K = hydraulic conductivity (ft/day), a = the distance between the pipe and the barrier (ft), b = the maximum allowable water table height measured from the barrier (ft), Q_d = recharge rate (ft^3/ft^2/day). (Note: Donnan's formula is valid in SI units also.)

For groundwater dewatering systems, neglecting landfill leakage, which is small compared to groundwater flow,

$$Q_d = ki \qquad (16.38)$$

Equation (16.37) changes to

$$L^2 = \frac{4(b^2 - a^2)}{i} \qquad (16.39)$$

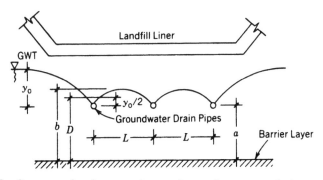

FIG. 16.15 Geometry for the groundwater dewatering system design—pipes above the barrier layer.

Equation (16.37) is used for the case in which the pipes are installed above a barrier (Fig. 16.15). For pipes installed on a barrier, $a = 0$ (Fig. 16.16). In a sensitivity analysis of several formulas for drain tile design, Slane (1987) found that (1) the pipe spacing and mound height depend on the permeability of the medium and recharge rate, and (2) the Donnan formula predicts the lowest mound height for a pipe spacing compared to other formulas. However, the Donnan formula is widely used for its simplicity. The pipe spacing obtained from it is within ± 20 ft of the spacing obtained using unsteady-state models (Bureau of Reclamation, 1978); hence a reduction factor of 0.8–0.9 is suggested to obtain a conservative estimate.

For pipes above a barrier the discharge from the pipes is given by (Bureau of Reclamation, 1978)

$$q_p = \frac{2\pi K y_0 D}{86,400L} \tag{16.40}$$

For pipe on the barrier it is given by

$$q_0 = \frac{4KH^2}{86,400L} \tag{16.41}$$

in which q_p or q_0 = the discharge from two sides per unit length of drain ($\text{ft}^3/\text{sec/ft}$), y_0 or H = the maximum height of the water table above the pipe invert (ft), K = the weighted average hydraulic conductivity of the soil between the maximum water table and the barrier or drain (ft/day), D = the average flow depth ($= d + y_0/2$) (ft), and L = the pipe spacing (ft).

The total discharge from a pipe is obtained by multiplying q_p or q_0 by the total length of the pipe. The pipe size can be calculated from the known discharge rate and from the velocity of water in the pipe using the Manning equation [Eq. (16.42)]:

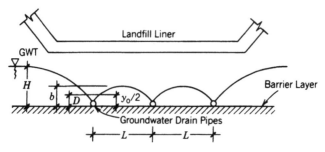

FIG. 16.16   Geometry for the groundwater dewatering system—pipes on the barrier layer.

$$V = \frac{1.486}{n_r} r^{2.3} s^{1/2} \qquad (16.42)$$

in which V = velocity in ft/sec, r = the hydraulic radius in ft, s = the slope of the groundwater collection pipe in ft/ft, and n_r = the coefficient of roughness. Example 16.1 enumerates the design steps. The minimum suggested pipe size is 10 cm (4 in.).

Example 16.1 (in F.P.S. units)   Design a groundwater collection system for a landfill in which the soil has a K of 2.83 ft/day (1×10^{-3} cm/sec) and a gradient of 0.1. The maximum permissible height of the water table is 10 ft above the barrier and the pipe invert is 5 ft above the barrier. The length of each pipe is proposed to be 1000 ft laid at a slope of 1%; $n = 0.015$ for plastic pipes. From Eq. (16.39)

$$L^2 = \frac{4(b^2 - a^2)}{i}$$

$$= \frac{4(10^2 - 5^2)}{0.1}$$

$$= 55 \text{ ft}$$

Using a reduction factor of 0.8, $L = 44$. From Eq. (16.40)

$$q_p = \frac{2\pi K y_0 D}{86,400L}$$

$$D = 5 + \tfrac{5}{2} = 7.5 \text{ ft}$$

$$q_p = \frac{2\pi \times 2.83 \times 5 \times 7.5}{86,400 \times 4}$$

$$= 0.00017 \text{ ft}^3/\text{sec ft}$$

The total discharge is $(0.00017/2) \times 1000 = 0.09$ cfs. From Eq. (16.41) assuming a 4-in. pipe flowing full,

$$V = \frac{1.486}{0.015} \times \left(\frac{4}{2 \times 12}\right)^{2/3} \times (0.01)^{1/2}$$

$$= 3 \text{ ft/sec}$$

The capacity of this pipe flowing full is

$$3 \times \pi(\tfrac{2}{12})^2 = 0.26 \text{ ft}^3/\text{sec}$$

which is more than the total discharge.

16.8   CHECK FOR LINER BLOWOUT

Blowout of the landfill liner must be checked if the landfill base is constructed below the water table, especially in a sandy environment. Dewatering wells can be used temporarily to lower the groundwater table during construction to avoid blowout. The analysis and example given below will help in determining whether there is a possibility of blowout of clay liners and when to turn off a dewatering system if one is used.

Land disposal facilities constructed under the water table encounter water pressure at the base and side of the site. An empty landfill, which is the worst condition (as far as blowout is concerned), should be assumed. Figure 16.17 shows a situation in which the clay liner may blowout because of a high water table. Analyses of both base and side liners for different situations are available.

When a composite liner is used, uplift of the synthetic membrane liner at the base and its slippage at the side wall due to seepage of water through the clay liner should be considered. Failure of the synthetic membrane liner due to such water seepage can be avoided if sufficient volume of waste is disposed of relatively quickly (e.g., within a year). This operational issue should be considered while designing composite liners below GWT. In general, for big landfills uplift of the synthetic membrane liner at the base is not likely because usually sufficient waste is disposed within a year. However, slippage of the synthetic membrane liner on a side slope due to reduced interfacial friction should be carefully evaluated. Failure can be avoided if the interfacial friction between the synthetic membrane and clay is high enough even under saturated condition. If necessary a layer of sand (30–60 cm thick) or geocomposite below the clay liner on the side slope should be included.

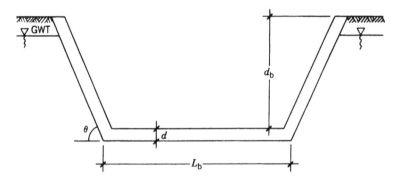

FIG. 16.17   Possible liner blowout situation.

16.8.1   Blowout of Base Liner

The following possible causes of blowout failure for the base liner were analyzed: (1) shear failure, (2) bending failure, and (3) punching shear failure (Bagchi, 1986b). The water head at which the stress due to water pressure equals the strength of the liner was defined as the critical head. It was observed that critical head expressions could be divided into two terms: one containing the shear or bending strength of clay and another containing the ratio of unit weights of soil and water. It was shown that high bending stress is the most probable cause of failure. Equation (16.43) gives the critical head (h_b):

$$h_b = \frac{4}{3} \frac{K_b C_u}{\gamma_w} \left(\frac{d}{L_b}\right)^2 + \frac{\gamma_s}{\gamma_w} d \qquad (16.43)$$

in which C_u = the unit shear strength of clay, K_b = constant (≈ 0.25), γ_w = the unit weight of water, γ_s = the unit weight of soil, d = the thickness of the liner, and L_b = the length of the landfill base. Based on a parametric study, it was concluded that the numerical contribution of the first term is negligible for practical design purposes (Bagchi, 1986b).

16.8.2   Blowout of Side Liner

Failure of a side liner below the water table or due to artesian pressure in a sand seam is discussed in this section. Failure at section $Y-Y'$ (Fig. 16.18) due to a high water table is discussed first. The possible causes of failure at

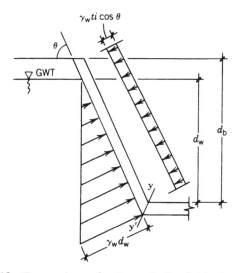

FIG. 16.18   Force scheme for the analysis of side liner blowout.

section $Y-Y$ include shear failure and bending failure. It was observed that in this case also, the critical depth of the water table above the liner can be subdivided into two terms: one containing the shear or bending strength of the clay and another containing the ratio of unit weights of soil and water. It was shown that high shear stress is the most probable cause of failure (Bagchi, 1986b). Figure 16.18 shows the force diagram for this case. Equation (16.44) gives the critical depth of water for shear failure at section $Y-Y$:

$$C_{DW} = \sqrt{\left(\frac{2C_u d \sin^2 \theta}{\gamma_w} + \frac{\gamma_s}{\gamma_w} d \sin^2 \theta \, d_b\right)} \qquad (16.44)$$

in which C_{DW} = the critical depth of the water table above the liner, D_b = the depth of the top of the base liner from the land surface, and other terms are as previously defined.

However, unlike the base liner blowout, the numerical value of the first term is not negligible compared to the numerical value of the second term for the range of values of different parameters usually encountered in the field.

Analysis of the blowout of a side liner due to a narrow water-bearing stratum (Fig. 16.19) is given by Oakley (1987). The factor of safety (FS) against failure is given by

$$FS = \frac{\gamma_T d \cos \alpha}{Z \gamma_w} + \frac{2d^2 f}{Z \gamma_w (T/\sin \alpha)^2} \qquad (16.45)$$

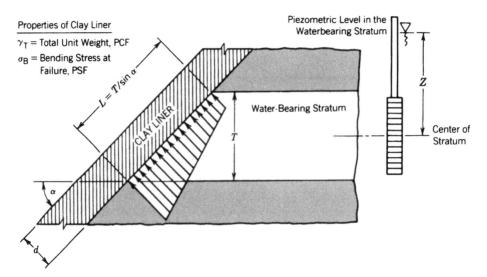

FIG. 16.19 Possible side liner blowout situation due to artesian pressure in sand seam. After Oakley (1987).

A factor of safety of 2–3 should be assumed. The following range of values for C_u is reasonable: 7425–9900 kg/m² (1500–2000 psf).

Example 16.2   Following are the dimensions of a square landfill, constructed using a dewatering system for which the possibility of liner blowout is to be checked: Assume C_u = 7500 kg/m², L_b = 50 m, d_b = 5 m, d_w = 3 m, d = 1.5 m, θ = 18.4°; unit weight of solid waste = 816 kg/m³, and γ_s = 1650 kg/m³.

Check for the Base Liner   From Eq. (16.43) and the discussion, the critical head for the base liner is

$$(1650/1000) \times 1.5 = 2.5 \text{ m}$$

Since the height of the GWT is (= $d + d_w$ = 1.5 + 3) 4.5 m above the bottom of the liner, blowout will occur if the operation of the dewatering system is stopped immediately after construction of the liner. One way to avoid blowout is to place a certain height of solid waste in the landfill before turning off the dewatering system.
   The head of water to be balanced by the solid waste is

$$4.5 - 2.5 = 2 \text{ m}$$

The equivalent height of solid waste that must be placed to resist blowout is

$$(2 \times 1000)/816 = 2.45 \text{ m}$$

Therefore, to prevent liner blowout, a minimum of 2.45 m of solid waste must be disposed of on the entire area of the lined landfill base, before the dewatering system can be shut off, or a thicker liner should be designed.

Check for the Side Liner   Failure near the base at section *Y–Y*. From Eq. (16.44)

$$C_{\text{DW}} = \sqrt{\left(\frac{2 \times 7500 \times 1.5 \times \sin^2 18.4}{10^3} + \frac{816 \times 1.5 \times 5 \times \sin^2 18.4}{10^3} \right)}$$
$$= 1.69 \text{ m}$$

Therefore, a side liner failure will occur at section *Y–Y* if the operation of the dewatering system is stopped immediately after construction of the liner. To prevent blowout either the liner should be thickened along the corner (section *Y–Y*) or a certain height of solid waste should be disposed in the landfill before turning off the dewatering system. Thickening of the corner is preferred because it will increase the shear resistance of the liner. Reliance on the shear resistance of the waste is not suggested.

Faulty construction may also lead to blowout of the base as well as the side liner. Liner blowout due to bad quality control cannot be analyzed. The best way to ensure proper quality control is to arrange on-site engineering supervision.

16.9   NATURAL ATTENUATION VERSUS CONTAINMENT LANDFILL

Because both NA-type and containment-type landfills have advantages and disadvantages, a designer has to consider various factors in choosing the correct type. A designer must verify whether any regulatory restrictions exist regarding siting of a natural attenuation type of landfill in a region. Some of the points discussed below are rather broad based and are sometimes beyond the control of an individual designer; however, they are discussed to help in developing a proper perspective and attitude toward each landfill type.

A NA-type landfill can be designed for a small waste disposal rate (nonhazardous wastes only) but a containment landfill cannot be designed and operated economically for a small waste volume because the cost of constructing and operating a containment landfill is high. The leachate has to be treated both during the active site life and after closure of a landfill, the cost of which must be obtained from the landfill customers. A small segment of liner cannot be constructed economically, and a larger liner may remain unused for several years causing severe deterioration due to exposure to the natural elements. If waste is disposed in a small area of a containment landfill, all precipitation water will be collected by the leachate collection system and must be treated, although the leachate strength is likely to be low. For these technical reasons a containment landfill for a small waste disposal rate is not the preferred option; in addition, operation of such a landfill could be very costly.

NA-type landfills cannot be used for a large waste volume because of unavoidable groundwater impact. A containment landfill is the only choice for large waste volumes. NA-type landfills put very little demand on the environment as a whole (only the landfill site is disturbed), whereas containment landfills disturb not only the areas in which they are sited but also the areas from which clay and sand is borrowed (especially clay-lined landfills). The total energy consumption for construction, operation, and maintenance of a containment landfill is much higher than for an NA-type landfill. However, this overall impact can be reduced only if the waste type is such that it can be easily disposed of in an NA-type landfill without the risk of groundwater contamination. But the individual designer has no control over waste type. Control on waste type is a waste management issue dictated largely by local and state regulations.

Thus, in summary, a designer should use judgment as to which type of landfill may be used in a particular situation. In general, if regulations allow, properly designed natural attenuation landfills are suitable for a small waste volume in some cases and are cheaper to operate, whereas containment landfills are suitable for larger waste volumes and may be the most economical option for certain communities. If the waste generation rate in a community is small but regulations do not allow the use of NA-type landfills, then collecting the waste in a container and shipping that container periodically to a nearby large landfill may be cheaper than constructing and operating a small containment landfill. This option should be investigated for small community landfills.

LIST OF SYMBOLS

a = distance between the pipe and the barrier
b = maximum allowable water table height measured from the barrier
C_{DW} = critical depth of water table to cause shear failure of side liner
C_u = unit shear strength of clay
d = thickness of liner
d_b = depth of the top of base liner from land surface
D = average flow depth
h = thickness of the rectilinear saturated slab at time t
h = critical head of waste to cause failure of base liner
h_{max} = height of the leachate mound at $x = p$
h_0 = initial thickness of the rectilinear saturated slab
H = leachate head over the liner at any distance x from the crest of the liner
I = leachate input rate
K = hydraulic conductivity
K_1 = permeability of drainage layer
K_2 = permeability of the liner
K_b = constant
L = drain spacing
L_b = length of landfill base
n = porosity of the drainage layer
n_r = coefficient of roughness
$2p$ = pipe spacing
q_p or q_0 = discharge from two sides per unit length of drain
Q_c = leachate collection rate
Q_d = recharge rate
Q_l = leakage rate
$(Q_x)_c$ = leachate flow rate toward the collection point, through a section located at a distance x
$(Q_x)_l$ = leachate leakage rate through the liner between the liner crest and a point x

r = hydraulic radius

s = slope of collection pipes

s = length of the rectilinear saturated slab at time t

S_0 = initial length of the saturated liner

t = time for the entire length of the leachate slab to drain into the collection line

v = saturated volume above liner at time t

v_1 = total volume of leachate collection

v_2 = total volume of leachate leakage

v_0 = total leachate volume

V = velocity

V_d = remaining volume of leachate in a day

V_i = volume of precipitation

y_0 or H = maximum height of water table above the pipe invert

$\overline{Y}$ = average thickness of water profile above the liner between the leachate collection pipe and the crest of a module

γ_s = unit weight of water

θ = slope of the liner

17

BIOREACTOR LANDFILLS

There is a long-standing debate regarding encapsulation of landfills at an early stage. In the 1980s landfill designers thought that constructing cover at an early date would minimize groundwater impacts. However, with additional information regarding rate of decomposition (Suflita et al., 1992), it was theorized that leachate will continue to contain significant amounts of contaminants many years (viz., more than 20 years) after a landfill is encapsulated. Many researchers also expressed skepticism regarding the longevity of the liner and the leachate collection system. According to them, encapsulation of landfills at an early date could create a perpetual threat to groundwater impact (Lee and Jones 1990; Harper and Pohland 1988). Subsequently, attempts were made to flush out the contaminants faster, using biodegradation of putrescible waste. Thus the goal of bioreactor landfills is to stabilize the readily and moderately decomposable organic waste constituents faster through controlled microbiological processes. In general, bioreactor landfills include a system for adding moisture because moisture content is probably the most important factor that enhances biodegradation. Because of the rapid biodegradation, the gas generation rate increases significantly. The generation rate also declines rapidly after closure. Figure 17.1 shows the comparison of gas generation rates between conventional and bioreactor landfills. Since the gas is generated at an early date, arrangements to collect the gas must be made to minimize air impact.

Both laboratory and field-scale research indicates that addition of moisture accelerates degradation of waste (EMCON Associates, 1975; Leckie et al., 1979; Barlaz et al., 1987; Doedens and Cord-Landwehr, 1989; Pohland, 1980; Reinhart and Townsend, 1998). Moisture availability is a key factor in sustaining bioreactor operation. Typical moisture content of municipal soild

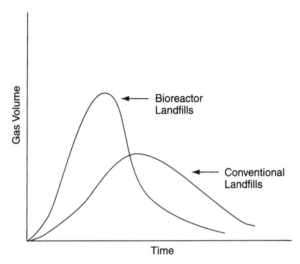

FIG. 17.1 Comparison of idealistic gas generation rates between conventional and bioreactor landfills.

waste in conventional landfills is approximately 20%. Minimum moisture content of 40% is necessary for a bioreactor landfill (Federal Register, 2002). As indicated above, the addition of water increases the microbial activities, which in turn increases the gas generation. In addition to gas generation, the following phenomena are also observed in bioreactor landfills:

1. The humuslike material generated due to the microbial activities filters some salts and metals from leachate.
2. The organic constituents in leachate decreases significantly.
3. The increase of pH leads to increase in metal concentration in leachate.

There are three types of bioreactor landfills: aerobic, anaerobic, and hybrid. In aerobic bioreactor landfills, both moisture and air is injected to promote aerobic bacterial activities. In anaerobic bioreactor landfills, only moisture is added to promote microbial activities. In hybrid landfills a sequential aerobic–anaerobic treatment is employed to enhance biodegradation for promoting early onset of methanogenesis. In hybrid bioreactor landfills, aerobic conditions are maintained in the top lifts for a short period of time (up to 3 months), after which the waste is covered to create anaerobic condition.

A field study indicated enhanced gas collection through the use of vacuum induced semi-aerobic (VSA) stabilization (NYSERDA, 2003). 12-inch (30-cm) HDPE pipes (structurally reinforced) were installed in horizontal trenches at 55-ft (1605-m) spacing. The pipes were installed 3 ft (90 cm) below the top of the waste. A vacuum of about 30–35 in. (76–89 cm) of water column was induced to collect the gas. The field study showed the system could create vacuum up to a distance of 110 ft (33 m). Significant amounts of landfill gas could be collected using the system. The study reported a 20% increase in air space.

17.1 MICROBIOLOGY OF LANDFILLS

Microbial activities play an important role in bioreactor landfills. The typical range of major components of municipal solid waste in the United States can be found in Table 14.1. Table 17.1 shows the percentages of methane-producing chemicals in municipal solid waste. About 60% of municipal solid waste is cellulose and hemicellulose, which accounts for about 90% of methane generation potential (Barlaz et al., 1990). Three groups of anaerobic bacteria decompose the cellulose and other polymers to produce methane. The first group consists of hydrolytic and fermentative microorganisms. These microorganism hydrolyze carbohydrates, fats, protein, and so on to form sol-

TABLE 17.1 Composition and Methane Potential of Municipal Refuse by Chemical Constituent

Chemical Constituent	Percent Dry Weight	Methane Potential[a]
Cellulose	51.2	73.4%
Hemicellulose	11.9	17.1
Protein[b]	4.2	8.3
Lignin	15.2	0
Starch	0.5	0.7
Pectin[c]	<3.0	—
Soluble sugars	0.35	0.5
Total volatile solids[d]	78.6	[e]

[a] Data expressed as a percentage of the total methane potential based on the cellulose hemicellulose, protein, sugar, and starch data. The methane contribution of pectin was not calculated because of the uncertainty associated with the pectin concentration in refuse. Methane potential was calculated from stoichiometry on the basis of 100% conversion of a constituent to carbon dioxide and methane. It should be recognized that 100% of the degradable constituents will not be mineralized, as some fraction is surrounded by lignin and not accessible for anaerobic degradation.
[b] Determined by multiplication of the total Kjeldahl nitrogen by 6.25. The actual protein content of refuse is probably lower than the value given here because some of the Kjeldahl nitrogen is actually nitrogen-containing humic materials and structural proteins that are not easily degradable.
[c] Actual value is probably less than 3%, but could not be quantified.
[d] An independent measure of volatile solids based on weight loss on ignition at 550°C.
[e] The volatile solids concentration is presented to illustrate that the chemical constituent analyses account for 110% of the volatile solids. No methane potential is calculated for volatile solids because this measurement includes both degradable and nondegradable carbon.
Reprinted with permission from Barlaz et al. (1990). Copyright CRC Press, Boca Raton, Florida.

uble sugars, amino acids, carboxylic acids, and glycerol. These chemicals are subsequently converted to form short-chain carboxylic acids, carbon dioxide, hydrogen, and acetate. The second group of microorganisms consists of obligate proton-reducing acetogens. These microorganisms convert the first-stage chemicals to acetate, carbon dioxide, and hydrogen. The third group of microorganisms are methanogens, which convert the hydrogen, carbon dioxide, and acetate to methane (Barlaz et at., 1990).

Several researchers reported formation of methane from refuse (Pohland et al., 1983; Barlaz et al., 1989a, 1990). According to Pohland et al. (1983), methane production goes through four phases. At present a fifth phase is included in which the methane production decreases due to the depletion of available food source. Based on a study of conventional municipal solid waste landfills, it is concluded that internal conditions of landfills changes over time primarily due to microbial activities and passes through the following five phases (USEPA, 2000):

Phase I: This is an aerobic phase, which lasts a few hours to one week. During this phase oxygen level decreases, and carbon dioxide level and temperature increase.

Phase II: This is a transition phase, which lasts 1–6 months. In this phase the landfill changes from aerobic to anaerobic condition. Volatile organic acids (VOA) are detected, and the chemical oxygen demand (COD) of the leachate increases signaling a shift to anaerobic metabolism.

Phase III: This is an anaerobic phase, which lasts from 3 months to 3 years. VOAs are rapidly formed leading to an increase of pH. The COD level of leachate peaks in this phase.

Phase IV: Gas generation is very high in this phase, which lasts 8–40 years or more. The acid compounds produced in earlier phases are converted to methane and carbon dioxide gas by methanogenic bacteria. The pH becomes more neutral, which reduces the metal concentration in leachate. Of the landfill gas released during this phase 50–60% is methane.

Phase V: In this phase methane production decreases significantly. The length of this phase is yet unknown. This is a phase of relative dormancy as biodegradable matter and nutrients become limiting. The concentration of leachate constituents reaches a stable state.

In aerobic bioreactor landfills, an attempt is made to sustain phase I. In anaerobic bioreactor landfills, an attempt is made to reduce the time of phase IV activities—to possibly 5–10 years.

17.2   POTENTIAL ADVANTAGES OF BIOREACTOR LANDFILLS

The increase in microbial activities due to the addition of liquid decomposes the waste mass faster, whereby the waste mass gets stabilized faster. However, note that all harmful chemicals are not flushed out due to liquid addition. Compared to conventional landfills, the following effects of liquid addition are observed in bioreactor landfills:

1. The waste mass is stabilized much faster.
2. The waste mass settles more and faster leading to 15–30% gain in air space during active landfill life.
3. There is a significant increase in gas generation.
4. Early stabilization decreases long-term environmental risk and reduces postclosure costs.
5. The disposal cost of leachate during active site life reduces significantly.

Note that aerobic bioreactors do not produce significant quantities of gas.

17.3 BIOREACTOR LANDFILL DESIGN

Several elements of bioreactor landfills should be designed carefully for environmental protection as well as to achieve the benefits. It may be noted that detail discussion regarding design and construction of most of the appurtenances are included in Chapters 16, 18, 19, and 20. Only the information relevant to bioreactor landfills is included in the following sections. Discussion regarding monitoring of bioreactor landfills is included in Chapter 22.

17.3.1 Liner

At a minimum a composite liner consisting of a 60-mil (1.5-mm) geosynthetic liner (usually HDPE sheets) placed over a 4–5 ft (1.2–1.5 m) clay layer is used for bioreactor landfills. The liner design should be discussed with the regulatory agency. Some U.S. states require multiple liners for bioreactor landfills.

Because of the increased possibility of slope failure in new phases of landfills, liner area at the base of a new phase should be bigger than what is normally built for conventional landfills. If failure occurs, this additional space will help in accommodating the waste mass resting on the slope.

17.3.2 Drainage Blanket

Because of increased microbial activities, and the possibilities of increased metal removal from the waste mass, chances of drainage blanket clogging are high in bioreactor landfills. The permeability of the drainage blanket should be higher than conventional landfills and a geotextile should be placed over the drainage blanket in bioreactor landfills (Koerner and Koerner, 1995). The drainage blanket material should be gravel, not sand. A monofilament woven geotextile, preferably treated with biocide, with an opening size of 0.5 mm and a relative area of 30% may also be used (Giroud, 1996). Since the volume of leachate may be higher than conventional landfills, excess leachate head buildup is possible. Therefore, the thickness of the gravel drainage blanket should be thicker compared to conventional landfills. The drainage blanket should be made thicker than the highest leachate estimated by using a suitable leachate apportionment model, as mentioned in Chapter 16.

17.3.3   Leachate Collection Pipe

The structural integrity of leachate collection pipes must be checked for bioreactor landfills. The density of waste mass can be 30% higher compared to conventional landfills (Phaneuf and Vana, 2000). In addition to structural strength, the leachate collection pipes should be designed for additional leachate flow and possible pipe clogging. The diameter of leachate collection pipes may need to be increased, and they may need to be cleaned more frequently to address these concerns. Discussions regarding leachate collection pipe design and cleaning are included in Chapter 19.

17.3.4   Liquid Introduction System

Liquid introduction system (LIS) is an important element of bioreactor landfills. There are several ways by which liquid can be introduced to bioreactor landfills. The most efficient system capable of distributing the additional liquid evenly over the entire waste mass should be chosen. In addition, the volume of liquid to be added for increasing the moisture content of the waste mass must be calculated. It may be noted that regulatory permission regarding addition of liquid, including recirculating leachate generated at the landfill, must be obtained where necessary. The volume of liquid to be added depends on the initial moisture content of the waste mass. Based on the literature, the average moisture content of MSW is 12–15%. Since the suggested average moisture content for bioreactor landfills is 40%, the additional volume of liquid to be added is 25–28%. The actual volume of liquid to be added is

$$V = M_p A T \qquad (17.1)$$

where M_p = additional percentage of moisture
  V = volume of liquid, L^3
  A = area of waste mass, L^2
  T = thickness of the waste mass, L

Equation (17.1) assumes a homogenous waste mass. However, in reality the waste mass is heterogeneous. Large materials such as carpet, wood logs, daily cover, and so on impedes uniform flow of liquids. Ideally, liquid movement in landfills is characterized by unsaturated flow. The suction head due to unsaturated condition in a landfill prevents downward movement of liquid.

The suction head or negative potential causes the liquid to move from an area of less negative potential to the adjacent area with higher negative potential until the waste mass reaches field capacity. Any liquid added beyond the field capacity of the waste will form leachate. Since the heterogeneity of the waste mass causes channeling, the waste mass is not wetted uniformly. In general, the additional liquid volume required to reach optimum moisture content in bioreactor landfills is more than the leachate volume generated in landfills. However, it is possible that leachate generated initially is sufficient to provide the optimum moisture. The leachate generation rate from conventional landfills can be estimated by any of the methods described in Chapter 13. If leachate generation data from nearby existing landfills is available, then those data may be used to predict additional liquid volume requirements by using the field-calibrated leachate apportionment model (Bagchi and Ganguly, 1990) described in Chapter 13. Theoretical estimation regarding additional liquid volume requirements is influenced by:

1. Actual moisture content
2. Heterogeneity
3. Compaction
4. Accuracy of the predictive model

Currently, field data regarding the volume of liquid added to reach optimum state is not available. Since moisture is a critical factor in bioreactor landfill design and operation, moisture content of waste mass should be monitored on site. Waste samples should be collected at well-defined horizontal and vertical locations regularly during active phases of a bioreactor landfill. In the absence of a regulatory requirement, the following schedule may be used to monitor moisture content:

1. Waste samples should be collected at 100 ft (30.5 m) grid points on the surface at 10 ft (3 m) vertical interval every week for 4 consecutive weeks.
2. If the moisture content values stabilize within the 4-week period, then the sampling frequency may be reduced to once every month.

After operating the first phase of the landfill, the data should be analyzed to obtain a realistic liquid recirculation schedule suitable for the site.

Design of the liquid introduction system must be such that the pipes and wells do not serve as points for air intrusion. The entire system should be easily accessible to ensure flow control and easy maintenance. In cold regions, the liquid introduction system should be buried below the frost depth to protect the system from freezing. It may be noted that microbial decomposition raises the temperature of the waste mass, which helps prevent freezing of the liquid introduction system. Odor is a problem, especially if leachate is recirculated. The installation and maintenance of the LIS can be costly if not designed properly. Various ways to introduce liquid in bioreactor landfills are discussed below. A comparison of these systems is included in Table 17.2.

17.3.4.1 Surface Spraying   Surface spraying is done by water tankers or spraying with a fire hose. The method can be used only in active phases. Surface spraying is a simple technique that provides uniform increase of moisture content. However, the method is labor intensive and cannot be used for

TABLE 17.2   Comparison of Various Liquid Introduction Systems

Recirculation Method	Disadvantages	Advantages
Prewetting	• Labor intensive • Blowing of leachate • Enhances compaction (may interfere with leachate routing) • Incompatible with closure	• Simple • Uniform and efficient wetting • Promotes evaporation
Vertical injection wells	• Subsidence problems • Limited recharge area • Interference with waste placement operations	• Relatively large volumes of leachate can be recirculated • Low-cost materials • Easy to construct during and following waste placement • Compatible with closure
Horizontal trenches	• Potential subsidence impact on trench integrity • Potential biofouling may limit volume • Inaccessible for remediation	• Low-cost materials • Large volumes of leachate can be recirculated • Compatible with closure • Unobtrusive during landfill operation
Surface ponds	• Collect stormwater • Floating waste • Odors • Limited impact area • Incompatible with closure	• Simple construction and operation • Effective wetting directly beneath pond • Leachate storage provided
Spray irrigation	• Leachate blowing and misting • Surface precipitation leads to decreased permeability • Cannot be used in inclement weather • Incompatible with closure	• Flexible • Promotes evaporation

closed landfills. Surface spraying may lead to the development of a solid hard pan due to the precipitation of leachate constituents (Robinson and Marib 1985) and odor problems. Surface spraying is ineffective when the surface is frozen during winter months.

17.3.4.2 Horizontal Trenches and Ponds Liquid may be introduced through horizontal trenches and ponds. Isolated ponds, 1–2 m in depth, are constructed at regular intervals throughout the landfill surface. Because ponds have a limited zone of influence, they need to be moved after a short interval to maintain uniform wetting. Horizontal trenches, 1–2 m in width and depth, are also used as liquid introduction system. The zone of influence depends on the rate of liquid introduction and hydraulic conductivity of waste. If the liquid injection rate is too high, then the waste mass influenced by the trench becomes saturated, creating an unfavorable condition for microbial activities. The trenches should be spaced (preferably 4–6 m apart) such that moisture content of waste mass increases uniformly. The trenches are usually filled with gravel or tire chips. As with the surface spraying, horizontal trenches and ponds create odor problems and are usable only in active landfills.

17.3.4.3 Horizontal Injection Liquid may be introduced by horizontal pipes embedded within the waste. Perforated polyethylene pipes are placed in 3 ft × 2 ft (1 m × 0.6 m) gravel [1 in. (2.5 cm) diameter, round] filled trenches (Fig. 17.2). The pipes are usually 6 in. (15 cm) in diameter and are placed at 6- to 8-m horizontal intervals and 2- to 3-m vertical intervals (Fig. 17.3). The spacing depends on the estimated injection rate. The horizontal pipes may also be installed in a dendritic pattern (Fig. 17.4). The cost of construction of horizontal pipes is low; these pipes are easy to construct except when a dendritic pattern is used. Liquid injection at various zones of the landfill can be controlled if control valves are installed on pipes. The horizontal injection system can be operated even after the landfill is closed.

17.3.4.4 Vertical Injection Liquid may be introduced using vertical wells. The wells are placed at predesigned grid points. The wells are constructed with 4-ft (1.2-m) diameter perforated concrete manholes. The bottom 6–9 ft

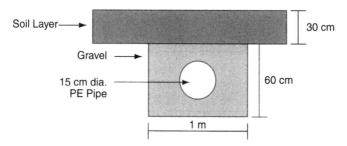

FIG. 17.2 Horizontal injection pipe (cross section) (not to scale).

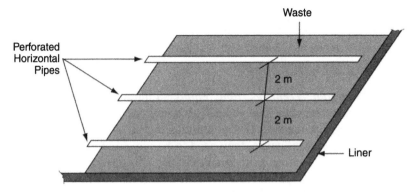

FIG. 17.3   Vertical spacing of horizontal injection pipes (not to scale).

(2–3 m) of the manhole is filled with gravel or concrete (Fig. 17.5). To increase the height of the well as the depth of waste increases, sections of the manhole are added vertically. The concrete-filled bottom portion provides stability, helps in better distribution of liquid within the waste, and avoids interference with the leachate collected on the liner. The radius of influence of vertical wells can be calculated by using the following formula (Reinhart and Townsend, 1998):

$$R = K_w / K_r \tag{17.2}$$

where R = radius of influence, L
    R = radius of recharge well, L
    K_w = permeability of media surrounding the well, L/T
    K_r = permeability of waste, L/T

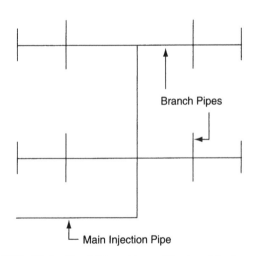

FIG. 17.4   Dendritic pattern of horizontal pipes.

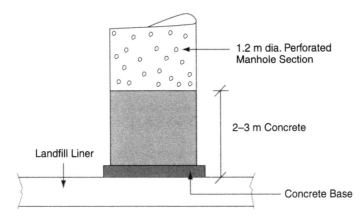

FIG. 17.5 Vertical well.

Note that Eq. (17.2) does not account for the liquid injection rate into the well.

Vertical injection wells can be used to introduce large volumes of liquid. The cost of construction of these wells is low and they are easy to install. Proper care needs to be taken to protect the well during the active phase of operation. The wells can be operated even after the landfill is closed.

17.3.4.5 Liquid Storage Provision for on-site liquid storage must be made for operating bioreactor landfills. The leachate generated in a bioreactor land-fill is, in most cases, not sufficient to provide the liquid necessary to raise the moisture content of the waste mass to the optimum level. In addition, the leachate generation rate (i.e., volume/day) may be more than or less than the required liquid addition rate. If the leachate generated in a period is more than what is needed for recirculation, then the excess leachate must be stored for future use. On the other hand, if the leachate generation rate is less than what is required for recirculation, then the additional liquid volume must be available on-site. Although a fair estimate can be made regarding the total yearly volume of liquid required for bioreactor landfills, it is a rather difficult task to estimate the daily volume of liquid required to raise the moisture to the optimum value. In the absence of site-specific modeling, the following approach may be used initially to roughly estimate the daily liquid volume requirements:

1. Obtain an estimate of monthly leachate generation rates, using the ap-portionment model mentioned in Chapter 16. Assume uniform generation to obtain a daily volume.
2. Estimate the required additional liquid volume by substituting the av-erage moisture content of the waste from the required optimum moisture content. The daily volume can be calculated by knowing the in-place waste and the average waste added each day.

3. The daily additional liquid requirement can be estimated from the difference of the available leachate volume and required liquid volume. The difference may be positive or negative. Positive values indicate excess volume of liquid available per day, and the negative values indicate the additional volume of liquid per day, which need to be supplied from outside sources.

4. Add all positive values for each month to estimate the total on-site liquid storage requirement.

5. Add all the negative values for each month to estimate the additional liquid required to be supplied to the landfill. If there is no source of water supply near the landfill, then arrangement must be made to store the liquid on-site.

The leachate and the noncontaminated liquid may be stored separately or together. If may be mentioned that leachate storage will require specially designed tanks or ponds to avoid groundwater impact. However, such special tank design is not necessary for storage of the noncontaminated liquid.

17.4   SLOPE STABILITY

Discussion on slope stability analysis for landfills is included in Chapter 19. Only the issues applicable to bioreactor landfills are discussed briefly in this section. The following issues should be considered while analyzing slope stability of bioreactor landfills:

1. The unit weight of waste mass increases by about 30% due to liquid addition. However, this increase in weight is rather gradual. So the slope analysis should be done for at least the following three different unit weight of the waste mass:
   a. No increase in unit weight
   b. Ten percent increase in unit weight
   c. Thirty to 40% increase in unit weight
   While only stability of internal working faces should be analyzed when top of the waste is below grade, both internal working face and external faces should be analyzed when top of waste mass elevation is above ground. The stability of the internal face as well as failure along the geomembrane on a side slope should be checked.

2. Possible leachate head within the landfill should be estimated properly. Slope stability analysis should include the effect of leachate mounding, particularly above the geomembrane surface on side slopes. Both slip circle and block failure analysis should be done.

3. Daily or intermediate cover, if not removed properly, can act as a barrier to liquid movement. Leachate mounding is very likely above such low permeability layers. In addition, a biosolid layer may develop on such surfaces (Fig. 17.6), which may act as a sliding surface. Although it is difficult to anticipate the development of such surfaces within a bioreactor landfill, stability analysis assuming development of such surface(s) at one or more arbitrary location(s) should be undertaken. Both slip circle and block failure analysis should be done.

17.5 POTENTIAL OBSTACLES TO BIOREACTOR LANDFILL DEVELOPMENT

There are several design, operational, monitoring, and financial issues that need to be addressed before developing bioreactor landfills. Additional studies are needed to address the following issues:

1. The impact of liquid addition on the leachate generation rate is to be addressed. A significant increase in leachate generation will increase the leachate head on the liner, leading to increased leakage through the liner.
2. Impact of liquid addition on the liner design needs further study. If the liquid addition increases the leachate head significantly, then the liner design may need to be revised (e.g., addition of another barrier layer along with a leachate collection system).
3. It may be noted that, while anaerobic decomposition can degrade recalcitrant chemicals, aerobic decomposition is not effective in degrad-

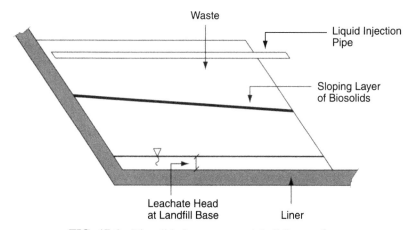

FIG. 17.6   Biosolids layer—potential sliding surface.

ing such chemicals. This issue is important for aerobic and hybrid bioreactor landfills. Steps to reduce the toxicity of the leachate (refer to Chapter 3 for discussion on toxicity reduction) should be undertaken to address this issue.

4. Impact of enhanced biodegradation on gas collection system design needs further study. Since gas generation in bioreactor landfills starts earlier, the gas collection system should be installed during the active phase of these landfills. Construction and maintenance of such collection systems in active landfills are difficult.

5. Impact of enhanced biodegradation on the air emissions needs further study. Because of the early start of gas generation, chances of violation of air emission standards is higher in active landfills.

6. Impact of biodegradation on long-term landfill settlement is not known. Landfill settlement will dictate the timing of the final cover construction. Additional data is necessary to assess the rate of settlement, the total settlement and the time to reach the total settlement.

7. Estimation regarding the additional liquid requirements is necessary for proper operation of bioreactor landfills.

8. Additional information is needed regarding the parameter and frequency of monitoring to control the microbiological processes.

9. Additional information is needed regarding the parameters and frequency of monitoring for environmental protection. It appears that, at a minimum, more frequent monitoring of leachate head, variation of leachate quality and quantity, and air emissions should be done.

10. Both short-term and long-term cost of bioreactor landfills should be calculated carefully. While recirculation of leachate will reduce leachate treatment cost, additional cost of installation and maintenance of liquid introduction system should be taken into consideration for cost estimation purposes.

11. The quality of leachate of matured bioreactor landfills will dictate the need for on-site pretreatment of leachate. The cost of construction and maintenance of such pretreatment plants should be included, if necessary, in the long-term care cost of biorecator landfills.

18

LINER MATERIALS

The different types of materials used to construct landfill liners and final covers fall into three categories: (1) clayey soil, (2) synthetic membranes or other artificially manufactured materials, and (3) amended soil or other admixtures. As discussed in Section 16.2, either only clay or a combination of two or all three types of materials is used for base liner and final cover construction. Each type has advantages and disadvantages that must be considered when choosing a particular liner material. A cost comparison must also be done prior to selecting an option. Material specifications, quality control tests and specifications, and minimum allowable thickness of the liner may vary from one state to another or in different countries. It is therefore essential that the local regulatory agency be contacted for acceptance criteria. The following discussions on each material type are provided as general guidelines and to provide a better understanding so that proper judgment regarding selection criteria can be developed. Inyang (1994) proposed a model to judge the long-term performance of waste containment systems.

At present geosynthetic clay liners are also being considered to replace clay liners. The purpose of landfill liners is to minimize or eliminate leakage of contaminants into groundwater. The transport of contaminants across liners can occur due to both advection and diffusion. In advection, movement of solutes is caused by a hydraulic gradient, whereas in diffusion the movement is caused by a difference in concentration of the solute. In media with high hydraulic conductivity advection is the dominant mode of solute transport, whereas in low hydraulic conductivity media diffusion is the dominant mode of solute transport. The solute transport through landfill liners is a combination of both advection and diffusion. Figure 18.1 shows the effect of hydraulic conductivity on transit times for various types of flow. The curve shows that

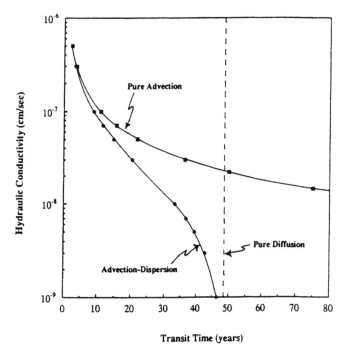

FIG. 18.1   Transit time (at c/co = 0.5) as a function of hydraulic conductivity. (From Shackelford, 1988.)

even with a hydraulic conductivity of 1×10^{-9} cm/sec the solute time for the concentration at the base of a liner to reach 50% of the input concentration under a hydraulic gradient of 1.33 for a 3-ft (0.9-m) liner (i.e., 1 ft (30 cm) of leachate head on top of a 3-ft (0.9-m) liner) is nearly 50 years, both under pure diffusion and advective-diffusive flow. An increase in the thickness of the liner will increase the transit time.

18.1   CLAY

Clayey soil is widely used for lining nonhazardous waste landfills. However, there is some debate regarding the definition of clay. Clay can be defined as the soil fraction that has particles equal to or finer than 0.002 mm (or 2 μm) or as the soil fraction that has particles equal to or finer than 0.005 mm (or 5 μm). Three commonly used grain size classifications are the unified soil classification system, the international classification system, and the MIT classification system (Means and Parcher, 1963). Both the international (proposed in 1927 and adopted by most countries except the United States) and MIT classification systems define clay by the 2-μm criterion. The MIT classification, which is simple, widely used in the United States, and easy to remember, uses the following particle size range:

Sand: 2–0.06 mm
Silt: 0.06–0.002 mm
Clay: 0.002 mm and less

Unless otherwise specified in this book, clay is defined by the 2-μm criterion.

The shape of the grain size curve is an important issue in obtaining a low permeability compacted soil liner. A soil with a grain size curve closely following the classical inverted S shape is expected to develop lower permeability.

18.1.1 Effect of Double Layer

An understanding of the role of the electrical double layer in controlling soil behavior is helpful. Clay surfaces usually have an excess negative charge. The space between two clay particles contains water with cations (positively charged ions) and anions (negatively charged ions). The density of cations near the negatively charged clay surface is high and becomes lower with increasing distance. The negatively charged clay surface together with the liquid phase adjacent to it is called the diffuse double layer. Based on the Gouy–Chapman theory, the thickness of the double layer is expressed as

$$\text{Th} = \left(\frac{DKT}{8\pi n_0 \varepsilon^2 v^2} \right)^{1/2} \tag{18.1}$$

in which Th = the thickness of the double layer, D = the dielectric constant, K = Boltzmann's constant, T = absolute temperature (in Kelvin), n_0 = ion concentration, ε = the unit electronic charge, and v = the valence of the ion.

Equation (18.1) indicates that the thickness is inversely proportional to the valence of the ion and square root of the ion concentration. The thickness of the double layer increases with the dielectric constant and temperature. The following conclusions were drawn from a parametric study of Eq. (18.1) (Mitchell, 1976).

1. The swelling behavior of clay depends partially on the electrolyte concentration within the double layer.
2. The addition of small amounts of di- or trivalent cations to a monovalent double-layer system can influence the physical properties significantly.
3. Double-layer thickness controls the tendency to flocculate and influences the swelling pressure of clays.
4. A decrease in double-layer thickness may cause shrinkage of clay. This is an important issue for waste disposal sites because the dielectric constant of leachate is expected to be different from water.
5. The effect of temperature change on the double layer is difficult to predict because the dielectric constant is also temperature dependent.

For water, the change in the value of $D \times T$ is not significant for a change of temperature from 0 to 60°C, which indicates that the double-layer thickness will not be influenced significantly due to a change in temperature in that range.

It should be noted that the Gouy–Chapman theory does not take into account the following important effects on the double-layer thickness: secondary energy terms (Bolt, 1955), superimposing electric field and structure of water on the double layer (Low, 1961; Mitchell, 1976), ion size in the double layer (Van Olphen, 1963), and pH (Mitchell, 1976). The Gouy–Chapman theory cannot explain the observed behavior of natural clay completely because natural clayey soils are mostly a mixture of a number of different clay minerals, each of which has a different double-layer thickness. Change in the electrolyte concentration in the diffused double layer causes a change in its thickness leading to a change in permeability and shear strength properties of clay.

18.1.2 Effects of Various Parameters on Clay Properties

The mechanical properties of clay depend on several interacting factors such as mineral composition, percentage of amorphous material, absorbed cation, distribution and shape of particles, pore fluid chemistry, soil fabric, and degree of saturation. The effect on the mechanical properties of soil due to a change in any of these factors can be predicted qualitatively using physicochemical theories. However, quantitative prediction of soil behavior based on the above factors and any improved double-layer theory is almost impossible, mainly because of the inadequacy of the available physicochemical theories and the difficulties in taking into account the effect of soil fabric and other *in situ* environmental factors. Several studies have been reported on the effect of change in pore fluid chemistry on the strength of clayey soil (Torrance, 1974; Yong, 1986; Yong et al., 1979) and on the permeability of clayey soil (Acar and Seals, 1984; Acar and Ghosh, 1986; Brown and Anderson, 1980; Brown et al., 1983, Brown and Thomas, 1985; Fernandez and Quigley, 1985; Green et al., 1981; Reeve and Tamaddoni, 1965; Foreman and Daniel, 1984; Daniel et al., 1984). Some of the above studies used pure solvents and high concentrations of chemical compounds and the rest used actual landfill leachate or diluted solvent.

Mitchell and Madsen (1987) summarized the literature on the effect of inorganic and organic chemicals on clay permeability. The summary indicates that although dilute solutions of inorganic chemicals may change clay permeability, dilute solutions or organic chemicals have virtually no effect. Significant changes in permeability of the clay samples were not observed when leached with diluted solvent or salt solutions (Carpenter and Stephenson, 1986; Acar and Seals, 1984; Brown and Thomas, 1985). Fang and Evans (1988) observed practically no difference in the permeability of clay samples permeated with tap water and landfill leachate. Whether a chemical species in the leachate will trigger a change in the permeability of a clay liner or not, it may leak through the liner. The rate of leakage will depend on the diffusion coefficient and permeability of the chemical through the liner (Daniel and Shackelford, 1988).

Most of the studies cited above are rather short term and so the effect on permeability due to long-term exposure is not readily known. Bowders et al. (1984) suggested a decision tree (Fig. 18.2), which can be used to decide

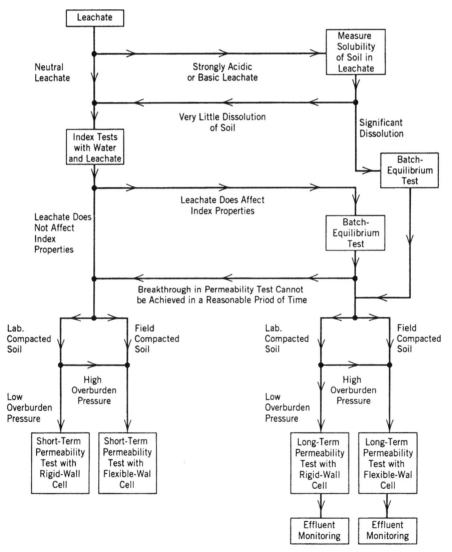

FIG. 18.2 Suggested decision tree as an aid in selecting a program of soil testing. After Bowders et al. (1984). Copyright ASTM. Reprinted with permission.

whether short-term or long-term permeability tests should be used. At present the general trend in the United States is not to use clayey soil as the primary liner in hazardous waste sites. Since pore fluid chemistry plays a significant role in changing the permeability of a soil, the effect of leachate from a proposed hazardous waste landfill on the chosen soil should be studied even if it is not required by the regulatory agency.

Because of the problem of running a long-term permeability test, one may use the decision tree shown in Fig. 18.2 to decide whether a long-term permeability test should be undertaken. As previously discussed, changes in pore fluid chemistry will influence all the mechanical properties of soil. It is postulated that the Atterberg limits reflect the strength and permeability characteristics of soil (Terzaghi, 1936).

The liquid limit of clay corresponds approximately to the water content at which the shear strength is between 2 and 2.5 kN/m^2 (Norman, 1958). The plastic limit indicates the lower boundary of the water content range below which the soil no longer behaves as a plastic. In other words a soil can be deformed without volume change or cracking above the plastic limit and will retain its deformed shape (Mitchell, 1976). The plastic limit is a reflection of the structure of water within the pore and the nature of interparticle forces (Terzaghi, 1936; Young and Warkentin, 1966). If the moisture content at which compaction is proposed (compaction moisture is usually 2–3% above optimum moisture) is less than the plastic limit, then the tendency to develop microcracks during liner construction is expected to increase. Therefore, it is essential to ensure that the plastic limit of soil is lower than the proposed compaction moisture content. Seed et al. (1962, 1964a,b) developed a theoretical relationship between liquid limit, plasticity index, and clay content (Fig. 18.3) that is similar to the plasticity chart used for classifying clay using the USCS method. Inorganic clays of medium plasticity are preferred, and caution must be exercised in using low plasticity inorganic clays for liner construction purposes. The ratio of the plasticity index and the percentage of clay fraction is termed activity (Skempton, 1953). Studies of artificially prepared sand–clay mineral mixtures indicate that the relationship between the plasticity index and percentage clay fraction does not necessarily pass through the origin but may intersect the percentage clay fraction axis between 0 and 10. Based on two studies Mitchell (1976) defined activity as

$$A_c = \frac{PI}{P_c - n} \tag{18.2}$$

in which A_c = activity, PI = plasticity index, P_c = percentage clay fraction, and n = constant (= 5 for natural soil and 10 for artificial mixtures).

The activity of smectites is highest and that of kaolinites is lowest. The higher the activity of a soil the higher its susceptibility to property changes due to a change in factors such as pore fluid chemistry and water content.

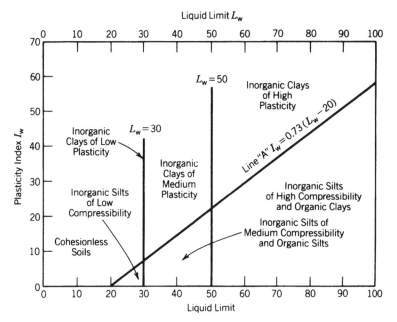

FIG. 18.3   Plasticity chart. After Mitchell (1976).

18.1.3   Effect of Compaction on Permeability of Clay

Compaction of soil during construction causes a change in soil fabric that influences mechanical properties, especially the permeability of soil. A dispersed soil fabric is expected to have lower pore diameter and higher tortuosity, which will influence permeability. A study of the Kozeny–Carman equation (Carman, 1956; Mitchell, 1976) is helpful in understanding the effect of pore size on soil permeability. The equation, although usable for sand, is considered inadequate in predicting permeability for clayey soil (Lambe, 1954; Michaels and Lin, 1954; Olsen, 1962). However, the equation is useful in understanding the effect of various *in situ* parameters on soil permeability. According to the theory, the permeability of saturated soil can be expressed as

$$K = \frac{1}{K_0 T_f^2 S_0^2}\left(\frac{e^3}{1 + e}\right)\left(\frac{\gamma}{\mu}\right) \tag{18.3}$$

in which K = the permeability of soil, K_0 = the pore shape factor, T_f = the tortuosity factor, S_0 = the specific surface per unit volume of particle, e = the void ratio, γ = the unit weight of the permeant liquid, and μ = the viscosity of the permeant liquid.

The equation shows the following:

1. Permeability will be reduced due to a reduction in the void ratio (e).
2. Permeability is dependent on the ratio of the unit weight and viscosity of the permeant liquid.
3. Permeability will decrease due to an increase in tortuosity (i.e., if a more zigzag path is followed by the liquid).
4. Permeability will decrease due to an increase in the specific surface area.

The permeability of soil is controlled mostly by large pores (Mitchell and Madsen, 1987), the distribution and size of which depend on the fabric of the soil. The following is a simplistic description of compacted clay soil fabric observable under the electron microscope; The clay particles aggregate together to form "miniclods" (or assemblages) that are packed tightly to form a larger clod. The sizes of the pores within the miniclods are smaller than the pores formed by the space between the miniclods. In addition to these two types of pores, a third type may exist in naturally occurring clayey soil due to cracks and fissures (Collins and McGown, 1974). The pores caused by cracks and fissures are the largest and transport the maximum amount of fluid. So, when compacting clay, measures should be taken to avoid the formation of cracks and fissures. The factors that influence the fabric of compacted clay are the water content during compaction, the method and effort involved in compaction, the clod size of clay, and the interlocking of layers.

Since dispersed fabric will increase the tortuosity and reduce pore size, the aim of field compaction should be to create a dispersed fabric. Kneading compaction induces high shearing strain, which breaks down flocculated fabric and helps create a dispersed fabric (Mitchell, 1976). Water content during compaction influences the rearrangement of clay particles, which greatly influences permeability. The method of compaction also influences permeability (Mitchell et al., 1965; Mitchell, 1976). The effects of water content and the method of compaction on permeability are shown in Figs. 18.4 and 18.5, respectively. Because of these reasons, a sheep's foot roller should be used, which will provide kneading compaction.

Current convention is to specify the degree of compaction and moisture content of the clay to achieve the desired permeability. Traditionally, to achieve the permeability demonstrated in the laboratory, the clay is compacted at 90–95% of the maximum modified proctor density at 1–4% wet of the optimum moisture. A study by Daniel and Benson (1990) showed that the acceptable permeability for clay liners (1×10^{-7} cm/sec) can be achieved even with less compactive effort compared to modified proctor compaction, provided the moisture content is above the optimum moisture. The acceptable zone for achieving a permeability of 1×10^{-7} cm/sec is shown in Fig. 18.6. The curve was generated using three different compactive efforts. In addition

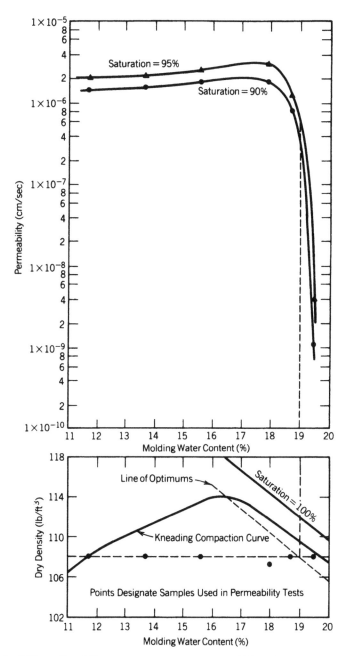

FIG. 18.4   Effect of molding water content on permeability of a silty clay (kneading compaction was used in preparing the samples). After Mitchell (1976).

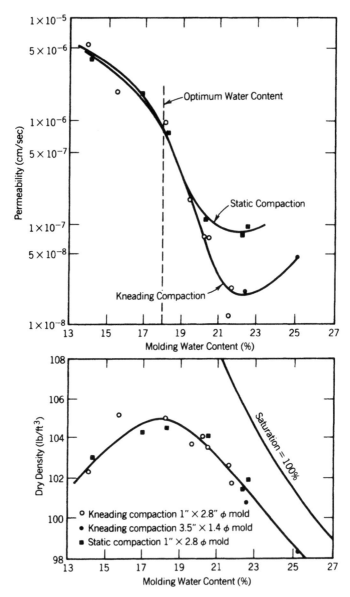

FIG. 18.5   Influence of the method of compaction on the permeability of silty clay. After Mitchell (1976).

to using the standard proctor compaction (ASTM D698) and modified proctor compaction (ASTM D1557), a third method termed the "reduced" proctor compaction was used for compacting soil. The reduced proctor compaction is identical to standard proctor compaction except that instead of 25 blows/ layer, 15 blows/layer are used. The following procedure to achieve a per-

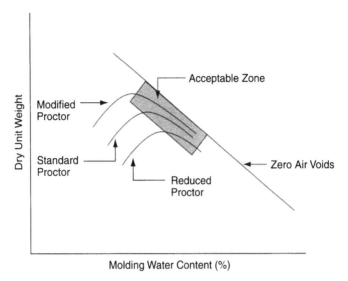

FIG. 18.6 Acceptable zone for permeability.

meability of 1×10^{-7} cm/sec or less has been suggested by Benson and Daniel (1990):

1. Compaction curves using modified standard and reduced proctor compaction procedures are to be generated using laboratory tests.
2. The saturated hydraulic conductivity of the specimens should be determined carefully.
3. The specimens with hydraulic conductivity of 1×10^{-7} cm/sec or less should be identified in the compaction curves (step 1). The acceptable zone for hydraulic conductivity is to be drawn encompassing all data point with a permeability of 1×10^{-7} cm/sec or less. Figure 18.6 shows a typical plot of the acceptable zone for hydraulic conductivity. Soil compacted within the acceptable zone is expected to achieve a permeability of 1×10^{-7} cm/sec. A field study by Benson et al. (1999) showed that a permeability of 1×10^{-7} cm/sec or less was achieved when clay liners were compacted at or above the line of optimum moisture.

18.1.4 Effect of Clod Size on Clay Permeability

Precompaction clod size may influence the structure of compacted clay and thereby the permeability of clay (Daniel, 1981; Barden and Sides, 1970). Influence of clod size on the hydraulic conductivity of a highly plastic soil (note: the soil was classified as CH per the USCS and had a clay fraction of 42%) was reported by Benson and Daniel (1990). For the clay compacted dry of optimum using standard proctor procedure, the hydraulic conductivity was 1×10^6 times higher than when the soil was prepared from large (19 mm) rather than small (4.6 mm) clods. Clod size did not influence the hydraulic conductivity when the soil was compacted wet of optimum using a modified proctor procedure.

It has been observed that the liquid limit of clay may be altered due to drying and rewetting, indicating a lack of rehydration to the original state (Sangrey et al., 1976). Therefore, an attempt should be made to maintain a natural moisture content whenever possible and sufficient time should be allowed to moisten the soil in the field. The weight of the compaction equipment should be high enough to break the clods.

18.1.5   Effects of Natural Elements on Clay Permeability

Changes of permeability due to exposure to the natural elements could be due to desiccation cracks and freeze–thaw degradation.

18.1.5.1 Desiccation Cracks.   Desiccation cracks in a liner may develop because compacted clayey soil liners are subjected to periods of drying usually immediately after construction. In a laboratory study the permeability of desiccated samples increased by one order of magnitude or more (Dunn, 1986). The depth of desiccation cracks can be as high as 1 ft, which is of concern for liner construction. Desiccation cracks are essentially due to shrinkage of the soil. Desiccation cracks increase with soil shrinkage (Kleppe and Olson, 1994). Soils with a high liquid limit are expected to shrink more. The degree of shrinkage (S_γ), defined in Eq. (18.4), can be used to classify the soil qualitatively (Jumikis, 1962):

$$S_\gamma = \frac{V_i - V_f}{V_i} \qquad (18.4)$$

in which V_i = the initial volume of the sample before drying and V_f = the final volume of the sample after drying. Standard shrinkage limit tests are used to find V_i and V_f.

Table 18.1 provides a soil grouping based on shrinkage. According to these criteria good soils shrink less.

TABLE 18.1   Soil Grouping Based on Shrinkage

Degree of Shrinkage (S_γ) (%)	Quality
Less than 5	Good
10–15	Poor
More than 15	Very poor

Daniel and Wu (1993) suggested the following procedure to establish the dry unit weight and water content range within which the volumetric strain is 4% or less and the hydraulic conductivity is 1×10^{-7} cm/sec or less:

1. Compaction curves using modified, standard, and reduced proctor procedure are to be generated using laboratory tests.
2. The volumetric strain of the compacted samples is to be determined using a standard test (ASTM D 427).
3. On the compaction curves the specimens with 4% or less volumetric strain are to be identified. The acceptable zone for the volumetric zone is to be drawn encompassing all data points having a volumetric strain of 4% of less. Figure 18.7 shows a typical plot of an acceptable zone for volumetric shrinkage.

It is essential that a clayey soil liner be protected from excessive drying after construction by spraying water if necessary. Disposal of waste will reduce the chance of development of desiccation cracks in liners. Desiccation cracks can be minimized by compacting relatively dry soil using high compactive effort (Daniel and Wu, 1993). Desiccation cracks of a final cover clay layer can be minimized by constructing a thick cover soil or by constructing a synthetic membrane overlaid by a layer of top soil.

18.1.5.2 Freeze–Thaw Degradation   Freeze–thaw degradation is due to alternate freezing and thawing of the clay liner exposed to subzero temperatures. Cracks in natural soils and an increase in vertical permeability due to the freeze–thaw effect have been observed by many (Chamberlain and Gow, 1978; Vallejo and Edil, 1981; Chamberlain and Blouin, 1977). Many researchers reported a two orders of magnitude increase in hydraulic conductivity of

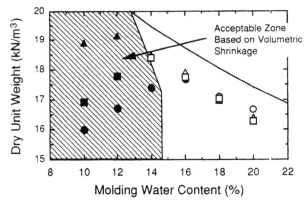

FIG. 18.7 Acceptable zone based on volumetric shrinkage considerations. Reproduced by permission of the publisher, ASCE, from Daniel and Wu (1993).

clay due to freeze–thaw in laboratory studies (Chamberlain and Blouin, 1976; Chamberlain and Gow, 1979; Chamberlain et al., 1990; Zimmie and La Plante, 1990; Othman and Benson, 1991; Kim and Daniel, 1992). Paruvakat (1993a) raised doubts regarding validity of laboratory data when a three-dimensional freezing technique is used in the laboratory. Although all laboratory tests showed a significant increase in hydraulic conductivity of frozen clay specimens, field studies reported by researchers indicated an increase in hydraulic conductivity only up to three-fourths of an order of magnitude (Starke, 1989; Paruvakat et al., 1990; Bagchi, 1993).

A layer of frozen clay develops at the surface of the liner when the ambient temperature falls below 0°C. Frozen soil develops a reticulate ice vein network that subdivides the soil into irregular blocks (Vallejo, 1980). The water present in the pores of the clay layer forms ice crystals. Since the volume of ice is 9% more than the volume of water, the ice crystals exert pressure on the surrounding soil causing a change in fabric and possibly a localized consolidation of the clay (Taylor and Luthin, 1978; Andersland and Anderson, 1978; Chamberlain and Blouin, 1976; Alkire and Morrison, 1982). This layer of frozen soil (known as frozen fringe) creates a suction causing water from lower portions of the layer to move upward. As a result, a localized hydraulic gradient develops within the soil that causes development of a nonuniform water content profile within the layer (Williams, 1986; Konrad and Morgenstern, 1980; Dirksen and Miller, 1966; Hoekstra, 1966). Migration of water to the frozen clay layer from the lower portion of a layer may result in vertical shrinkage cracks depending on the plasticity index of the clay and the water content during compaction. If sufficient moisture is available during freezing, then a reticulate ice vein structure may develop. An equilibrium develops within the compacted clay liner whereby the penetration of the frozen fringe stops at a depth that is related to ambient temperature and the temperature of the compacted clay liner. The clay liner temperature depends on solar radiation, soil moisture, evaporation rate, heat storage, and insulation on the liner (e.g., snow cover). The depth of the frozen fringe depends primarily on the duration and degree of the subzero temperature. When the ambient temperature rises above 0°C, the ice lenses thaw. This thawing may increase the size of pores and its interconnectedness. Thus a layer of degraded clay forms that could be as deep as the frozen fringe penetration. When the ambient temperature falls below freezing again, the upper layer of the liner provides less resistance to frost penetration.

The presence of desiccation cracks at the liner surface, an almost unavoidable phenomenon, probably triggers the degradation due to freeze–thaw. The presence of interconnected pores that is dictated by compaction during construction is another key issue in freeze–thaw degradation.

Thus, the depth of liner degradation due to freeze–thaw depends primarily on the duration and degree of subzero temperature, available solar radiation, and thickness of cover on the liner. However, the degree of liner degradation or, in other words, increase in hydraulic conductivity depends primarily on the number of freeze–thaw cycles and the size and distribution of interconnected pores. The size and distribution of interconnected pores are a reflection of compactness of the liner, which depends on soil type, compaction equipment and effort used, and the quality control program used during liner construction.

The necessary conditions for the formation and growth of ice veins are a low overburden pressure, a slow freezing rate, a sufficient supply of moisture to keep the soil saturated at the beginning and during the freezing process, and soil with 3–10% clay (Martin, 1958; Vallejo, 1980; Linell and Kaplar, 1959; Anderson et al., 1978). Several field studies indicated that the depth of liner degradation due to exposure in one winter in Wisconsin (note: in Wisconsin the winter temperature ranges from −9.5 to −1°C and there are usually two or three freeze–thaw cycles) in one winter is approximately 30 cm (Starke, 1989; Paruvakat et al., 1990; Bagchi, 1993). The focus of laboratory studies was to gain an insight into the effect on intrinsic property of permeability of compacted clay samples due to freeze–thaw cycles. A direct use of the data for predicting field behavior of compacted clay liner due to winter exposure is bound to lead to an extremely conservative design. Discrepancy between laboratory and field data is often reported in the technical literature. Caution must be exercised in extrapolating laboratory data to predict field behavior. Field studies have clearly shown that the effect of freeze–thaw on clay liner permeability is not as severe as predicted by laboratory studies and, most importantly, the depth of degradation is not expected to exceed 30 cm (for winter conditions similar to Wisconsin) if proper care is taken during and after liner construction.

The following recommendations are made for protecting clay liners from freeze–thaw degradation:

1. Clay liners should not be exposed to freeze–thaw for two or more consecutive winters. Sufficient frost protection material should be placed over the liner before the onset of the second winter.
2. Good quality clay (CL or CL-ML type of soil per USCS) and a proper QA/QC program for liner compaction should be used to reduce the effect of freeze–thaw on clay liners (note: see Chapter 20 for a QA/QC program).
3. Appropriate care (e.g., sufficiently, thick layer of sand or waste) for protecting clay liners from freeze–thaw degradation should be taken from the first winter if the average infield hydraulic conductivity of the liner is 1×10^{-7} cm/sec or more.
4. Appropriate care should be used in protecting clay liners from freeze–thaw degradation if the thickness of the liner is 60 cm or less. A minimum liner thickness of 120 cm is recommended in all cases.

18.1.6 Permeability Test

Proper permeability tests should be used to assess the permeability of liners. The terms permeability and hydraulic conductivity are used interchangeably in the literature. When the term hydraulic conductivity is used it specifically refers to the conduction property related to water, whereas the term permeability is used to refer to a conduction property related to any liquid, which also includes water. Although the use of laboratory permeability tests is met with skepticism because of macrofabric features (Dunn, 1986; Day and Daniel, 1985), current practice continues to be to obtain several undisturbed samples and to test them in the laboratory.

18.1.6.1 Laboratory Permeability Test Peirce et al. (1986) advocated the use of a consolidation test to determine hydraulic conductivity of clay in the laboratory because in their opinion the test will simulate field conditions. However, others prefer a flexible wall permeameter, with back pressure, over fixed wall permeameter, or a consolidometer (Bagchi, 1987d; Daniel et al., 1984). Standard test method for measuring hydraulic conductivity of fine grained soil using a flexible wall permeameter (ASTM D5084-90) has since been adopted by ASTM. The gradient during permeation, the chemistry of the permeating fluid, the degree of saturation, and confining pressure influence the permeability of a soil sample (Olsen, 1962; Mitchell, 1976). All these factors must be carefully monitored to simulate field conditions as far as practicable. Trainor (1986) reported a dedicated flexible wall permeameter that is equipped with back pressure to saturate soil samples before performing a permeability test using leachate (Fig. 18.8). It is essential that 90–100% saturation be achieved during a permeability test. A low confining pressure and low gradient (maximum 10) are preferred during these permeability tests to simulate field conditions.

There are several sources of error in a laboratory permeability test (Olson and Daniel, 1979; Zimmie et al., 1981). Sources of error also exist in estimating field permeability using laboratory test results (Daniel, 1981). Most of the error sources can be eliminated if proper quality control is exercised during construction and laboratory testing. Use of a triaxial permeameter and landfill leachate as the permeant is preferred. It may be difficult and time consuming to run a permeability test using landfill leachate as permeant, especially if the bacterial population is high in the leachate (i.e., leachate obtained from putrescible waste). Such a special test procedure would mean special laboratory arrangements and therefore higher costs.

The following approach may be used to minimize the total cost yet maintain a good quality control on tests. A small number of samples (10–15%) is tested under stricter test conditions (i.e., testing in triaxial cells with low confining pressure and hydraulic gradient) and then the results are compared with those obtained by the standard falling head permeability test. The degree of saturation of each of the samples of the second set must be determined at

FIG. 18.8 Apparatus for the clay permeability test using triaxial pressure. (Courtesy of RMT Inc., Madison, WI.)

the beginning and end of the test. If the degree of saturation is 90% or more and the results of the two sets are close, then the results of the falling head permeability test may be acceptable. The degree of saturation should be obtained carefully because it greatly influences the permeability of soil samples (Johnson, 1954).

18.1.6.2 In Situ Permeability Test   Several *in situ* permeability test methods are available (Daniel, 1989). Each method has its advantages and disadvantages, which are included in Table 18.2. The methods can be subdivided into four major categories: borehole tests, porous probe, underdrain, and infiltrometer. The following is a brief description of each of the methods.

Borehole Tests   Two different methods are available in this category: Boutwell permeameter and constant head permeameter.

BOUTWELL PERMEAMETER   This is a two-stage borehole test. The test schematic is shown in Fig. 18.9. A hole is drilled in the compacted liner in which the casing is placed; the annular space is grouted. Falling head tests are performed. Next the hole is deepened by augering or by pushing a thin-walled sampling tube into the compacted liner; the smeared soil is removed from the hole. The test is performed again after assembling the device. Both horizontal

TABLE 18.2 Advantages and Disadvantages of *in Situ* Permeability Test Methods

Type of test	Device	Advantages	Disadvantages
Borehole	Boutwell permeameter	Low equipment cost (<$200 per unit)	Volume of soil tested is small
		Easy to install	Unsaturated nature of soil not properly taken into account
		Hydraulic conductivity is measured in vertical and horizontal direction	Testing times are somewhat long (typically several days to several weeks for hydraulic conductivities $<10^{-7}$ cm/sec)
		Can measure low hydraulic conductivity (down to about 10^{-9} cm/sec)	
		Can be used at great depths and on slopes	
	Constant head permeameter	Low equipment cost (<$1000 per unit)	Volume of soil tested is small
		Easy to install	The hydraulic conductivity that is measured is primarily the horizontal value (in some applications, the value in the vertical direction is desired)
		Unsaturated nature of soil taken into account relatively rigorously	The device is not well suited to measuring very low hydraulic conductivities (less than 10^{-7} cm/sec)
		Relatively short testing times (a few hours to several days)	
		The hydraulic conductivity that is measured is primarily the horizontal value (which is an advantage if this is the desired value)	
		Can be used at great depths	
Porous probe	BAT permeameter	Easy to install	High equipment cost (>$6000)
		Short testing times (usually a few minutes to a few hours)	Volume of soil tested is very small
		Probe can also be used to measure pore-water pressures	Soil smeared across probe during installation may lead to underestimation of hydraulic conductivity
		Can measure low hydraulic conductivity (down to about 10^{-10} cm/sec)	The hydraulic conductivity measured is primarily the horizontal value (in some applications the value in the vertical direction is desired)
		The hydraulic conductivity that is measured is primarily the horizontal value (which is an advantage if this is the desired value)	The unsaturated nature of the soil is not properly taken into account
		Can be used at large depths	

Method	Type	Advantages	Disadvantages
Infiltrometer	Open, single-ring infiltrometer	Low cost (<$1000) Easy to install Very large infiltrometer can be used to test a large volume of soil Hydraulic conductivity in the vertical direction is determined	Low hydraulic conductivity (<10^{-7} cm/sec) is difficult to measure accurately Must eliminate, or make a correction for, evaporation May need to correct for lateral spreading of water beneath infiltrometer Testing times are relatively long (usually several weeks to several months for hydraulic conductivities <10^{-7} cm/sec) Must estimate wetting-front suction head Cannot be used on steep slopes unless a flat bench is cut
	Open, double-ring infiltrometer	Low equipment cost (<$1000) Hydraulic conductivity in the vertical direction is determined Minimal lateral spreading of water that infiltrates from inner ring	Low hydraulic conductivity (10^{-7} cm/sec) is difficult to measure accurately Must eliminate or make a correction for evaporation Testing times are somewhat long (usually several days to several months for hydraulic conductivities <10^{-7} cm/sec) Must estimate wetting-front suction head Cannot be used on steep slopes unless a flat bench is cut
	Closed, single-ring infiltrometer	Low equipment cost (<$1000) Hydraulic conductivity in the vertical direction is measured Can measure low hydraulic conductivity (down to 10^{-8}–10^{-9} cm/sec)	Volume of soil tested is somewhat small because diameter of ring is < 1 m Need to correct for lateral spreading of water if wetting front penetrates below the base of the ring Testing times are long (usually several weeks to several months) Must estimate wetting-front suction head Very difficult to use on steeply sloping ground

TABLE 18.2 (*Continued*)

Type of test	Device	Advantages	Disadvantages
	Sealed, double-ring infiltrometer	Moderate equipment cost (<$2500) Hydraulic conductivity in the vertical direction is determined Can measure low hydraulic conductivity (down to about 10^{-8} cm/sec) Minimal lateral spreading of water that infiltrates from inner ring Relatively large volume of soil is permeated	Testing times are relatively long (usually several weeks to several months) Must estimate wetting-front suction head Cannot be used on slopes unless a flat bench is cut
	Air-entry permeameter	Modest equipment cost (<$3000) Relatively short testing times (a few hours to a few days) Hydraulic conductivity in the vertical direction is measured Can measure low hydraulic conductivity (down to 10^{-8}–10^{-9} cm/sec) Wetting-front suction head is estimated in second stage of test	A relatively small volume of soil is permeated because the wetting front usually does not penetrate more than a few centimeters into compacted clay Cannot be used on slopes unless flat bench is cut Several important assumptions are required
Underdrain	Lysimeter pan	Low cost The hydraulic conductivity in the vertical direction is measured Large volumes of soil can be tested Few experimental ambiguities No disturbance of soil	Must install underdrain before the liner is constructed Relatively long testing times (usually several weeks to several months for hydraulic conductivities less than 10^{-7} cm/sec) Must collect and measure seepage from underdrain, which usually necessitates a sump and a pump.

After Daniel (1989). Reprinted by permission from ASCE.

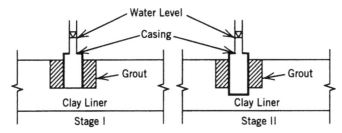

FIG. 18.9 Boutwell permeameter.

and vertical hydraulic conductivity of the compacted liner can be calculated using this test. It is reported that the field hydraulic conductivity measured using this method compared well with laboratory test; this test also indicated that horizontal hydraulic conductivity is typically 5–10 times vertical hydraulic conductivity for compacted clay soils (Daniel, 1989). It takes several days to several weeks to complete this test.

CONSTANT HEAD BOREHOLE PERMEAMETER   In this method (Fig. 18.10) a constant head of water is maintained using a Mariotte system (Olson and Daniel, 1981; Reynolds and Elrick, 1985). The rate of flow needed to maintain two constant water levels is measured, which is then used to calculate the hydraulic conductivity. *In situ* hydraulic conductivity obtained using a borehole permeameter was reported to be an order of magnitude higher than laboratory values (Stephens et al., 1988). The test can be completed in a relatively short time—a few hours to a few days.

Porous Probes   In this method either a constant or falling head test is performed using a porous probe pushed into the soil (Fig. 18.11). BAT permeameter is similar to a porous probe (Tortensson, 1984). In a BAT permeameter a chamber is lowered into a borehole created by the probe and brought into contact with the porous probe using a hypodermic needle and a septum. The

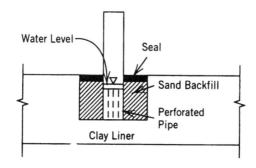

FIG. 18.10 Constant head borehole permeameter.

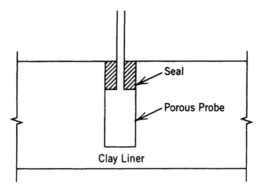

FIG. 18.11   Porous probe.

chamber contains both air and water. The mixture is either pressurized or evacuated causing water to flow out of or into the chamber. A pressure transducer records the change of air pressure within the chamber due to the flow of water. The hydraulic conductivity is measured from the time required for a change in air pressure. *In situ* hydraulic conductivity of compacted clay soil compared well with laboratory values (Chen and Yamamoto, 1987). The time taken to perform the test is only a few minutes to a few hours.

Air Entry Permeameter   The air entry permeameter (AEP) consists of a sealed ring, about 60 cm in diameter, embedded approximately 10 cm into the soil (Fig. 18.12). The test is done in two stages. In the first stage the rate of infiltration is determined from falling or constant head test until the wetting front penetrates to the base of the ring (note: this may take several weeks for

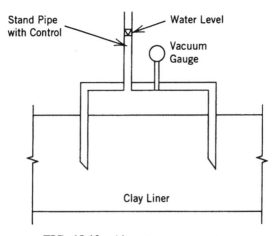

FIG. 18.12   Air entry permeameter.

compacted soil). In the second stage the valve that allows flow of water is closed. This closing of the valve creates a negative pressure, which is measured with a gauge. The negative pressure develops because of the soil suction, which tries to suck water out of the AEP. When the vacuum gauge reaches a peak, the AEP is disassembled and the depth to the wetting front is measured by finding out the variation of water content with depth. The hydraulic conductivity is calculated from the water entry suction pressure and air entry suction pressure; these values are obtained from a plot of the suction pressure with volumetric water content. *In situ* hydraulic conductivity obtained by using AEP was about one-half order of magnitude higher than laboratory values (Daniel, 1989). The test takes several weeks to complete.

Lysimeter A lysimeter (Fig. 22.5) is essentially a pan constructed with material of very low permeability (e.g., synthetic membrane) and back-filled with highly permeable material (e.g., sand); the water collected in the pan is drained to a stand pipe or similar device (see Section 22.3.1 for details). Hydraulic conductivity is calculated by Darcy's law and the measured flow rate in the pan. These are installed below the compacted clay liner and are used mainly for landfill monitoring purposes. However, use of the lysimeter to calculate the *in situ* hydraulic conductivity of compacted clay liners has been reported, which indicated good agreement with other *in situ* tests (Day and Daniel, 1985; Rogowski et al., 1985). Lahti et al. (1987) reported good agreement with laboratory values. Time taken to complete the test is several weeks to several months.

Infiltrometers Four different types of infiltrometers are available. Of these the sealed double-ring infiltrometer (SDRI) appears to provide representative *in situ* values of compacted soil liners (Sai and Anderson, 1991).

OPEN, SINGLE-RING INFILTROMETER In this method a ring is embedded in the compacted soil (Fig. 18.13). The ring is filled with water. Hydraulic conductivity of the soil is calculated by measuring the rate of change of the water head within the ring. *In situ* hydraulic conductivity of compacted clay liners

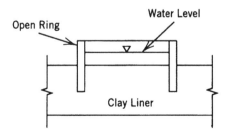

FIG. 18.13 Open single-ring infiltrometer.

was reported to be in good agreement with values obtained by using other methods (Daniel, 1984; Day and Daniel, 1985). Time taken to complete the test is several weeks to several months.

SEALED, SINGLE-RING INFILTROMETER   The apparatus is similar to the open single-ring infiltrometer (Fig. 18.14) except that the ring is sealed, which minimizes loss due to evaporation from an open ring. Hydraulic conductivity is calculated from the rate of change of level in the stand pipe. Testing time using this method varies from several weeks to several months.

OPEN, DOUBLE-RING INFILTROMETER   Two rings or boxes are embedded in soil, which are filled with water; the rings are covered to minimize evaporation (Fig. 18.15). Hydraulic conductivity is calculated using the rate of change of water head in the inner ring. The purpose of the outer ring is to limit lateral spreading of water originating from the inner ring. *In situ* hydraulic conductivity of compacted soil compared well with the values obtained using other methods (Daniel, 1984; Day and Daniel, 1985). Testing time using this method varies from several weeks to several months.

SEALED, DOUBLE-RING INFILTROMETER   In this method two circular or square rings are embedded in the compacted soil (Fig. 18.16). A small flexible bag is attached to the inner ring, which is sealed. Water is siphoned into the inner ring up to a depth of 25 mm. The outer ring is filled with water until the water level is 10 cm above the top of the inner ring. Flow of water from the inner ring is replaced automatically by the flexible bag. Volume of water through the inner ring is calculated by finding the loss of weight of the flexible bag. The infiltration rate is calculated from the weight loss and the area of the inner ring. The test is continued until the infiltration rate becomes steady or reaches a specified value. While Daniel and Trantwein (1986) and Chen and Yamamoto (1987) reported that the *in situ* hydraulic conductivity obtained using this method was up to one order of magnitude higher than laboratory values. Elsbury et al. (1988) (cited in Daniel, 1989) reported excellent agree-

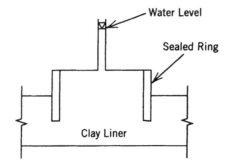

FIG. 18.14   Sealed single-ring infiltrometer.

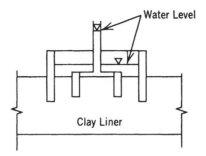

FIG. 18.15 Open double-ring infiltrometer.

ment between values obtained using SDRI and lysimeter. Testing time using this method varies from several weeks to several months. A standard test method for SDRI is available (ASTM D5093-90).

18.1.7 Density Tests

Knowledge about two types of density testing is required for clay liner construction purposes: (1) modified Proctor's density test (ASTM D1557-78) performed in the laboratory to find the maximum density and optimum moisture content relationship and (2) field density and moisture content test for quality control purposes.

18.1.7.1 Laboratory Test The dependency of soil density on the moisture content of soil was first investigated by Proctor (1933). The compaction procedure was later standardized to develop the commonly known standard Proctor's method. The weight of the hammer in the standard Proctor's test is 5.5 lb (2.49 kg), having a free fall of 12 in. (30.48 cm); the sample is compacted

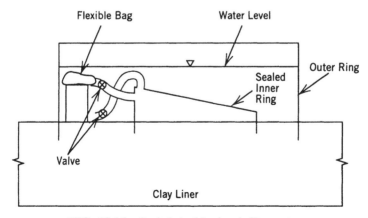

FIG. 18.16 Sealed double-ring infiltrometer.

in a 4-in. (10.16-cm) mold in three layers, each layer receiving 25 blows. In a modified Proctor's test the soil is compacted in five layers in a 4-in. (10.15-cm) mold using a 10-lb hammer (4.54 kg) with a 18-in. (45.7-cm) free fall. Each layer is compacted with 25 blows. Bulk density and moisture content of the compacted soil obtained from several samples compacted at different moisture contents is used to develop a moisture–density relationship (Fig. 18.17). The optimum moisture is the moisture content at which the soil exhibits maximum density. The modified Proctor's test is expected to simulate compaction offered by a sheep's foot roller.

18.1.7.2   Field Test   Two types of field tests are available: (1) direct measurement using a sand cone, drive cylinder, or rubber balloon and (2) indirect measurement using a nuclear gauge.

Direct Methods   The sand cone test (ASTM D1556) consists of digging out a sample of the soil, measuring the volume of soil by filling it with sand, and then finding the dry weight of the removed sample. The dry unit density is obtained by dividing the weight by the volume of sand. The rubber balloon method (ASTM D2167) is the same as the sand cone method except that the volume of the hole is measured by a water-filled balloon under constant pressure. The drive cylinder test (ASTM D-2937) consists of driving a standard tube into soil, trimming the soil flush with the ends of the tube, and finding the dry weight of the soil in the tube. Dry unit weight is obtained from the weight of the soil and the known volume of the cylinder. The moisture content of the soil is obtained by oven drying a field soil sample (ASTM D2216-80). A microwave oven may also be used for drying the soil samples (Gee and Dodson, 1981).

Indirect Method   A nuclear gauge is used to determine both density and moisture content (ASTM D2922). The method is much faster and is widely used currently for quality control purposes. The nuclear gauge essentially

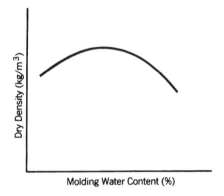

FIG. 18.17   Typical dry density versus water content relationship for clayey soils.

consists of a gamma ray source and a detector. The reflected emission sensed by the detector is inversely proportional to the density of the soil. The density can be measured by any of two operation modes: back scattering and direct transmission. In the back scattering mode the gauge is kept at the top of the compacted surface. Any air void between the gauge and the liner will lead to a lower density reading. In the transmission mode the source of the gamma rays is lowered into a hole (5–30 cm deep). The error in density measurement is minimal in this mode. It is essential to calibrate the gauge for each project and on a regular basis thereafter against a direct method of density measurement (e.g., sand cone method) because the composition of the soil and handling may affect results.

A nuclear moisture gauge consists of a fast neutron source and a neutron detector. The water content is proportional to the hydrogen atom concentration in the medium. It is essential to calibrate the gauge against oven-dried samples on a regular basis. Proper precautions must be taken while handling nuclear meters because the devices use radioactive material.

18.1.8 Shear Strength

The bearing stress on clay liners of big landfills can exceed 100 psi (667 kPa). Daniel and Wu (1993) proposed the use of a procedure similar to the procedures described earlier (see Sections 18.1.3 and 18.1.5) for determining the acceptable zone for unconfined compressive shear strength. Unconfined compression or unconsolidated–undrained triaxial tests should be run on the specimens.

18.1.9 Specifications

The primary purpose of clay liners is to provide a low hydraulic conductivity (1×10^{-7} cm/sec or less) barrier layer, either at the base or in the final cover. The procedures to develop acceptable zones, which related dry unit weight to low hydraulic conductivity, low volumetric shrinkage (4% or less), and adequate strength are discussed in Sections 18.1.3, 18.1.5, and 18.1.8, respectively. All three criteria are superimposed in Fig. 18.18, which shows the acceptable zone meeting all criteria. The procedure is not widely used at present.

The criteria for choosing a clay is primarily based on the recompacted permeability achievable under field conditions. A clay that can be compacted to obtain a low permeability (1×10^{-7} cm/sec or less) sample when compacted to 90–95% of the maximum Proctor's dry density at wet of optimum moisture is chosen for landfill liner construction. Note that clay with a high liquid limit (LL) tends to develop more desiccation cracks, clay with a very low plasticity index (PI) or plastic limit (PL) is less workable, and a well-graded soil is expected to develop low permeability when compacted properly. Therefore, the PI, LL, and some minimum requirements regarding grain size distribution should be specified. Inorganic clays of medium plasticity (Fig.

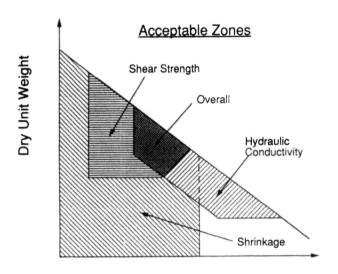

FIG. 18.18 Acceptable zone based on low hydraulic conductivity, low desiccation induced shrinkage, and high unconfined compressive strength. Reproduced by permission of the publisher, ASCE, from Daniel and Wu (1993).

18.3) are best suited for liner construction. Usually a soil with the following specifications would prove suitable for liner construction.

LL greater than or equal to 30%
PI greater than or equal to 15%
0.074 mm and less fraction (P200) greater than or equal to 50%
Clay fraction greater than or equal to 25%

The minimum percentage compaction (usually 90–95% of the maximum modified Proctor's density) should be specified. To obtain better kneading action and lower permeability, all compaction must be done at wet of optimum moisture. Note that the shape of the grain size distribution curves should be studied to confirm that they are close to an "inverted S" shape. The specifications regarding Atterbergs limits and grain size percentages are also helpful in quality control during construction (refer to Section 20.2 for more details).

In reality it may be difficult to obtain clay that will satisfy all the specifications described above. Field experience indicates that clayey soil with the following values can also be compacted to obtain a low permeability liner:

PI between 10 and 15%
LL between 25 and 30%

0.074 mm or less fraction between 40 and 50%

Clay content between 18 and 25%

However, it is prudent to perform field trials for such marginal clayey soils.

18.1.9.1   Field Trial for Marginal Clays   Necessary laboratory test data described in Section 18.1.6 through 18.1.9 must be obtained prior to the field trial. Then the soil should be spread in thin lifts [20–25 cm (8–10 in.)] before compaction in two or more plots and compacted at two or more different percentages using the equipment proposed for use for the actual liner construction (for more guidance on construction refer to Sections 20.1 and 20.2). Undisturbed soil samples should be collected per the recommendations provided in Section 20.2 and tested for permeability. The actual moisture content and percentage compaction that will provide the desired permeability are decided based on the field test data. Field trials should be undertaken for all major landfill projects.

18.2   SYNTHETIC MEMBRANE

There are several polymers that are combined with different additives to form a thermoplastic widely known as geosynthetic membrane (or geomembrane). The molecules that make up plastics are very large and are called polymers. The factors that influence the physical properties of polymers include the size of the molecules, the distribution of different molecular sizes within a polymer, and the shape and structure of individual molecules. Different additives are added to the polymers to improve manufacturing and the usefulness of the final product.

From the numerous polymers and additives available, thousands of possible formulations can be developed. However, in practice only a few are used and these are named after the major polymers. The following polymers are in common use: butyl rubber, chlorinated polyethylene (CPE), chlorosulfonated polyethylene (CSPE), ethylene-propylene rubber (EPDM), high-density polyethylene (HDPE), medium, low-, and very low-density polyethylene (MDPE, LDPE, and VLDPE), linear low-density polyethylene (LLDPE), and polyvinyl chloride (PVC). Each has certain advantages and disadvantages, which are summarized in Table 18.3. It may be noted that all membranes in the same category do not have the same composition. Thus, testing of each lot of synthetic membrane is essential to determine whether the one proposed for a project conforms to certain standard physical properties. Discussions on factors considered for choosing a membrane are included in Sections 18.2.1–18.2.4 and a discussion on polyethylene is included in Section 18.2.5.

TABLE 18.3   Advantages and Disadvantages of Commonly Used Synthetic Membranes

S1 Number	Synthetic Membrane	Advantages/Disadvantages
1	Butyl rubber	Good resistance to ultraviolet (UV) ray ozone, and weathering elements
		Good performance at high and low temperatures
		Low swelling in water
		Low strength characteristics
		Low resistance to hydrocarbons
		Difficult to seam
2	Chlorinated polyethylene (CPE)	Good resistance to UV, ozone, and weather elements
		Good performance at low temperatures
		Good strength characteristics
		Easy to seam but poor seam reliability
		Moderate resistance to chemicals, acids, and oils
3	Chlorosulfonated polyethylene (CSPE)	Good resistance to UV, ozone, and weather elements
		Good performance at low temperatures
		Good resistance to chemicals, acids, and oils
		Good resistance to bacteria
		Low strength characteristics
		Minor problem during seaming
4	Ethylene-propylene rubber (EPDM)	Good resistance to UV, ozone, and weather elements
		High strength characteristics
		Good performance at low temperatures
		Low water absorbance
		Poor resistance to oils, hydrocarbons, and solvents
		Poor seam quality
5	Low-density and high-density polyethylene (LDPE and HDPE)	Good resistance to most chemicals
		Good strength and seam characteristics
		Good performance at low temperatures
		Poor puncture resistance

TABLE 18.3   (*Continued*)

S1 Number	Synthetic Membrane	Advantages/Disadvantages
6	Medium, very low, and linear low-density polyethylene (MDPE, VLDPE, LLDPE)	Good to excellent chemical resistance Good seam quality (Note: These products are newly introduced in the market; properties should be reviewed before use)
7	Polyvinyl chloride (PVC)	Good workability High strength characteristics Easy to seam Poor resistance to UV, ozone, sulfide, and weather elements Poor performance at high and low temperatures

18.2.1   Damage by Soil Microbes, Rodents, and Vegetation

In general, the resins in membranes are resistant to microbiological attack, but the additives are not (Connolly, 1972; Kuster and Azadi-Bokhsh, 1973; Potts et al., 1973). Since microbial attack is expected for some waste types, care should be taken to choose a proper additive. The two following tests may be used to determine microbial resistance: ASTM G-21 and ASTM G-22. Both are short-term tests and at present no long-term test is available.

Insects and burrowing rodents may severely damage plastics (Connolly, 1972). PVC is more susceptible to attack by rats than polyethylene. Hoofed animals such as deer may puncture a synthetic membrane.

Certain grass species may germinate and penetrate through synthetic membrane. To prevent such damage, use of a herbicide prior to synthetic membrane installation is recommended (Schultz and McKias, 1980). However, study of canals lined with synthetic membranes indicate that roots of vegetation are not likely to penetrate synthetic membranes (Hickey, 1969).

18.2.2   Workability

The cost of installation will depend on the ease with which the membrane can be handled. Thicker membranes [1.5 mm (60 mils) and above] have the advantage of having more tolerance to handling abuse. A thicker membrane is less likely to be weakened by the seaming process. The main disadvantage of a thicker liner is that it is heavier and may require special equipment to handle it. Synthetic membrane liners will be exposed to temperature variation prior to and during installation. Therefore whether the membranes stick together at anticipated field temperatures should be investigated. Hypalon may be susceptible to this problem of sticking or "blocking" (Lee, 1974; Woodley, 1988). ASTM D-1893 and D-3354 can be used to test blocking properties of synthetic membranes.

Ease of seaming is an important consideration in the installation of a synthetic membrane. In general, it is more difficult to seam a membrane that is more chemically resistant (Forseth and Kmet, 1983). The long-term durability of the seam must be considered. The chemical compatibility of the solvent may vary considerably (Haxo, 1982). The problem of seaming a membrane after long-term exposure to waste would also be considered. Membrane aging may affect the ability to repair a damaged area or seam a new membrane to an old portion when expanding a landfill. Hypalon is known to lose seaming ability within a year even when protected by a layer of soil (Forseth and Kmet, 1983).

18.2.3   Compatibility with Waste

The chemical compatibility of the synthetic membrane with the waste leachate must be determined. Koerner (1986) reported a chemical resistance chart indicating chemical resistance of many common synthetic membranes against several generic chemicals. Waste–membrane compatibility tests typically involve immersing a membrane coupon in a waste leachate or waste slurry. Membrane samples are exposed to actual or synthetic leachate either on both sides (immersion) or on one side only (tub or pouch encapsulation). The samples are removed after a period of time (up to 120 days) and tested for several physical properties (Koerner, 1986). Because of the magnitude of possible error, attempts are being made to develop a test based on diffusion parameters (Lord and Koerner, 1984). At present, however, the 9090 test, which is an immersion-type test proposed by the U.S. EPA, is commonly used in landfill projects in the United States. A designer should check with the regulatory authority as to whether a compatibility test should be undertaken and, if so, an acceptable test procedure for the type of membrane selected. In addition to exposure to waste, synthetic membranes will be exposed to soil. Therefore compatibility with the on-site soil should also be evaluated. Naturally occurring oxides of metals, chlorides, and sulfur compounds and some organic compounds may react with the membrane; high or low pH soil can also degrade the membrane (Connolly, 1972). Short-term burial tests such as ASTM D-3083 can be used to determine short-term compatibility with the on-site soil. Table 18.4 includes suggested values of property change that may be allowed after exposure to the chemical or waste.

18.2.4   Mechanical Properties

The mechanical properties of concern and corresponding test methods are listed in Table 18.5. The details of test procedures can be found in the relevant ASTM manual or in the cited references. Resistance to degradation due to

TABLE 18.4 Suggested Limits of Different Test Values for Incubated Synthetic Membranes Used as Liner

Property	Resistant
Permeation rate	<0.9 g/m^2-hr
Change in weight (%)	<10
Change in volume (%)	<10
Change in tensile strength (%)	<20
Change in elongation at break (%)	<30
Change in 100 or 200% modulus (%)	<30
Change in hardness	10 points

After Koerner (1990). Reprinted with permission from the Industrial Fabrics Association International.

leachate exposure is also a physical property needing assessment. It is not included in Table 18.5 because it has already been discussed in Section 18.2.3. The name of each test clearly indicates the property under investigation. Most of the tests are related to installation; a detailed discussion regarding their need can be found in Section 20.2. It should be mentioned that the permeability of synthetic membranes is extremely low and doubt exists regarding the use of Darcy's equation in predicting membrane permeability (Giroud, 1984).

18.2.5 Discussion on Polyethylene

Although many polymers have been used in the past for landfill liner construction, it appears that HDPE and LDPE are the preferred choices in many

TABLE 18.5 Standard Tests Used in the United States for Testing Physical Properties of Synthetic Membranes

S1 Number	Physical Properties	Standard Test Method
1	Tensile strength	ASTM D638
2	Tear resistance	ASTM D1004
3	Puncture resistance	ASTM D4833
4	Low-temperature brittleness	ASTM D746 procedure B
5	Environmental stress crack resistance	ASTM D1693 condition C
6	Permeability	Refer to Section 7.2.5.1
7	Carbon black percent	ASTM D1603
	Carbon black dispersion	ASTM D2663
	Accelerated heat aging	ASTM D573, D1349
8	Density	ASTM D1505
9	Melt flow index	ASTM D1238

instances. A comparison of physical properties of some synthetic membranes and their performance as liners can be found elsewhere (Fong and Haxo, 1981; Cadwallader, 1986). It should be mentioned that new polymers are also being developed (Brookman et al., 1984; Wollak, 1984) that may provide alternatives to HDPE and LDPE (e.g., VLDPE, LLDPE). Polyethylene is the general name for many polymer resins. Proper selection of polyethylene resin is important. Resins used for pipe manufacturing are ideal for manufacturing quality liner material. Polyethylene is a crystalline polymer and hence tends to have a regular crystal lattice. The molecules crystallize by folding of the polymer chains forming lamellae or platelike polymer crystals (Fig. 18.19). These lamellae are arranged to form larger aggregates known as spherulites (Fig. 18.20). The physical properties of polyethylene are greatly influenced by the size, shape, and arrangement of the spherulites. The smaller the diameter of the spherulites the greater is the resistance to stress cracking. Brittleness in polyethylene increases due to the increase in spherulite dimension. The molecular weight distribution also plays an important role in stress cracking. Low-molecular-weight species occupy the space between high-molecular-weight species thereby resisting crack propagation. Both crystalline and amorphous material are present in a polymer. Cracks tend to propagate through amorphous material. There are essentially three amorphous materials or zones that link the crystalline zones (Fig. 18.21). (1) Tie molecules: these are chains that tie two lamellae; (2) loose loops: these are closed chains that · loosely hang from a lamellae but do not interconnect lamellae; (3) celia: these are open chains loosely hanging from lamellae. Thus, tie molecules form a bridge between lamellae and spherulites. Polyethylenes containing few tie molecules are brittle in nature. However, with a high number of tie molecules, a polyethylene will be highly ductile but not very stiff (Lustiger and Corneliussen, 1988). Thus, a balance must be obtained in the number of tie molecules. The three parameters that control tie molecule numbers are molecular weight, comonomer content, and density.

Molecular Weight   A low melt flow index indicates longer average polymer chains (or high molecular weight). The distribution of molecular weight is also important to obtain better control over tie molecules.

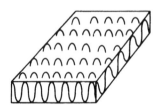

FIG. 18.19   Polymer crystal with folded chain that forms lamellar crystal plates.

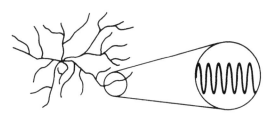

FIG. 18.20   Pattern and interconnection between lamellae and spherulite.

Comonomer Content   Comonomers are long-chain olefins (1-butene or 1-hexene) that inhibit crystallinity and cause short branches. Short-chain branching increases entanglements of tie molecule.

Density   This is an indirect measure of crystallinity. The more crystalline the polymer, the fewer the number of tie molecules. A typically good value for density is 0.95 g/cm³ and for melt flow index is 0.22 g/10 min (Cadwallader, 1986).

18.2.5.1   Permeability of Polyethyene (PE)   In general, the hydraulic conductivity of PE liners is extremely low (of the order of 1×10^{-10} cm/sec) and hence the leakage of water is negligible when used as landfill liners. Water vapor transmission (WVT) through a synthetic membrane is calculated using the following equation (Koerner, 1986):

$$\text{WVT} = \frac{m \times 24}{t \times a} \text{ g/m}^2/\text{day} \qquad (18.5)$$

in which m = weight loss (g), t = time interval (hr), and a = area of the specimen (m). In the water vapor transmission test, a specimen is sealed over an aluminum cup with water in it. A controlled relative humidity is maintained

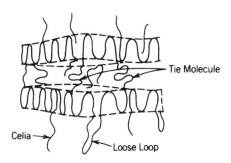

FIG. 18.21   Typical structure of polyethylene lamellae.

on either side of the cup. The entire assembly is placed in a chamber in which the relative humidity is controlled. The weight change of the cup is monitored over a length of time. The test time varies between 3 and 30 days. Permeance is defined as change of water vapor transmission due to the difference in relative humidity, which is mathematically expressed as follows:

$$\frac{WVT}{\delta_p} = \frac{WVT}{S(R_1 - R_2)} \tag{18.6}$$

in which δ_p = the vapor pressure difference across the membrane (in millimeters of mercury), S = the saturation vapor pressure at test temperature (in millimeters of mercury), R_1 = the relative humidity within a cup, and R_2 = the relative humidity outside the cup. The permeability is calculated by the following equation:

$$\text{Permeability} = \text{permeance} \times \text{thickness} \tag{18.7}$$

As mentioned earlier in this section, the hydraulic conductivity of synthetic membranes is extremely low. However, the permeation of other chemicals, especially organic species, is not as low. Synthetic membrane liners exhibit permselective properties, that is, rate of permeation through the synthetic liner depends on the chemical specie. Table 18.6 shows permeation rates of a mixture of hydrocarbons through a 1-mm (40-mil)-thick HDPE. The transport of nonelectrolytic contaminants through synthetic liners should be determined experimentally.

ASTM E96 test method may be employed by replacing the water in the cup with the solvent of interest (Koerner, 1986). Additional relevant standard tests are (1) ASTM D1434 for gas vapor transmission and (2) ASTM D814 for organic vapor transmission. Telles et al. (1988) suggested a method called the "Permachor method," in which the permeability coefficient for a given polymer/permeant system is calculated using apparent permeation energy (E_p) at temperature T using the following equation:

TABLE 18.6   Permeation Rates of a Mixture of Hydrocarbons through HDPE

Hydrocarbon	Permeation Rate ($g/m^2/day$)
Trichloroethylene	9.4
Tetrachloroethylene	8.1
Xylene	3.0
iso-Octane	0.8
Acetone	1.4
Methanol	0.7

After Telles et al. (1988).

$$\ln P = \ln P_0 - [E_p/(R_0 - T)] \tag{18.8}$$

in which P_0 and R_0 are universal gas constant and preexponential factor, respectively. Several researchers have reported their findings on permeation of liquid through various polymers (Salame, 1961; Nasim et al., 1972; Haxo et al., 1984). The concentration of organic species in leachates from most nonhazardous waste landfills (e.g., municipal waste) is not high and, hence, special test(s) to find the permeation rate of such chemicals need not be undertaken. However, if the concentration of organic species is expected to be relatively high for a particular landfill, then permeation test(s) should be performed. Small reductions in densities below 0.93 g/cm^3 result in large increases in permeability of polyethylene (Telles et al., 1988).

Leakage through holes in synthetic membrane, though not an intrinsic property like permeability, need to be considered for estimating leakage through a liner. The number of holes or flaws per square meter of a liner, arial distribution of such holes, and their sizes are difficult, if not impossible, to predict. However, even under strict quality control, few holes are expected to be present either at pre- or postconstruction stage. The range of such defects is between 1 and 30 holes per acre (2.5 to 75 holes per hectare). Most defects are usually less than 0.1 m^2 (Giroud and Bonaparte, 1989). Leakage rate through a synthetic liner depends on the permeability of the material immediately below the membrane and the permeability of the drainage blanket above the membrane. Leakage rate in three possible situations are discussed:

1. Flow of liquid through a hole in the synthetic membrane may be considered as free flow if it is overlain and underlain by high permeability material (e.g., gravel).
2. Flow of liquid through a hole in the synthetic material is less than case 1 if it is underlain by material of moderate permeability (e.g., silty soil).
3. Flow of liquid through a hole in the synthetic material is significantly low if it is overlain by a high to moderate permeability material (e.g., sand) and underlain by a low permeability material (e.g., compacted clay). This scenario models a composite liner. The following empirical equations may be used for this case (Bonaparte et al., 1989):

For good contact:
$$Q = 0.21a^{0.1}h^{0.9}k_s^{0.74} \tag{18.9}$$

For poor contact:
$$Q = 1.15a^{0.1}h^{0.9}k_s^{0.74} \tag{18.10}$$

in which Q = steady-state leakage rate through a hole in the synthetic liner (m^3/sec), a = area of the hole (m^3), h = head of liquid on the synthetic membrane at the hole location (m), and k_s = hydraulic conductivity of the underlying low permeability layer (m/sec).

The good and poor contact conditions are defined as follows (Bonaparte et al., 1989):

Good contact: A synthetic membrane installed with very few wrinkles on a smooth well compacted clay surface.

Bad contact: A synthetic membrane with several wrinkles; the underlying surface is not well compacted and appears to be rough.

18.2.6 Types of Commonly Available Synthetic Membranes

Four types of membranes are usually available.

Nonreinforced   These are manufactured in plants with an extrusion or callendering process. The ranges of available width and thickness when the extrusion process is used are width, 4.85–10 m (16–33 ft), and thickness, 0.25–4 mm (10–160 mils). The ranges of available width and thickness when the callendering process is used are width, 1.5–2.4 m (5–8 ft), and thickness, 0.25–2 mm (10–80 mils).

Reinforced   Woven or nonwoven geofabrics are coated with polymeric compounds. The typical thickness of reinforced synthetic membranes ranges between 0.75 and 1.5 mm (30 and 60 mils). The width of the membrane depends on the width of the geofabric.

Laminated   A nonwoven geofabric is callendered to a synthetic membrane (usually nonreinforced). The thickness and width depend on the thickness and width of the geofabric and synthetic membrane.

Textured   Both HDPE and LLDPE are available with a textured surface. The textured surface increases interface friction between the underlying or overlying material (e.g., soil, geofabric). Usually both edges of textured geomembrane sheets have a 6-in. (15-cm) width of nontextured band for facilitating welding.

18.3   GEOSYNTHETIC CLAY LINER

The geosynthetic clay liner (GCL) is gaining popularity. Geosynthetic clay liners are manufactured by sandwiching a uniform layer of dry bentonite between two geotextiles or attached to a synthetic membrane with an adhesive. The bentonite is kept in place by using a water-soluble adhesive. When a geosynthetic clay liner comes in contact with water, it swells, forming a continuous layer of bentonite 12–25 mm in thickness. There are four types of GCL available in the United States: geotextile-enclosed, adhesive-bonded

GCL; geotextile-encased, stitch-bonded GCL; geotextile-encased, needle-punched GCL; and geomembrane-supported adhesive-bonded GCL.

18.3.1 Hydraulic Conductivity of GCL

As discussed in Section 18.4.2 the hydraulic conductivity of bentonite may increase due to permeating fluid chemistry. The hydraulic conductivity of geotextile-encased bentomite increased by an order of magnitude when permeated with $CaSO_4$ and $CaCl_2$ (Shan and Daniel, 1991). Significant change of hydraulic conductivity was observed when GCLs were permeated without proper hydration (Daniel et al., 1993; Ruhl and Daniel, 1997). Significant decrease in hydraulic conductivity of GCLs due to increased effective continuing stress have been reported by several researchers (Shan and Daniel, 1991; USEPA, 1993; Petrov et al., 1997). Boardman and Daniel (1996) reported that alternate wetting and drying has no significant effect on the hydraulic conductivity of GCLs. Several researchers have reported that freeze–thaw cycles do not influence the hydraulic conductivity of GCLs (Hewitt and Daniel, 1997; Kraus et al., 1997). LaGatta et al. (1997) reported that GCLs maintained a hydraulic conductivity 1×10^{-7} cm/sec or less even under high differential settlement.

The hydraulic conductivity of GCLs is sensitive to permeant, hydration, and stress. Therefore the hydraulic conductivity needs to be assessed on a case by case basis. The hydraulic conductivity of GCLs increased up to an order of magnitude when permeated with real and synthetic municipal solid waste (MSW) leachate (Kerry, 1998).

Table 18.7 includes a comparison between synthetic clay liners and compacted clay liners. The hydraulic conductivity of synthetic clay liners varies between 1×10^{-7} and 1×10^{-9} cm/sec, which may decrease by almost an order of magnitude due to a 10-fold increase in effective stress (James Clem Corporation, 1992). The interface shear values of synthetic clay liners depend on the contact material on which they are laid. It is preferred that an appropriate test be undertaken for determining the shear strength parameters. However, the following values may be assumed in the absence of actual test data (James Clem Corporation, 1992):

Smooth HDPE	$\phi = 11$	$C = 0$
Textured HDPE	$\phi = 27$	$C = 1.92$ kPa
Wet clay	$\phi = 24$	$C = 1.44$ kPa
Sand	$\phi = 24$	$C = 3.84$ kPa

It may be noted that the shear strength parameters will depend on the type of geotextile used in the manufacture of the synthetic clay liner. Both stitch-bonded and needle-punched GCLs are suitable for use in low normal stress

TABLE 18.7 Comparison between Synthetic Clay Liners and Compacted Clay Liners

Synthetic Clay Liner	Compacted Clay Liner
Easy to install	Difficult to construct correctly
Light equipment can be used for construction	Heavy equipment is necessary for construction
Choice of equipment is not critical for construction	Choice of equipment used for compaction is critical
Majority of the quality control tests are done in factory; very little quality control test is necessary during installation	Some of the quality control tests are done at the borrow source; substantial quality control tests are necessary during construction
Construction time is short	Construction time is long
Construction is interrupted due to rain	Construction is interrupted due to rain
Installation at low temperature (up to $-5°C$) is permissible	Construction at or below freezing temperature not permissible
Effect of desiccation and freeze–thaw is low	Effect of desiccation and freeze–thaw is pronounced
Has relatively high tensile and shear strength properties	Has low tensile and shear strength properties
Can be installed easily at relatively steeper slopes (up to 2H:1V)	Cannot be constructed easily in steeper slopes
Adjusts to differential settlement	Performance is poor in case of a differential settlement
Does not consume significant airspace	Consumers relatively higher airspace
Cannot be used in direct contact with most leachate	Can be used in direct contact with most leachate
Overall cost is low in most cases	Overall cost depends primarily on haul distance and liner thickness

situations (e.g., final cover). Only needle-punched GCLs are suitable for bottom liners (Fox et al., 1998).

18.4 AMENDED SOIL AND OTHER ADMIXTURES

Results of studies on three types of amended soil [the amenders used are bentonite, asphalt, and cement and two admixes (which are asphaltic concrete)] are available. Bentonite-amended soil has proved to be more useful in landfill projects; hence, the discussion is divided into two categories: nonbentonite mixes and bentonite-amended soil.

18.4.1 Nonbentonite Mixes

Fong and Haxo (1981) reported findings in which the following four admix materials were exposed to municipal solid waste leachate for a fairly long time (56 months):

Paving Asphalt Concrete   Concrete made of a hot mix of 7% asphalt (60–70 penetration grade) and granite proportioned to meet a 0.25-in. maximum gradation for dense graded asphalt concrete.

Hydraulic Asphalt Concrete   A concrete made of a hot mix of 9% asphalt (60–70 penetration grade) and granite proportioned to meet an 0.25-in. maximum gradation for dense graded asphalt concrete.

Soil Asphalt   This was a mixture of 100 parts of soil with 7 parts of liquid asphalt, grade SC-800.

Soil Cement   This was a mixture of 9.5 parts of soil, 5 parts of kaolinite, 10 parts of portland cement, and 8.5 parts of water.

Although the permeability and other physical properties of the asphaltic mixtures did not deteriorate much, a significant loss in compressive strength due to absorption of water or leachate was observed. The soil cement probably gained strength and decreased permeability; however, a caveat was issued regarding the cracking of soil cement as observed in highway projects. The study did not favor the use of any of these mixtures in lining a landfill. Additional studies (Haxo et al., 1985) indicate that except for soil cement all other mixes performed poorly when exposed to different industrial waste leachate.

No study is available regarding the use of these admixtures in landfill final cover construction. Use of the admixtures as a layer in a multiple layered cap (i.e., as a layer below a synthetic membrane) may be studied for chemical compatibility and retention of structural integrity under differential settlement conditions. Since asphaltic concrete is not as flexible as soil, it is highly likely that it will not be able to withstand stresses induced by differential settlement. A detailed report on the subject is available (Haxo et al., 1985).

18.4.2   Bentonite-Amended Soil

Bentonite-amended soil is used in different civil engineering projects. Bentonites are essentially clay minerals of the smectite group. The basic structure of a smectite consists of repeated stacking of layers; each layer consists of an octahedral sheet sandwiched between two silica sheet, as schematically shown in Fig. 18.22. The bond between the layers is weak and has an unbalanced charge. Water is easily absorbed between the layers whereby the basal spacing increases causing swelling of the clay. The basal spacing can vary from 9.64 Å to complete separation (Mitchell, 1976). Because of the unbalanced charge, smectite exhibits a high cation-exchange capacity whereby ions are absorbed in the interlayer spaces.

Montmorillonite, the most common mineral of the smectite group, is formed essentially from the substitution of aluminum by magnesium in the octahedral sheet. Naturally occurring bentonite is a variety of montmorillonite. The interlayer swelling of bentonite and its fabric is influenced by the composition of the pore liquid. A change in the pore liquid composition may

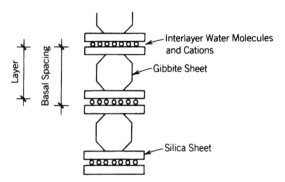

FIG. 18.22   Schematic diagram showing structure of bentonite.

trigger isomorphous substitution whereby the swelling may increase or decrease, which may cause a change in the fabric. Bentonite swells heavily due to the addition of water and forms a low permeability fabric. However, if the pore fluid is changed, the fabric as well as the basal spacing may be reduced causing an increase in hydraulic conductivity. The effect of pore fluid on the permeability of bentonite can be reduced by treating the bentonite with special polymer(s). Since the pore fluid chemistry has significant effect on permeability of bentonite, the compatibility of leachate (expected from the landfill) with the bentonite-amended soil must be studied. Changes in the permeability of the bentonite mix due to a change in the permeating fluid is well documented (D'Appolonia, 1980; Alther, 1982).

18.4.2.1   Mixing Process   Central plant mixing (Fig. 18.23) is reported to be more effective than in place mixing using agricultural equipment (Bagchi, 1986a; Goldman et al., 1986; Lundgren, 1981). For constructing final cover over a papermill landfill in Wisconsin a well-graded sand with 15% of 0.074 or less fraction (P200) content was mixed with a 4% commercially available powered bentonite (a polymer-enriched natural bentonite) to develop a low permeability (1×10^{-7} cm/sec or less) mix (Bagchi, 1986a). The use of small truck-mounted concrete batch plants for mixing bentonite is not documented and is worth studying.

The quality of the mix must be checked to ensure uniformity and correctness of the bentonite percentage. The methylene blue test (ASTM C-837) and the sand equivalent test (ASTM D-2419) were found to be useful in estimating the bentonite percentage in the mix (Alther, 1983; Bagchi, 1986a; Barvenic et al., 1985). A minimum of five trial runs should be made to check the quality of the mix visually and using grain size analysis. The bentonite percentage should be checked using either of the tests mentioned above during the trial runs. The permeability should also be checked using the field mix compacted in the laboratory. Quality control during construction of the liner is discussed in Section 20.2. The following steps may be followed to determine the suitability of the bentonite amendment for a project.

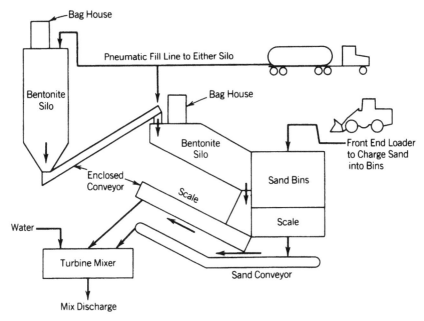

FIG. 18.23 Flow diagram for sand–bentonite mixing. After Bagchi (1986a).

1. Find the percentage of bentonite to be added to well-graded sand (use the sand borrow proposed for the project) that will develop a low permeability mix. A proper permeant must be used for the permeability tests. It is recommended that the sand contain a minimum of 10–15% of 0.074 mm or less fraction (P200) soil. The percentage of bentonite may very between 3 and 15%, which depends on the sand and the permeating liquid. Only powdered bentonite should be used for mixing because complete hydration of pellets cannot occur in the field.

2. Perform trial runs using the mixing equipment proposed for the project to check quality and permeability of the field mix.

3. Run a sand equivalent or methylene blue test on a regular basis during actual construction to check the percentage bentonite added.

18.5 COMPOSITE LINER

A composite liner is a geomembrane underlain by a compacted clay liner or GCL. The leakage through composite liners takes place through defects in geomembrane and permeation of water vapor through the geomembrane. The leakage rate depends on the quality of contact (Section 18.2). To minimize leakage the geomembrane must be placed in good contact (often called "intimate contact") with the underlying clay liner or GCL. Field studies show the leakage rate through composite liners is significantly low compared to compacted clay liners (Melchior and Miehlich, 1994). Foose et al. (2002)

reported a mathematical model to compare solute transport in three different composite liners. The first composite liner consisted of 2 ft (61 cm) of compacted clay liner overlain by a 60-mil (1.5-mm) HDPE liner called a D liner (composite liner required by USEPA for municipal waste landfills). The second composite liner consisted of 4 ft (122 cm) of compacted clay liner overlain by a 60-mil (1.5-mm) HDPE liner called a Wisconsin NR 500 liner (composite liner required by Wisconsin Deptartment of Natural Resources for municipal waste landfills). The third composite liner consisted of a 6.5-mm-thick GCL overlain by a 60-mil (1.5-mm) HDPE liner. Cadmium was used

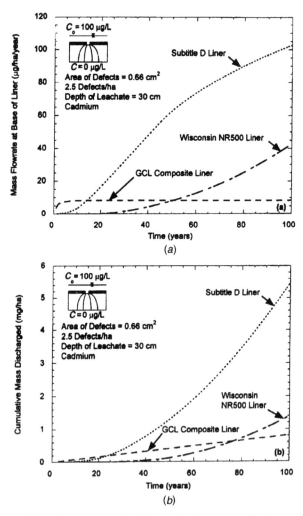

FIG. 18.24   (a) Mass flow rate and (b) cumulative mass discharge for cadmium. Reproduced by permission of the publisher, ASCE, from Foose et al. (2002).

to represent inorganic leachate constituents, and toluene was used to represent organic leachate constituents. The mass flow rate and sorptive capacity for cadmium for the three liners varied within an order of magnitude. However, the mass flux for toluene through GCL was estimated to be two to three orders of magnitude greater compared to clay composite liners. The sorptive capacity for toluene in clay liners was estimated to be one order of magnitude greater than the GCL composite liner. Figure 18.24 shows the mass flux and cumulative mass discharge for cadmium. Figure 18.25 shows the mass flux and cumulative mass discharge for toluene. Diffusive transport depends on both concentration and temperature. Some organic contaminants (e.g., volatile

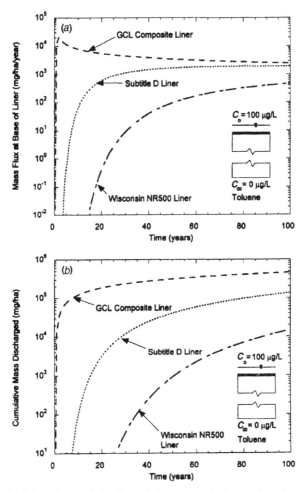

FIG. 18.25   (*a*) Mass flux and (*b*) Cumulative mass discharge for toluene using semi-infinite bottom boundary condition. Reproduced by permission of the publisher, ASCE, from Foose et al. (2002).

organic compounds, VOCs) can readily diffuse through HDPE geomembranes, and may migrate through thin (60 cm or so) composite liners within a decade (Kerry, 1998).

LIST OF SYMBOLS

A_c = activity of clay
a = area of specimen
a = area of hole
C = cohesion
D = dielectric constant
E_p = apparent permeation energy
e = void ratio
h = head
K = Boltzmann's constant
K = permeability of soil
K_s = hydraulic conductivity of the underlying layer
K_0 = pore shape factor
m = weight loss
n = constant (= 5 for natural soil and 10 for artificial mixtures)
n_o = ion concentration
P_c = percentage clay fraction
PI = plasticity index
P_0 = universal gas constant
Q = steady-state leakage
R_1, R_2 = relative humidity
R_0 = preexponential factor
S = saturation vapor pressure at test temperature
S_r = degree of shrinkage
S_0 = specific surface per unit volume of particle
T = absolute temperature
T_f = tortuosity factor
Th = thickness of double layer
t = time
V_f = final volume of sample after drying
V_i = initial volume of sample before drying
v = valence of ion
WVT = water vapor transmission
δ_p = vapor pressure difference
ε = unit electronic charge
γ = unit weight of the permeant liquid
ϕ = angle of internal friction
μ = viscosity of the permeant liquid

19

DESIGN OF LANDFILL ELEMENTS

A detailed design of several landfill elements, in addition to liners, is necessary. The proper functioning of each of these elements is essential to construct and maintain a landfill in an environmentally sound manner. All the elements mentioned in this chapter may not be necessary for all landfills. Engineering designs for most of the elements are already available; however, in some cases changes need to be made to adapt them to a landfill situation.

19.1 LEACHATE COLLECTION SYSTEM

The leachate collection system consists of a drainage layer, leachate trench, and pipe, leachate line clean-out ports, a leachate collection pump and lift station, and a leachate storage tank. The leachate storage tank may not be necessary for a site in which the leachate is discharged directly to a sewer. Permission from the proper authority is necessary before landfill leachate can be discharged into a sewer.

19.1.1 Leachate Collection System Failure

A knowledge about causes of leachate collection system failure is essential to appreciate the design of various elements of the system. A leachate collection system can fail due to the malfunctioning of one or more of the system elements. Kmet et al. (1988) and Bass (1985) discussed several causes of failure. The pipe may fail because of clogging, crushing, or faulty design.

19.1.1.1 Clogging   The pipe may clog because of the buildup of fines, the growth of biological organisms, or the precipitation of chemicals (Bass, 1985). Buildup of fines can result from sedimentation from the leachate or migration of fines from the trench (if the liner material is clay). To minimize the possibility of soil buildup, it is a good idea to use geotextile or filter fabric in the trench, as discussed in Section 19.1.2. Migration of surrounding soil into the trench will not occur if the pore space of the filter layer is small enough to hold the 85% size of soil (Cedergren, 1977). Biological clogging occurs because of the presence of microorganisms in the leachater. Factors that may contribute to biological clogging include the carbon–nitrogen ratio in the leachate, nutrient availability, concentration of polyuromides, temperature, and soil moisture (Kristiansen, 1981). Bioslime layer may form within 1–4 years of exposure to landfill leachate. The major types of bacteria found in the slime include sulfate-reducing bacteria, denitrifying bacteria, methanogenic bacteria, iron-related bacteria, and slime-forming bacteria. The generic types include *Pseudomonas, Klebsiella, Enterobacter, Micrococcus,* and *Bacillus* (Fleming et al., 1999). Clogging resulting from chemical precipitation could be caused by chemical or biochemical processes. Factors that control chemical precipitation are change in pH, change in partial pressure of CO_2, or evaporation (Bass et al., 1983). The precipitates produced due to biochemical processes are generally mixed with slime consisting of bacterial colonies that adhere to the pipe wall. Polymers and additives are susceptible to biodeterioration from microorganisms (Klausmeir and Andrews, 1981). Cracking of polyurethanes caused by fungal growth has also been reported (Hamilton, 1988). Incorporation of an antimicrobial additive to the plastic composition can delay or reduce the extent of the formation of bioslime (Hamilton, 1988). Laboratory tests indicate that sand drainage blankets clog faster than gravel drainage blankets. Researchers also suggest use of disinfectants and pretreatment of municipal waste prior to disposal for reducing clogging potential (Turk et al., 1997).

19.1.1.2 Crushing of Pipe   Crushing of a pipe may occur if the strength of the pipe chosen for the landfill is insufficient. Plastic pipes are considered flexible. The structural design of these pipes are discussed in Section 19.1.3.

19.1.1.3 Faulty Design   A leachate collection pipe may also fail because of faulty design. In general, the leachate flow rate is very low ($\sim$0.5–1.0 cm³/ min); however, in some landfills the flow can be significantly higher for an accidental runon due to failure of the diversion structure. The size of the leachate collection pipe may be insufficient to effectively manage such situations, although these situations are not common for most landfills. Sizing the leachate collection pipe to drain such an exceptionally high water volume is not standard practice. The pipe may also fail due to uneven settlement, especially near exit points from the landfill and at manhole entry points. In designing a leachate collection system, all these causes of failure must be taken into consideration. A leachate collection system may also fail because of the malfunctioning of a joint. Hence each joint must be designed carefully.

19.1.2   Drainage Layer

The drainage layers are constructed with gravel, sand, or geocomposites (also called "drainage composites"). As indicated in Section 19.1.1 gravel drainage blankets are less prone to clogging than sand drainage blankets. The hydraulic conductivity of the drainage layer should be at least 1×10^{-2} cm/sec to maintain leachate flow.

Geocomposites are manufactured by bonding (using heat fusion) geotextile on one or both sides of a geonets. Geonets are synthetic nets manufactured using parallel ribs placed in layers. Geonets can be biplanar or triplanar. Biplanar geonets have two parallel sets of ribs placed one over the other. Triplanar geonets have three parallel sets of ribs placed in three layers. The in-plane hydraulic conductivity of geonets can be performed using ASTM D4716. The flow rate of geonets depends on the pressure on the geonet (USEPA, 1989c). The allowable flow rate for geonets should be calculated for each site.

The flow rate determined in the laboratory is reduced using the following formula to arrive at a site-specific value (Koerner, 1998):

$$q_{allow} = q_{ult}/(RF_{IN} \times RF_{CR} \times RF_{CC} \times RF_{BC}) \qquad (19.1)$$

where q_{allow} = allowable flow rate for final design purposes
$\quad\quad\quad q_{ult}$ = flow rate obtained from ASTMD 4716 using water
$\quad\quad\quad RF_{IN}$ = reduction factor for elastic deformation, or intrusion, of the adjacent geosynthetics into the geonet's core space
$\quad\quad\quad RF_{CR}$ = reduction factor for creep deformation of the geonet and adjacent geosynthetics into geonet's core space
$\quad\quad\quad RF_{CC}$ = reduction factor for chemical clogging and/or precipitation of chemicals in the geonet's core space
$\quad\quad\quad RF_{BC}$ = reduction factor for biological clogging in the geonet's core space

The reduction factors vary for different uses. The landfill-related values of different factors are included in Table 19.1. the allowable flow rate should be reduced by a factor of 2–3 to obtain the field flow rate capability of the geocomposite. The leachate production rate is expressed as volume per unit time. The volume rate should be divided by maximum horizontal distance perpendicular to the leachate collection pipe for rectangular cells to obtain the volume of leachate that is required to be collected per unit length of pipe. This required flow rate must be less than or equal to the flow rate capability of the geocomposite.

TABLE 19.1 Recommended Reduction Factors for Determining Allowable Flow Rate in Biplanar Geonets

Application Area	Reduction Factor			
	RF_{IN}	RF_{CR}[a]	RF_{CC}	RF_{BC}
Drainage blankets	1.3–1.5	1.2–1.4	1.0–1.2	1.0–1.2
Surface water drains	1.3–1.5	1.1–1.4	1.0–1.2	1.2–1.5
Primary drainage layer	1.5–2.0	1.4–2.0	1.5–2.0	1.5–2.0
Secondary drainage layer	1.5–2.0	1.4–2.0	1.5–2.0	1.5–2.0

[a]These values are sensitive to the density of the resins used in the geonet's manufacture. The higher the density, the lower the reduction factor; creep of the covering of the geotextile(s) is a product-specific issue.
After Koerner (1998).

19.1.3 Design of Leachate Trench and Pipe

Leachate pipes are generally installed in trenches that are filled with gravel. The trenches are lined with geotextile to minimize entry of fines from the liner into the trench and eventually into the leachate collection pipe. Typical trench details are shown in Fig. 19.1 and 19.2. Usually the design shown in Fig. 19.1 is used in landfills in which liner material is clay and the design shown in Fig. 19.2 is used in landfills in which the primary liner material is synthetic membrane. It is essential to have a deeper excavation below the collection trench so that the liner has the same minimum design thickness even below the trench.

19.1.3.1 Leachate Trench The gravel used in the trench should be mounded as shown in Figs. 19.1 and 19.2 to distribute the load of compaction machinery and thereby provide more protection for the pipe against crushing. The geotextile, which acts as a filter, should be folded over the gravel. Alternatively, a graded sand filter may be designed to minimize the infiltration

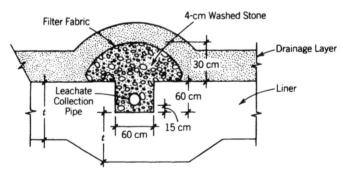

FIG. 19.1 Leachate collection trench detail.

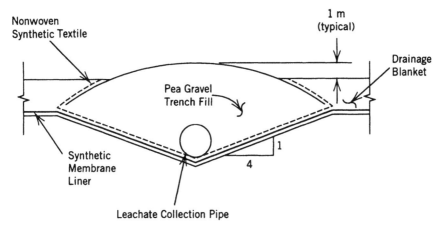

FIG. 19.2   Leachate collection trench for synthetic membrane liner.

of fines into the trench from the waste. Turk et al. (1997) argue that filter fabric over the leachate collection pipes should be avoided to minimize slime growth. Designs for both soil and geotextile filters are discussed below.

Soil Filter   Although several criteria are available for the design of granular filters, actual variations among them are minimal (Koerner, 1986). The following approach for the design of a soil filter may be used (Cedergren, 1977).

FIRST CRITERION

$$\frac{D_{15} \text{ of the filter}}{D_{85} \text{ of the overlying soil}} < 4\text{--}5 \tag{19.2}$$

SECOND CRITERION

$$\frac{D_{15} \text{ of the filter}}{D_{15} \text{ of the overlying soil}} > 4\text{--}5 \tag{19.3}$$

in which D_n = the particle size of which $n\%$ of the soil particles are smaller. The first criterion is aimed at preventing migration of overlying soils into the filter layer, and the second criterion is aimed at ensuring sufficient hydraulic conductivity of the filter layer to maintain proper drainage. A critique of several design criteria can be found in Sherard et al. (1984a,b) along with their suggestions for alternative criteria.

Geotextile Filters   Filter design criteria for geotextiles have been frequently discussed in the literature (Cedergren, 1977; Chen et al., 1981; Giroud, 1982; Horz, 1984; Lawson, 1982; Carroll, 1983).

The approach for filter fabric design primarily consists of comparing the soil particle size characteristics with the apparent opening size (AOS; also called the equivalent opening size or EOS) of the filter fabric. The following simple procedure is recommended by Koerner (1986)

1. For soils in which ≤50% of the particles pass through a 0.074-mm sieve (P200), the AOS of the filter fabric should be ≥0.59 mm (US No. 30 sieve).

2. For soils in which >50% of the particles pass through a 0.074-mm sieve (P200), the AOS of the filter fabric should be ≥0.297 mm (US No. 50 sieve). The criteria proposed by Carroll (1983) are more restrictive and the criteria proposed by Giroud (1982) are probably the most conservative.

The AOS or EOS of the filter fabric is found by sieving glass beads of known size through the fabric (Corps of Engineers, 1977). The test has several problems but is widely used for lack of a better test (Koerner, 1986). The EOS values for several commercial filter fabrics are available elsewhere (Corps of Engineers, 1977).

19.1.3.2 Leachate Pipe   As indicated in Section 19.1.1, a leachate pipe may fail due to clogging, crushing, or faulty design. Design and maintenance of leachate pipes for each of these situations are discussed below.

Clogging   Cleaning the pipe on a regular basis is an effective way of minimizing clogging due to biological or chemical processes. A clean-out port along with an exit point design detail is included in Fig. 19.3. Essentially two methods are available for cleaning leachate lines: mechanical and hydraulic. At present all the cleaning equipment mentioned below are not used for leachate line cleaning but are used for sewer line cleaning. However, they may be adapted (redesigned) to suit leachate line cleaning.

Three different types of mechanical equipment are available for cleaning purposes: rodding machines, cable machines, and buckets. In rodding machines, a series of rigid roads are joined together to make a flexible line that is pushed or pulled through the line. In cable machines, an attachment is spinned at the end of a cable that is pushed or pulled through the line. Both machines use a rotary motion of various attachments to clean the line. Several attachments for both types of machines are available (Bass, 1986). The disadvantage of the rodding/cable machine is that it cannot remove the dislodged debris, and large quantities of water are required to flush the debris after rodding. Rodding/cable cleaning is advantageous for situations in which hydraulic methods are not capable of dislodging chemical or biological buildup or jetting may damage the pipe. Rodding/cable cleaning is less expensive than hydraulic methods in some cases. Machines fitted with a bucket that

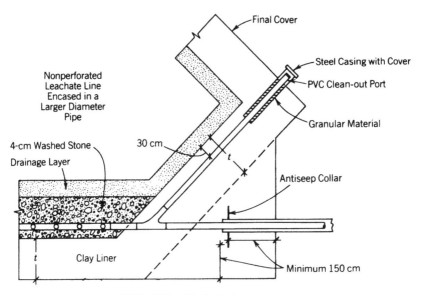

FIG. 19.3   Typical clean-out port.

opens when pulled from one direction but closes when pulled from the other direction may be used to remove large quantities of debris from collection pipes. The bucket is pulled through a line in between two manholes, by a cable connected to a power winch (Hammer, 1977). Several accessories are available to remove materials not cleaned by a bucket (Foster and Sullivan, 1977). One disadvantage of the use of a bucket is that a manhole at each end of the pipe must be available to run the equipment. Buckets are available for use in 15-cm (6-in.) pipes and can be used to clean lines up to 230 m (750 ft) long (Foster and Sullivan, 1977).

Hydraulic equipment is almost exclusively used for cleaning leachate pipes. Two different types of hydraulic equipment are available for leachate line cleaning: jetting and flushing. In jetting, water is pumped through a nozzle that is self-propelled. Water pressure of up to 13.8 MPa (2000 psi) can be generated normally. It was observed that a high-pressure jet may damage the drainage layer (Ford, 1974); however, only a 5-cm (2-in.)-thick drainage layer was used for the experiment. The nozzle loses its self-propelling motion at low pressure. An on-site experiment may be performed to determine the maximum allowable pressure that will not damage the drainage layer. The section used for the experiment must have sufficiently thick overburden so that a realistic equivalent pressure is exerted on the drainage layer. A low pressure (in excess of the self-propelled pressure) may be specified for the first few years to reduce the risk of damaging the filter layer. However, use of extremely high pressure is not recommended. Jetting is effective in cleaning most types of clogs, is easy to use, and needs access from only one end of the pipe. However, jetting may damage the filter layer and collection pipe, and may not be effective in removing heavy debris. A vacuum device may need to be used to clean debris dislodged by the jet. The nozzle size and type should be based on size, length, and expected clogging condition in the pipe. Jetting can clean up to a maximum length of 303 m (1000 ft); however, the capability of jetting equipment may vary (Foster and Sullivan, 1977). A field study using different pipe diameters undertaken by Babcock and Mossien (1993) indicated that (1) 303 m (1000 ft) of 10-cm (4-in.)-diameter pipes will take 4–6 hr to clean whereas the same length of 15-cm (6-in.) or 20-cm (8-in.)-diameter pipe can be cleaned in less than an hour; (2) negotiating a bend in a 10-cm-diameter pipe is relatively difficult compared to 15- or 20-cm-diameter pipes; (3) 20-cm-diameter pipes should be used if the length of the collection line is more than 303 m; (4) it is preferable to limit leachate collection pipe length to 303 m; and (5) it is preferable to use 20.7 MPa (3000 psi) of jet pressure for easy and effective cleaning.

Collection lines can also be cleaned using a hose connected to a high-pressure water source (e.g., a fire hydrant). Sewer balls (inflatable rubber balls that reduce the flow area so that water flows around the ball at higher velocity, increasing the cleaning ability) and sewer scooters (or hinged-disk cleaners) are two attachments that may be used with the hose for better cleaning of the line (WPCF, 1980). The debris are washed downstream and should be removed using a vacuum device. Although access to only one end of the pipe is needed for flushing, access from both sides of the pipe is preferable for proper cleaning. Access from both ends of a pipe is required if a sewer ball or sewer scooter is used. Flushing is simple and is useful only in the initial years or in certain waste types (e.g., foundry waste) landfills, where heavy biological or chemical buildup is not expected.

Crushing   Leachate collection pipes may be crushed during the construction of or during the active life of the landfill. To safeguard against crushing, leachate pipes should be handled carefully and brought on the liner only when the trench is ready. Running of heavy equipment over a pipe must be avoided. A pipe can be installed in either a positive or negative projection mode. Every effort should be made to install it in a negative projection mode (Figs. 19.1 and 19.2), although at times it may be necessary to install a pipe in a positive projecting mode (Fig. 19.4). The strength of a pipe must be checked to ascertain whether it will be able to withstand the load during both pre- and postconstruction periods. Usually two types of pipes are used, PVC and HDPE. The pipes are considered as flexible pipe. The design approach consists of calculating the deflection of the pipe, which should not exceed 5% [Uni-Bell Plastic Pipe Association (Uni-Bell), 1979]. Two approaches are available for checking pipe strength: (1) approach using Iowa formula and (2) approach developed by Paruvakat (1993b).

Either of the following formulas, commonly known as modified Iowa formulas, can be used to estimate pipe strength:

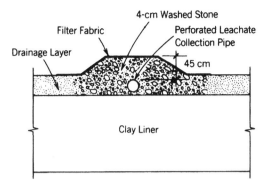

FIG. 19.4 Leachate collection pipe in a positive projection mode.

$$\% \frac{\Delta}{D_{\mathrm{p}}} = \frac{DBP(100)}{0.149(E/\Delta Y) + 0.061E'} \tag{19.4}$$

$$\% \frac{\Delta}{D} = \frac{DBP(100)}{[2E/3(\mathrm{DR} - 1)^3] + 0.061E'} \tag{19.5}$$

in which D = the deflection lag factor, B = the bedding constant, P = the pressure on the pipe (psi), D_{p} = the pipe diameter (in.), Δ = deflection, E = the modulus of elasticity of the pipe material (psi), E' = the modulus of the soil reaction (psi), DR = the dimension ratio = outside diameter/wall thickness (both in inches), $F/\Delta Y$ = the pipe stiffness (psi). Equation (19.4) is used when the pipe stiffness ($F/\Delta Y$) is known and Eq. (19.5) is used when DR is known. Note that these formulas are in F.P.S. units so all metric dimensions must be converted before using the formula. Pipe stiffness is measured according to ASTM D-2412 (ASTM) (Standard Test Method for External Loading Properties of Plastic Pipe by Parallel-Plate Loading). The modulus of elasticity of the pipe material depends on the compound used. The values of the variables in Eqs. (19.4) and (19.5) can be found in the UniBell handbook (Uni-Bell, 1979), from experiment, and from the published literature. The approximate range of values of some variables are included in Table 19.2, which provides a general guideline. Accurate values of these variables should be used for deep (20 m or more) landfills. The following formula may be used to calculate pipe stiffness (Uni-Bell, 1979):

$$\frac{F}{\Delta Y} \simeq 0.559E \left(\frac{t}{r}\right)^3 \tag{19.6}$$

in which t = the wall thickness (in.) and r = the mean radius of the pipe (in.). For pipes with SDR numbers (e.g., SDR 35), the following formula may also be used for pipe stiffness (Uni-Bell, 1979):

TABLE 19.2   Approximate Range of Values of Different Variables of Eqs. (19.4) and (19.5)

Variable	Range	Remarks
B	0.08–0.1	Pipes embedded in gravel or sand
D	1.5–2.5	If the soil in the trench is not compacted, then the higher value of D should be used
E' (in psi)		If the soil in the trency is not
Crushed rock	1000–3000	compacted, then the lower
Sandy soil and rounded gravel	~100–400	value should be used
Stiffness		
Schedule 40	~129	Values are for 6-in-diameter pipes
Schedule 80	~700	
SDR 35	46	
SDR 26	115	

$$\text{Pipe stiffness} = 4.47 \, \frac{E}{(DR - 1)^3} \qquad (19.7)$$

Laboratory and field tests on buried HDPE pipe showed that the Iowa formula or its variations highly overpredicted pipe deflection (Watkins, 1990). Based on laboratory study Watkins (1990) concluded that for pipes enveloped in select pipe-zone backfill (PZB), incipient failure condition will not occur up to a ring deflection of 18–21% (with angle of internal friction of the PZB between 30° and 35°) irrespective of the height of the landfill. The PZB is typically placed above the pipe to a height of 30–60 cm (Fig. 19.5). When

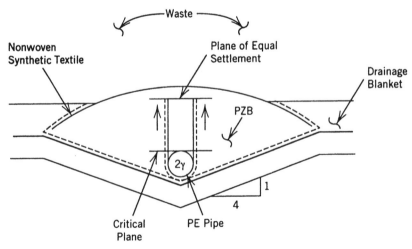

FIG. 19.5   Settlement of PE pipe under the waste load.

waste is placed over the PZB, the pipe will deflect. This deflection of the pipe will cause a redistribution of stresses within the PZB adjacent to the pipe.

Paruvakat (1993) proposed a new approach for designing leachate collection pipes that considers the stress redistribution; his approach essentially combines the Iowa formula and Marston's formula. Paruvakat's (1993b) method can be used for both positive and negative projection modes.

Marston's formula assumes that relative movements take place along imaginary vertical planes extending upward from the sides of the pipe (Spangler and Handy, 1982). For the landfill situation, Marston's formula

$$W_c = \gamma (C_c)(B_c)^2 \tag{19.8}$$

$$C_c = \frac{e^{\pm 2kM(H_e/B_c)} - 1}{\pm 2kM}$$

$$+ \left[\frac{H}{B_c} - \frac{H_e}{B_c} \right] e^{\pm 2kM(H_e/B_c)} \tag{19.9}$$

in which W_c = load per unit length of the pipe, γ = unit weight of soil above the pipe, B_c = outside width of the pipe, H = height of fill above pipe, H_c = height of plane of equal settlement above the critical plane, k = earth pressure coefficient, M = tan ϕ, ϕ = angle of internal friction, and e = base of natural logarithm.

Figure 19.6 shows the positive projection mode where the relative stiffness of the pipe is smaller than the surrounding backfill and, therefore, shortening of the vertical height of pipe is more than the deformation of the adjacent backfill. Thus, an incomplete ditch condition will occur provided the plane of equal settlement is within the PZB. The height of the plane of equal settlement above the critical plane can be calculated from Eq. (19.10); all relevant terms are depicted in Fig. 19.6:

$$\pm r_{sd}(P_p) \left[\frac{H}{B_c} \right] = \left[\frac{1}{2kM} + \left\{ \frac{H}{B_c} - \frac{H_e}{B_c} \right\} \pm \frac{r_{sd}}{3} \frac{P_p}{} \right] \frac{e^{\pm 2kM(H_e/B_c)}}{\pm 2kM}$$

$$\pm \frac{1}{2} \left[\frac{H_e}{B_c} \right]^2$$

$$\pm \frac{r_{sd} P_p}{3} \left[\frac{H}{B_c} - \frac{H_e}{B_c} \right] e^{\pm 2kM(H_e/B_c)}$$

$$- \frac{1}{2kM} \left[\frac{H_e}{B_c} \right] \mp \left[\frac{H}{B_c} \right] \left[\frac{H_e}{B_c} \right] \tag{19.10}$$

in which r_{sd} = settlement ratio and P_p = ratio of pipe projection above the natural ground surface to the pipe diameter; r_{sd} is given by

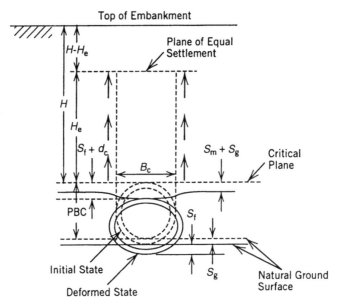

FIG. 19.6 Case of an incomplete ditch condition for a positive projecting conduit. After Paruvakat (1993b). Reprinted with permission from the author and Industrial Fabrics Association International.

$$r_{sd} = \frac{(S_m + S_g) - (S_f + d_c)}{S_m} \tag{19.11}$$

in which S_m = compression deformation of the backfill column adjacent to pipe of height PBC, S_g = settlement of natural ground surface adjacent to the pipe; S_f = settlement of the pipe into the liner, and d_c = shortening of the vertical height of pipe.

(Note: In Eqs. (19.9) and (19.10), for ± signs, use the upper sign when r_{sd} is positive and use the lower sign when r_{sd} is negative.) For PE pipes in landfill liners, S_g and S_f are the same whereby r_{sd} reduces to

$$r_{sd} = \frac{S_m - d_c}{S_m} \tag{19.12}$$

and d_c is larger than S_m because the PZB materials are stiffer than the pipe. So r_{sd} is negative, indicating an incomplete ditch condition. Equations (19.9), (19.10), and (19.12) need to be solved simultaneously to find the deflection d_c for the design load of W_c. The following is a step-by-step procedure for calculating d_c (Paruvakat, 1993b):

STEP 1   Assume a value of r_{sd}.

STEP 2   Calculate S_m based on the estimated vertical stress using the following formula:

$$S_m = \frac{\text{(vertical stress) (mean pipe diameter)}}{E_s} \qquad (19.13)$$

in which E_s = deformation modulus of the PZB material [732.6 MPa (15.3 × 10^6 psf) approx].

STEP 3   Using the assumed value of r_{sd} and the calculated value of S_m, find d_c from Eq. (19.12).

STEP 4   Using the calculated value of d_c, find W_c from the following, which is a variation of the Iowa formula:

$$d_c = \frac{DBW_c}{EI/r^3 + 0.016E'} \qquad (19.14)$$

in which D = deflection lag factor, B = bedding factor, W_c = load per unit length of pipe, E = modulus of elasticity of the pipe material, I = moment of inertia of the pipe wall, E' = modulus of soil reaction, and r = mean radius of pipe.

STEP 5   Using Eq. (19.8) and W_c, calculate C_c.

STEP 6   Using Eq. (19.9) and C_c, calculate H_e/B_c.

STEP 7   Use the calculated value of H_e/B_c in Eq. (19.10) to find r_{sd}. Steps 1–7 should be repeated until the calculated and assumed values are equal.

STEP 8   Verify that PZB extends up to at least H_e to satisfy the original assumption made in deriving the equations. The percent deflection ($d_c/2r$ in percent) can be checked against allowable percent deflection, or the actual stress on the pipe wall due to W_c can be checked against the allowable stress to calculate the factor of safety, which should be 2 or more.

   In general, for shallow landfills (20 m for municipal or other wastes with similar unit weight and 10 m for sandy-type waste), the Iowa formula may be used. For deeper landfills, Paruvakat's method described above should be used because it will provide a realistic pipe deflection.

Example 19.1 (in F.P.S. units)   Determine the suitability of 6-in. schedule 40 PVC pipe placed in a trench (Fig. 19.1) filled with rounded washed gravel (dumped in the trench). The landfill is 70 ft deep and has a waste-to-cover ratio of 5:1. The final cover will be 4 ft thick and the drainage blanket 1 ft thick. The unit weights of waste and soil are 50 and 110 lb/ft³, respectively. The maximum leachate head is expected to be 3 ft.

$$\text{Thickness of daily cover} + \text{final cover} = (70 - 40) \times \frac{1}{5 + 1} + 4 = 15 \text{ ft}$$

Pressure on the pipe $(70 - 15) \times 50 + 15 \times 110$

$$+ 3 \times 62.4 + 1 \times 110 = 4697.2 \text{ psf} = 32.6 \text{ psi}$$

Using Eq. (19.4), the percentage deflection can be calculated:

$$\% \frac{\Delta}{D_p} = \frac{2 \times 0.09 \times 32.6 \times 100}{0.149 \times 129 + 0.061 \times 200} = 18.6\%$$

Since the percent deflection is more than 5%, a thicker pipe should be used. The percentage deflection for a schedule 80 pipe will be 5.03%. The maximum percentage deflection allowable is usually 5, however, manufacturers of PVC pipe may allow a value of up to 7.5%. Use of the lower value of 5% deflection is recommended if approximate values of variables are used in Eq. (19.4) or (19.5).

Pressure due to live load should be added to the pressure P to account for compaction machinery running in the landfill. The effect of live load is maximum for the first lift. Deflection of the pipe may be checked due to pressure of a compactor running on 1-ft (30-cm) lift of waste. The effect of live load will be reduced with increases in the depth of the landfill. the Boussinesq equation (Sowers and Sowers, 1970) may be used to determine the effect of live load at various depths. The approximate pressure of equipment at various depths (Fig. 19.7) may be obtained by multiplying by the appropriate factor given in Table 19.3. In Fig. 19.7, W_p is the width of the compactor wheel in contact with the soil, q is the contact pressure of the equipment, and z is the waste thickness.

Example 19.2 (in F.P.S. units)   Assume $q = 20$ psi and $W_p = 12$ in. Find the total pressure on the pipe when the equipment is compacting the first lift of refuse 1 ft thick (use appropriate values from the landfill in Example 19.1):

$$0.1 \times 20 + \frac{1 \times 110}{144} + \frac{1 \times 50}{144} = 3 \text{ psi} \tag{19.15}$$

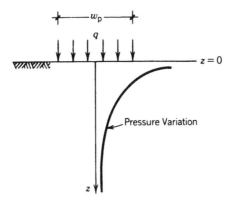

FIG. 19.7   Variation of pressure with depth due to a load at the surface.

This value of $P = 3$ should be added to the value of P in Example 19.1 to estimate percentage deflection. This should be checked, especially for landfills in which lower stiffness pipes (e.g., 6 in SDR-35) are proposed. It may be noted that the Boussinesq equation assumes an elastic soil condition, whereas most waste types are nonelastic; hence the values given in Table 19.2 are conservative.

Faulty Design   The bends in the leachate line should be smooth. Cleaning equipment cannot negotiate sharp bends. Criss-crossing of leachate lines should be avoided. Sometimes secondary leachate lines are connected to a header line to carry the entire leachate generated in a landfill. The diameter of the header pipe should be designed to handle total peak flow of leachate. T-joints should not be used to connect header pipes to a secondary line. A smoother 45° or lesser bend should be used to facilitate cleaning activities.

A minimum number of manholes should be used in a landfill. A flexible connection should be used for entry and exit points of leachate lines (Fig.

TABLE 19.3   Approximate Values of Pressure Coefficients at Various Depths

Depth	Coefficient
W_p	0.4
$2W_p$	0.1
$3W_p$	0.055
$4W_p$	0.03
$5W_p$	0.02
$6W_p$	0.015

19.8). The manholes may settle, particularly if they are more than 7.5 m (25 ft) in height. The settlement should be estimated for deep manholes. A lightweight material (other than concrete) should be considered for deep manholes.

The holes in a leachate line should be made as shown in Fig. 19.9. The pipe should be laid such that the holes are at the lower half of the pipe. Holes close to the springing line reduce the strength of the pipe and hence should be avoided. Slotted pipe may also be used for leachate collection.

19.1.4   Leachate Line Clean-Out Port

A typical clean-out port is shown in Fig. 19.3. The clean-out pipe must be well guarded at the exit point. A shallow concrete manhole may be constructed to provide additional safety against runover. Usually the pipe is laid along the side slope. However, if it is laid on a nearly vertical slope, a smooth bend should be used to connect it to the leachate line.

19.1.5   Leachate Collection Pump and Lift Station

The pump capacity must be calculated carefully for proper functions. In choosing a pump, both suction and delivery head must be considered; it should be noted that the density of leachate is somewhat higher than water. A typical detail of a lift station and pump is shown in Fig. 19.10. Usually automatic submersible pumps are used in a lift station. Positioning of the starting and shut off switches should be such that the pump can run for a while; frequent start and stop may damage the pump. The shut off switch must be located at least 15 cm (6 in.) below the leachate line entry invert.

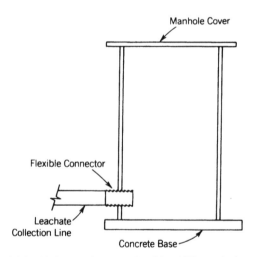

FIG. 19.8   Typical detail of landfill manhole.

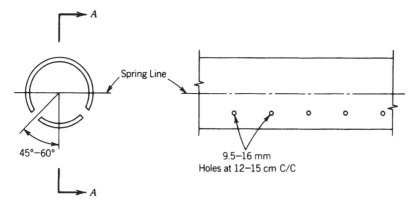

FIG. 19.9   Perforations in leachate collection pipes.

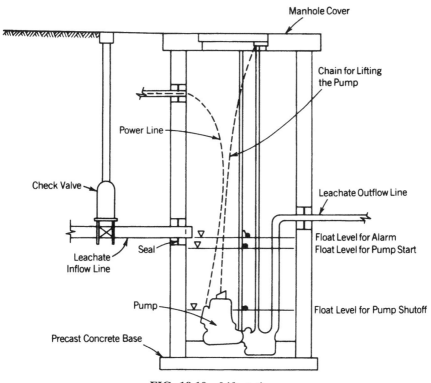

FIG. 19.10   Lift station.

Guide rails must be provided so that the pump can be lifted and lowered easily for maintenance. For larger landfills arrangements should be made for a standby pump, either available on-site or easily accessible. A valve operable from the ground surface should be installed on the incoming leachate line(s). This is useful during periodic maintenance of the pump. If the lift station is constructed outside the landfill, then it should be encased in clay or synthetic membrane to minimize the potential of leachate leakage into the ground. The lift station must be made leakproof from inside. The settlement of the lift station should be checked; in addition to the self-weight of the lift station, the weight of the pump and guide rail and of the leachate ponded inside must be taken into consideration in calculating settlement. The leachate collection line entry connection should be made flexible.

Most pumps are efficient if run continuously; however, since the leachate generation rate varies, intermittent operation of the pump becomes essential. A 12-min cycle is considered satisfactory (Bureau of Reclamation, 1978), an issue which should be verified from the prospective pump manufacturers. The sump size is estimated based on the volume of leachate inflow in one-half cycling time; this would mean equal on and off time. the storage capacity (or sump size) (S_s) is given by

$$S_s = V_c t \tag{19.16}$$

in which V_c = the rate of leachate collection per minute and t = one-half the cycling time. The pumping rate (P_1) is determined by

$$P_1 = \frac{V_c t + V_c t}{t} \tag{19.17}$$

$$= 2V_c \text{ volume/min}$$

The maximum volume of leachate expected in a day should be used while estimating the storage capacity and pumping rate. The maximum leachate level in the sump (for starting the pump) should be 15 cm (6 in.) below the inlet pipe invert. The minimum leachate level in the sump (at which the pump will stop) should be 60–90 cm (2–3 ft) from the bottom of the sump. The difference D_f between the maximum (pump start level) and minimum (pump shut off level) elevation should be kept low [60–90 cm (2–3 ft)] so that the sump dimension is reasonable. The cross-sectional area A_r of a sump is calculated by assuming a value of D_f and from the known value of S_s:

$$A_r \times D_f = S_s \tag{19.18}$$

From the known difference in head, h_d, between suction and delivery, the pump break horsepower (BHP) can be calculated as follows:

$$\text{BHP} = \frac{\gamma_1 h_d P_1}{550E} \tag{19.19}$$

in which γ_1 = the leachate density (note: if the density of leachate is higher than water, a 10–15% higher value may be used for a conservative estimate) and E = pump efficiency. Both the delivery and suction head should be taken into consideration when choosing a submersible pump.

Example 19.3 (in F.P.S. units) Design a circular sump for a leachate collection system for the following case: V_c = 1.25 ft³/min, cycling time $(2t)$ = 12 min, D_f = 2 ft, and h_d = 16 ft. Assume a pump efficiency of 80% and the leachate density to be 10% higher than water.

$$\text{Sump storage capacity} = 1.25 \times 6 = 7.5 \text{ ft}^3$$

For a sump of diameter d

$$\frac{\pi d^2}{4} \times 2 = 7.5$$

$$d = 2.1 \text{ ft}$$

Use a 2-ft-diameter sump:

$$\text{Pumping rate} = 2 \times 1.25 = 2.5 \text{ ft}^3/\text{min}$$

$$\text{BHP} = \frac{62.4 \times 1.1 \times 16 \times 2.5}{550 \times 0.8} = 6.24$$

19.1.6 Leachate Holding Tank

Leachate tanks should have enough volume to hold leachate for a period of time (usually 1–3 days) during the peak leachate production season. The regulatory agency should be contacted to find out if a minimum holding capacity is mandated. The holding volume will depend on frequency of pumping out and maximum allowable discharge rate to a treatment plant.

Both double-wall and single-wall leachate holding tanks may be used. Single-wall leachate tanks should be installed within a clay or synthetic membrane encasement (Fig. 19.11). Arrangement should be made to monitor the inside of the encasement. The monitoring well will detect tank leakage at an early date. The well should be monitored once a month for indicator parameters. This type of encasement is not needed for double-wall tanks; however, provision for monitoring the space in between the two walls should be available.

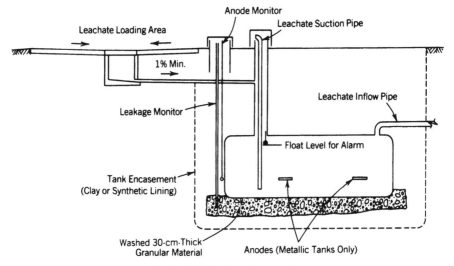

FIG. 19.11   Leachate tank.

Both metallic and nonmetallic tanks may be used. Metallic tanks must be protected against corrosion. The inside of the tanks must be coated with suitable material so that leachate does not damage the tank. The tank(s) should be pressure tested before installation to check for leaks. Manufacturer's guidelines regarding leak testing and installation should be followed. A long-term performance warranty should be obtained for the tank(s) from the manufacturer.

Tank(s) should be installed properly. Improper handling or backfilling may damage tanks during installation. Tanks should be anchored properly with the base concrete if the water table is expected to rise above the base level of tank. Anchoring is important for tanks installed at shallow depth below the water table. Figure 19.12 shows a strapping arrangement; the strapping inter-

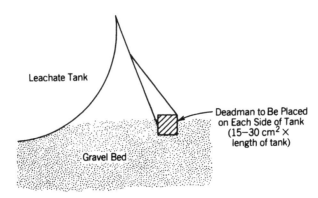

FIG. 19.12   Deadman to hold down tank.

val should be 1.8 m (6 ft) or per the manufacturer's design. The tank trench should be backfilled with pea gravel or crushed stone (3–15 mm size). A manhole is constructed above the tank that houses the pump (Fig. 19.13). A guiderail should be provided so that the pump can be lifted easily for maintenance and repair. The pipe exiting the manhole should be encased in another pipe of larger diameter to provide secondary containment. Some tankers are equipped with a pump so that the pump in the manhole is not needed. The capability of the tanker should be checked to find out whether a pump is necessary. A concrete loading pad should be constructed away from the tank. The pad should be sloped toward a sump connected to the tank so that any spill during loading will go back to the tank.

19.1.7 Leachate Removal Systems

Leachate from a landfill may be removed either by gravity flow or by using a side slope riser. Either system has its advantages and disadvantages. Each system is discussed separately in the following sections.

19.1.7.1 Gravity Flow   This system of leachate removal is used where the landfill base is at grade or at shallow depth. This is a popular system used in clay-lined landfills. The header pipe, which connects all the leachate from the landfill, exits through the side. The number of header pipes should be minimized to minimize liner penetration. The header pipe either discharges directly into a sewer line or into a sump with a lift station (see Section 19.1.5).

An antiseep collar should be constructed on all leachate line exit points to ensure that leachate does not seep through the hole. Figure 19.14 shows a typical antiseep collar design. The soil around the collar should be hand

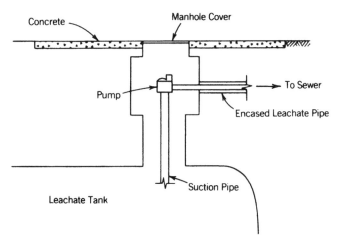

FIG. 19.13   Leachate tank manhole with outflow pump.

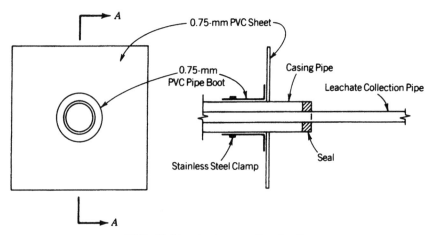

FIG. 19.14   Antiseep collar detail.

compacted. Care should be taken when compacting berm around a leachate exit point.

The advantages of gravity flow are (1) the method is economical, especially if direct discharge to a sewer line can be made; (2) the operational cost of a gravity flow is very low; (3) not much maintenance of the system is required; and (4) the system is in use for more than a decade, so that design and construction details are well established.

The disadvantages of the gravity flow system are (1) the chance of leakage at the exit point is relatively high, especially in landfills where synthetic membrane liners are used, and (2) the integrity of the antiseep collar cannot be tested and leakage through the liner cannot be monitored.

19.1.7.2 Side Slope Riser   This system of leachate removal is used primarily where the landfill base is deep (in excess of 20 m). In most cases this system is used mostly for landfills with synthetic membrane liners. Figure 19.15 shows a typical side slope riser detail. A 30–45 cm (12–18 in.) diameter HDPE pipe is used that houses the leachate withdrawal pipe. A sump pump is installed at the lower end that pumps the leachate. A side slope riser is installed at the end of each leachate line. Thus the need for a header pipe within the landfill is eliminated; however, the leachate from each side slope riser is discharged into a header pipe that runs outside of the landfill. Thus it eliminates T-joints within the landfill; T-joints are a major source of problems in leachate line cleaning. The leachate from the collection line is collected in a sump within the landfill that is withdrawn by the pump. The sump is filled with gravel. Enough redundancy should be built into a side slope riser. In addition to a steel chain attached to the pump, a hook should be attached to the pump to retrieve it if the chain is broken. In addition, the diameter of side slope riser pipe should be large enough (45 cm is suggested) so that a second pump can be installed in case the first one fails. In addition, a steel plate

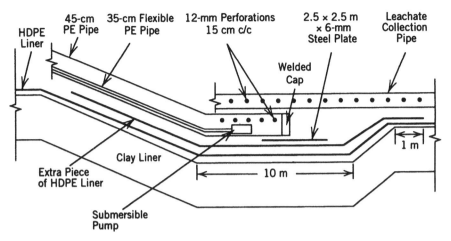

FIG. 19.15 Side slope riser (not to scale).

should be placed in the sump area. The purpose of this steel plate is to provide a guide in case a cassion needs to be installed in the event of side slope riser failure. the failure of a side slope riser means that leachate from the entire cell will be ponded, which in turn will increase leakage. Failure of the leach-ate collection line in a cell is also a possibility.

The advantages of the slope riser are (1) it can be used for deep landfills; (2) the chances of leakage at the exit point is minimal; (3) it eliminates the need for a header pipe within a landfill; (4) additional redundancy can be built into the design; and (5) leachate removal from only one cell will be affected in case of pump failure in a side slope riser. The disadvantages of the side slope riser are (1) this system is difficult and costly to construct; (2) leachate removal from a cell will stop if a pump fails; and (3) no direct access to the collection point is possible, making maintenance of the pump and electrical system difficult.

19.2 STORMWATER ROUTING

Routing of stormwater on and around the landfill is essential to reduce leach-ate generation. All runon water should be diversified away from the landfill by constructing drainage ditches. Usually landfills are not located on natural drainage ways with few exceptions. Precipitation falling on a landfill also needs to be routed to natural drainage swales or toward a sedimentation basin. The following sections mainly address routing or precipitation falling on a landfill; similar approaches can be used to route runon water.

19.2.1 Design of Stormwater Ditch

The design of a stormwater ditch (also called a drainage swale) uses principles of open channel flow. There may be several ditches running over and around a landfill; in many instances one or more secondary ditches are connected to a primary drainage, which carries the entire runoff from the landfill area. In designing these ditches care should be taken to estimate proper volume of runoff water flowing through each section. Ditches running over the landfill should have low base slope to minimize erosion (note: the recommended maximum is 10%). Even though short-term maintenance is expected, long-term maintenance of drainage ditches cannot be ensured. The Manning formula is used to design a channel section:

$$V = \frac{1.486}{n_r} r_h^{2/3} S^{1/2} \tag{19.20}$$

in which V = the mean velocity of water (fps), r_h = the mean hydraulic radius (ft, obtained by dividing the cross-sectional area by the wetted perimeter), S = slope of the energy line (= slope of channel for small slopes), and n_r = the roughness coefficient.

The greatest difficulty in applying the Manning formula is to choose the proper value of n_r. Typical values of n_r based on experience and observations are available (Chow, 1959). Some restrictions should also be imposed on the allowable velocity to minimize scour. Table 19.4 provides some suggested velocity ranges and n_r values based on information obtained from several sources (Chow, 1959; SCS Engineers, 1976; Schwab and Manson, 1957; Bureau of Reclamation, 1978). The initial values are to be used for estimating the channel dimensions, which should be checked for adequacy in the long term. Chow (1959) provides a detailed procedure for designing a grassed channel. However, for most landfills a simpler approach shown in the example below may be used for designing drainage swales on a landfill. Usually either a trapezoidal or triangular section is used for landfill drainage swales. A typical drainage swale arrangement is shown in Fig. 19.16, and a typical cross section of a primary drainage ditch is shown in Fig. 19.17. Additional erosion protection measures such as lining with an erosion mat or riprap or construction of check dam(s) should be undertaken if higher velocity is used. In some cases drop inlets (Figs. 19.18a and b) may be needed to route surface water from the landfill surface to a nearby surface drainage swale.

TABLE 19.4   Recommended Values of n_r and V for Design of Drainage Swales in Landfills

Variable	Values
n_r	Initial: 0.02–0.03
	Long term: 0.1–0.14
Maximum permissible V in fps	Initial: 3–5
	Long term
	Clay: 4–5
	Sandy loam: 1.5–2.5

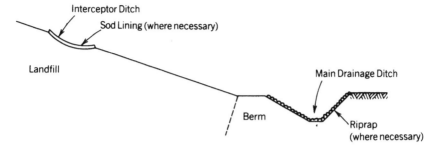

FIG. 19.16 Typical drainage swale arrangement in landfills for surface water routing.

Example 19.4 (in F.P.S. units)   Design a drainage swale to carry a flow of 5.8 ft³/sec (0.15 m³/sec) from a landfill.

For all drainage designs, a trial-and-error method is used to find the dimensions of the section. For most cases a slope of the base is assumed and kept constant throughout the trial-and-error process. A triangular cross section with 3 : 1 side slope and 1% base slope is assumed for the example.

For the initial design:

1. Using the Manning formula, find the dimensions of the section capable of allowing the design flow (use initial n_r values given in Table 19.3).
2. Check whether the velocity is within the maximum permissible value. The trial computations are given in Table 19.5. The long-term design consists of verifying whether the channel is large enough to route 1.5 times the design flow in the long term. The retardance to flow will increase due to growth of vegetation and hence the high n_r values suggested in Table 19.4 should be used. The entire channel depth of 1.8 ft is used for the check, which obviously reduces free board. However, such a reduction in free board is acceptable only for secondary swales on the landfill. For primary swales the final freeboard should be at least 80% of the initial freeboard. (Thus, for the example the revised depth of channel could be 1 not 1.8 ft.)

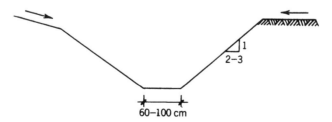

FIG. 19.17 Typical cross section of drainage ditch.

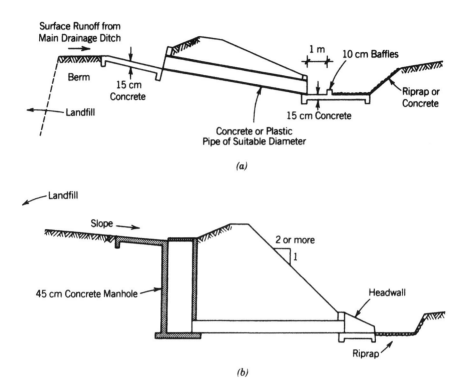

FIG. 19.18   Drop inlets for surface water routing: (*a*) inclined and (*b*) vertical.

The r_h value for a 1.8-ft-deep triangular section is 0.85 ft; V = 0.95 ft/sec; and Q = 9.25 ft³/sec. Since the capacity of the ditch is more than 1.5 times the design flow the swale dimension is acceptable.

19.2.2   Design of Culvert

Circular or rectangular culverts are used to drain water below a road. The culvert inlet and outlet should provide a smooth transition to minimize erosion at entrance and exit points; concrete should be used for entrance and exit. In

TABLE 19.5   Trial Computations for the Example on Drainage Swale Design

Trial	Depth (ft)	r_h (ft)	V (ft/sec)	Flow ft³/sec
1	0.6	0.29	2.6	2.8
2	0.8	0.37	3.06	5.87
3	1.8	0.47	3.59	10.77
	Depth of channel = 0.8 + 1 (free board) = 1.8 ft			

many instances, maintenance of the culvert in the long term is not envisioned. Therefore, a concrete culvert is preferable over a metal culvert, which needs to be replaced more often. As an alternative, arrangements should be made to replace the culvert with an open section by cutting the road at the end of the long-term care period (usually 30–40 years after closure of landfill); this alternative should be resorted to only if lack of maintenance after the long-term care period is envisioned. A culvert should be overdesigned because in most cases long-term maintenance is not expected. A culvert can flow full or partly full. The flow characteristics depend on inlet geometry, slope, size, roughness, approach, tailwater condition, and so on. Although the use of nomographs is suggested for high design flows, 45- to 50-cm (18- to 20-in.) circular section culverts with a minimum base slope of 1% can be safely used for flows up to 0.28 m³/sec (10 ft³/sec).

19.2.3   Design of Stormwater Basins

Sedimentation basins may need to be constructed before allowing the surface water to enter a natural flowage. The purpose of the sedimentation basin is to reduce the total dissolved solids (TDS) from the surface water. The sediment collected at the bottom of the basin should be cleaned periodically and disposed of in the landfill. Settling velocity of a particle is calculated using Stoke's law:

$$V_s = \frac{g(\rho_s - \rho_w)d^2}{18\mu}$$ (19.21)

in which V_s = the settling velocity of a particle, g = the gravitational constant, ρ_s = the density of the particle, ρ_w = the density of the water, d = the diameter of the particle, and μ = the absolute viscosity of water. Ideally the surface area of a basin (A) capable of settling all particles with the settling velocity of V_s is given by

$$A = \frac{Q}{V_s}$$ (19.22)

in which Q = the surface loading or overflow rate. In reality the sedimentation basin does not perform according to the above theory and the two following factors influence basic performance: currents and particle interactions. Four types of currents are identified: surface currents induced by wind, convection currents arising from temperature differences within and outside the basin, density currents that develop due to different densities of the incoming water and basin water, and eddy currents produced by the incoming water. Three different particle interactions are identified: settling is increased due to flocculation of two or more particles, settling of a particle is hindered due to the upward movement of the water particle displaced by the particle, and settling of a particle is hindered due to high concentrations of sediment near the bottom of the basin. The design of inlet and outlet structures also influence sedimentation. Particles of 100–10 μm (0.0039–0.00039 in.) and above are usually removed from the runoff water. The areas of the basin for 1 m^3/sec (35.937 ft^3/sec) flow for different particle sizes are included in Table 19.6. Performance of a sedimentation basin depends very little on the depth of the basin. Normally the depth of basin used is 1.5 m (5 ft). The design approach discussed above (and enumerated in the example below) is considered adequate for most landfill designs. Charts and nomographs are also available (Gemmell, 1971; Fair and Geyer, 1954) for more detailed design. Figure 19.19 shows a typical plan and cross section of a sedimentation basin. Usually the length-to-width ratio of 2 : 1 is used for basin sizing. The size of the basin depends on particle size distribution and total suspended solids (TSS) of the runoff water. Usually removing the particles that are 40 μm and above provides an acceptable TSS of the effluent. However, regulations may dictate removal of lower particle size.

Example 19.5   Design a sedimentation basis for removing particles 40 μm and above for a landfill in which the expected peak flow is 1.5 m^3/sec (54 ft^3/sec).

From Table 19.6 the required base area of the basin is

$$1.5 \times 476 = 714 \text{ m}^2$$

Assuming the width to be A and length-to-width ratio as 2 : 1:

$$2A^2 = 714$$

or

$$A = 18.9 \text{ m}$$

Use a basin size of 38 $\times$ 19 $\times$ 1.5 m.

TABLE 19.6   Sedimentation Basin Surface Area for 1 m^3/sec Flow

Particle Size (mm)	Area (m^2)
0.1 (100 μm)	125
0.06 (60 μm)	263
0.04 (40 μm)	476
0.01 (10 μm)	6.7×10^3
0.001 (1 μm)	6.7×10^5

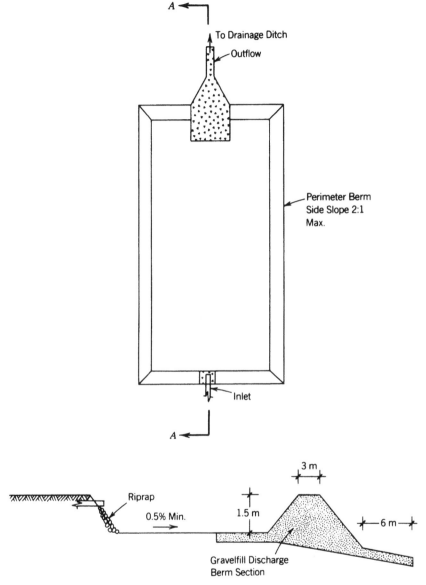

FIG. 19.19 Typical sedimentation basin detail.

19.3 GEOSYNTHETIC MEMBRANE

Several design issues are discussed. A simple approach using fundamental principles of mechanics are used in deriving the equations. The values used in the examples are approximate and may be different for a product used in a site. New products with better physical properties are marketed from time to time, hence physical properties for the synthetic membrane should be ob-

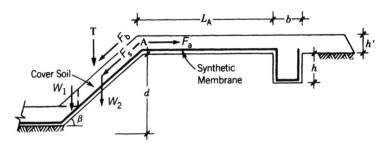

FIG. 19.20 Synthetic membrane design variable.

tained from the manufacturer or from experiments whenever necessary. Dimensions of the trench and so on shown in Fig. 19.20 have been used for deriving equations in this section.

19.3.1 Runout and Anchor Trench Design

A trench is constructed on the berm, and the geomembrane from the side slope of the landfill is runout to the trench (Fig. 19.20). The depth and width of the trench is dictated by construction equipment. Usually the anchor trenches are 2 ft (60 cm) deep and 2–3 ft (60–90 cm) wide. The runout length (L_A) is calculated to provide sufficient friction so that the geomembrane does not slip down when the cover soil is placed. The allowable pull (F_s) on the geomembrane (per unit width) is given by

$$F_s = \sigma_a t \tag{19.23}$$

The total resistive force (per unit width) is generated due to the sum of friction of the geomembrane with soil on friction of the geomembrane with soil on the runout (F_r) and friction of the geomembrane with soil on the trench vertical (F_{tv}) and trench base (F_{tb}):

$$F_r = L_A h' \gamma_s \tan \delta \tag{19.24}$$

where δ = friction angle between geomembrane and soil.

$$F_{tv} = K_0 \gamma_s (h' + h/2) \tan \delta \tag{19.25}$$

where K_0 = earth pressure coefficient at rest.

$$F_{tb} = \gamma_s (h + h')b \tan \delta \tag{19.26}$$

The resistance provided by the geomembrane on the right side of the trench is neglected, equating all forces in the horizontal direction:

$$F_s \cos \beta = F_r + F_{tv} + F_{tb}$$

Or

$$\sigma_a t \cos \beta = L_A h' \gamma_s \tan \delta + K_0 \gamma_s (h' + h/2) \tan \delta + \gamma_s (h + h') b \tan \delta$$

$$\tag{19.27}$$

$$L_A = (\sigma_a t \cos \beta / h' \gamma_s \tan \delta) - K_0 (1 + h/2h') + (1 + h/h') \quad \text{(19.28)}$$

Example 19.6 Find the runout length for a 40-mil (1.016-mm) HDPE liner ($\sigma_y t = 1440$ kg/m); the cover soil is 60 cm thick and the anchor trench is 60 cm wide and 60 cm deep; γ_s is 1700 kg/m^3; $\delta = 20°$; the side slope is 3 : 1 (i.e., $\beta = 18.46°$); $K_0 = 0.5$ from Eq. (19.26):

$$L_A = \frac{1440 \times \cos 18.46}{0.6 \times 1700 \tan 20°} - 0.5 \left(1 + \frac{1}{2} \right) - (1 + 1) = 0.93 \text{ m}$$

19.3.2 Allowable Weight of Vehicle

To avoid pullout, rupturing, and undesirable elongation leading to failure, heavy vehicles cannot be allowed on synthetic membranes. The maximum allowable weight of the vehicle (T) can be calculated using Eq. (19.29). A cover soil should be applied before allowing any vehicle into the landfill:

$$T \sin \beta + W \sin \beta = W \cos \beta \tan \delta + F_{rc} + \frac{\sigma_y t}{FS} \tag{19.29}$$

in which T = the allowable weight of the vehicle, W = the total weight of the soil on the slope (= $\gamma_h h' / d / \sin \beta$), β = the slope angle of the liner, and FS = the factor of safety.

Example 19.7 Calculate the maximum weight of a vehicle for the case in Example 19.6. The landfill is 6 m deep and has a side slope of 3.5 : 1 ($\beta = 15.9°$).

From Eq. (19.29)

$$T \sin 15.9 + 1.7 \times 10^3 \times 0.6 \times 6$$
$$= 1.7 \times 10^3 \times 0.6 \times (6/\sin 15.9) \times \cos 15.9 \times \tan 18$$
$$+ 1.7 \times 10^3 \times (0.6 + 0.6) \times 1.2 \tan 18$$
$$+ 1.7 \times 10^3 \times 0.6 \times 1 \times \tan 18 + 1440/1.3$$
$$= 11{,}253 \text{ kg}$$

19.3.3   Check for Sliding of Cover Soil

The cover soil may slide down the slope if adequate friction is not provided by the membrane. Sometimes a geotextile is placed below or above the membrane for protective purposes. In such cases, the friction angle between the synthetic membrane and geotextile (which are usually lower) or the lowest friction angle between the materials should be used for design purposes. Williams and Houlihan (1986) studied the friction angle for three separate interfaces: soil to synthetic membrane (range: 27°–17°), soil to geotextile (range 30°–23°), and geotextile to synthetic membrane (range 23°–6°). The study showed that the friction angles depend on the materials involved. Even though the soil to synthetic membrane and soil to geotextile friction angles were somewhat close, the geotextile to synthetic membrane friction angles are drastically low; HDPE friction angles were lowest in all cases. In the absence of experimental data, the following values may be used for sandy cover soil and HDPE liners:

Soil to synthetic membrane = 17°. Soil to geotextile = 23°–25° (depends on type of geotextile). Synthetic membrane to geotextile = 6°–8°.

Proper friction angle should be used for designing a combined layer of geotextile and synthetic membrane. The factor of safety against sliding (FS) is given by (see Fig. 19.21)

$$FS \times \frac{W \cos \beta \times \tan \delta}{W \sin \beta} = \frac{\tan \delta}{\tan \beta} \tag{19.30}$$

Example 19.8   Calculate the factor of safety against sliding of a 60-cm soil cover laced over a geotextile; the geotextile is in turn placed over an HDPE synthetic membrane. The unit weight of soil = 1.6 g/cm³ and the slope of the liner = 18.4°. The following friction angles may be assumed: geotextile and soil, 23°; synthetic membrane and soil, 18°; synthetic membrane and geotextile, 10°.

In this case the geotextile is most vulnerable to sliding over the synthetic membrane, hence

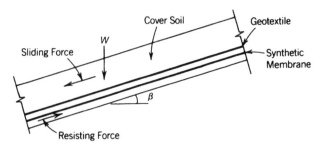

FIG. 19.21   Force schematic for synthetic membrane design.

$$FS = \tan 10/\tan 18.4 = 0.53$$

To avoid failure the liner slope may be reduced to $5:1$ (not a very practical approach though); the use of geotextile may be avoided and the slope is reduced to $3.5:1$ (the $FS = \tan 18/\tan 15.9 = 1.1$); an anchor trench is designed to withstand a higher pull from the geotextile ($\delta = 10°$ should be used in the design) and the strength of the geotextile and synthetic membrane should be checked [Eq. (19.29) can be used (assuming $T = 0$ and $FS = 1$) to find the stress and Eq. (19.28) can be used for the anchor trench design].

19.3.3.1   General Equation for Cover Soil Sliding   An equation correlating effects of slope angle, strength properties and thickness of cover soil, seepage forces and equipment load with the properties of the synthetic membrane has been proposed by Druschel and Underwood (1993). The maximum allowable pull (F_p) is given by (see Fig. 19.20)

$$F_p = F_b + F_s + \frac{(W_e + W_2)\sin(\beta - \delta)}{\cos \delta} - \frac{W_1 \sin \phi}{\cos(\beta + \phi)} \qquad (19.31)$$

in which F_b = equipment braking force, F_s = seepage force, W_e = weight of vehicle on the side slope, W_1 = weight of soil anchorage at base, W_2 = weight of cover soil on the side slope, and β = slope angle of berm or liner. F_b can be assumed to be 30% of W_e; thus

$$F_b = 0.3W_e \qquad (19.32)$$

and

$$F_s = \frac{\rho_w h_s^2 \tan \phi}{2 \sin \beta \cos \beta} + \frac{\rho_w h_s}{\sin \beta}\left(d - \frac{h_s}{2 \cos \beta}\right)\cos \beta \tan \delta \qquad (19.33)$$

in which h_s = seepage thickness (this will be maximum after a rain event or during spring thaw when the entire side slope soil will be wet), ρ_w = unit weight of water. W_1 and W_2 can be calculated using the following equations:

$$W_1 = \frac{\gamma_s(h')^2}{2 \sin \beta \cos \beta} \qquad (19.34)$$

$$W_2 = \frac{\gamma_s h'}{\sin \beta}\left(d - \frac{h'}{2 \cos \beta}\right) \qquad (19.35)$$

19.3.4   Check for Uneven Settlement

Uneven settlement may occur in landfill cover. However, the size of the set-tlement is an unknown that cannot be predicted. The maximum permissible size of curvilinear settlement may be estimated using the theory of cables. However, synthetic membranes are into elastic material, and the stress devel-oped due to a deflection is expected to be somewhat less than that predicted by Eq. (19.36). The shape of a cable deflected due to a uniform load is parabolic. The maximum stress would occur at M and N (Fig. 19.22) and its magnitude is given by (Ogden and Ripperger, 1981)

$$\sigma_y = H(1 + 16S_r^2)^{1/2} \tag{19.36}$$

in which

$$H = \frac{(w + \gamma_s)h'L^2}{9d_f}$$

S_r = the sag ratio = d_f/L, L = the diameter of a depression, and d_f = max-imum deflection.

Equation (19.36) can be used to estimate the permissible size of a depres-sion; or, if the depression size is known, that σ_y can be estimated. Every effort should be made to avoid the formation of a depression when using a synthetic membrane.

19.4   BERM DESIGN

Berms around a landfill should be checked for stability. Various approaches are available for checking stability of a berm, which essentially deals with slope stability. One or more chapters will be needed to do justice to the topic, which makes it somewhat beyond the scope of this book. A basic concept and a few references are cited that should be considered as a complement to this section. Designers dealing with a difficult situation should consult the current literature on slope stability analysis.

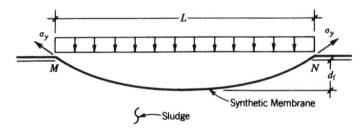

FIG. 19.22   Force schematic for the deflection of synthetic membrane.

Failure along a circular arc is assumed for most analysis (Fig. 19.23). The berm within the circular arc is subdivided into slices and analyzed by equilibrium (Bishop, 1955; Bishop and Morgensterm, 1960; Skempton, 1964). A graphical method known as the friction circle method is also available (Taylor, 1948). The method of slices using a modified Bishop's method (Bishop, 1955) provides good accuracy. Analysis can be made easier by using a hand calculator or computer (Whitman and Bailey, 1967; Little and Price, 1958). A computer disk for use in a personal computer is also available (Bosscher, 1987). Summary discussions on the topic can be found elsewhere (Morgenstern, 1992; Lambe and Silva-Tulla, 1992). The height of the berm, climatic condition, effective angle of internal friction (ϕ), effective cohesion (C'), and unit weight of the berm material influence the stability of the side slope. The effect of earthquakes on the structural stability of a berm constructed in earthquake-prone regions should also be investigated (Marcuson et al., 1992). Although rigorous analyses using correct material properties is suggested for checking the stability of berms, a 2–2.5 horizontal (H) to 1 vertical (V) side slope may be considered structurally safe for 3–4 m (10–13 ft) high berms made of sandy soil (not applicable for earthquake-prone regions).

Apart from ensuring the structural stability of inboard and outerboard side of a berm, the outboard side of a berm should be protected from erosion due to wind and water. Freeze–thaw also causes deterioration of the slope face. Soil erosion depends on rainfall intensity, length (along the slope) and gradient of the slope, status of vegetation on the berm, and soil type. The length and slope of a berm should be designed to minimize erosion. The erosion of bare soil is much higher for a 3H : 1 V compared to a 4H : 1 V slope (Soil and Conservation Service, 1972). The length of the slope may be reduced by constructing a horizontal bench on the (outboard side) berm and by constructing drainage ditches. These measures should be undertaken if the slope length exceeds 20 m (66 ft). Growth of vegetation retards soil erosion (Gray and Leiser, 1982). The effect of freeze–thaw is prominent in berms with no

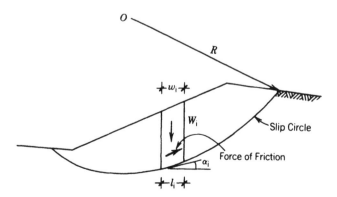

FIG. 19.23 Geometry and forces in "method of slice" analysis.

vegetation. To account for soil erosion the outboard side of berm slopes should be made less steep than what is calculated by structural stability analysis. A 3V : 1H slope for a 3–4 m (10–13 ft) high berm (made of sandy soil) may be used in the absence of a detailed analysis. Note that the suggested side slope is less steep than mentioned in the above paragraph; the reduction in slope is to account for the erosion losses.

19.5   LANDFILL STABILITY

The stability of new landfills as well as during lateral and vertical expansions should be assessed for both static and dynamic conditions. This section includes discussion of static conditions. Landfill stability under dynamic conditions due to seismic forces are discussed in Section 19.6. Dynamic forces generated due to blasting should also be considered if necessary. Landfills may become unstable during construction, operation, and after closure. The following are the primary failure modes, which cause loss of landfill stability (Mitchell and Mitchell, 1992; USEPA, 1999):

1. Failure of landfill berms (see Section 19.4)
2. Failure of drainage layer and geosynthetic membrane on the side slope (see Sections 19.3.1, 19.3.2, 19.3.3, and 19.3.3.1)
3. Failure of landfill foundation
4. Failure of waste slope
5. Failure due to human activities in the vicinity of landfills
6. Failure due to collapse of sink holes

Landfills may become unstable due to both soil properties as well as properties of synthetic materials used for landfill construction. Both rotational and translational slides (Fig. 19.24) are involved in landfill failures.

Landfill foundation may fail due to the presence of a soft layer below the liner. Failure could be due to excessive settlement or due to lack of shear strength. Both trench type (Figs. 11.1 and 11.2) and at grade (Fig. 11.4) landfills should be checked for foundation failure if a soft soil layer is encountered during subsoil investigation. A check for both settlement and bearing capacity failures should be made.

Waste face can fail due to lack of stability of the slope. Sliding of the waste mass as a whole may also take place. Case histories of landfill failure due to sliding of the waste mass have been reported in the literature (Byrne et al., 1992; Stark, 1999). Both short-term and long-term analysis, should be undertaken to check stability of the waste face (Howland and Landva, 1992). Stability of the waste face should be checked both from an operational as well as a design standpoint.

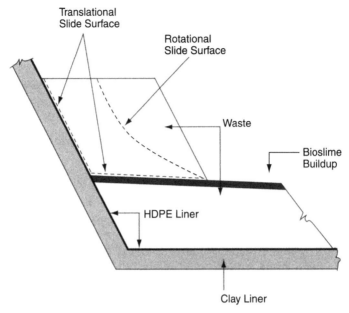

FIG. 19.24   Probable rotational and translational slide surfaces.

In a canyon-type landfill (Fig. 11.3), in addition to checking stability of the waste face during operation and at closure, the stability of the canyon itself should also be checked.

The problem in analyzing a waste slope is to obtain proper shear strength parameters for the waste and to develop an applicable analysis. For relatively homogeneous waste, such as foundry sand, the standard tests may be used to obtain the value of ϕ' and C'. However, for most putrescible waste types the standard tests are not applicable for the following reasons: (1) many sludge-type wastes (e.g., papermill sludge) release gas during the test, which will influence the test results; (2) putrescible wastes may deteriorate in the landfill leading to a change in the shear properties; (3) some wastes are extremely heterogeneous (e.g., municipal garbage) and hence small samples cannot provide the shear properties of the waste as a whole; and (4) it is difficult to compact a waste in the laboratory.

Stability failures in bioreactor landfills may occur during operation. The failure surface may be above or below the geomembrane. While analyzing the slope stability of bioreactor landfills, reduction of pressure due to liquid buildup within the waste mass should be taken into consideration. As mentioned in Chapter 17, slime buildup along a sloping daily or intermediate cover may lead to sliding failure of the waste mass above the plane. A block analysis should be used for analyzing these situations.

Since most of the wastes are heterogeneous, strictly speaking the theories mentioned in Section 19.4 are not applicable. Thus, analyzing the stability of

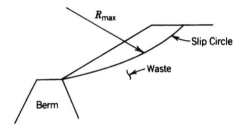

FIG. 19.25   Maximum allowable slip-circle for analyzing a waste slope.

a waste slope is somewhat difficult. A combination of theoretical analysis and field study should be undertaken to investigate the stability of waste slopes (Bagchi, 1987a). The stability analysis theories mentioned in Section 19.4 may be used if a higher factor of safety (1.5–2) or lower values of ϕ' and C' are used in arriving at a stable slope angle. While checking the final waste face care should be taken to see that the entire slip circle passes through the waste only and no part intercepts the berm (Fig. 19.25). The shear strength properties (ϕ' and C') obtained from standard laboratory tests should be reduced by 15–25%. Observations regarding stable waste slopes of existing landfills for the same waste type (if available) should be taken into consideration. Several papers discussing various issues of this topic can be found elsewhere (ASTM, STP1070, 1990). The following waste slopes have been observed to be stable:

Municipal waste: 3H : 1V to 4H : 1V

Sludge-type waste with minimum 40% solid content: 8H : 1V

Fly-ash-treated sludge: 7H : 1V to 6H : 1H

Sandy foundry waste and fly ash: 3V : 1H to 4V : 1H

Usually a two-dimensional (2D) approach is used for slope stability analysis. A 2D analysis assumes that the slopes are infinitely wide. In general, a 2D analysis provides a conservative estimate of a factor of safety (Duncan, 1992). However, a three-dimensional (3D) analysis should be used for situations involving complicated topography, nonuniform shear strength, uplift pressure due to perched liquid condition, and gas pressure. In these situations the factor of safety is influenced by the direction of movement. A 3D slope stability analysis can combine the effects of directional variation of slope geometry and shear strength (Stark and Eid, 1998). Such 3D analysis is very helpful for bioreactor landfills and landfill expansions involving two different types of liner system.

The following issues should be taken into consideration when selecting a slope stability software package:

1. Whether the software can analyze both rotational and transnational failure
2. Whether nonlinear failure envelops can be used for shear strength characteristics
3. Whether perched liquid head can be analyzed
4. Whether the analysis can address gas pressure conditions
5. Whether both 2D and 3D analysis can be done

Periodic field reconnaissance survey of active and closed landfills, especially in those landfills where stability analysis was not done, can help avoid a slope failure. Cracks near the head of a slide are usually normal to the direction of the horizontal movement, whereas the cracks along its edges are usually parallel to it (Stark, 1999). In landfills, cracks can appear due to both settlement and slope instability. In general, reappearance of cracks in the same location within a short period of time indicates slope instability. Settlement cracks take a long time to reappear because the biological processes that cause landfill settlement require a long time. In rotational failure walls of cracks are slightly curved and concave toward the direction of movement. In addition, for rotational failure the crack is horseshoe shaped in plan (Varnes, 1978). The field reconnaissance survey may include the following:

1. Inspection of toe of landfills for unplanned excavation (applicable to active landfills), erosion, uplift, lateral displacement, or seepage. Bulging of the toe may also point toward a near-term slope failure.
2. The top surface of landfills for cracks and unusual settlements. The shape, locations, and depth of cracks should be noted carefully; the information is helpful for predicting instability.
3. Inspection of the area surrounding the landfill to identify any human activities (e.g., deep excavation near a landfill), which may impact the stability of landfill slopes.
4. Assessment of liquid withdrawal rate (groundwater, natural gas and petroleum) from any shallow aquifer, which runs below the landfill.

Cracks should be instrumented following detection to assess rate and direction of crack propagation, if any. Catastrophic slope failure can be avoided if signs of failures are detected at an early stage.

The regulatory agency should be contacted to check whether the rules call for a mandatory maximum permissible waste slope. Proper geotechnical investigation should be undertaken at all sites before siting a landfill (refer to Chapter 12 for details).

19.6 COMMENTS ON SEISMIC DESIGN OF LANDFILLS

Seismic design of landfill elements is an emerging field in which standards of practice are evolving rapidly. The analytic techniques used for involved and are somewhat beyond the scope of this book. Although it appears that seismic design of landfill elements is undertaken in some cases, there is a lack of literature dealing with the topic. The purpose of this section is to familiarize readers with the current trends in design approaches rather than providing methods of analyses. A designer needs to address the following issues that are associated with seismic analysis:

1. Selection of appropriate design ground motion
2. Evaluation of the effect of selected ground motion on the proposed site
3. Estimation of the dynamic properties and shear strengths of waste fills, liner materials, and subgrade materials
4. Checking the seismic stability of the landfill elements due to the selected ground motion

A regulatory agency should be consulted to choose the proper ground motion. In general, a landfill should not be located on or near (within 60 m) a known fault. In an earthquake-prone zone seismic investigation should be undertaken to ensure the absence of a fault under the proposed site. Although primary faults are mapped, secondary faults are not well mapped. USEPA requires that landfills located in seismic impact zones should be designed to resist the maximum horizontal acceleration (MHA) for the area. A seismic impact zone is defined as an area that has 10% or more probability of exceeding MHA of $0.1g$ in 250 years. The dynamic response analysis should be performed to check the stability of subgrade strata and/or natural slope faces. The estimation of dynamic properties of soil materials can be easily done using available laboratory tests. Pseudostatic stability analyses (Newmark, 1965; Makdisi and Seed, 1978) is generally used to check dynamic stability. Use of a computer program (Schnabel et al., 1972) for this purpose has also been reported (Seed and Bonaparte, 1992).

Although Singh and Murphy (1990) reported dynamic characteristics of waste fill, in general there are few data on this important issue. Seed and Bonaparte (1992) suggested the use of back analysis from field recordings of waste fills to obtain reliable data. Anderson et al. (1992) reported a case history using this approach.

Seismic design of landfills involves checking the stability of both the waste face and the liner. Stability of the liner (especially synthetic liners) dictates whether a failure will occur. Use of same shear strength properties of the waste mass for analyzing both static and dynamic loadings is reasonable, although the approach is conservative. The performance of 10 natural attenuation type of landfills after an earthquake in California (acceleration range $0.1-0.45g$) was studied by Orr and Finch (1990). Relatively minor surface damage was observed, indicating good attenuation and damping properties of waste fills in general. However, subtle failure such as localized rupture of clay or synthetic membrane liner or final cover can lessen the integrity of the landfill.

The mode of failure due to the ground motion during an earthquake will induce a sliding block type of failure along the liner–waste interface (Singh, 1992). This drag will cause a strain on the clay liner resulting in rupture. For synthetic membrane liners the dynamic properties of interface materials can be established using conventional laboratory tests. However, for clay one needs to find the limiting permanent deformation beyond which the clay will crack or rupture. The development of cracks, in clay will depend on the magnitude of the sliding shear force and stress–strain characteristics of the compacted clay. The differential settlement of subsoil due to an earthquake will make a clay liner act like a beam, resulting in tension cracks at the bottom of the liner. Thus from a seismic design standpoint, use of a more plastic clay in constructing the clay liner appears to be preferable. It should be noted, however, that highly plastic clay is difficult to compact and chances of developing desiccation cracks are also higher. Table 19.7 provides a summary of possible earthquake design problems and suggested solutions for avoiding failure of the landfill.

19.7 ACCESS ROAD DESIGN

Roads, both within and outside of a landfill, are important in maintaining the smooth operation of a landfill. The road within the landfill should be designed

TABLE 19.7 Selected Techniques to Improve Landfill Stability in Seismic Impact Zones

Problem Area	Suggested Solutions

Liquefaction of foundation soils	Densify foundation soils
	    Dynamic compaction
	    Vibrocompaction
	    Compaction grouting

Inadequate slip resistance in liner/leachate collection system	Increase strength of soil liner components (use different soil, revise compaction specification) Increase interface strength (geomembranes) Reduce grading Change overall bottom geometry
Potential slip in waste	Alter operations to increase strength of waste (increase compaction or soil bulking) Alter cell configuration Institute interior soil berms Institute or increase size and/or strength of perimeter dikes
Slip of landfill cover	Increase strength of soils/materials Increase interface strength (geomembranes) Reduce landfill slopes

After Weiler et al. (1993). Copyright 1993 by National Solid Waste Management Association. Reprinted with permission.

so that dumping vehicles can move in and out easily. A typical cross section of a road within the landfill is shown in Fig. 19.26. Both geometric and structural design of road(s) outside of the landfill area should be undertaken. Sufficient sight distance, merging, and an exit lane to and from the primary public road should be provided. The vehicles entering and exiting from the landfill do not maneuver at high speed so a rigorous design calculation is not needed. A typical design for a landfill entrance is shown in Fig. 19.27. Caution sign(s) should be posted on the public road regarding the existence of a landfill and/or heavy vehicles entering the road.

Some arrangements should be made to minimize spreading of waste on the public road and off the landfill area, which sticks to the wheels of dump trucks. Washing of tires may be needed in hazardous waste landfills. Running the trucks over a rough gravel road or constructing a grate or a series of bumps in a portion of the exit road are some of the approaches used with partial success to reduce waste spreading.

The structural design drawing of the access road should indicate the thickness of the base and subbase coarse material and the type of surfacing. Flexible pavements are used for the access road. Palmer and Barber (1940) and Barber (1946) developed a method for designing pavement thickness using Boussinesq's equation (Kansas State Highway Commission, 1947). The pavement thickness (t_p) is given by

$$t_p = \left[\sqrt{\left(\frac{3 A_t mn}{2 \pi E_s \Delta} \right)^2 - a^2} \right] \left[\sqrt[3]{\left(\frac{E_s}{E_p} \right)} \right] \tag{19.37}$$

in which A_t = the total weight of the vehicle, m = the traffic coefficient, n = the saturation coefficient, E_s = the modulus of elasticity of the subgrade, Δ = the allowable deflection (= 0.1 in.), a = the radius of contact of the tire, and E_p = the modulus of elasticity of the pavement or surface course. In the absence of experimental values the following approximate values (in F.P.S. units) may be used for the design: m = 0.5, n = 0.5 for an average annual rainfall of 15.0–19.9 in. and 1.0 for a rainfall of 40–50 in. n increases by 0.1 for each 5-in. increase in average annual rainfall (i.e., n = 0.6 for an average annual rainfall of 20–24.9 in.); E_s = 5000 psi for gravel and 1500 psi for clay, E_p = 15,000 psi, and a = 6 in.; a granular base coarse material reduces frost heave.

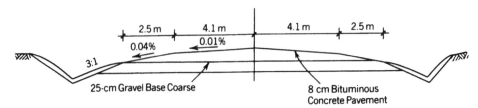

FIG. 19.26 Typical structural design of a landfill access road.

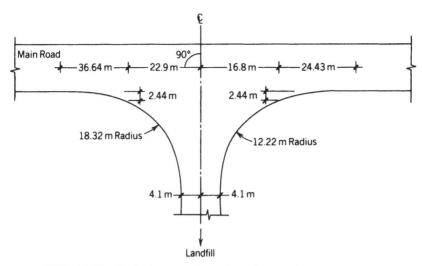

FIG. 19.27 Typical geometric design of a landfill entrance road.

If a granular base coarse material is to be used for a portion of the bituminous mat, then the thickness of the base coarse material (t_b) can be calculated as

$$t_b = (t_p - t_s) \sqrt[3]{\frac{E_p}{E_b}} \qquad (19.38)$$

in which t_s = thickness of wearing surface (usually 5–15 cm (2–6 in. thick) and E_b = modulus of elasticity of the base coarse material. Several other design approaches are available for the design of flexible pavements (Yoder, 1967).

Example 19.9 (in F.P.S. units)   Design a landfill access road for a region having an annual average rainfall of 30–34.9 in. and a single wheel load of 20,000 lb. The modulus of elasticity of a base coarse gravel is 5000 psi and that of the subgrade clayey soil is 500 psi.
   From Eq. (19.37)

$$t_p = \left[\sqrt{\left(\frac{3 \times 20,000 \times 0.5 \times 0.8}{2\pi \times 500 \times 0.1} \right)^2 - 6^2} \right] \left[\sqrt[3]{\left(\frac{500}{1500} \right)} \right]$$

$$= 11.5 \text{ in. of bituminous mat directly on the clayey subgrade}$$

To include a 2-in. surface and 6-in. of gravel base, the thickness of subgrade coarse material is calculated as follows from Eq. (19.38):

$$\text{Thickness of base coarse gravel} \times (11.5 - 2) \sqrt[3]{\left(\frac{1500}{5000}\right)}$$

$$= 13.7 \text{ in.}$$

Hence, the required

$$\text{Thickness of subgrade} = (13.7 - 6) \sqrt[3]{\left(\frac{5000}{1500}\right)}$$

$$= 11 \text{ in.}$$

Thus the road section will consist of 2 in. of wearing surface, 6 in. of base coarse gravel, and 11 in. of subgrade.

19.8   LANDFILL COVER DESIGN

The purpose of a landfill cover is to minimize infiltration of water into the landfill. A multilayer configuration is used for a cover in which each layer has a task to perform (Fig. 19.28). The first layer, called the grading layer, should consist of coarse-grained material and is usually 15–60 cm (6–24 in. thick). The thickness depends on the stability of the waste and gas collection system design. For an unstable waste surface, a layer of bark or a geotextile

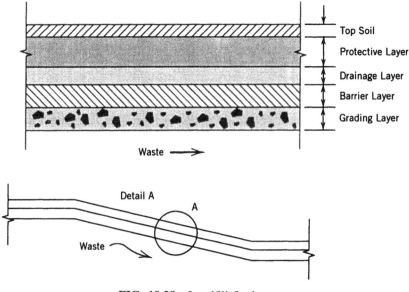

FIG. 19.28   Landfill final cover.

may be used below the grading layer. The purpose of the grading layer is to provide a stable surface on which the low permeability layer can be constructed and to facilitate landfill gas venting. The second layer, called the barrier layer, provides a barrier for water infiltration. The material used can be clay, synthetic clay liner, or synthetic membrane. This is an important layer that must be constructed carefully. A low permeability layer must be constructed below a synthetic membrane layer. The purpose of the third layer, called the protective layer, is to protect the barrier layer from freeze–thaw and desiccation cracks and to provide a medium for root growth. The thickness of this layer should be sufficient to perform both the above tasks; the thickness of this layer should be between 30 and 105 cm (1 and 3.5 ft) depending on geographic location. Although incorporation of this layer in the final cover may increase infiltration, it is considered as an important element for the tasks it is supposed to perform.

As indicated above, a low permeability layer (i.e., either a clay or synthetic clay layer) should be used below a synthetic membrane in the barrier layer. Usually a 60-cm-thick clay layer or a layer of synthetic clay is used for this purpose. However, only a 60-cm-thick clay layer may be used as the barrier layer. If the liner is a composite liner, then the barrier layer should also be a composite layer.

The drainage layer may be constructed using coarse sand (5–15 cm thick) or synthetic net. The drainage layer is an important component of the final cover if a synthetic membrane is used at the top of the barrier layer. The purpose of this drainage layer is to provide better drainage of the protective layer so that the interface of the protective layer and the synthetic membrane does not become saturated. In general, the friction angle of the fine-grained soil–synthetic membrane interface decreases due to saturation of the fine-grained soil. A decrease in the interfacial friction angle may result in unstable soil conditions leading to failure or increased erosion of the protective layer. If the permeability of the protective layer is high (1×10^{-3} cm/sec or more), then it will drain quickly, whereby reduction in the interfacial friction angle due to saturation is not expected. So use of the drainage layer to avoid a reduction in the interfacial friction angle due to saturation in landfills with a high permeability protective layer is not justifiable. However, use of a layer of sand or synthetic net will protect the synthetic membrane from damage during placement of the protective layer. If the soil used for the protective layer is sandy, then a separate drainage layer may not be essential. If the soil used for the protective layer has relatively low permeability and a synthetic membrane is used below the protective layer, then a drainage layer should be used.

Example 19.10 (in F.P.S. units)   Check the stability of the various components of the final cover configuration shown in Fig. 19.28. The thicknesses of the various layers are as follows: (1) top soil, 6 in.; (2) protective layer of silty sand, 2 ft; (3) barrier layer, 60 mils of textured HDPE over a 2-ft compacted clay layer; (4) grading layer, 6 in. of sand; (5) drainage layer is a geocomposite net (a synthetic net with a synthetic textile on either side). The interfacial friction angles are (1) protective layer/geocomposite, 25°; (2) geocomposite/textured HDPE, 32°; (3) textured HDPE/clay layer, 30°; (4) clay layer/grading layer, 34°. Undrained cohesion and friction angles of the clay samples compacted to 90% modified Proctor were 500 psf and 15°, respectively. The maximum slope of the final cover is 3.5H : 1V. The length of the slope is 200 ft. The average unit weight of the soil is 130 pcf. The undrained friction angle of the silty sand is 28° and the cohesion is zero.

For drained conditions the factor of safety (FS) of the final cover components can be calculated using the infinite slope method. The infinite slope method neglects passive resistance near the toe. This method is valid for long slopes:

$$FS = \frac{\tan R}{\tan A} \tag{19.39}$$

in which R = the required friction angle, and A = the slope angle = 16° for a 3.5H : 1V slope. Usually an FS between 1.25 and 1.5 is used to calculate stability. The lower value is used when confidence on the data (e.g., interfacial friction angle) is high and the consequences of failure are not catastrophic. For importance structures a high FS is used. Using an FS of 1.4, R is calculated from Eq. (19.39) as 21.2°. Since all the interfacial friction angles are higher than 21.2°, the final cover components are stable under drained conditions.

Undrained conditions may occur either in the protective layer or in the clay layer. The shear strength of soil for this case is given by

$$S_u = \sigma_n \tan \phi_u + C_u \tag{19.40}$$

in which S_u = the undrained shear strength of the soil layer, ϕ_u = the undrained friction angle of the soil, and σ_n = normal pressure. The FS for this situation is given by

$$FS = \frac{S_L \times S_u}{S_L \times \gamma_s t \sin A}$$

$$= \frac{S_u}{\gamma_s t \sin A} \tag{19.41}$$

FS = 1.4, t = 2.5 ft (2 ft protective layer + 6 in. top soil), A = 16°, and γ_s = 130 pcf.

$$1.4 \times 130 \times 2.5 \sin 16 = 2.5 \times 130 \cos 15° \tan \phi_u + C_u$$

or

$$125 = 312 \tan \phi_u + C_u$$

Table 19.8 is developed for various combinations of ϕ_u and C_u. Since the undrained friction angle of the silty sand protective layer is 28°, a failure is not expected.

To check the failure of the clay layer, a t value of 4.5 needs to be used. In this case from Eq. (19.41)

$$S_u = 1.4 \times 4.5 \times 130 \sin 16 = 226$$

now

$$S_u = \sigma_n \tan \phi_u + C_u$$

$$= 4.5 \times 130 \cos 16 \tan 15 + 500$$

$$= 650.7$$

Since the actual S_u is more than the required S_u, the clay layer will be stable. (Note: Usually the interfacial friction angle between a synthetic material and a soil is 90% of the internal friction angle of the soil.)

Finally, 10–15 cm (4–6 in.) or organic soil should be spread on top of the protective layer to facilitate seed germination and help the growth of vegetation. Necessary lime and nutrient should be applied at least for the first 5 years to help vegetative growth. Vegetation is helpful in reducing soil erosion, increasing structural stability of the cover, and reducing infiltration by increasing evapotranspiration. A horticulturist should be consulted to help choose proper species, which can grow vigorously without cultivation.

TABLE 19.8 Combination of ϕ_u and C_u Values for Example 19.10

ϕ_u (in degrees)	C_u (in psf)
0	125
4	103
8	81
12	58
16	35
21.8	0

19.8.1 Clay or Amended Soil as Barrier Layer

The properties of these materials are discussed in Chapter 18 and construction-related issues are discussed in Chapter 20. The thickness of this layer depends on its ability to provide a low permeability layer to reduce infiltration. From principles of soil mechanics it can be determined that a thin barrier layer is enough to provide overall low equivalent permeability of the cover. However, a thicker layer is needed for construction-related issues. A 60-cm (2-ft)-thick layer for clay and a 30-cm (1-ft) layer for bentonite-amended soil is considered to be the minimum. A synthetic clay liner may also be used as a barrier layer. A thicker layer will perform better in case of differential settlement of the waste.

Installation of a filter layer, to minimize migration of fines from the clay layer to the underlying coarse-grained soil, should be considered, especially if the layer is thin (30 cm). The soil filter criteria mentioned in Section 19.1.1 may be used to design this layer. As an alternative, a geotextile layer may be laid over the grading layer where needed.

19.8.2 Synthetic Membrane as Barrier Layer

The properties of these materials are discussed in Chapter 18 and construction-related issues are discussed in Chapter 20. A thickness of 1 mm (40 mils) is considered minimum but 1.5–2 mm (60–80 mils) is preferable. Provision should be made in the membrane regarding installation of a gas extraction system in the future (refer to Section 20.2). If the soil used for the protective layer has a significant amount of 150-mm (6-in.) particles, then use of a geotextile is recommended. In such cases, the side slope should be reduced or a special synthetic membrane with high surface friction should be used. A low permeability layer must be constructed below a synthetic membrane liner to minimize infiltration in case of a leak in the membrane.

19.8.3 Alternative Covers

Cover designs, other than the configuration shown in Fig. 19.28 may be used for arid and semiarid regions. These alternative cover designs are based on utilization of storage capacity and unsaturated flow characteristics of fine-grained soil, and evapotranspiration. Two alternative cover designs have been reported in the literature: capillary barrier and monolayer barrier.

19.8.3.1 Capillary Barrier   This alternative cover design (Fig. 19.29) consists of a fine-grained layer overlying a coarse-grained layer (Benson and Khire, 1995). The contrast in particle size restricts downward flow of water due to the difference in matrix suction at the interface of the two layers. The fine-grained top layer becomes saturated before allowing any downward migration of water. For low annual precipitation part of the water is stored in

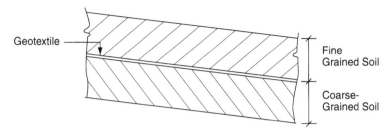

FIG. 19.29 Capillary barrier.

the upper fine-grained layer, and part is diverted laterally in the surface layer; the water stored in the upper layer is depleted through evapotranspiration (Khire et al., 1999; Stormont, 1995). For areas with low precipitation, the upper fine-grained soil layer will act as a reservoir for moisture, and the underlying layer will act as a barrier to water percolation by not allowing moisture from the upper layer to pass through it. Both pilot-scale experiments and field studies (Nyhan et al., 1990; Benson and Khire, 1995) have shown that the two-layer design is effective in arid and semiarid regions. Performance of capillary barriers are better if the hydraulic conductivity of the upper layer is low and the lower layer is of a uniform coarse material (Benson and Khire, 1995; Stormont and Anderson, 1999). Clayey silt, silty sands, and sandy clays are expected to be good soil types for the upper layer. Although a three-layer capillary barrier has been suggested for regions where potential for shrinkage cracks is high (Yeh et al., 1994), in reality shrinkage cracks will allow percolation of significant precipitation volume. Inclusion of a biota barrier layer is helpful in protecting the upper layer from burrowing animals and woody plants (Benson and Khire, 1995).

A geotextile or fiberglass geotextile may be used to protect the lower layer from intrusion of fines from the upper layer (Qian et al., 2002). To minimize percolation through the final cover during high precipitation, more layers with contrasting grain size may be used; these layers are expected to divert infiltration via lateral flow (Nyhan et al., 1990; Yeh et al., 1994).

19.8.3.2 Monolayer Barrier Monolayer barriers consist of a thick layer of fine-grained soil with a top soil (Fig 19.30). The top soil must be well vegetated. The thick fine-grained lower layer stores water that is used by the vegetation for evapotranspiration (Benson and Khire, 1995). The literature reports a field study demonstrating the effectiveness of the monolayer (Qian et al., 2002). Since in monolayer barriers evapotranspiration is the primary way of minimizing percolation, existence of vegetation year round is an important issue. Thus it will not be effective in cold regions or in extremely hot regions where maintaining good vegetative growth year round is not possible. In addition, proper postclosure maintenance programs must be included to maintain the cover.

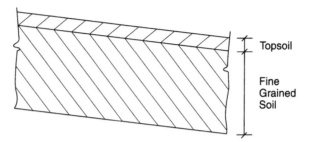

FIG. 19.30   Monolayer barrier.

19.9   GAS VENTING SYSTEM DESIGN

Only venting systems are discussed in this section; design of systems for collection and use of landfill gas is not included although the extraction system design included in Section 19.8.2 can provide some input. It is important to decide at the beginning whether an active or passive system will be utilized for venting the gas. Contrary to popular belief, a passive venting system cannot be converted to an efficient active venting system simply by connecting the vents with a header pipe and by installing a blower at the end of the header pipe. Only a series of deep gas extraction wells, each connected to a header pipe and placed suitably in the landfill, can provide efficient withdrawal of gas from a landfill. The following issues need to be considered for choosing one system or the other:

1. Landfill design: chances of gas migration are higher from natural attenuation type of landfills than from containment-type landfills.
2. Type of soil surrounding the landfill: gas migration can occur more easily through sandy soil than through clayey soil.
3. Distance of usable closed space (homes, warehouses, etc.) near the landfill. Landfill gas can migrate 150 m (500 ft) or more. Any usable closed space within 300 m (1000 ft) of a landfill should be monitored for methane gas concentration.
4. Possibility of future use of the landfill.
5. Regulatory mandate: the regulatory agency may mandate the type of gas venting system to be used in a landfill.
6. Waste type: gas generation depends on waste type. Gas venting is essential for putrescible waste.

19.9.1   Passive Venting System

Such systems are installed where gas generation is low and off-site migration of gas is not expected. Essentially passive venting is suitable for small municipal landfills (40,000 m³) and for most nonputrescible containment-type landfills. The system may consist of a series of isolated gas vents (Fig. 19.31). No design procedure is available to calculate the number of vents required, but one vent per 7500 m³ (~10,000 yd³) of waste is probably sufficient. Sometimes these isolated vents are connected by a perforated pipe embedded in the grading layer (Fig. 19.32).

19.9.2   Active Venting System

An active venting system consists of a series of deep extraction wells con-nected by a header pipe to a blower that either delivers the gas for energy reuse purposes or to an on-site burner or simply releases it to the atmosphere. Microturbines may also be used for energy recovery from landfill gas (Whe-less and Wiltsee, 2001). Whether the gas can be released to the atmosphere without burning depends on the following:

1. *Chemical constituents of the gas.* If hazardous air contaminants such as vinyl chloride or benzene are present, then burning the gas is the pre-ferred option. If such contaminants are absent, releasing the gas to the atmosphere may be acceptable in some (but not all) situations. In ad-dition the regulatory agency should be contacted to determine whether burning landfill gas is mandatory.

2. *Landfill location.* If the landfill is located near/within a community, then burning is necessary because methane has an odor that may create a nuisance condition.

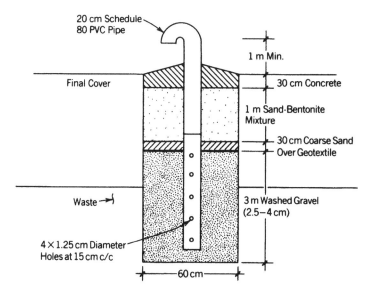

FIG. 19.31   Typical detail of an isolated gas vent.

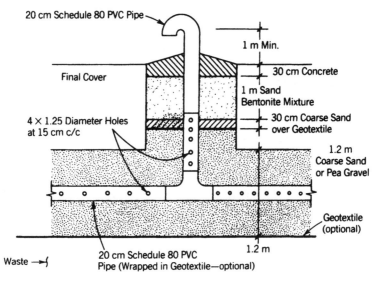

FIG. 19.32   Typical detail of a passive gas venting system with a header pipe.

A typical layout of an active venting system is shown in Fig. 19.33, which includes the elements of the system. Designs of these elements are included in the following paragraphs. A detail design is included in Example 19.11.

Extraction Well   Spacing of extraction wells is a key issue in extracting landfill gas efficiently. They should be spaced such that their zone of influence overlaps. As shown in Fig. 19.34, a 27% overlap can be obtained by installing the extraction wells on the corners of an equilateral triangle of side $1.73R$ and a 100% overlap can be obtained by installing the extraction wells on the corner of a regular hexagon of side R. A square array would provide a 60% overlap. Thus, spacing of extraction wells is given by

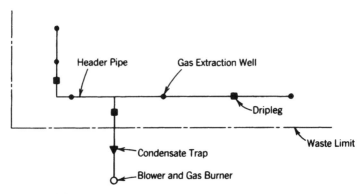

FIG. 19.33   Typical layout of active gas venting system elements.

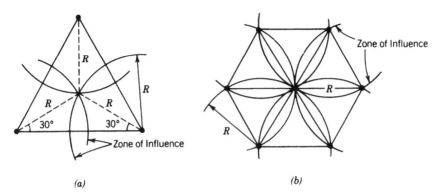

FIG. 19.34 Positioning of gas extraction well for complete overlap: (a) triangular array and (b) hexagonal array. Solid circles indicate locations of gas extraction wells.

$$\text{Spacing} = (2 - O_1/100)R \tag{19.42}$$

in which R = the radius of influence of gas extraction wells and O_1 = the required overlap.

The zone of influence of a gas extraction system should be determined from actual field study. An extraction well should be installed within the landfill with gas probes at regular distances from the well (Fig. 19.35). Short-term and/or long-term testing is done to design an efficient withdrawal system. Short-term extraction tests usually run for 48 hr to several days. A short-term test is sufficient where the intention is to design an extraction system to minimize landfill gas migration.

A long-term test is used to simulate full recovery project conditions. The suggested probe spacing is shown in Fig. 19.35. The extraction wells should penetrate 80–90% of the refuse thickness and lower 70–80% of the well should be perforated. The well should be pumped for at least 48 hr and then pressure at all the probes should be monitored for 3 consecutive days, at least twice a day. The probes nearest to the well show highest negative pressure, which drops rapidly with distance. The radius of influence is that radius at which the negative pressure is nearly zero. In the absence of test data a radius of influence between 45 and 67 m may be used. The radius of influence depends on the cover type, depth of landfill, age, and composition of waste. The lower value is used for relatively shallow landfills [15 m (50 ft) or less] where the final cover does not have a synthetic membrane barrier layer. Figure 19.36 shows details of a typical gas extraction well. A well with arrangements for separating leachate is used where a relatively high leachate head buildup is expected [1.2 m (4 ft) or more].

Header Pipe Nonperforated 15- to 20-cm diameter plastic pipes are used to connect the extraction wells to a blower. The diameter of the pipe may be increased to reduce head loss due to friction. These pipes are embedded in

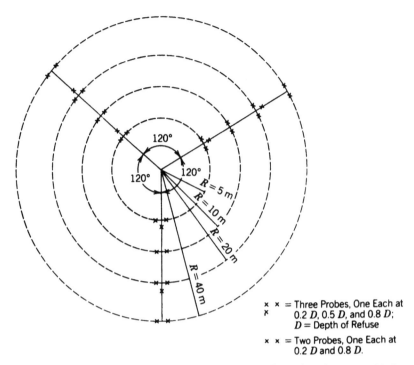

FIG. 19.35   Gas extraction well and probe cluster configuration for zone of influence determination.

trenches filled with sand (Fig. 19.37). If leachate is extracted from the landfill along with gas, then the leachate discharge pipe can also be embedded in the same trench. PVC or HDPE pipes are used for header pipes. The header pipe must not be perforated. The head loss through these holes is high and a significantly high capacity blower will be needed without much increase in extracted gas volume.

Blower   The blower should be installed in a shed at an elevation slightly higher than the end of the header pipe to facilitate condensate dripping. The size of the blower is designed based on total negative head and volume of gas to be extracted. Note that a three-phase electrical connection is needed for most motors with horsepower of 5 or more. If a three-phase electrical connection is not available on-site, then the system should be designed with multiple blowers each requiring less than a 5-hp motor. An approach for calculating the blower size is enumerated in the example at the end of this section.

Condensate Removal   Landfill gas is saturated, which forms condensate due to a reduction of temperature while moving through the header pipe. Arrangements need to be made to remove the condensate before the gas enters the

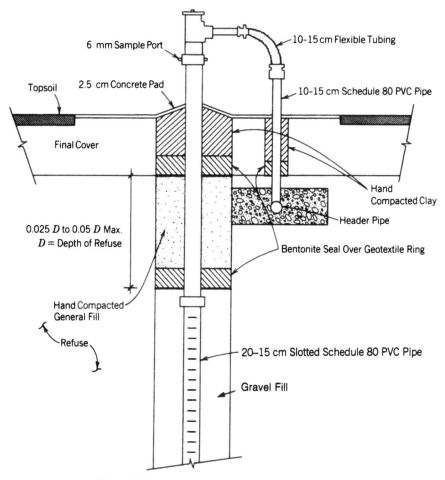

FIG. 19.36 Typical detail of gas extraction well.

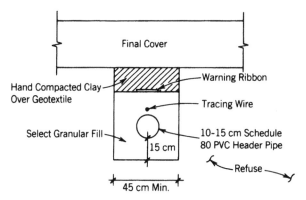

FIG. 19.37 Typical detail of gas extraction header pipe and trench.

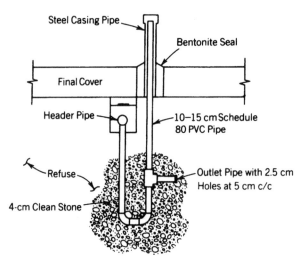

FIG. 19.38   Typical detail of a drip leg.

blower. A typical drip leg used for condensate removal is shown in Fig. 19.38. Usually drip legs are spaced at 150–230 m (500–750 ft). Condensate may be allowed to drip back into the landfill or collected and treated with leachate. A study by Cook et al. (1991) indicated that collection of condensate from a (natural attenuation type) landfill gas extraction system reduces volatile organic compounds (VOC) concentration in surrounding groundwater. Thus condensate removal will decrease VOC concentration in leachate. It should be noted, however, that the concentration of several parameters in condensate usually exceeds the hazardous waste concentration limit; thus the condensate may need to be handled as hazardous waste if dictated by the regulatory agency. The drip leg exit should be designed in such a way that the condensate can be directed to the leachate tank or collected in a separate tank if needed. This option would allow the collection of condensate if the reduction of VOC in the leachate becomes necessary. The drip leg should be designed such that the seal is maintained at all times.

Burner   Landfill gas may have enough methane to burn once ignited, but for complete burning of hazardous air contaminants an additional methane supply may be needed. The operating temperature and residence time in a burner should be enough to destroy the contaminants completely. An operating temperature of 815–900°C (1500–1650°F) and a residence time of 0.3–0.5 sec is needed to destroy most hazardous air contaminants. A flame arrester should also be installed on the landfill gas line to arrest flames going back into the blower.

Example 19.11 (in F.P.S. units)   Design an active gas venting system for a landfill of 30,000-ton waste volume using HDPE pipes for the layout shown in Fig. 19.39. Assume a gas production rate of 0.3 ft³/lb/year and equal

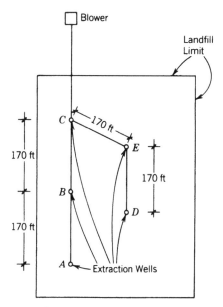

FIG. 19.39   Layout of active gas collection system for Example 19.11.

production rate from each well. A minimum negative head of 2 in. of water column is to be maintained at the farthest well. (Note: For formulas in this example, refer to Perry, 1976.)

$$\text{Total volume of gas generated from the landfill} = \frac{30,000 \times 2000 \times 0.3}{365}$$

$$= 45,315 \text{ ft}^3/\text{day}$$

$$\text{Reynolds number} = \frac{D_i V \rho_g}{\mu} \tag{19.43}$$

where D_i = inside diameter of the pipe = 5.7 in. = 0.475 ft
  V = velocity = 0.58 ft/sec
  ρ_g = density of gas = 0.0808 × 1.05 = 0.08484 lb/ft³
  μ_g = viscosity of gas = 0.093 lb/ft-hr

(Note: Values of ρ_g and μ_g are estimates based on a gas of molecular weight 30 and specific gravity of 1.05. Use of laboratory values is suggested for major/critical projects.)

The Fanning friction factor is a function of the roughness of the pipe and Reynolds number. For plastic pipes that may be categorized as drawn pipe, a roughness factor (rf) of 0.000005 may be used, thus (rf)/d = 0.000005/0.475 = 0.0001. An appropriate friction factor for each segment of the header pipe was found. The loss of pressure (Δp) in each pipe was calculated using

$$\Delta p = \frac{32\rho_g f L_p q^2}{\pi^2 g_c D_i^5} \tag{19.44}$$

in which f = the Fanning friction factor, L_p = the length of the pipe (ft), q = the volumetric flow rate (ft³/sec), g_c = the dimensional constant (= 32.17), and D_i = the internal diameter of the pipe (= 0.475 ft).

Loss if pressure in each fitting and valve can be calculated by knowing the velocity and the friction loss factor (K). In the absence of better estimates for K, the following values can be used: 45° elbows, 0.35; 90° bends, 0.75; tee, 1.0; gate valve (½ open), 4.5; Eq. (19.45) is used to estimate the pressure loss in fittings and valves:

$$\Delta p = \rho_g K \frac{V^2}{2g} \tag{19.45}$$

in which V = the velocity at the fittings (ft/sec) and g = the gravitational constant.

The velocity at each fitting must be calculated. Frictional losses at different pipe sections and fittings are included in Table 19.9. Loss of head at each extraction well is given by

$$\Delta p_s = \frac{\mu_g F_g D_r}{2K_g g} \left[R^2 \ln \left(\frac{R}{r_w} \right) + \frac{\gamma_w^2}{2} - \frac{R^2}{2} \right] \tag{19.46}$$

in which μ_g = the gas viscosity (= 0.0293 lb/ft-hr = 8.14×10^{-6} lb/ft-sec), F_g = the gas production rate (= 0.3 ft³/lb/year = 9×10^{-9} ft³/lb/sec), D_r = the density of the refuse (= 1000 lb/yd³ = 37.03 lb/ft³), K_g = the intrinsic permeability of the refuse (permeability $\times \mu/\rho_w = 1 \times 10^{-7}$ cm² = 1.08×10^{-10} ft²), R = 100 ft, and r_w = the radius of the extraction well (assume 2 ft).

TABLE 19.9   Friction Loss in Different Sections of the Header Pipe and Fittings

Pipe Section	Friction Loss (psi)
DE	0.00006
EC	0.00019
AB	0.00006
BC	0.00019
CF	0.00094
Bend at E	0.00397
Bend at C	0.0159

Putting proper values in Eq. (8.27) $\Delta\rho_w = 0.093$ psi

Total loss of head in five extraction wells $= 5 \times 0.093 = 0.465$ psi

Total loss of head $= 0.465 + 0.02131$

Total loss of head $= 0.48631$ psi

It is common practice to maintain a minimum negative head of 2 in. of water ($= 0.0722$ psi) in the header pipe. Thus the

Total suction head required $= 0.48631 + 0.0722$

$= 0.55831$ psi

$$\text{Horsepower of the suction motor} = \frac{\text{pressure in psf} \times \text{flow rate in ft}^3/\text{sec}}{550}$$

$$= 0.08 \text{ hp} \qquad (19.47)$$

Brake horsepower with an 80% efficiency $= 0.08/0.8 = 0.1$ hp

An alternative design procedure for designing gas extraction systems is discussed below:

1. Estimate gas flow from each well; assume 1 cfm of gas/vertical foot of the extraction well (note: use the entire well depth, not the slotted length).

2. Estimate the flow in each segment of the header pipe. The gas flow in the header pipe increases in the direction of flow due to addition of flow from each well in the line. For example, the flow in section EC (Fig. 19.39) is the sum of flows from extraction wells D and E.

3. The header pipe diameter is calculated using Spitzglass formulas (Spitzglass, 1912) given in Eq. (19.48). Choose a diameter of the pipe such that the maximum velocity in the header pipe is 50 ft/sec and the maximum head loss/100 ft of pipe is 1 in. of water column:

$$Q = 3550K \left(\frac{\Delta_p}{SL_p} \right)^{1/2} \qquad (19.48)$$

in which Q = flow rate (ft^3/hr) and S = specific gravity of the gas.

$$K = \frac{D_i^5}{[1 + (3.6/D_i) + (0.30D_i]^{1/2}} \qquad (19.49)$$

4. For a looped header pipe, the pipe diameter may need to be varied to balance head loss at the point where the loop meets (e.g., point *C* in Fig. 19.39). Start with an external pipe diameter of 6 in.

5. To the head loss in the header pipe add the head loss in each bend and pipe fittings and each well as described in Example 19.11. To this head loss add the head loss due to the elevation difference between the header pipe and the blower and the head loss at entry to the blower.

6. Use the total head loss and flow rate to choose a blower.

7. From the total gas flow rate and an average temperature range of 100–70°F (note: gas temperature may vary depending on landfill age, waste composition, and ambient temperature) estimate condensate volume. To estimate condensate volume find the vapor content of the saturated gas at the maximum and minimum anticipated temperature using a standard gas chart. Condensate volume is the difference in water vapor of saturated gas at maximum and minimum temperature.

Manufacturers' charts relating flow rate, head loss, diameter and rpm of the wheel, and outlet diameter may be used to choose a suction pump. A higher pump rating should be used to account for uncertainties in the refuse, higher than estimated friction loss due to a change in gas properties, and so on.

Usually a larger blower is installed to compensate for the reduction in efficiency through time, higher gas density due to saturation, and reduction in performance during cold weather.

Performance Monitoring   After an active system has been installed, its performance should be monitored to determine whether the system is working as designed. For this purpose the pressure and gas composition in each extraction well and the off-site gas probes are monitored twice a day for 2–3 days. The monitoring is done 7 days after the shakedown. During the shakedown period, the opening of valves in the extraction wells needs to be adjusted to arrive at the design pressure at the farthest well. Any serious leakage/blockage in the header pipe or malfunctions of extraction well valves and the blower assembly can be detected through this performance monitoring.

19.9.3   Temporary Gas Collection Systems

Temporary gas collection systems are used to mitigate odor problems in landfills. In some waste type (e.g., sludge) odor is a significant operational problem that might affect health (e.g., nausea) of workers or of the population living close to a landfill. If a change in operational practice, such as covering the waste with soil at the end of the day, is not a practical alternative or if odor persists even after instituting such an operational practice, then temporary gas collection should be undertaken.

In this system the exposed waste is covered with one or more synthetic membranes seamed together. Attempts should be made to minimize the exposed area by covering it with 15 cm of soil. The floating cover of membrane is connected to a system of flexible pipe manifold that will extract gas from under the membrane (Fig. 19.40). A low radius of influence of 15–20 m should be assumed in spacing the collection points. The laterals are connected to a header pipe, which in turn is connected to a blower of sufficient capacity. The blower may be sized by using the method described in Section 19.9.2. The system is not efficient, so loss of head will be rather high. For design purposes 10 cfm of gas from each extraction point may be assumed. A drip leg should be used on the header pipe before it enters the blower. The leachate/condensate from the drip leg may be directed back to the landfill or collected separately and treated. The membrane should be secured in place

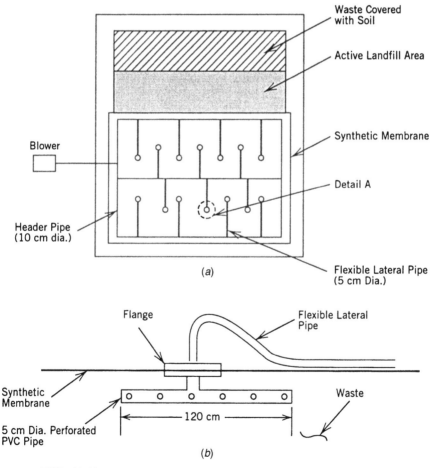

FIG. 19.40 Temporary gas collection system (*a*) plan (*b*) detail A.

with a sufficient number of sand bags. The membrane and the flexible pipe assembly should be designed such that it can easily be moved within the landfill to accommodate active disposal area. Usually a 40-mil-thick synthetic membrane is used. The membrane should be disposed in a landfill after use.

19.10   CONVERTING EXISTING NATURAL ATTENUATION LANDFILLS TO CONTAINMENT LANDFILLS

Many times existing natural attenuation landfills have to be converted to containment landfills. Before proceeding with such a design, it is necessary to confirm that groundwater contamination is nonexistent at the site. The purpose of such conversion is to retrofit old landfills for a change in regulatory rules or to reduce future groundwater contamination potential. Retrofitting must not be confused with remedial actions, which are undertaken when groundwater contamination has been detected or is highly likely to occur. Retrofitting may not be possible in all cases. It should be attempted only at sites having suitable geology and hydrogeology. Two approaches are normally used: construction of a base leachate collection system, and construction of a perimeter leachate collection system. A third option is to use leachate extraction wells within a landfill that will reduce groundwater impact.

19.10.1   Base Collection System

If a sizeable area of the landfill has not been used, then a regular base collection system can be constructed on that part and covering the existing landfill with a barrier layer as shown in Fig. 19.41. A 60- to 90-cm (2- to 3-ft) layer of sand compacted to 85% relative density on the inside of the existing landfill will provide a stable support for the liner. If needed, a 30-cm (1-ft) layer of gravel may be compacted below the sand layer to provide additional strength. The waste slope on which the liner is to be constructed should be

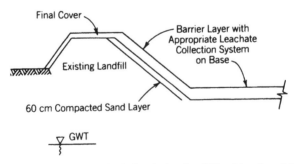

FIG. 19.41   Retrofitting of partially full existing landfill with a basal leachate collection system.

checked for stability. Groundwater should be below the subbase of the new portion of the landfill.

19.10.2   Perimeter Collection System

If the entire base area is already utilized and if a suitable low permeability layer exists at reasonable depth below the existing landfill, then a perimeter collection system may be constructed as shown in Fig. 19.42. A cutoff wall, which should extend at least 90 cm (3 ft) into the low permeability layer, has to be constructed. Then a perimeter leachate collection line should be installed inside of the cutoff wall. A submersible pump needs to be used to pump out leachate from the trench. The design can significantly reduce leachate seepage into the groundwater.

19.10.3   Leachate Extraction Wells

To minimize groundwater impact leachate extraction wells may be installed within natural attenuation type of landfills or in containment-type landfills in which leachate collection systems for one or more cells have failed. Use of these wells is successful if leachate is ponded within the landfill or if the waste is saturated (e.g., sludge-type waste). Usually 10-cm-diameter wells with slotted screens are installed with the waste. The well is installed within a gravel pack similar to a gas extraction well (see Fig. 19.36). The yield from a well depends on the amount of leachate present and on the pump capacity. A radius of influence between 10 and 20 m may be used if installation of more than one well is planned. The leachate pumped from each well is collected by a header pipe (in case of multiple wells) and discharged into a collection tank or directly into a sewer. The approach used for groundwater well design may be used for designing these wells. The optimum extraction schedule should be determined by recovery rate of each well; and 10–15

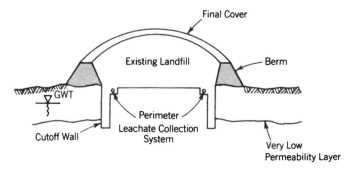

FIG. 19.42   Retrofitting of existing landfill with a perimeter leachate collection system.

cycles per day may be used for optimum performance. A detail draw-down test may be undertaken to optimize extraction (Hentges et al., 1993).

LIST OF SYMBOLS

A_γ = area of sump
a = radius of contact of tire
A = surface area of sedimentation basin
a_t = total weight of vehicle
B = bedding constant
b = trench width
B_c = outside width of pipe
D = deflection lag factor
d = depth of landfill
d = diameter of particle
D_f = difference between maximum and minimum elevation
d_f = maximum deflection
D_i = inside diameter of header pipe
D_n = particle size of which $n\%$ of the soil particles are smaller
D_p = pipe diameter
D_r = density of refuse
DR = dimension ratio = outside diameter/wall thickness
E = modulus of elasticity of pipe material
E' = modulus of soil reaction
E_p = modulus of elasticity of pavement or surface course
E_s = deformation modulus of PZB material
E_s = modulus of elasticity of subgrade
f = Fanning friction factor
$F/\Delta y$ = pipe stiffness
F_b = equipment braking force
F_g = gas production rate
F_p = maximum allowable pull
F_R = total force resisting pullout
F_{rc} = total force resisting pullout when cover soil is applied
F_s = seepage force
g = gravitational constant
g_c = dimensional constant
H = height of fill above pipe
h = trench depth
h' = thickness of landfill cover
h_d = delivery head of sump
H_e = height of plane of equal settlement above critical plane
h_s = seepage thickness
I = moment of inertia of pipe wall
k = earth pressure coefficient

K = friction loss factor in fittings and valves
K_g = intrinsic permeability of refuse
L = diameter of depression
L_p = length of pipe
M = tan ϕ; ϕ = angle of internal friction
m = traffic coefficient
n = saturation coefficient
n_r = roughness coefficient
O_1 = overlap radius of influence
P = pressure on pipe
P_1 = pumping rate
P_p = ratio of pipe projection above natural ground surface to pipe diameter
q = contact pressure
Q = floor rate
Q = surface loading or overflow rate
q = volumetric flow rate
r = mean radius of pipe
r = pipe radius
R = radius of influence
r_h = mean hydraulic radius
r_{sd} = settlement ratio
r_w = radius of extraction well
S = slope of energy line (= slope of channel for small slope)
S = specific gravity of gas
S_f = settlement of pipe into liner
S_g = settlement of natural ground surface
S_m = compression deformation of backfill column adjacent to pipe of height P_{BC}
S_r = sag ratio, = d_f/L
S_s = storage capacity of sump
T = maximum allowable weight of vehicle
t = one-half the cycling time of pump
t = thickness of synthetic membrane
t = wall thickness of pipe
t_b = thickness of base coarse material
t_p = total pavement thickness
t_s = thickness of wearing surface
V = mean velocity of water
V = velocity of gas in header pipe
V_c = volume of leachate collection
V_s = settling velocity of particle
W = total weight of soil on slope
w = weight per square area of synthetic membrane
W_1 = weight of soil anchorage at base
W_2 = weight of cover soil on side slope

W_c = load per unit length of pipe
W_e = weight of vehicle on side slope
W_p = width of compactor wheel in contact with soil
z = waste thickness
β = slope angle of liner
Δ = deflection
Δp = loss of pressure in header pipe
Δp_w = loss of head at each extraction well
γ = unit weight of soil
γ_1 = leachate density
γ_s = unit weight of soil
δ = friction angle between membrane and soil
μ = absolute viscosity of water
μ_g = viscosity of gas
ρ_g = density gas
ρ_g = density of particle
ρ_w = density of water
ρ_w = unit weight of water
σ_y = yield stress of synthetic membrane

20

LANDFILL CONSTRUCTION

This chapter includes construction and quality control tests for liners and final cover construction. Details of many landfill appurtenances and their construction have already been included in Chapter 19. The functional success of liner and final cover elements depends on the correct choice of material and proper quality control tests during construction. Discussions regarding properties of various liner materials are included in Chapter 18, which will be helpful in making the correct choice of material for liner or final cover construction relative to waste type.

Two types of construction specifications are usually used. In a "work-type" specification the contractor is told what to do and how to do it. A performance specification on the other hand requires a specific end result. The bid for the work-type specification will be lower compared to performance specification because the contractor does not have to spend time in researching how to do the job or hire a technician/engineer for quality control purposes. As far as liner construction goes it may be somewhat risky to use a detailed work-type specification because all the construction and quality control issues are not yet standardized. For example, the number of passes for compacting a clay liner lift specified in the contract may or may not be sufficient to obtain the desired density. Thus, this type of detail work-type specification should be avoided in preparing a bid document. It is essential that independent quality control personnel are utilized for maintaining construction quality.

20.1 SUBBASE CONSTRUCTION

The subbase for a liner system refers to the ground surface on which the liner is constructed. Compaction and grading of the subbase are necessary so that the actual liner can be constructed easily. If the subbase is not compacted, then it becomes difficult to compact the first one or two lifts to 90 or 95% modified Proctor's density. The subbase should be compacted to the same degree as the actual liner. If the subbase material is sandy, then it should be compacted to 85–90% relative density. It should be noted that the Proctor density versus moisture relationship does not hold good for coarse-grade materials (ASTM D698-78). So the percentile relative density should be specified and the standard relative density test (ASTM D4253-83) should be used. The density should be checked at 30-m (100-ft) grid points. A smaller or larger grid may be chosen depending on the soil properties, and reliability of the contractor. A vibratory roller may be used for compacting sandy soil. However, if the sand is already at a dense state, application of vibration will loosen the sand. Therefore, the *in situ* density of the sand should be checked prior to choosing compaction equipment. A nuclear device or other standard method (e.g., sand cone) may be used to determine the in-place density. Consolidation and rebound characteristics of the subbase should be studied during design of the site. If the subbase is predominantly sandy, then checking consolidation/rebound is not as important as when the subbase is clayey. Based on a national survey Peirce et al. (1986) reported that the maximum possible overburden pressure in hazardous waste landfills can be 370 kPa (7575 psf). Settlement of the subbase for such high pressure should be checked. Rebound of subbase due to excavation of overburden should be checked if the subbase is clayey or if a thick clayey stratum is identified at a short distance below the subbase grade. Rebound of the subbase may cause uneven heave of the liner resulting in failure of the leachate collection system. When estimating rebound, the weight of the liner may be considered as a weight-restricting rebound if quick construction of the liner is planned; however, the total weight of the waste should not be taken into account because sufficient time elapses between liner construction and disposing of waste up to the final grade, which may be enough for total rebound to take place. This issue should be investigated during landfill design where necessary.

20.2   LINER CONSTRUCTION

As mentioned in Chapter 18, four types of materials are used for liner construction: clay, bentonite-amended soil, synthetic clay liner, and synthetic membrane. The techniques used for constructing a liner using clay or bentonite-amended soil are very similar. Therefore, construction and quality control issues for these two materials are grouped together.

20.2.1   Construction of Clay and Amended Soil Liner

Because the construction and quality control used for both base liner and the barrier layer in the final cover system are the same, the construction of the barrier layer in the final cover will not be discussed separately.

20.2.1.1 Construction Laboratory studies have shown that clod size influences the permeability of the clay (Daniel et al., 1984; Benson and Daniel, 1990). So it is important to reduce clod size. Clod sizes recommended by engineers vary between 4.6 and 25 mm and are dependent on the thickness of the lift (Goldman et al., 1986). The clod size becomes a more important issue if the preconstruction moisture content of the soil is less than the wet of optimum moisture at which compaction is proposed. Wetting of clay takes time. If the clod size used in the field for permeability testing is larger than what was used in the laboratory, then the entire clay mass may not attain the same moisture content within the short time period usually allowed in the field. The type of equipment used for compaction also plays a role in developing low permeability. It is recommended to start compaction with a small clod size of approximately 2.5 cm (1 in.). Tiller or disks may be used for breaking up clay, however, high-speed pulvi mixtures, which are used for breaking asphalt pavement, are expected to reduce clod size of clayey soil to less than 3.8 cm (1.5 in.) after two passes (Goldman et al., 1986).

Compaction of clay must be done at "wet of optimum" moisture. Adequate time must be allowed to ensure uniform distribution of moisture. Nonuniform moisture distribution in the clay liner can be due to larger clod size, uneven water distribution by sprinklers, and insufficient "curing time" allowed for moisture penetration (Ghassemi et al., 1983). The moisture content of compacted portions of the liner should be maintained during inactive periods to prevent drying (which may lead to desiccation cracks) or overwetting (which will need long periods to dry). Moisture can be maintained by "proof rolling" the liner with a rubber tire or smooth steel drum vehicle or covering the liner with plastic. Once started, a clay liner project should be finished entirely. If construction is halted due to onset of winter (in cold regions), then scarification and recompaction should be done for the top 30 cm (1 ft) of the liner.

Three main categories of rollers are available for compaction: sheep's foot roller (self-propelled or towed), vibratory (smooth drum or sheep's foot), and rubber tire. A nonvibratory-type sheep's roller is recommended for constructing a clay liner because it can provide kneading compaction and has the ability to break down clods further. A vibratory-type roller or rubber tire rollers may not provide the right type of particle orientation desired for constructing a low permeability liner. Vibration may increase pore pressure within clay clods temporarily, which increases its shear strength; as a result more pressure will be needed to construct a uniform, homogeneous well-compacted layer (Bagchi, 1987c). Static compaction produces a flocculated structure with large pores (Mitchell et al., 1965). At wet of optimum the particle groups are weaker; high shear strain will help in breaking down the flocculated structures (Seed and Chan, 1959). Comparisons of the effect of kneading versus static compaction for various water content on soil permeability show that the lowest permeability can be achieved by using kneading compaction at wet of optimum moisture (Mitchell et al., 1965). The use of a heavy-weight sheep's foot roller is preferred by many for compacting clay (Hilf, 1975; Department of the Navy, 1971). The size of the landfill should also be considered in choosing equipment. Because large compaction equipment needs a large turn-ing radius, suitable smaller equipment should be selected for small landfills. However, small equipment may not provide the pressure required to achieve a low permeability liner. Thus, construction of a small containment-type clay-lined site poses a construction-related problem, which should be considered during design of such sites.

The slope of the sidewall should also be considered when selecting equip-ment. Two types of construction techniques are used to construct a liner on a side slope: flat and stepped construction (Fig. 20.1*a* and *b*). Usually stepped construction is used for slopes steeper than 3:1. Another approach is to con-struct a much flatter slope to facilitate construction (Fig. 20.2). The final slope of 3:1 is maintained by constructing a much thicker sand drainage layer. In addition to the ease of compaction using a sheep's foot roller, the thicker sand drainage layer provides better protection for the clay liner on the side wall from the effects of freeze–thaw and desiccation. However, a larger area is required to construct a landfill using a flatter side slope.

20.2.1.2 Quality Control   There are several issues involved in providing proper quality control during liner construction using clay or bentonite-amended soil: quality control before and during clay liner construction, qual-ity control personnel, and documentation report. Each item is important in providing proper quality control for the successful construction and function-ing of a landfill. Construction of containment-type landfills involves a sub-

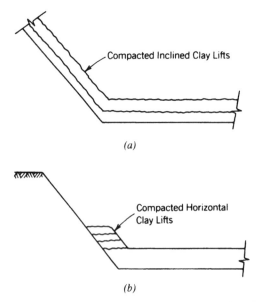

(*a*)

(*b*)

FIG. 20.1   (*a*) Inclined side liner construction. (*b*) Stepped side liner construction.

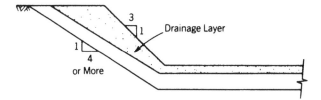

FIG. 20.2   Flatter sidewall construction.

stantial amount of money. In addition, failure of a landfill would contaminate the groundwater; groundwater cleanup may cost millions of dollars. Therefore sufficient importance should be given to quality control during construction of such structures. Spending at least 10–20% of the cost of the project for quality control purposes is routine for all civil engineering projects.

Quality Control before and during Clay Liner Construction The steps involved in choosing the borrow source for a clay liner, discussed in Section 12.5.2, must be followed. The quality of soil, percentage compaction, and moisture content at compaction are thus known before liner construction is undertaken. Proper testing is required to implement the proposed compaction during construction. The items that need attention are lift thickness, number of equipment passes, frequency of testing compaction density and moisture content, and permeability of the compacted liner.

Usually a lift thickness of 20–25 cm (8–10 in.) before compaction [which becomes 15–20 cm (6–8 in.) after compaction] is recommended. This lift thickness provides enough kneading for the entire lift and provides good bonding with the lift below. Tieing consecutive lifts is an important issue for those side liners that are constructed using horizontal lifts (Fig. 20.1*b*). An improperly tied lift may exhibit higher permeability in the horizontal direction along the lift boundary. A sheep's foot tends to mesh the boundary between successive lifts (Johnson and Sollberg, 1960) and thus is a better choice of equipment for compaction.

The total number of passes depends on the compactive effort, which is expressed as kg-m/m^3 (ft-lb/ft^3). Foot contact pressure is an important issue for achieving the desired permeability at the proposed density and moisture content. Two approaches are available. In the first approach the foot contact pressure and number of passes are specified; in the second approach the numbers of passes are calculated based on compactive effort, drawbar pull, and so on. DM-7 (Department of the Navy, 1971) specifies four to six passes of sheep's foot rollers for fine-grained soil, foot contact pressure of 250–500 psi (1722–3445 kPa) for PI greater than or equal to 30 and 200–400 psi (1378–2756 kPa) for PI less than 30. (Note: Foot contact pressure should be regulated such that the soil is not sheared on the third or fourth pass.) The specified foot contact area varies between 5 and 14 in.^{2} (32.25 and 90.3 cm^2). The second approach is to calculate the number of passes based on the following formula:

$$CE = (P \times N \times L)/(R_W) \times (L_T) \times (L) \qquad (20.1)$$

in which CE = the compactive effort, P = the draw bar pull, N = the number of passes, L = the length of the pass, R_W = the roller width, and L_T = the lift thickness. The usual field approach is to measure the density more frequently at the beginning of a compaction project. The number of passes required for a particular equipment to achieve the specified density at wet of optimum is standardized based on the initial test runs. (Note: Use of the DM-7 guideline for foot contact pressure is suggested.) Usually the foot shears the soil initially, but after a few passes the foot "rides on the clay" (i.e., no longer shears it). If the foot contact pressure is excessively high and the layer continues to be sheared even after seven or eight passes, then foot contact pressure should be reduced by reducing the weight of the drum. However, if the foot does not shear the clay lift to at least two-thirds its thickness from the beginning, then the weight of the drum should be increased. The weight of the drum can be adjusted by adjusting the volume of sand within the drum.

The number of density, moisture content, and permeability tests to be done for quality control purposes must be specified. Clay liners are a heterogeneous medium in a microscale but may be considered homogeneous if the scale of observation is enlarged. A statistical approach using representative elementary volume (REV) has been attempted to determine the number of quality control tests necessary during clay construction (Rogowski and Richie, 1984). The REV for a hydrologic property is defined as the smallest volume above which the decrease in variance of property of the medium is not significant (Bear, 1972). REV is difficult to define in the real world, however, according to Benson (1990) a 1500 cm^2 area is large enough to incorporate the variability. So when small samples of soil liners are measured to predict the overall permeability of the liner, they are assumed to be representative of the entire compacted volume. The probable variability is highest for permeability, average for water content, and lowest for compacted density (Rogowski and Richie, 1984). The frequency of the quality control tests specified by design engineers varies widely (Goldman et al., 1986). Usually a large number of density and moisture content tests are specified during construction to ensure uniformity. Fewer permeability tests (laboratory or field) are usually specified. A detailed study of compacted density, moisture content, and permeability of a clay liner indicated some variability of these properties within the liner (Rogowski et al., 1985), which were within reasonable limits. Although a shelby tube sample represents a small liner area, proper compaction technique and quality control can ensure uniformity whereby sample size becomes a redundant issue. Significant difference in permeability was not observed between shelby tube samples and large 30-cm-diameter samples collected from Wisconsin landfills (Bagchi, 1993). However, depending on the confidence in the quality control tests and the importance of a project, *in situ* permeability tests may need to be performed. Refer to Section 18.1.6.2 for a detail discussion on *in situ* permeability tests.

In the absence of a specified testing frequency the following testing frequency may be used: Dry density and moisture content tests on every 30 m (100 ft) grid points for each lift subject to a minimum of 12 tests per hectare (5/acre) per lift. Density and moisture content testing frequency should be increased for confined areas in which equipment movement is restricted or hand compaction is necessary. Grain size analysis, liquid limit, plasticity index, and permeability should be performed on samples collected at 25% of the 30 m (100 ft) grid points subject to a minimum of 5 tests per hectare (2/acre) per lift permeability tests must be performed on undisturbed samples. The grain size analysis and Atterberg limit tests are necessary to document that the soil originally proposed is being used for liner construction. In addition, the grain size distribution, Atterberg limits, and Proctor's curve should be checked during construction for every 3800 m^3 (5000 yd^3) of soil placed to check the variability of the source. The color of the soil should also be checked in every truck load; a significant change in soil color usually indicates a change in soil properties.

Usually clay is required to be compacted to 90% of maximum modified Proctor's density at wet of optimum (2–3% more than optimum) moisture. Daniel and Benson (1990) argue that this method may not ensure the specified permeability if the compactive effort is varied in the field. They suggested a new approach for controlling permeability during liner construction. The recommended procedure involves (1) compaction of 5–6 specimens each with modified, standard, and reduced Proctor (note: a reduced Proctor compactive effort procedure is similar to a standard Proctor except the number of blows is reduced to 15) compactive efforts in the laboratory; (2) determination of saturated hydraulic conductivity of each specimen; and (3) specifying the acceptable zone of density and moisture content on a dry unit weight versus moisture content plot. While plotting the density versus moisture content data, a different symbol should be used for the specimens whose hydraulic conductivities were equal to or less than the maximum allowable value. A typical plot with acceptable zone specified by this method and traditional acceptance zone is shown in Fig. 20.3. Because the zone of acceptance is larger, construction operations are expected to accelerate. See Section 18.1 for additional discussion.

Quality Control Personnel The presence of technical personnel with some experience and knowledge about clay compaction projects is essential during landfill liner construction. The person(s) must be capable of advising the field crew, performing the quality control tests in the field (e.g., compaction tests, moisture content tests, and collection of undisturbed shelby tube samples for

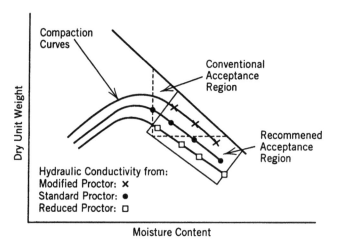

FIG. 20.3 Typical compaction curves showing conventional recommended acceptance regions. Based on Daniel and Benson (1990).

permeability tests), and surveying. The field technician(s) must be under the direct supervision of a qualified engineer capable of supervising the technician(s). Many regulatory agencies require that the quality control tests are signed off by certified or registered engineers. In those situations it is essential that the supervising engineer be certified or registered by an authority acceptable to the regulatory agency. Although the construction contractor may arrange for on-site technical supervision, the owner should hire independent personnel for quality control purposes. A conflict of interest will exist if the quality control personnel are employees of the construction contractor or have an interest in the construction firm. The owner should realize that the landfill owner is ultimately responsible for any faulty construction.

For controlling the quality of a clay construction project the quality control technician(s) should be present at all times when compaction is done. The homogeneity of the compacted clay cannot be ensured by performing quality control tests alone, no matter how many or what type of tests are done. In addition, the supervising engineer should arrange for routine and surprise visits during the construction of the landfill. Although the interest of the contractor and the quality control team are in conflict, a good professional relationship is essential for the smooth running of the project. A preconstruction meeting attended by representatives from the contractor, quality control team, owner, and regulatory agency is suggested. In many cases the contractor insists on using compaction equipment other than the sheep's foot roller because the required compaction can be easily obtained by using a rubber tire vehicle. The contractor and the quality control team must be convinced that compaction and moisture content specifications are not the goal but are only indirect tests to achieve a low permeability clay liner. Perhaps a technical presentation arranged by the owner (or the regulatory agency) regarding landfill clay liner construction and quality control tests is an effective means of bringing all parties to a common understanding and mutual cooperation.

The quality control tests are usually arranged by the owner and performed with equipment (and the laboratory) owned by the quality control team or their firm. For major projects, especially large municipal waste and hazardous waste sites, the owner should arrange to perform at least 10–20% of the tests by an independent consulting firm not owned by the quality control team or the contractor. Sometimes this type of "split sampling" is performed by the regulatory agency. The owner should more than welcome such a split sampling program because this will provide additional quality control for the project at no extra cost. At no time should the quality control tests be performed in a laboratory owned by the contractor.

Documentation Report All the items of construction must be well documented. Clear and concise documentation provides construction details, notes any departure from the original proposal, and discusses reasons for such departures; it is also helpful in case of any litigation. So due importance should be given to the documentation report. The location of any test must be clearly indicated. Since compaction, moisture content, permeability, and other tests are done on different lifts, it is a good idea to draw each lift separately and then show the appropriate test locations. The drawings and narrative should be clear and concise. A daily log of construction activities and quality control tests should be maintained at the site by the technician(s). Such field logs are important documents, especially in case of a dispute. Writing over and erasing must be avoided in the log book. In case of an error, the mistake should be simply crossed out. The report should be reviewed in detail by all the parties, that is, the owner, contractor, and the quality control team, before submitting it to the regulatory agency if directed. Usually the consulting firm that designed the site has responsibility for quality control and documentation. If detailed documentation requirements are spelled out by a regulatory agency, then those must be followed to obtain a permit.

20.2.2 Construction of Synthetic Membrane Liner

Although the construction activities for synthetic membrane installation are not as time consuming as clay liner construction, the quality control tests are intensive. Extreme care is necessary during synthetic membrane liner installation.

20.2.2.1 Construction The subbase must be properly prepared for installation of synthetic membrane. The subbase should be compacted as per design specifications (usually 85–90% relative density for sand or 90% modified Proctor density for clay or amended soil). The subbase must not contain any particles greater than 1.25 cm (0.5 in.). Larger particles may cause protuberance in the liner, especially due to freeze–thaw effects during winter (Katsman, 1984). A herbicide may be used on the subbase below the synthetic membrane to inhibit vegetative growth. Use of herbicide can be avoided if the time interval between subbase construction and synthetic membrane installation is not large (e.g., one month). The panel layout plan should be made in advance so that travel of heavy equipment on the liner can be avoided. In no case should a vehicle be allowed on a completed liner. Only the seaming equipment, seam testing equipment, and necessary minimum number of personnel should be allowed on the liner. Seaming of panels within 0.9 m (3 ft) of the leachate collection line location should be avoided if possible; this issue can be finalized during the layout plan. The subbase must be checked for footprints or similar depressions before laying the liner. The seaming equipment tends to get caught in such small depressions, causing burnout and subsequent repair. A small piece of the synthetic membrane placed below the membranes that are being seamed (this piece is moved forward along with the seaming equipment) may reduce burnout due to small depressions. The crew should be instructed to carry only the necessary tools and not to wear any heavy boots (tennis shoes are preferred). Laying of the synthetic membrane should be avoided during high winds [24 kmph (15 mph) or more]. Seaming should be done within the temperature range specified by the manufacturer.

Synthetic membrane absorbs heat very easily because of the carbon black and becomes overheated very quickly during sunny days. In higher latitudes construction should be done within a temperature range of 4–38°C (40–100°F). In lower latitudes the higher limit should be lower than 38°F to avoid overheating. Wrinkles may develop due to seaming at various times of the day, especially if the daytime and nighttime temperature differential is high. Typically, seaming is done (especially extrusion type) after the dew evaporates from the liner during the early hours after sunrise. Welding cannot be done in the rain or if the membranes are wet after a recent rain. So the membranes to be seamed must be dried using sponge before seaming. In case it becomes essential to continue seaming during rain (e.g., only a few minutes of seaming left) a portable protective structure may be used to cover the area being seamed. Such portable protective structures should be kept ready in the equipment shed at all times for emergency use.

Settlement hubs may need to be installed in the final cover over some waste types that are expected to settle (refer to Section 22.7 for numbers, etc.). The construction details of a settlement hub are included in Fig. 20.4. Proper care must be exercised in connecting the boot with the membrane. Gas vent risers are also needed in most landfills. Figure 20.5 shows details of a gas vent riser used in synthetic membrane final covers. In some instances (mostly for putrescible waste) active gas venting may become necessary at a future date. A passive gas-venting system cannot be retrofitted to act as an active gas-venting system. If it is felt that an active gas-venting system needs to be installed at a future date, then provision for installing deep extraction wells should be made. A preliminary design regarding extraction well location

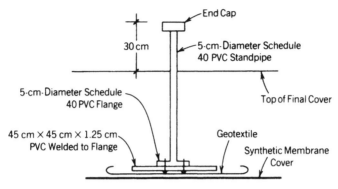

FIG. 20.4 Typical settlement hub on synthetic.

should be made so that holes on the membrane are made at proper locations during installation. A 15- to 20-cm-diameter riser pipe (closed at the top) should be installed at the proposed extraction well locations; connection details of such risers are the same as the details shown in Fig. 20.5. Punching holes and connecting the membrane with the extraction well may be extremely difficult, if not impossible, task in the future.

Transportation of the membrane must be done carefully, both from the factory to the site and within the site. A carefully prepared panel layout plan can reduce handling. The panel length and mark should be handed over to the factory prior to their manufacturing. Necessary overlap [15–23 cm (6–9 in.)] should be used in preparing the panel layout plan. The seaming machines

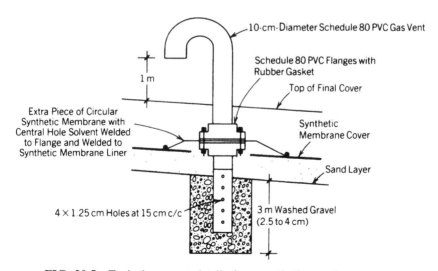

FIG. 20.5 Typical gas vent details for a synthetic membrane cover.

capability and seaming technique used dictate the overlap length. The marked panel can then be placed at a location where it can be laid without much movement.

Synthetic membranes deteriorate due to ultraviolet rays. They should be covered if kept on site, with a nontransparent sheet. If the estimated time of installation is high, then the rolls of synthetic should be stored in a shed to avoid exposure to ultraviolet rays. If the ambient temperature is expected to be high, then the shed should be well ventilated so that the temperature inside is not excessive. Manufacturer's guidelines on storage must be followed at all times.

Several types of seaming methods are available. The following are some of the commonly used seaming techniques: thermal-hot air, hot wedge fusion, extrusion welding (fillet or lap), and solvent adhesive. The manufacturer usually specifies the type of seaming to be used and in most cases provides the seaming machine. Manufacturer's specifications and guidelines for seaming must be followed. Seaming is more of an art even with the automatic machines. Only persons who are conversant with the machine and have some actual experience should be allowed to seam. For HDPE, hot wedge fusion and extrusion welding type of seaming are commonly practiced. A detailed study on strength and durability of seams was conducted by Morrision and Parkhill (1985). The results indicate the following:

1. No direct correlation exists between the seam shear and peel strength.
2. The peel strength of a seam should be tested to evaluate the quality of the seam.
3. The dead load peel test is not a valid procedure for evaluating the quality of a seam.
4. Short-term (6 months or less) chemical immersion tests may not be enough to determine chemical compatibility of some seams.
5. Of the three field seaming methods (thermal, extrusion, and mechanical) used for seaming HDPE membranes, the extrusion lap weld produces the highest shear and peel strength.

Synthetic membranes must be covered with soil and or waste as soon as possible. The results of the quality control field tests are available immediately; however, results of quality control tests done in the laboratory are not available immediately. The turnaround time for these tests must be minimized by proper planning and project scheduling. A testing time should be reserved in the laboratory (owned by the manufacturer or independent third party), where the quality control tests are to be done so that the waiting period is minimized. Enough volume of soil should be stockpiled near the site so that it can be spread on the finished membrane as soon as the tests results are available and the final inspection is over. Synthetic membranes can be damaged by hoofed animals. Bare membrane should be guarded against such damage by fencing the area or by other appropriate methods.

Usually liners are constructed in phases to minimize the length of time the liner is exposed to the elements. Such phasing reduces liner deterioration due to freeze–thaw and desiccation. Thus liners built in subsequent phases need to be connected with the previous phase. The splice detail for connecting liners is shown in Fig. 20.6 (note: Fig. 20.6 shows splice details for both synthetic and clay liners).

At least a foot of sand or similar soil should be spread on the membrane. The soil should be screened to ensure that the maximum particle size is 1.25 cm (0.5 in.) or less. The traffic routing plan must be carefully made so that the vehicle(s) does not travel on the membrane directly. Soil should be pushed gently by a light dozer to make a path. Dumping of soil on the membrane should be avoided as much as possible. One or two main routes with 60–90 cm (2–3 ft) of soil should be created for use by heavier equipment for the purposes of soil moving. The damage of the membrane due to traffic can be severe, but will probably remain undetected. Even the utmost precaution and quality control during installation will be meaningless if proper care is not taken when covering the membrane. As a guideline, it may be assumed that the time required to cover the membrane with soil is 50–100% of the time spent on installation. "Slow" and "carefulness" are the key words during the soil spreading process. It is recommended that a maximum of two parties (each party consisting of one small dump truck and a light dozer) be allowed for soil spreading. Minor wrinkles form in synthetic membranes (note: this effect is relatively more pronounced in polyethylene membranes) due to expansion during the day and contraction during the night resulting from a temperature differential. Sand should be hand shoveled to the back of minor wrinkles to prevent them from being pushed ahead of the sand and becoming a "large air bubble" (Fig. 20.7). The first lift of waste should be spread and

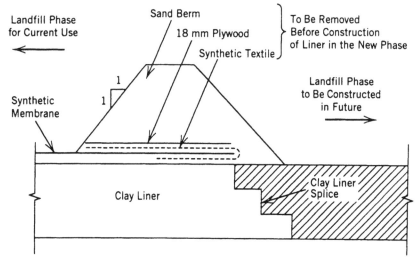

FIG. 20.6 Splice details for synthetic and clay liners.

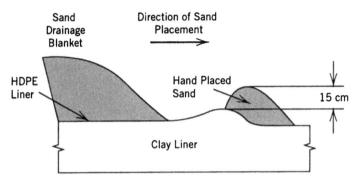

FIG. 20.7   Hand placement of sand to reduce formation of air bubble.

compacted with light vehicles. It is preferable not to compact the first foot of waste. No bulky items should be dumped in the first lift.

The bid specification should include warranty coverage for transportation installation and quality control tests. The cost of a project may increase due to the warranty. A survey of available manufacturers' warranties may be made by the owner to ensure that the owner is making the best choice. Unlike clay or amended soil, the manufacturer's warranty is an important issue in synthetic membrane bids. The manufacturers usually provide a warranty regarding strength and durability, field and factory seams, safe transportation of rolls from factory to site, and product quality. Comparison of warranties from different bidders should be made along with the cost. The experience of the company (both in manufacturing and installation), quality control during manufacturing and installation, physical installation should be asked in the bid so proper comparison among different bidders can be made.

20.2.2.2   Quality Control   The quality control issues are the same as for liner construction. However, since comments on quality control personnel will be similar to those made for clay liner projects, they will not be repeated in this section. Only quality control before and during membrane installation and documentation report are discussed here.

Quality Control before and during Membrane Installation   Tests of several physical properties of the membrane must be performed before installation. Usually most of these tests are performed at the time of manufacturing in the manufacturer's laboratory. The owner may arrange for an independent observer to oversee the tests, conduct the tests in an independent laboratory, or use a split sampling technique. This issue of responsibility for preinstallation quality control tests must be clearly mentioned or resolved during the biding process. The following are tests (with the relevant test designations indicated with parentheses) used for quality control purposes: sheet thickness (ASTM D751), melt index (ASTM D1238), percentage carbon black (ASTM D1603), puncture resistance (ASTM D4833), tear resistance (ASTM D1004), dimensional stability (ASTM D-1204), density (ASTM D729), low-temperature brittleness (ASTM D746), peel adhesion (ASTM D4437), and bonded seam strength (ASTM D4437). The quality control tests that are performed during installation include the following:

1. Inspection of subbase grade. Both density and smoothness of the surface should be checked. As previously indicated, the subbase should not have any depressions that tend to create more burnouts during seaming. The installation contractor and the quality control team should inspect the entire site.

2. Checking delivery tickets. The delivery tickets of all rolls delivered to the site must be inspected to verify that the site received proper shipments.

3. Verification of the proposed layout plan. The rolls may be marked at the factory or at the site. Usually the width of the roll is fixed but the length varies based on the layout plan.

4. Checking roll overlap. The overlap of each roll must be checked during and after sheet placement. Usually a 15-cm (6-in.) overlap is preferred, however, the seaming machine and the type of seaming dictate overlap width.

5. Checking anchoring trench and sump. The anchoring trench and sump dimensions must be checked prior to installation. No sharp objects or rocks should be present in any trench or sump. Material passing through a 1.25-cm (0.5-in.) sieve should be used to fill a trench or sump.

6. Testing of all factory and field seams using proper techniques over full length. Usually the vacuum box test (Fig. 20.8) is the most successful test for thick [0.75 mm (30 mils) and above] HDPE membranes. Air lance testing is not suitable for thick membranes. A dual hot wedge seam is tested by pressurizing the space in between the parallel seams. A lowering of pressure indicates leaking within the test section. Two newer techniques, the ultrasonic impedance plane (UIP) technique (also known as the wave resonant frequency technique) and the ultrasonic pulse echo technique, were also found to be useful for HDPE membranes (Morrison and Parkhill, 1985). These techniques are still at the development stage and may be available for field use in the near future. It should be noted that these nondestructive tests check bonding between membranes, not seam strength. Only destructive testing for verifying seam strength is available currently.

7. Destructive seam strength test. Test seams at least 60 cm (2 ft) long for each seaming machine should be made twice a day: once at the beginning of each work period. Each seamer should perform at least one test seam each day, if more than one person is handling one seaming machine. A portion of each test seam should be tested in the field and another in a laboratory. The manual field peel test is known to have produced erroneous results (Katsman, 1984) and hence should be avoided. The peel adhesion tests performed in the field can be standardized by specifying specimen size and a set displacement of the testing equipment; the displacement may be calibrated to reflect a

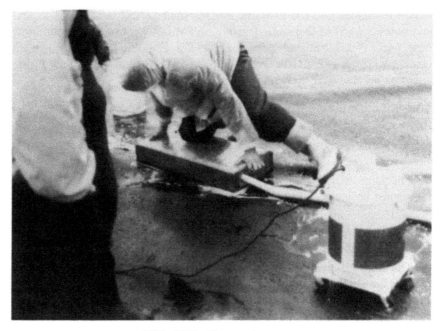

FIG. 20.8   Vacuum box test.

tensile force. As an alternative a tensile testing machine may be used in the field. The test seam samples sent to the laboratory should be tested for peel adhesion (ASTM D4437) and bonded seam strength (ASTM D4437).

8. Patch up repair. Damage to synthetic membranes is unavoidable. Synthetic membranes can be damaged as a result of bad seams or burnouts, manufacturing defects, shipping problems, accident puncturing (e.g., dropping sharp object), wind damage during installation, and wrinkles. Every damaged area must be repaired (Fig. 20.9). Each patch up repair must be tested for proper seaming. There is no set standard or number of allowable repairs. Based on experience and literature data (Katsman, 1984) it appears that an average of one or two repairs per 930 m² (10,000 ft²) of installed synthetic membrane area is reasonable. This estimate assumes that the quality control team is very vigilant. Lapses on the part of the quality control team may show a lower repair rate but actually may represent a worse job. A detailed field report of all activities is very essential. The repair location, cause of failure (i.e., bad seam or manufacturing defect), time of detection, time of repair, and time of retesting must be recorded meticulously. Without such a detailed record, it is not possible to identify any problem in the future. A surprise visit by the supervising engineer should be arranged to ensure better quality control.

Documentation Report   The documentation report is in general similar to the documentation report discussed in Section 20.2.1 for clay construction. The report should include field and laboratory test results, all repair locations,

FIG. 20.9   Application of patch to repair a damaged area.

warranties from the installer and manufacturer where appropriate, and a detailed field log of construction activities.

20.3   BERM CONSTRUCTION

Sandy soil is usually used to construct landfill berms. The compaction equipment most suitable for coarse-grained material includes rubber tire rollers, smooth wheel rollers, and a crawler tractor. It is suggested that lift thickness before compaction should be 25–30 cm (10–12 in.) and tire pressure should be 413–551 kPa (60–80 psi). The number of passes should be standardized in the field based on design relative density (usually 90%). Three to six passes are necessary to achieve the design density. Water may be added during compaction, but it is necessary to know whether the dry or wet method (ASTM D4253) was used in developing the relative density specifications and tests. Berm is constructed in horizontal lifts. The outer face and top of the berm should be covered with topsoil and vegetated as soon as possible. Surface erosion may be caused by a heavy rainstorm or freeze–thaw, resulting in failure of the berm. Vegetation helps in stabilizing slopes (Gray and Leiser, 1982).

Berms for natural attenuation type of landfills may include a clay core and/ or a riprap on the interior face. The design drawing should be studied carefully to develop a working drawing for construction of clay core. For vertical extension of the berm, the upper berm should be keyed to the lower berm by a 90 × 90 cm (3 × 3 ft) to 150 × 150 cm (5 × 5 ft) key as shown in Fig. 20.10. The top surface of the existing berm must also be scarified. The key and the first lift should be compacted carefully. A power pamper or rammer may be used for constructing the key. A vibratory compactor may loosen the already compacted lower berm near the top. (Note: A dense sand loosens due to vibration.) Therefore, it is preferable that nonvibratory compacting equipment be used for constructing the upper berm. The junction of the lower and upper berm may provide a channel for leachate to seep out, especially for natural attenuation type of landfills. A 0.5- to 0.75-mm (20- to 30-mil) synthetic membrane may be placed on the inside face of the berm to minimize chances of leachate seepage.

Quality control tests should include density tests (any of the methods described in Section 18.1.3 may be used) at a regular frequency. The following frequency of testing is recommended: density tests should be performed on each compacted lift at every 30 m (100 ft) for lift widths less than 4.5 m (15 ft). For lift widths greater than 4.5 m (15 ft) the density test should be checked at every 23 m (75 ft). The test location should follow a zigzag pattern so that density is tested both near the ends and at the middle of the berm. The grain size analysis of the berm material should be performed for each 765–1900 m^3 (1000–2500 yd^3) of material used; the higher limit is for a uniform borrow source and the lower limit is for a nonuniform borrow source. The material used for berm construction should not be too gravelly. The subbase grade on which the berm is constructed should be compacted. A 15-cm (6-in.) layer of 2-cm (0.75-in.) size gravel may be used to provide a firm base. The base of the berm should be at least 15 cm (6 in.) below the existing ground surface on the outer face to minimize chances of leachate seepage at the toe of the berm. If the geotechnical investigation indicates a soft clay lens below the site and the design calls for a preloading of the area before actual construction, then necessary arrangements should be made to preload the area. If preloading is undertaken, then additional quality control tests, such as the settlement history of the berm, should also be recorded during the preloading and postconstruction periods.

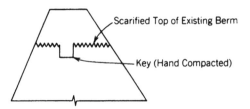

FIG. 20.10   Details for berm extension.

20.4 SAND DRAINAGE BLANKET CONSTRUCTION

Construction of a sand drainage blanket is not as difficult as construction of a liner, but more is involved than just dumping sand on the finished liner. The sand should be pushed carefully so that vehicles do not travel on the liner directly. A light dozer should be used for this operation. If the drainage layer is constructed on a synthetic liner, then additional care must be taken to protect the liner. A sand drainage blanket on the side liner may erode due to heavy rainfall. Pea gravel is a more stable material on the side liner.

The 0.074 or less fraction (P200) content of the drainage blanket material should not be more than 5%. A clean coarse sand is the preferred material for the drainage blanket, however, pea gravel may also be used for this purpose. A layer of gravel should never be used as a drainage blanket; the fines from the waste may migrate and clog the blanket. A filtering medium design approach may be used in designing a graded filter over a gravel drainage blanket. The material specified in the design should be followed strictly when constructing the drainage blanket.

The quality control tests include tests for grain size analysis and permeability. Usually one grain size analysis for each 765 m^3 (1000 yd^3) and one permeability test for each 1900 m^3 (2500 yd^3) of material used is sufficient. For smaller volumes a minimum of four samples should be tested for each of the above properties. The permeability of the material should be tested at 90% relative density.

20.5 LEACHATE COLLECTION TRENCH CONSTRUCTION

The location of the leachate collection line, as shown in the design drawing, should be strictly followed. The pipe spacing is a critical item for minimizing leakage through the liner. A leachate collection pipe may be placed in a trench. When the collection pipe is to be installed in a trench, care must be taken to ensure that the trench has the design slope (minimum of 0.5%) toward the collection manhole. A ditch must be excavated in the subbase (see Figs. 19.1 and 19.2) along the proposed leachate trench location. As previously indicated, seaming of the synthetic membrane liner should be avoided within 90 cm (3 ft) of the leachate collection trench. The gravel used in the trench and the leachate collection pipes should be brought in carefully. The collection pipes should be connected near the trench site so they do not have to be dragged on the liner. Dragging may rip the synthetic membrane. It is recommended that the drainage blanket be spread over the synthetic liner first, on the entire liner except 15 cm (6 in.) away from the collection trench. This will allow movement of light vehicles on the liner for placement of gravel and collection pipes.

When the leachate collection pipe is not installed within a trench, there is no need to thicken the liner below the collection pipe location. Typical details for this type of collection pipe installation are shown in Fig. 19.4.

All leachate collection lines that penetrate a liner should have an antiseep collar. A minimum of 1.5 m (5 ft) of compacted clay should be placed around the collar in all directions. Leachate transfer lines located outside the lined area should be encased in at least 60 cm (2 ft) of clay or a double-cased pipe should be used.

Quality control tests for a leachate collection pipes and trenches should include the following:

1. Density testing (for nonsynthetic liners only) at 30 m (100 ft) centers.
2. A 0.074 mm or less fraction (P200) content testing of the gravel for every 76 m^3 (100 yd^3) of material use. The gravel should have a uniformity coefficient of less than 4, a maximum particle diameter of 5 cm (2 in.) [a particle size of 3.7 cm (1.5 in.) is recommended for a synthetic membrane liner], and should be subangular. Limestone and dolomite should not be used because they may dissolve in acidic leachate and clog the collection pipes. The integrity of the gravel for specific leachate types (especially from hazardous waste) may be checked in case of doubts.
3. Stiffness and strain of leachate collection pipes, in accordance with ASTM D2412. The test results should conform to specified standards (Uni-Bell, 1979).
4. Checking invert of collection trench. The invert of the leachate collection trench should be checked at 9.5-m (25-ft) intervals using a level instrument with good accuracy so that the 0.5% slope of the collection line can be checked. In addition, a dye test may be done to ensure the slope of the pipe. In a dye test a dye is injected into the collection line at a low rate (so that it does not leak out through the perforations) until it comes out through the other end. At steady state the rate of injection should be equal to the rate of outpour. This test may be done if there is any doubt regarding the survey work and in important landfills.
5. Cleaning of leachate collection lines. The leachate collection pipes should be cleaned using a water jet, sewer line cleaning equipment, or high-velocity gravity flow immediately after finishing construction. Water jetting will clean all construction debris inside a pipe and will detect any severe damage (e.g., crushing) to the pipe. The dye test should be performed after cleaning.

20.6   DOUBLE- OR MULTIPLE-LINER CONSTRUCTION

The construction and quality control are guided by each liner material. Discussions on the construction of clay liners and synthetic membrane liners are included in earlier sections of this chapter. The most difficult construction is the case in which both liners are synthetic membranes. As indicated in Section 16.2, the lower synthetic membrane is highly prone to damage, which is difficult or almost impossible to detect. The main source of damage is probably vehicular traffic, which is unavoidable during construction of the primary liner. Extreme care should be taken with traffic movement on a synthetic liner.

20.7   GROUNDWATER DEWATERING SYSTEM CONSTRUCTION

Usually this system is installed in a sandy environment. The construction of a dewatering system involves excavating trenches in which the groundwater pipes are to be laid, laying perforated groundwater collection pipes, and connecting them to a manhole or sump. Details of a groundwater collection trench is shown in Fig. 20.11. Usually the groundwater collection pipes are drained by gravity wherever possible, and so the water is allowed to be discharged in surface water bodies via a ditch. However, for hazardous waste sites the groundwater collected may not be allowed to be discharged to surface water because of fear of contamination. Arrangements need to be made to direct the pipe to either a separate or a dual manhole.

In some landfills the groundwater collection system may need to be installed within a sand bed. Typical details of this type of installation are shown in Fig. 20.12. A geotextile should be laid over the sand bed. This will minimize migration of fines from the clay liner. It will also help the construction of the synthetic membrane liner.

20.8   LYSIMETER CONSTRUCTION

Lysimeters are constructed below liners. Typical details of lysimeters are given in Section 22.3.1. The critical construction events are laying the synthetic membrane and the boot on the collection pipe. The subbase should be cleared of all stones larger than 1.3 cm (0.5 in.). Overexcavation and recom-

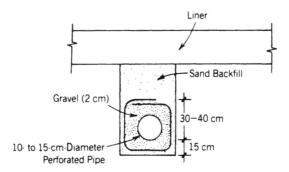

FIG. 20.11   Typical details of a groundwater collection trench and pipe.

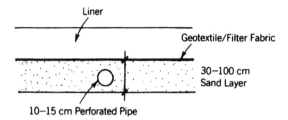

FIG. 20.12 Typical details of a groundwater collection pipe laid in a sand layer.

paction of the subbase with soil passing U.S. no. 4 sieve is suggested. Leak testing of seams should be done very carefully. If possible, a membrane without a seam should be installed; otherwise the seam should be completed and tested in the factory to provide better quality control. The synthetic membrane must be laid on a low permeability layer (e.g., 60-cm clay layer, synthetic clay liner).

The lysimeter is usually tested for leakage. The lysimeter is filled with water and the top is covered as tightly as possible and left undisturbed for 24–48 hr. The water elevation is measured at the beginning and end of the test. The test can successfully detect serious leakage problems but cannot detect small holes. The compatibility of the synthetic membrane with the leachate should also be tested.

20.9 LANDFILL COVER CONSTRUCTION

Landfill cover construction and quality control issues are similar to those for liner construction and therefore will not be discussed here. The layer below the low permeability layer, referred to as the grading layer or gas-venting layer, should be constructed using poorly graded sand (SW- or SP-type sand per USCS classification). A grain size analysis for every 400 m³ (520 yd³) of material use is recommended for quality control purposes. The layer should be compacted to 85–90% relative density to provide a firm subbase for the low permeability layer above. The density should be tested at 30 m (100 ft) grid points.

The layer above the low permeability layer, referred to as the protective layer, should be SM, SC, or ML (per USCS classification). A finer soil mixed with sand may be used if silty soil is not available. A horticulturist or soil scientist should be consulted if such mixing is planned. A grain size analysis for every 765 m³ (1000 yd³) of material use is recommended for quality control purposes. This layer should not be compacted to help root penetration.

Laying of the topsoil layer should be done as soon as the protective layer construction is finished. Heavy construction equipment should not be allowed on the finished surface. The nutrient and liming requirements for the topsoil should be assessed from a competent agricultural laboratory. In the absence of a regulatory recommendation/requirement regarding seed mix, a horticulturist or soil scientist should be consulted. A combination of grass and bush-type vegetation capable of surviving without irrigation water should be planted. At least five samples of topsoil per hectare (2.4 acres) should be tested for nutrient and liming requirements. Nutrient and seed mix application rates should be supervised on site for quality control purposes.

20.10   MATERIAL PROCUREMENT, CONSTRUCTION SCHEDULING, AND SO ON

All construction material should be procured/ordered well in advance. Landfill construction must not be scheduled until specifications of all materials required for construction are approved by the proper authority and the contractor/supplier signs a specific delivery agreement. Construction scheduling is an important issue, especially in cold regions. Enough waste [usually 1–1.5 m (3–5 ft) lift] should be disposed of on a landfill liner before the onset of the winter/dry season to protect the liner from the natural elements. In general, because soil is compacted to a higher density in clay liner projects than their natural states, the shrinkage factor must be estimated. The following formula is used:

$$\text{SF} = (F_d - N_d)/N_d \tag{20.2}$$

in which SF = the shrinkage factor, F_d = the proposed field density in kg/m³ (lb/ft³), and N_d = the natural density of the soil at the borrow source. The borrow source should have a volume of $(1 + \text{SF}) \times F_v$, in which F_v = the required volume of soil at the site. A certain percentage for spillage and so on should be allowed, varying from 2 to 10% of the fill volume (lower value for larger projects). The minimum required soil volume (R_v) at the borrow source with a 5% spillage is

$$R_v = (1 + \text{SF})F_v + 0.05F_v = F_v(1.05 + \text{SF}) \tag{20.3}$$

Example 20.1   Find the required volume of soil at a borrow source for a project in which a 5% spillage factor may be assumed. The following values are known: F_d = 1680 kdg/m³ (105 lb/ft³), N_d = 1360 kg/m³ (85 lb/ft³), and F_v = 100,000 m³.

20.11   EROSION CONTROL DURING LANDFILL CONSTRUCTION

An erosion control plan should be prepared before landfill construction is started. This will minimize erosion of soil from the construction site and sediment load in the surface water body (e.g., lake, stream). If planned properly the area disturbed during construction can be kept to a minimum. Figure 20.13 shows a typical erosion control plan. Straw bales and silt fences are used in drainage swales to minimize off-site migration of fines. Straw bales must be properly placed in trenches without any gaps between adjacent bales. The straw bales should be anchored in place using stakes. Typical details for straw bales and silt fences are shown in Figs. 20.14 and 20.15. The straw bales and silt fences should be inspected at least once a week to ensure that they are still functional. Use of a sedimentation basin is recommended if the disturbed area is more than 2 acres. The disturbed area should be vegetated as early as possible after completion of construction. The straw bales and silt fences should be maintained until vegetative cover is established on the disturbed areas. The regulatory agency should be consulted to find out if there are any specific requirements regarding erosion control during construction.

20.12 CONSTRUCTION ON LANDFILLS

In the past construction on landfills were generally avoided. Very little consideration, if any, was given to the postclosure use of landfills. Since the early

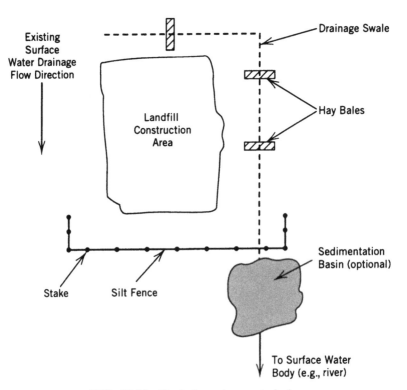

FIG. 20.13 Typical erosion control plan.

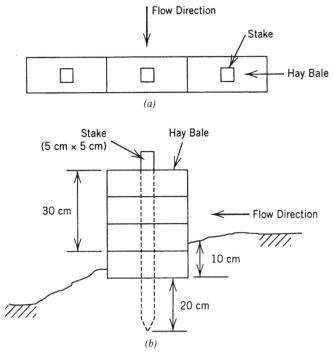

FIG. 20.14 Details of hay bales (*a*) plan (*b*) cross section (not to scale).

1980s attention was given to the postclosure maintenance, monitoring, and use of landfills. Initially, the usual practice was to designate closed landfill surface as open space for parks, golf courses, and so on. However, as developable space in urban areas are becoming scarce, postclosure development of landfills are becoming increasingly common. In Japan, because of scarcity of

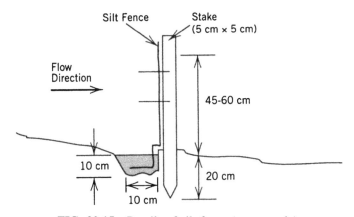

FIG. 20.15 Details of silt fence (not to scale).

developable land, 67% closed landfills have been reused for various projects (Shimizu, 1997). Construction on closed landfills and cleaned up lands (which were contaminated previously) is also rising in the United States. In some European countries, preprocessing of waste is widely endorsed in part to enhance postclosure development potential (Bouazza and Kavazanjian, 2001).

Both engineering and environmental challenges are faced while developing a structure on closed landfills. Engineering challenges primarily include foundation design and utility alignment. Environmental challenges include mitigation of explosive gas, health risk due to probable exposure to harmful chemicals, impact on soil, and impact to groundwater and air. There are two types of use of closed landfill surfaces: hard use and soft use. Hard uses include commercial and residential buildings, roadways, bridges, and so on. Soft uses include golf courses, athletic field, parks, and so on.

There are basically four types of wastes on which development is undertaken: inert waste [e.g., construction and demolition (C&D) debris], nonhazardous waste (e.g., foundry sand), putrescible waste (e.g., municipal solid waste), and hazardous waste. Although development for hard use and soft use is undertaken on all closed landfills, development on hazardous waste landfills is not common.

20.12.1 Development on Inert Waste Disposal Sites

Development is probably easiest on inert waste and nonhazardous industrial waste. The foundation on these types of waste can be constructed using conventional geotechnical techniques. Possibilities of environmental impacts due to construction are usually minimum except in a few cases. Because of lack of regulations in the past, hazardous wastes were disposed in some of these landfills. Therefore proper site investigation should be done prior to undertaking construction on landfills containing these types of wastes. In addition to assessing geotechnical properties (e.g., bearing capacity) of the site, an attempt should be made to collect the following information: the disposal history of the site, records of waste types disposed, list of site users, type of equipment used for operating the site, and postclosure use envisioned during or prior to starting operation. In addition to C&D waste, inert waste landfills may contain whole or shredded tires, asbestos, excess soil excavated from construction sites, and contaminated soil. Proper protection about health and safety of workers should be undertaken, especially if asbestos, contaminated soil, or hazardous waste are expected to be encountered during construction.

20.12.2 Development on Hazardous Waste Disposal Sites

Hazardous waste disposal sites may include contaminated soil and wastes from commercial and industrial processes. Because of environmental concerns, and health and safety issues, development on this type of disposal sites is not common. Since hazardous waste landfills are not subject to the same degree of settlement or gas generation as municipal waste landfills, the engineering challenge associated with development on hazardous waste disposal sites is not as difficult. Although the degree of toxicity associated with hazardous waste is high, it is possible to develop these site for both soft and hard uses (Bouazza and Kavazanjian, 2001). Discussion about remediation of contaminated sites is included in Chapter 9. Serious consideration should be given regarding the potential effect of the chemicals present in the waste (e.g., corrosion) and of the foundation materials (e.g., concrete, steel). Additional discussion about piles is included in Section 20.12.3. Deep pile foundations are not undertaken in landfills with liners because the pile may serve as a conduit for contaminants to reach the groundwater.

Open spaces and golf courses were developed on landfills containing highly acidic (pH less than 1.0) waste (Collins and Ramanujam, 1998). Because of the low bearing capacity of the waste, two geogrid were placed below the cap in the golf course areas. The foundation layer also included a gas extraction system with a blower connected to an activated carbon treatment unit. The cap was tied into a soil–bentonite slurry wall encircling the site (Hendricker et al., 1998).

20.12.3 Development on Municipal Solid Waste Landfills

Development on closed municipal waste landfills includes athletic fields, parks, commercial developments, office buildings, shopping centers, highways, high rise buildings, and so on (Evans et al., 1998; Hinkle, 1990; Gifford et al., 1990; Rollin and Fournier, 2001, Shimizu, 1997). Engineering design for construction of foundation, gas migration control, utility corridors, and so on were needed for the above projects. Environmental challenge for such projects includes hazards of landfill gas migration, downward migration of contaminants into groundwater and so on. Landfill gas poses a major problem due to the risk of fire and explosion and the health and safety of the workers during construction.

Which structures are allowed on these landfills depends on age and composition of the waste, degree of compaction, liner design (if any), and final cover design. Waste settlement influences both soft and hard use projects. Waste settlement also impacts design of shallow and deep foundation, drainage grades, site utilities, access roads, and other ancillary features. Both total and differential settlement poses an engineering challenge. Significant differential settlement should be expected because waste depth varies across landfills. Conventional mechanical ground improvement techniques (e.g., surcharging or dynamic compaction) may delay onset of consolidation settlement, but the long-term settlement due to decomposition is not effected by these techniques. However, ground improvement techniques have been used effectively to densify older landfills (Van Impe and Bouazza, 1996) because settlement due to decomposition is not a major issue in older landfills. Settlement due to dynamic compaction is higher in younger landfills (Lewis and

Langer, 1994) compared to older landfills (Van Impe and Bouazza, 1996). Additional discussion regarding settlement of putrescible waste landfills can be found in Chapter 14.

20.12.3.1 Development for Hard Use   Both bearing capacity and settlement need to be assessed for foundation design. In addition, proper gas control systems should be installed to ensure that the structure poses no significant risk of fire and explosion or to human health and safety. It is prudent to avoid construction on these landfills until the waste mass stabilizes. The following criteria should be considered before undertaking development on putrescible waste landfills:

1. Construction should be undertaken 10 years after closure of a landfill.
2. The waste depth.
3. Installation cost of gas control systems.
4. The water table beneath the landfill should be stable and low.

It may be noted that the above set of criteria are not strictly followed while undertaking development on landfills (Bouazza and Kavazanjian, 2001).

Comments on Foundation Design   Both total and differential settlements usually govern the choice of foundation type. In general, shallow foundations are used to support light structures and deep foundations are used to support heavy structures. Deep foundations are usually used for construction on older landfills without a bottom liner. Table 20.1 shows the relative advantages and disadvantages of shallow and deep foundations on landfills. Thick final covers provide substantial bearing capacity for shallow foundation. Thin final covers transmit the load to the waste, causing significant settlement. In these cases total settlement dictates the design of utility connections and differential settlement dictates the structural design of the foundation. Therefore, raft foundations rather than isolated column footings are used in these cases. If needed, one or more layers of geogrid or geotextile may be used to support the foundation.

TABLE 20.1   Relative Advantages and Disadvantages of Shallow and Deep Foundations

	Deep Foundations	Shallow Foundations
Bearing capacity	Excellent	Limited to two stories
Relative settlement	Poor	Good
Differential settlement	Excellent	Acceptable
Building protection	Poor	Excellent
Maintenance	High	Low

After Bouazza and Kavazanjian (2001).

Piles are used for deep foundations. Since the waste itself does not have enough strength, the load is transferred to a deeper strata with adequate capacity to carry the load. Depending on the geotechnical characteristics of the media, the piles are designed either to carry the load through friction or end bearing. Because of settlement of the waste, negative friction is very likely. The down-drag force may be 10% of the weight of the overlying waste (Oweis and Khera, 1990). Based on waste shear strength a down-drag of 15–20% of pile design capacity should be used (Gifford et al., 1990). The lateral earth pressure at rest (k_0) for waste is necessary to calculate the down-drag using frictional shear strength. Preliminary investigation indicates that k_0 (0.26 ≤ k_0 ≤ 0.4) decreases with an increase in fibrous constituents, and that k_0 is maximum when decomposable fibers cease to exist in the waste (Landva et al., 2000). The interface friction strength between household waste and concrete has been reported to be 30 kPa (Rine et al., 1994, cited in Bouazza and Kavazanjian, 2001).

Field pull-out tests on a number of piles should be done to estimate the down-drag load. The down-drag may be reduced by using friction-reducing coatings on piles, use of double piles, or filling a predrilled oversize hole with bentonite slurry (Dunn, 1995). Note that the effect of the relatively high temperature of decomposing materials (60°C) may impact the friction-reducing coating. This issue should be considered when choosing the coating. A degradation of the coating will increase the down-drag, which may lead to foundation failure. Potential corrosion of steel piles should also be considered. Deep pile foundations, penetrating the landfill bottom, may serve as a conduit for contaminants to reach the groundwater. Conical or wedge-shaped driving tips are often used to avoid downward migration of contaminants. Usually deep foundations are not used for constructing structures on containment-type landfills (Bouazza and Kavazanjian, 2001).

Decomposition of waste and deposition methods can induce horizontal movements inside landfills. Large deformations within landfills is possible, which can produce both horizontal and vertical loads (Bouazza and Kavazanjian, 2001).

Comments on Gas Control   The methane and carbon dioxide produced in landfills can find its way into buildings and accumulate to explosive concentration. Methane is explosive within a concentration range of 5–15% by volume. Organic compounds such as benzene and vinyl chloride are also found in landfill gas. Some of the constituents of landfill gas are carcinogenic in trace concentration. Therefore, long-term exposure, even to a low level of these constituents, can cause serious health effects. Landfill gas migrates through the final cover primarily through advection. As indicated in Chapter 22, falling barometric pressure and high winds create suction, which pulls landfill gas upward. Explosion due to landfill gas migration can occur if proper venting is not undertaken. There is a wide range of gas protection methods that can be used to provide necessary safety precautions. Table 20.2 provides a list of suggested protection measures for residential, commercial,

TABLE 20.2 Suggested Protection Measures for Residential, Commercial, and Industrial Structures[a]

Limiting CH_4 con. (% by vol)	Limiting CO_2 con. (% by vol)	Limiting Borehole Gas Volume of CH_4 or CO_2 (l/h)	Residential Building	Office/Commercial/Industrial Development
<0.1	<0.1	<0.07	No special precautions	No special precautions
<1.0	<1.5	<0.7	Well-constructed ground or suspended floor slab, geomembranes sealed around penetrations, passively underfloor subspace and wall cavities	Reinforced cast in situ ground slab. All joints and penetrations sealed. Possibly geomembrane. Granular layer below slab passively vented to atmosphere with interleaved geocomposite strips or pipes.
<5.0	<5.0	<3.5	Well-constructed suspended or ground slab. Gas-resistant geomembrane and passively ventilated underfloor subspace	Reinforced concrete cast in situ ground slab. All joints and penetrations sealed. Waterproof/gas-resistant geomembrane and passively ventilated underfloor subspace.
<20	<20	<15	Well-constructed suspended or ground slab. Gas-resistant geomembrane and passively ventilated underfloor subspace, oversite capping and in ground venting layer	Reinforced concrete cast in situ ground slab. All joints and penetrations sealed. Gas-resistant geomembrane and passively ventilated underfloor subspace.
<20	<20	<70	Specific gas-resistant geomembrane and ventilated underfloor void, oversite capping, and in ground venting layer and in ground venting wells	Reinforced concrete cast in situ ground slab. All joints and penetrations sealed. Gas-resistant geomembrane and passively ventilated underfloor subspace. In ground venting wells.
<20	<20	<70	Not suitable unless gas regime is reduced first and quantitative assessment carried out to assess design of protection measures in conjunction with foundation design	Reinforced concrete cast in situ ground slab. All joints and penetrations sealed. Gas-resistant membrane and actively ventilated underfloor subspace, with monitoring. In ground venting wells.

[a]Con. = concentration
After Bouazza and Kavazanjian (2001).

and industrial structures for various gas concentration levels. Figure 20.16 shows a conceptual scheme for controlling landfill gas. An alarm system to provide warning about gas accumulation may be placed within a building. The key to successful gas control is the long-term maintenance of the protection measures. All the controls must be maintained properly until the landfill gas production is stopped.

The landfill gas should be monitored at several locations to determine the end point of gas production. Therefore, in addition to structural costs, the long-term maintenance and monitoring costs should be taken into consideration for hard use projects. The long–term maintenance plan should be well documented. The plan must be followed as long as recommended by the developer.

20.12.3.2   Development for Soft Use   Soft uses generally include parks, golf courses, athletic fields, and so on. Both settlement and gas migration are also important issues for soft uses. If no ancillary facilities are built (e.g., clubhouse, office space), then the risk of gas accumulation is eliminated. However, slow venting of hazardous constituents of landfill gas can still pose risks. Landfill gas also affect vegetation, which is important for parks and golf courses. Therefore, adequate gas venting should be undertaken to ensure health and safety of users.

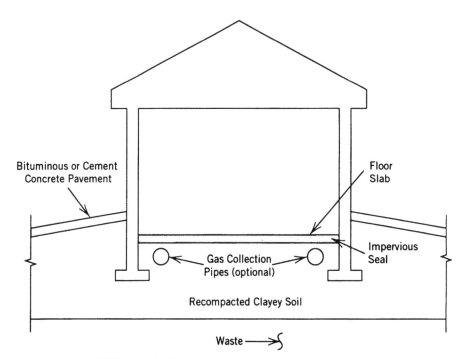

Bituminous or Cement
Concrete Pavement

Floor
Slab

Gas Collection
Pipes (optional)

Impervious
Seal

Recompacted Clayey Soil

Waste

FIG. 20.16   Typical detail for buildings on landfill.

Both total and differential settlements can impact utility lines, access roads, and ancillary facilities. Even relatively small differential settlement can affect soft use projects (e.g., athletic fields). Therefore careful assessment of settlement should be done for soft use projects as well.

Soft use demands good to excellent vegetative covers. Significant amount of water needs to be sprayed to maintain vegetation. Such water addition may increase infiltration. Increased infiltration is expected to increase gas production, settlement, and groundwater impact. The final cover should be designed or upgraded to minimize infiltration and venting of gas. However, an improved final cover, which reduces gas venting through the cover, may increase gas migration along the landfill perimeter and bottom. Chances of such migration is high for natural attenuation type of landfills (see Chapter 15). So, the aim of gas migration should be to control migration through the perimeter, bottom, and final cover.

Long-term monitoring and maintenance of the gas control system and final cover is important. Therefore, like hard use projects, the long-term cost of maintenance and monitoring should be considered while planning a soft use project.

LIST OF SYMBOLS

CE = compactive effort
F_d = proposed field density
F_v = required volume of soil at the site
L = length of pass
L_T = lift thickness
N = number of passes
N_d = natural density of the soil
P = drawbar pull
R_v = required soil volume at the borrow source
R_w = roller width
SF = shrinkage factor

21

LANDFILL REMEDIATION

Many old landfills were constructed without any liners or leachate collection systems and/or were closed with an inadequate final cover. These landfills are likely to pollute groundwater due to leakage of leachate through the landfill base. In addition, an inadequate final cover of a landfill may be eroded over time exposing the waste. Precipitation falling on such exposed waste can dissolve contaminants and subsequently pollute surface water bodies. Improperly designed and constructed landfills may pollute groundwater, surface water, as well as air. In many cases reliable groundwater monitoring data are not available from these landfills. Proper site assessment is necessary before undertaking remedial actions. If there are many such old or abandoned landfills, then it becomes essential to develop a system that can help in prioritizing the landfill sites where remedial action is needed. This approach of developing a priority list is also known as a hazard ranking system. Although improperly designed, constructed, or closed landfills can pollute groundwater, surface water, and air, in reality the potential for air pollution due to hazardous air pollutants is extremely low from nonhazardous solid waste landfills. In addition, chances of fire, explosion, and potential harm due to direct contact with the waste are also very low. Therefore, for practical purposes the site assessment and hazard ranking system for nonhazardous waste landfills need to consider only the potential for groundwater contamination. However, a more comprehensive hazard ranking—taking into consideration migration of pollutants in groundwater, surface water, and air, as well as potential harm from fire, explosion, and direct contact—should be undertaken for landfills where hazardous waste has been disposed. A detail approach for such a comprehensive hazard ranking system can be found elsewhere (USEPA, 1992a). The site assessment and hazard ranking system suggested is applicable to nonhazardous solid waste landfills only. (Note: The approach is based on the hazard ranking system adopted by U.S. EPA).

For existing landfills, it is essential to find out whether a landfill has potential for groundwater contamination. If there are too many landfills that are known to have impacted groundwater already or have potential for such impact and enough resources (both financial and human) are not available, then a priority action list needs to be developed for remediating the worst ones first. A priority action list is developed based on the hazard ranking system. The priority action list may be developed at a national, state, or regional level. It should be mentioned that the hazard ranking system can also be utilized to assess whether a natural attenuation type of landfill has potential for impacting groundwater.

Once a landfill is chosen based on the priority list, then a remedial action plan is developed for the site. Detailed site investigation and risk analysis are done to choose a suitable remediation method. This chapter primarily addresses the issues related to the development of a priority list. Although issues related to site-specific remediation are discussed briefly in this chapter, refer to Chapter 9 for a discussion on site investigation and remediation method and to Chapter 10 for risk analysis.

21.1   SITE INVESTIGATION

Site investigation is done prior to undertaking any remedial action. Monitoring data of active or recently closed landfills may indicate the need for remedial action. However, site assessment is needed to ascertain groundwater impact. Site assessment can be done either directly through groundwater monitoring or indirectly through the study of site features. If the site has a well-designed and properly operated groundwater monitoring system, the adverse impact on the groundwater can be detected through the study of the monitoring data. Comments regarding groundwater monitoring system design and data analysis can be found in Chapter 22. However, chances of the existence of a well-designed and operated groundwater monitoring system are low for older sites. If monitoring data is not available, then the site features are studied to assess the possibilities of groundwater impact from the site.

Knowledge about both the surface and subsurface features of the site are needed for the assessment. The surface features include the condition of vegetation on the cover, average slope of the cover, and type of soil used to construct the cover. The information is obtained through field survey and direct observation. The soil type used in the cover can be assessed by collecting soil samples from the cover at 100-ft (30-m) grid points. The subsurface features include depth to groundwater and the permeability of the subsoil. The information is obtained through subsoil investigation around the site. At a minimum, three monitoring wells should be installed around the site to find out the depth to groundwater. The boreholes used for collecting soil samples may be used for installing the monitoring wells. Detailed discussion regarding site investigation can be found in Section 9.5.

21.2 DEVELOPING A PRIORITY LIST

Probability of groundwater impact is high in natural attenuation landfills, especially if the volume of the landfill exceeds 45,000 yd^3 (35,000 m^3). Improperly constructed or closed containment landfills may also pollute groundwater. If there are several landfills where remediation is needed, and if the total funding available for undertaking remedial action at all landfills is not sufficient, then a list is developed for prioritizing the action. A systematic approach, known as hazard ranking, is used to develop a priority list. In the hazard ranking method, a site is scored based on the possibility of impact on groundwater and surface water. The scoring is done based on contaminant migration possibility and strength of leachate. Impact on groundwater from a landfill site depends on:

1. Depth to groundwater
2. Infiltration potential
3. Permeability of unsaturated zone
4. Physical state of the waste
5. Containment design
6. Leachate strength
7. Groundwater use
8. Service value

The following sections include information for calculating the score. The score for individual items mentioned above are estimated first. Then the individual scores are combined to find a total score. If a contamination is assessed through direct evidence or in other words through groundwater monitoring data, then a score of 45 (which is the highest score) is assigned to the site. However, qualitative evidence such as objectionable taste or smell in a nearby down-gradient well may be used as direct evidence only if it can be confirmed that it results from release at the site. Waste quantity also influences contamination potential of a site. However, since the waste volume in most landfills is greater then 2000 yd^3 (1500 m^3), a score of 8 will be used to account for the waste quantity.

21.2.1 Depth to Groundwater

The depth to the groundwater is measured vertically from the lowest point of the landfill to the highest seasonal groundwater level. Depth to groundwater greatly influences the length of time a pollutant can migrate to the ground-

TABLE 21.1   Depth to Groundwater

Depth to Groundwater	Assigned Value
>150 ft	0
76–150 ft	1
21–75 ft	2

After Wisconsin Department of Natural Resources (2001).

water body. Values for depth to groundwater are shown in Table 21.1 To add to the importance of depth to groundwater the value obtained from Table 21.1 is multiplied by a factor of 2.

21.2.2   Infiltration Potential

Infiltration potential is a measure of the site characteristics that facilitate water accumulation on the landfill cover and its movement through the waste, which causes the generation of leachate. Infiltration potential depends on slope of the final cover, type of cover soil, status of vegetative cover, and precipitation at the site. The infiltration score is obtained from Table 21.2 after adding the values from Tables 21.3 and 21.4, and the value assigned for net precipitation.

Net precipitation value is obtained from Table 21.5 based on annual net precipitation for the site location. Annual net precipitation is the sum of monthly net precipitations. Net precipitation values for sites located in the United States can be found elsewhere (USEPA, 1992a). Net monthly precipitation is calculated by subtracting the monthly potential evapotranspiration from the monthly precipitation. A net precipitation value of zero is assigned if for a month the potential evapotranspiration exceeds the precipitation. Average monthly precipitation for each month can be obtained from the nearest weather station. Approximate rate of transpitation can be obtained from the pan evaporation rate, provided plants continue to get sufficient water. Discussion regarding experimental estimation of evapotranspiration is included in Sections 13.4.1 and 13.4.2.1.1. The following formulas may also be used for obtaining potential evapotranspiration:

TABLE 21.2   Infiltration Potential

Infiltration Potential	Infiltration Score	Assigned Value
Low	0–6	0
Moderately low	7–11	1
Moderately high	12–17	2
High	18–22	3

After Wisconsin Department of Natural Resources (2001).

TABLE 21.3 Values for Combined Effect of Slope and Vegetative Cover

	Site Surface Slope			
Vegetative Cover	<3%	3–5%	5–8%	>8%
None	9	7	6	5
Poorly established < (sparse, root zone 6″)	8	6	5	4
Established (good, root zone 6–12″)	6	4	3	2
Well established > (lush, root zone 12″)	4	2	1	0

After Wisconsin Department of Natural Resources (2001).

$$E_i = 0.6F_i(10T_i/I)^a \qquad (21.1)$$

where E_i = monthly potential evaportranspiration for month i, in inches
F_i = monthly latitude adjusting value for month i (see Table 21.6)
T_i = mean monthly temperature for month i, in °C

$$I = (T_i/5)^{1.514} \qquad (21.2)$$

$$a = 6.75 \times 10^{-7}I^3 - 7.71 \times 10^{-5}I^2 + 1.79 \times 10^{-2}I + 0.49239 \quad (21.3)$$

The annual net precipitation is obtained by summing the monthly net precipitation values.

21.2.3 Permeability of the Unsaturated Zone

Permeability of the unsaturated zone between the bottom of the landfill and the top of the highest seasonal groundwater level influences the travel time

TABLE 21.4 Infiltration Values for Various Types of Cover Soil

Soil Score	Infiltration
Surface soil type	Value
Sand	8
Silty sand	7
Sandy loam	6
Silty loam	5
Peaty topsoil	4
Clay loam	3
Silty clay	2
Clay	1

After Wisconsin Department of Natural Resources (2001).

TABLE 21.5   Net Precipitation Factor Values

Net Precipitation (in.)	Assigned Value
0	0
>0–5	1
>5–15	3
>15–30	6
>30	10

After USEPA (1992a).

of contaminants from the landfill. Table 21.7 provides the score for permeability of the unsaturated zone for various soil types.

21.2.4   Physical State of the Waste

The physical state of the waste influences the percolation of precipitation water through the waste. In general, the physical state of municipal solid waste is not consolidated like inorganic waste. Therefore, a score of 1 may be considered as reasonable for the physical state of waste.

21.2.5   Containment Design

Containment is a measure of how good the landfill design is in minimizing leachate infiltration to the groundwater. Both leachate collection system de-

TABLE 21.6   Monthly Latitude Adjusting Values[a]

Latitude[b] (deg)	Month											
	Jan.	Feb.	Mar.	Apr.	May	June	July	Aug.	Sep.	Oct.	Nov.	Dec.
≥50 N	0.74	0.78	1.02	1.15	1.33	1.36	1.37	1.25	1.06	0.92	0.76	0.70
45 N	0.80	0.81	1.02	1.13	1.28	1.29	1.31	1.21	1.04	0.94	0.79	0.75
40 N	0.84	0.83	1.03	1.11	1.24	1.25	1.27	1.18	1.04	0.96	0.83	0.81
35 N	0.87	0.85	1.03	1.09	1.21	1.21	1.23	1.16	1.03	0.97	0.89	0.85
30 N	0.90	0.87	1.03	1.08	1.18	1.17	1.20	1.14	1.03	0.98	0.89	0.88
20 N	0.95	0.90	1.03	1.05	1.13	1.11	1.14	1.11	1.02	1.00	0.93	0.94
10 N	1.00	0.91	1.03	1.03	1.08	1.06	1.08	1.07	1.02	1.02	0.98	0.99
0	1.04	0.94	1.04	1.01	1.04	1.01	1.04	1.04	1.01	1.04	1.01	1.04
10 S	1.08	0.97	1.05	0.99	1.00	0.96	1.00	1.02	1.00	1.06	1.05	1.09
20 S	1.14	0.99	1.05	0.97	0.96	0.91	0.95	0.99	1.00	1.08	1.09	1.15

[a] Do not round to nearest integer.
[b] For unlisted latitudes lower than 50° north or 20° south, determine the latitude adjusting value by interpolation.
After USEPA (1992a).

TABLE 21.7   Values for Hydraulic Conductivity of Various Unsaturated Zone Materials

Type of Material	Approximate Range of Hydraulic Conductivity	Assigned Value
Unfractured clay, cemented till, shale; unfractured metamorphic and igneous rocks	10^{-7} cm/sec	0
Silt, loess, silty clays, silty loams, clay loams; less permeable limestone, dolomites, and sandstone; moderately permeable till; fractured clay	10^{-5}–10^{-7} cm/sec	1
Fine sand and silty sand; sandy loams; moderately permeable limestone, dolomites, and sandstone (no karst); moderately fractured igneous and metamorphic rocks, some coarse till	10^{-3}–10^{-5} cm/sec	2
Gravel, sand; highly fractured igneous and metamorphic rocks; permeable basalt and lavas, karst limestone and dolomite	10^{-3} cm/sec	3

After Wisconsin Department of Natural Resources (2001).

sign, and surface water diversion structures are taken into consideration when obtaining a score for containment. Table 21.8 includes scores for containment design.

21.2.6 Leachate Strength

Leachate strength influences the groundwater impact potential of the site. Chances of groundwater impact rise with leachate strength. Although leachate contains many pollutants, chemical oxygen demand (COD) is used as a surrogate parameter. As indicated in Table 13.14 the COD of municipal waste various from 6.6 to 99,000. COD for various waste types are given in Tables 13.10 through 13.15. However, for hazard ranking purposes, the COD of leachate sample obtained from the site should be used. The values assigned for COD are included in Table 21.9.

21.2.7 Groundwater Use

Use of groundwater within 3 miles (4.8 km) of the site dictates the allowable impact to groundwater. If the groundwater is used for nondrinking purposes (e.g., industrial use), the hazard due to use of polluted groundwater is con-

TABLE 21.8   Values for Containment Effectiveness

Liner Type	Assigned Value
Essentially nonpermeable liner, liner compatible with waste, and adequate leachate collection system.	0
Essentially nonpermeable compatible liner, no leachate collection system, and landfill surface precludes ponding.	1
Moderately permeable, compatible liner, and landfill surface precludes ponding.	2
No liner or incompatible liner; moderately permeable compatible liner; landfill surface encourages ponding; no runon control.	3

After Wisconsin Department of Natural Resources (2001).

sidered low. However, if the groundwater is used for drinking purposes and no alternative sources of groundwater are available, then the hazard due to groundwater pollution is high. Table 21.10 includes assigned values for groundwater use. To add to the importance of groundwater use, the value obtained from Table 21.10 is multiplied by a factor of 3.

21.2.8   Service Values

Service value is a combined factor based on the distance to the nearest well and population at risk. The probability of impacting a groundwater well decreases with its distance from the landfill site. Although groundwater contamination has been observed at distances more than 3 miles away, in general, the possibility of impact becomes zero at a distance of 3 miles from the landfill site. Assigned values for the well distance are included in Table 21.11.

   Risk to population served by a groundwater source increases with increasing population living within a 3-mile radius of the landfill site. When counting the population, residents as well as other regular users of the groundwater source (e.g., workers in factories and offices) should also be included. Either an actual head count or a statistically based population estimate is done to

TABLE 21.9   Assigned Values for Leachate Strength

Leachate COD in mg/L	Assigned Value
1,000–10,000	2
10,000–20,000	4
20,000–30,000	6
30,000–40,000	8
>40,000	10

After Wisconsin Department of Natural Resources (2001).

TABLE 21.10 Assigned Values for Groundwater Use

Groundwater Use	Assigned Value
Commercial, industrial, or irrigation; and another water source presently available; groundwater not used, but usable for drinking water	1
Drinking water with municipal water from alternate unthreatened sources presently available (i.e., minimal hookup requirements); or commercial, industrial, or irrigation with no other water source presently available	2
Drinking water; no municipal water from alternate unthreatened sources presently available	3

After Wisconsin Department of Natural Resources (2001).

arrive at a population count. If aerial photography is used to count the number of houses, then an average number of residents per dwelling is used. The average number depends on the population density for the region/country. In the United States, the number of residents/dwelling is 2.8. If the groundwater is used for irrigation, then an average number per acre (hectare) of land is to be used. In the United States, the average number of population/acre is 1.5. The total population within 3 miles of the landfill site is used. The values corresponding to population are shown in Table 21.12. The estimated population value and the well distance value is combined to find the service values given in Table 21.13.

21.3 SELECTION OF REMEDIAL METHODS AND RELATED ISSUES

There are various issues that need to be resolved before selecting a remedial method. Risk assessment should be done prior to undertaking a remedial action. Discussion regarding risk assessment is included in Chapter 9. Detailed discussion on physical, chemical, and biological remediation methods

TABLE 21.11 Assigned Values for Nearest Well Distance

Distance to Well	Assigned Value
>3 miles	0
2–3 miles	1
1–2 miles	2
2000 ft to 1 mile	3
<2000 ft	4

After Wisconsin Department of Natural Resources (2001).

TABLE 21.12   Assigned Values for Population at Risk

Population	Assigned Value
0	0
1–100	1
101–700	2
701–1500	3
1501–5000	4
>5000	5

After Wisconsin Department of Natural Resources (2001).

are included in Chapter 9. Only the *in situ* remediation methods (Section 9.6) are used for landfill remediation. Although landfill mining (Section 23.12) has been reported in the literature for salvaging recyclables, it is not undertaken for remediation purposes. It may be mentioned that *ex situ* remediation methods (Section 9.7) in conjunction with landfill mining may be used for cleanup purposes. In many cases a treatment train (active removal followed by natural attenuation or engineered biodegredation) is used for cleanup purposes.

A remedial action may be undertaken either by the own/operator of the site or by a governmental body (e.g., state environmental regulatory body). The primary issue of importance is to establish the cleanup goal, which consists of the target concentration and the time frame within which to reach the target concentration of various contaminants. This is an important step in choosing a remedial method, especially if the remedial action is undertaken by the owner/operator of a site. Currently, there are three approaches available to setup a target concentration. Detailed discussions regarding the three approaches are included in Section 9.8. The cleanup goal depends heavily on groundwater use (refer to Section 21.2.7) and service value (refer to Section 21.2.8). If the contaminated aquifer is the only source of drinking water for

TABLE 21.13   Service Values Based on Population and Well Distance

Value for Population Served	Value for Distance to Nearest Well				
	0	1	2	3	4
0	0	0	0	0	0
1	0	4	6	8	10
2	0	8	12	16	20
3	0	12	18	24	30
4	0	16	24	32	35
5	0	20	30	35	40

After Wisconsin Department of Natural Resources (2001).

the affected population, then the target concentration must be based on human toxicological data. In addition, the allowable time frame to finish cleanup should reflect the least possible time based on available technology. Arrangements are made to supply the necessary amount of clean water to the affected population until the water use source is cleaned up to the required standard. This is an added cost for the remedial action. Therefore, the total cost of the project would include not only the cost of remedial method but also the cost of supplying water to the affected population. Thus, when choosing a remedial method, the technological feasibility, the total cost of the project, and the human and environmental needs must be considered. It is possible that the total cost of a project will be reduced by choosing a costlier option as the remediation method. Funding sources and the amount of funds available at different stages of the project must also be taken into consideration when choosing a remedial method.

Monitoring plays an important role in remediation actions. Monitoring of the site both before and after the cleanup is essential to establish the effectiveness of the remediation method. Discussion on a groundwater monitoring network is included in Section 22.4. The groundwater monitoring wells may deteriorate over time. Discussion regarding maintenance and rehabilitation is included in Section 9.4. The total cost of a remedial action must be estimated carefully. Both the initial cost and long-term monitoring and maintenance costs are important. The initial cost includes cost of design and construction of the chosen remediation method. The long-term costs include the cost of monitoring and maintenance of the monitoring network. The cost of providing drinking water to the affected population, if any, is also included in the long-term costs. Proper estimate of cash flow for running the project from start to finish is important. Discussion regarding project cost estimation is included in Section 9.9. Discussions on long-term cost estimates for a project and cash flow estimates are included in Sections 26.2 and 26.4, respectively.

Proper source of funding is critical for undertaking a remedial action project by the owner/operator of a landfill. Ideas regarding funding a project are included in Section 26.5. Financial assistance may be provided by state or federal agencies. Proper investigation of all funding sources should be undertaken.

22

PERFORMANCE MONITORING

There are essentially two purposes in monitoring a landfill: to find out whether a landfill is performing as designed and to ensure that the landfill meets all the regulatory standards. The performances of some of the monitoring instruments are well established and are well accepted, although many new instruments are being developed and are awaiting widespread acceptance. Usually hydrogeologists are assigned the task of developing a performance-monitoring program for groundwater around landfills; engineers conversant with air monitoring are entrusted with ambient air monitoring around landfills. However, landfill design engineers need to understand the basics of monitoring so that they can interact with the persons responsible for developing a monitoring program for a landfill and develop a proper perspective for overall design and management of a landfill. Although this chapter is meant mainly for engineers, hydrogeologists will also benefit from the information.

The following items are usually monitored in a landfill: leachate head within the landfill, head in the dewatering system (if installed), leakage through the landfill base, groundwater around the site, gas in the soil and the atmosphere around the landfill, head and quality of the leachate in the collection tank, and stability of the final cover. All of these items need not be monitored in every landfill. The following issues are considered when developing a monitoring program: (1) what equipment to use, (2) where to install the monitoring equipment(s), (3) how often monitoring should be done, and (4) what chemical constituents should be monitored. Mainly the first three issues are discussed in this chapter. Guidelines for item 4 are included in Chapter 14.

In addition to the above-mentioned issues the following three "management-type" issues have to be resolved for an overall management of performance monitoring:

552

1. *Identification of a laboratory for testing purposes.* A laboratory capable of providing correct detection limits should be used. For example, let us assume that regulatory action is mandated for a certain chemical constituent at a concentration of 0.001 ppm. The laboratory chosen must be capable of detecting the chemical species at 0.001 ppm or less. In addition, the laboratory must follow standard procedures for storing and testing samples. So an inspection of the laboratory should be undertaken prior to contracting for a project.

2. *Data acquisition and storing.* Standard forms should be used that identify the chemical constituents to be tested for each monitoring point. This form should also indicate the highest allowable concentration of each chemical constituent, which are usually mandated by the regulatory agency. After obtaining the results from the laboratory, the data should be stored properly. A computer may be used to store and retrieve data, especially if the total amount of data to be handled is quite large.

3. *Analysis of data.* A set procedure should be used to analyze the data. A statistical method or other methods should be used to judge whether a medium (e.g., groundwater) has been impacted by the landfill.

22.1 LEACHATE HEAD MONITORING

The leachate head is monitored in a containment-type landfill; leachate head monitoring is optional for a natural attenuation type of landfill. Essentially two types of leachate head well design are available (Gear, 1988): horizontal and vertical. Typical designs of both well types are included in Figs. 22.1 and 22.2. Use of a wide concrete pad or stainless steel plate will allow reconstruction of another well if the first one is damaged. If a head well needs to be installed in the future after waste placement, then the exact location of the concrete pad/steel plate must be recorded in the drawing. Table 22.1 indicates the advantages and disadvantages of each well type. The leachate head within a landfill is expected to vary both with location and time. It is

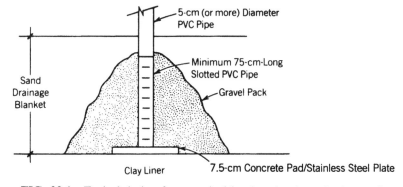

FIG. 22.1 Typical design for a vertical leachate head monitoring well.

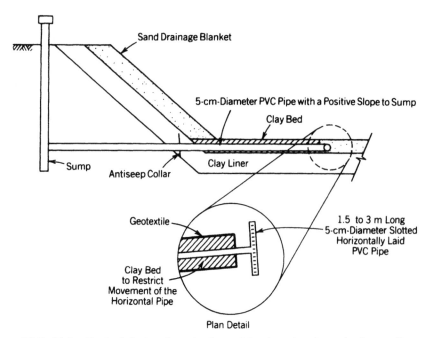

FIG. 22.2 Typical design for a horizontal leachate head monitoring well.

expected to be highest near the crest of the liner and lowest near the collection pipes. At a minimum the leachate head level should be monitored near these two locations.

The leachate head level varies with time. Therefore monitoring of the leachate head level should be done frequently to see if the landfill is performing as designed. A weekly monitoring for the first 3–4 years of operation and monthly monitoring thereafter is suggested.

TABLE 22.1 Advantages and Disadvantages of Vertical- and Horizontal-Type Leachate Head Wells

Sl Number	Vertical Head Well	Horizontal Head Well
1	Can be installed before, during, or at completion of waste placement	Can be installed before waste placement
2	Must be extended periodically if installed before or during waste placement	No extension is necessary
3	Chance of damage during waste placement is high	Chance of damage during waste placement is low
4	Hard to protect from waste shifting and settling, especially in sludge landfills	Waste shifting and setting has no effect

22.2   MONITORING HEAD IN THE GROUNDWATER DEWATERING SYSTEM

Head within the groundwater dewatering system should be monitored to check whether the system is performing as designed. The head is also expected to vary both in terms of location and time. Only horizontal monitoring wells are used to monitor the head. The design of the head well is similar to that for the horizontal leachate head well. Monthly monitoring to study the seasonal variation is suggested for the first 3–4 years. Thereafter the frequency may be reduced to only the few months of the year in which seasonally high levels were observed prior to construction of the landfill.

22.3   LEAKAGE MONITORING

The unsaturated zone between the liner and the groundwater table (if the liner is completely above the seasonal high groundwater table) has to be monitored to detect leaks. The literature identifies two approaches:

1. Installation of instrumentation that can collect leachate exfiltrate. These are termed direct leakage monitors.
2. Installation of instruments that can detect water percolation, and so on. These are termed indirect leakage monitors.

Bumb et al. (1988) and McKee and Bumb (1988) had proposed models that can be used to design a leak detection monitor network for hazardous waste landfills in sensitive environments. It should be mentioned that although early leak detection provides an early warning, the results may not be used to enforce any legal remedial action if the laws/rules are applicable to groundwater. Water in the vadose zone is not normally considered groundwater unless the legal definition explicitly includes it. This is more of a legal issue, and hence a lawyer should be consulted as to how the results may be used for enforcement purposes. However, the use of direct or indirect leakage monitors should not be restricted because of legal constraints. Remedial action costs less if a problem is detected at an early stage. A leakage monitor should be considered as an integral part of landfill design.

22.3.1   Direct Leakage Monitors

Two types of direct leakage monitors are available: suction lysimeters and basin lysimeters. The location and number of lysimeters depend on landfill design. Installation of ore than one lysimeter below the subbase is preferable because performance of the landfill can be continued to be monitored even if one lysimeter fails. The lysimeters should be installed near the edge of a landfill so that the length of transfer piping is minimum; horizontal bends on transfer pipe should be avoided. Lysimeters can be installed almost anywhere below a natural attenuation landfill. However, for containment-type landfills the following recommendations should be followed in selecting a lysimeter location. The leachate head is highest at the crest of the base liner, so this is the location where maximum leakage is expected. Minimum leakage is expected below the leachate collection trenches. If one lysimeter is to be installed per phase in a landfill, then it should be installed below a crest of the base liner. If two lysimeters are to be installed per phase, then they should be located at the following points: one below a crest of the base liner and the other below the middle of a module. A third may be installed below the collection trench if necessary (Fig. 22.3). There should be at least one lysimeter below each phase of a containment-type landfill.

22.3.1.1 Suction Lysimeter   Suction lysimeters should be installed below the liner area. To protect these lysimeters during landfill construction and operation, they should be installed several feet below the subbase grade. These lysimeters do not function very well if the moisture content of the vadose zone is extremely low.

Three different designs for suction lysimeters are available: vacuum operated, vacuum-pressure operated, and vacuum-pressure samplers with check valves (Everett, 1981). These have been used (in nonlandfill projects) for a number of years, both in laboratory and field studies (Parizek and Lane, 1970; Apgar and Langmuir, 1971; Johnson and Cartwright, 1980). Suction lysimeters may provide valuable information if installed and operated properly. Typically a series of suction lysimeters are installed at several depths in the same or immediately adjacent boreholes. A typical suction lysimeter is shown in Fig. 22.4.

22.3.1.2 Collection Lysimeter   These are also called basin-type lysimeters. These have been used successfully in many landfills (Kmet and Lindorff, 1983). The lysimeter is to be constructed in the field. Figure 22.5 shows a

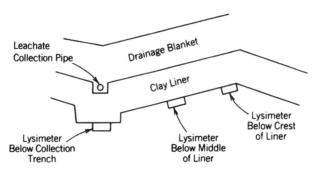

FIG. 22.3   Suggested location of lysimeters in order of preference.

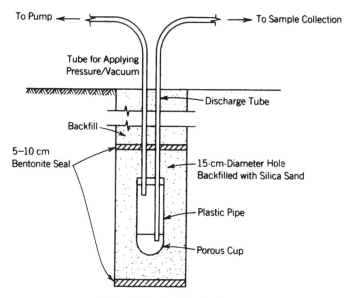

FIG. 22.4   Suction lysimeter.

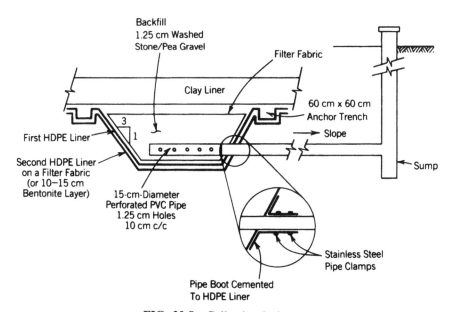

FIG. 22.5   Collection lysimeter.

collection lysimeter design. The exfiltrate collected can be monitored for rate of leakage and quality. The basin should be sufficiently large so that it collects an adequate volume of sample necessary for determining the quality of leachate. Usually 2 liters of leachate is necessary for analytical work. A testing laboratory should be contacted to verify the volume requirement.

The transfer pipe drains to a standpipe from which the leachate is withdrawn, either by pump or manually, on a regular basis. The standpipe should be sized in such a way that it can hold leakage in between withdrawals. The standpipe should be large enough so that it can be cleaned occasionally.

The leachate head in the standpipe should be monitored more frequently (the suggested interval is 15 days) for the first 2 years initially. Leachate must be removed whenever the level reaches 15 cm (6 in.) below the invert of the transfer pipe. The frequency of head monitoring in the standpipe and withdrawal may be reduced when a stabilized leakage rate is observed for at least 2 subsequent years. The transfer pipe should be cleaned periodically (the suggested frequency is once every 2 years).

Example 22.1   Determine the area for a basin lysimeter installed below a 1.3-m- (4.3-ft)-thick clay-lined site. Assume 100% spillage, and that sampling frequency = 30 days, liner permeability = 1×10^{-7} cm/sec (note: liner permeability is usually reduced due to clogging, etc., within 2–3 years of construction), and average leachate head over the liner = 30 cm.

$$\text{Leakage/sec} = (1 \times 10^{-8}) \times (130 + 30) \times A/(130)$$

$$= (1.23 \times 10^{-8}) \times A$$

$$\text{Leakage/30 day} = 0.03 \times A \ (= \text{basin area})$$

There should be at least 4 liters of liquid in a 30-day period (assuming a 100% spillage). Therefore

$$0.03A = 4 \times 1000$$

or

$$A = 125386 \text{ cm}^2$$

Assuming a length-to-breadth ratio of 2, the dimensions of the basin should be $(2B \times B = A = 125{,}386)$ 250 cm $\times$ 500 cm (8 ft $\times$ 16 ft). The size should be increased by 50–100% to account for lower permeability and additional spillage.

22.3.2   Indirect Leakage Monitors

Many researchers are attempting to develop new instruments or to adapt an existing instrument for landfill leakage detection purposes (Davies et al., 1984; Maser and Kaelin, 1986; Hwang, 1987; Burton, 1987; Daniel et al., 1981; Christel et al., 1985; Everett et al., 1982; Phene et al., 1971a,b; Thiel et al., 1963; Wilson, 1981, 1982, 1983). In most cases indirect leakage monitors were tested in an experimental setup or a conceptual use was discussed.

The indirect monitors can be subdivided into two groups: (1) instruments that detect changes in moisture content of the vadose zone and (2) instruments that detect a change in the chemical concentration (salinity in most cases) in the vadose zone. These monitors detect changes of moisture content or chemical concentration in the vadose zone but not the quality at any particular time. Since some leakage is expected through a clay-lined landfill, these monitors will provide information at the initial stage when changes are taking place. But after several years, when steady state is reached in the vadose zone, most of these monitors will probably fail to provide any valuable information. However, for landfills that are designed for zero or negligible leakage (double lined for landfills lined with synthetic membrane), use of these types of monitors may provide valuable information if a serious leak develops. Brief descriptions of several indirect monitors reported in the literature follow. In general, indirect monitors are not used for routine landfill monitoring but it appears that some of these monitors have potential for future use.

22.3.2.1 Neutron Moisture Meter   The meter essentially provides a profile of the moisture content of the soil (Wilson, 1980) below the landfill. These meters can be installed below a landfill or moved through a borehole (Fig. 22.6) next to the landfill (Daniel and Kurtovich, 1987).

22.3.2.2 Gamma Ray Attenuation Probes   These are used for detecting changes in moisture content (Schmugge et al., 1980). The degree of attenuation of gamma rays depends on the bulk density and water content of the

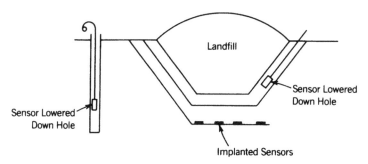

FIG. 22.6   Neutron moisture meter placement.

medium. Two different methods are available: transmission and scattering. In the transmission method two wells are installed at a known distance. A single well is used in the scattering method. The method is usually limited to shallow depth because of difficulties in installing parallel wells.

22.3.2.3   Electrical Resistance Blocks   These also measure changes in water content. Electrode blocks embedded in porous material are installed in the soil. The electrical properties of the blocks change with the changing water content of the vadose zone (Schmugge et al., 1980). The water content is calculated from calibration curves after determining changes in the electrical properties of the blocks.

22.3.2.4   Thermocouple Psychrometers   These meters are used to detect changes in moisture content. The meter relies on cooling the thermocouple junction by the peltier effect (Merill and Rawlins, 1972). Two types of meters are available: wet bulb and dew point. The dew-point method is more accurate (Wilson, 1981). The relationship between soil water potential and relative humidity is used in reading the water content of the medium.

22.3.2.5   Tensiometers   Tensiometers have been used for many years by soil scientists to measure the matric potential of soil. (Richards and Gardner, 1936; Marthaler et al., 1983). Tensiometers measure the negative pressure (capillary pressure) that exists in unsaturated soil. Many researchers (Christel et al., 1985; Long, 1982; Klute and Peters, 1962; Thiel et al., 1963) reported the use of an electrical pressure transducer instead of a conventional mercury manometers. Figure 22.7 shows a sketch of a typical tensiometer.

22.3.2.6   Heat Dissipation Sensors   These units also can monitor water content by measuring the rate of heat dissipation from the block to the surrounding soil. Calibration curves relating matric potential and temperature difference for the on-site soil need to be developed in the laboratory.

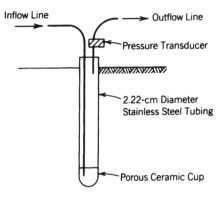

FIG. 22.7   Tensiometers.

22.3.2.7 Electrical Probes   These probes can be used to determine the salinity of the vadose zone. Four probes are installed in an array (Wenner array) so that conductivity of the soil can be measured (Rhoades, 1979; Rhoades and Van Schilfgaarde, 1976). A calibration curve relating conductivity and soil salinity is used for predicting soil salinity.

22.3.2.8 Salinity Sensors   As the name indicates, these sensors can be used to monitor soil salinity. Electrodes attached to a porous ceramic cup are installed in the soil (Oster and Ingualson, 1967). Calibration curves relating specific conductance and salinity are used to interpret readings.

22.3.2.9 Soil Gas Probes   These probes monitor volatile organic compounds (VOC) in the soil. The gas probe consists of a galvanized pipe that can be driven into the ground easily. A typical soil gas probe is shown in Fig. 22.8. The gas may be analyzed *in situ* using a portable gas chromatograph or tested in a laboratory after absorbing it in charcoal (Silka, 1988; Daniel and Kurtovich, 1987).

22.3.2.10 Wave Sensing Devices   Use of both seismic and acoustic wave propagation properties for leak detection in experimental setup has been reported in the literature (Maser and Kaelin, 1986; Davis et al., 1984). In the seismic wave technique, the difference in travel time of Rayleigh waves between the source and geophones is used to detect leaks. In the acoustic emission monitoring (AEM) technique, sound waves generated by blowing water from a leak are utilized in leak detection.

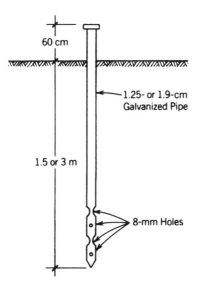

FIG. 22.8   Soil gas probe.

22.3.2.11   Time Domain Reflectrometry (TDR)   This technique is based on the difference in dielectric properties of water and soil. TDR measures wide-frequency bandwidth and short-pulse length that are sensitive to the high-frequency electrical properties of the material (Davis et al., 1984). TDR is reported to be insensitive to soil type, density, temperature, and pore liquid quality but is sensitive to changes in the water content of soil (Topp et al., 1980).

22.4   GROUNDWATER MONITORING

Two approaches are available to detect groundwater contamination: a direct method using groundwater monitoring wells and an indirect method using geophysical techniques. However, groundwater contamination detection using geophysical techniques is normally used for remedial actions but not for routine monitoring purposes. Hence only the direct method is discussed. The following are issues that should be considered for the design, installation, and sampling of a groundwater monitoring network: (1) coordinates of sampling points, (2) minimum number of sampling points required, (3) design and installation of wells, (4) groundwater quality status prior to landfilling, (5) frequency of sampling, and (6) collection and preservation of samples. The first two issues are discussed in some detail and the other issues are briefly addressed in the following section.

22.4.1   Coordinates and Number of Sampling Points

An approximate leachate plume configuration needs to be visualized for the design and installation of a monitoring network prior to disposal of waste in a landfill. This is a difficult task, especially if the soil below the groundwater table has extensive stratification, lenses of soil of different permeability, fractures, and so on. Such nonuniformity is a rule rather than an exception in glaciated regions.

A thorough site characterization is essential for the successful design of a monitoring well network. Some guidelines for site characterization are discussed in Chapter 12. When designing a monitoring network, it is important to recognize that contaminant migration is a three-dimensional phenomenon. Thus, the vertical depth at which a well screen is placed is as important as its grid location. The three major factors that influence monitoring well location include chemical characteristics of the contaminants expected to seep into the aquifer (this item is waste specific), design of the landfill (i.e., whether the site is lined and if so the liner type; this item is landfill design specific), and the hydrogeological characteristics of the site (this item is site specific).

The movement of a contaminant in the subsurface depends partly on its solubility in water and on its diffusivity in and reactivity with the geologic material beneath the site. The mobility of different leachate constituents will be different. Light immiscible contaminants tend to float while transport of heavy miscible contaminants is governed by its viscosity and density. Non-reactive species (e.g., chloride, low-molecular-weight organics) are expected to be quite mobile, whereas heavy metals and high-molecular-weight organics are much less mobile (Technical Enforcement Guidance Document, 1986). Thus, attenuation mechanisms of leachate constituents (discussed in detail in Chapter 15) should be given due consideration when designing a monitoring network.

The plume configuration is influenced by the velocity with which the ex-filtrate (leachate after percolating through the landfill subbase) enters the aquifer (Paschke, 1982). This entry velocity is higher for natural attenuation landfills than for containment landfills. Therefore the leachate constituents that leak primarily due to advective transport are expected to penetrate deeper into the aquifer for natural attenuation landfills than for containment landfills. However, landfill design is not expected to have a significant influence on plume for leachate constituents that leak primarily through diffusion.

The hydrogeologic characteristics of a site can be heterogeneous or homogeneous. A monitoring network for a homogeneous situation is easy to understand, and the concept can be helpful in visualizing monitoring networks in heterogeneous situations. A three-dimensional plume geometry for a homogeneous aquifer is discussed in the following paragraphs, which is helpful in developing a concept for a three-dimensional array of monitoring points.

It is simpler to locate monitoring points in uniform aquifers. The following generalization regarding plume geometry is applicable to deep sandy aquifers. The discussion will provide some insight into a monitoring network design. It is recognized that the soil stratigraphy used is very simple and is not usually encountered in the real world. It is introduced here with the hope that developing a concept for the design of a monitoring network in a simple case will help in visualizing a monitoring network in more complicated real-world situations. The general trend of concentration distribution of contaminants (CDC) and the shape of the plume (based on theory, laboratory experiments, and field observation) are summarized in this section (Bruch and Street, 1967; Paschke, 1982; Paschke and Hoppes, 1984; Kimmel and Braids, 1974; Nicholson et al., 1980; Bear, 1969; Bouwer, 1979; Cherry et al., 1981; Fattah, 1974; Pickens and Lennox, 1976). A circular landfill base is assumed for easy depiction and interpretation. A plume for a circular landfill base will have an elliptical cross section (Paschke and Hoppes, 1984); if the concentrations for a particular cross section are plotted, then all the points will ideally lie on a bell-shaped surface (Fig. 22.9). The highest concentration (*MM*) at a particular section occurs slightly below $0.5d$, where d is the depth of the lower boundary of the plume at that section measured from the water table. The concentration *MM* will be higher for sections closer to the landfill. The spread of contamination increases with distance from the landfill. To visualize the concentration variation within a landfill plume, one may imagine that the ordinates of the bell-shaped surface change with distance from the landfill; the height of the bell decreases and the base area increases as the distance from the landfill

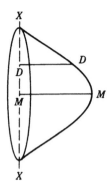

FIG. 22.9   Concentration distributon of contaminants.

increases. In Fig. 22.10 the dotted lines at $X-X$, $Y-Y$, and $Z-Z$ illustrate the concentration distribution in a vertical plane within the leachate plume at these locations.

Let $(DD)x$ represent the concentration of a contaminant at section $X-X$ and at some depth from the top of the plume, while $(MM)x$ represents the highest concentration of a contaminant at that section. At a subsequent section (e.g., $y-y$), $(MM)y$ will be less than $(MM)x$ and $(DD)y$ will be less than $(DD)x$. However, depending on the relative location of the planes $x-x$ and $y-y$, $(MM)y$ could be less than, equal to, or greater than $(DD)x$. Hence, depending on the

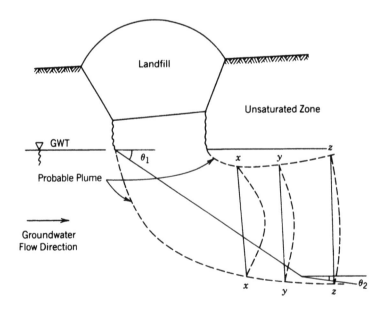

FIG. 22.10   Cross section of a landfill plume in a thick homogeneous sandy aquifer.

three-dimensional coordinates of a monitoring point, it is possible to find higher concentrations of a contaminant at a down-gradient monitoring point than in a relatively up-gradient monitoring point.

As shown in Fig. 22.10, the leachate plume for sandy aquifers can be enclosed approximately by straight lines. The angle in Fig. 22.10 is termed the divergence angle, which is the angle made by a straight line (with the horizontal) obtained by joining the far end of the landfill and the boundary of the plume at an arbitrarily fixed distance [a distance of 91 m (300 ft) was used]. The suggested plume geometry is based primarily on field plumes (Kimmel and Braids, 1974; Nicholson et al., 1980). The soil stratigraphy for the two field plumes was as follows: the base of the unlined NA-type landfill was near the groundwater table, which occurred within a sandy aquifer consisting of coarse to fine sand with little or no fines. The groundwater flow was nearly horizontal in the vicinity of the landfills. For containment-type sites the divergence angle is expected to be lower because the leachate entry velocity into the groundwater table will be lower.

Discussions in the preceding paragraphs can be used to design a network in homogeneous aquifers. Although homogeneous aquifers do exist, heterogeneity in aquifers is more common. Two common types of heterogeneity are interbedding (or alternate layers of sandy and clayey/silty materials) and clay lenses in a predominantly sandy stratum. General monitoring approaches for these two settings are discussed. Knowledge about the geology of the site is important in designing a monitoring network. Although contaminant transport modeling can be used to develop the probable plume configuration for heterogeneous subsoil conditions, in practice such modeling is not undertaken. In general, monitoring wells are placed on potential migration paths.

Interbedded Aquifer   Groundwater flow direction can be different in the upper and lower stratum. There may be different horizontal and vertical gradients in each stratum; there is a possibility of existence of vertical gradients in the upper strata in most cases. Proper assessment of the flow direction and gradients (horizontal and vertical) is the key to a successful monitoring network design. Figure 22.11 shows an example monitoring network for a natural attenuation type of municipal waste landfill sited in an interbedded aquifer. Note that background water samples are collected from all three strata, and monitoring points are well dispersed in each stratum because the plume is expected to penetrate into the lower sandy aquifer. For a containment landfill the plume configuration is expected to be different and hence the monitoring network should be adjusted accordingly. If light immiscible compounds are present in the leachate, then additional monitoring points should be installed near the water table in the upper aquifer.

Aquifer with Clay Lenses   The size of the lenses is an important issue for such sites. Sometimes the lenses are quite large and may have perched water. Whether the perched water should be monitored depends on the groundwater rules applicable for the site. (Note: In some states in the United States perched

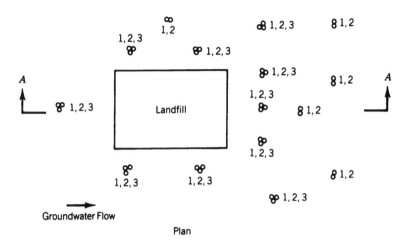

Plan

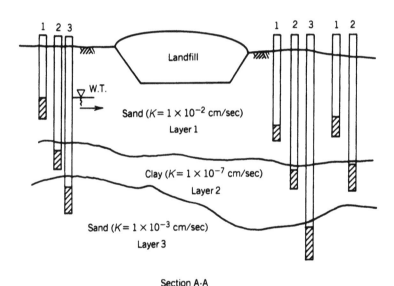

Section A·A

FIG. 22.11   Monitoring of an interbedded aquifer.

water monitoring is required, whereas in others only the "usable aquifer" needs to be monitored.) If a big lens is present within an aquifer, it should be monitored. Figure 22.12 shows an example of a containment landfill for papermill waste sited in this type of aquifer. Note the background water is sampled at two depths and only the large clay lens is monitored separately.

Thus, a three-dimensional array of monitoring points is needed for proper monitoring of the groundwater down-gradient of landfills. The soil stratigraphy beneath the landfill base and the hydrogeology of the site will influence

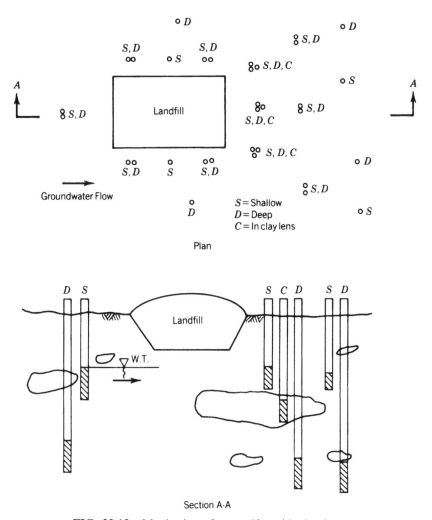

FIG. 22.12   Monitoring of an aquifer with clay lenses.

the plume geometry. Monitoring points for each landfill should therefore be custom designed. Data from down-gradient monitoring wells could be anomalous and are sometimes almost impossible to interpret without data from a background well that should be installed at a sufficient up-gradient distance from the facility (Bagchi et al., 1980). It should be mentioned that leachate plumes do not always sink to the bottom of an aquifer, although leachate is denser than groundwater. The plume geometry depends on the leachate entry velocity, groundwater velocity, permeability of the aquifer, soil stratigraphy below the landfill, density of the leachate, *in situ* diffusion–dispersion coefficient of leachate constituents, and the thickness of the aquifer (Paschke and Hoppes, 1984). Therefore, data from a number of deep monitoring wells will not necessarily represent the worst conditions of groundwater contamination from a landfill. A three-dimensional array of monitoring points intercepting the plume is necessary to determine the status of groundwater contamination due to landfill siting. The most difficult part is to predict the plume geometry.

22.4.2   Design of Wells and Sampling Frequency

Once the sampling point coordinates are fixed, properly designed monitoring wells need to be installed at these points. The design and installation of background wells and the status of groundwater quality are critical issues.

Usually an up-gradient well is installed to monitor background water quality. Monitoring at regular frequencies is needed to judge the change in quality of the groundwater down-gradient of the landfill. The quality of background water does change and it may be impacted due to a pollution source up-gradient of the landfill. If a known pollution source is absent, then monitoring at a single point is enough. However, if a known pollution source exists up-gradient, then several monitoring points will be necessary to judge the background water quality. Usually a long-screen well (Fig. 22.13*a*) is used for monitoring background data. Multilevel monitoring is probably a better approach because this will indicate the CDC if one exists. However, several piezometers (Fig. 22.13*b*) installed at different depths can also be used. The above well designs are also used to monitor down-gradient water quality. It may be noted that in addition to collecting water samples at different depths, the depth of the water table must also be monitored for interpreting data.

The material for well casing depends on the chemical parameters monitored. PVC pipes can be used to monitor inorganic constituents. However, metallic pipe coated with a nonreactive chemical or stainless-steel pipe is recommended for monitoring organic constituents. For shallow wells 5-cm-(2-in.)-diameter schedule 40 PVC casing may be used; however, for deeper wells 31-m (100-ft or more) schedule 80 PVC pipe should be used. The well screen material should also be nonreactive with the expected leachate exfiltrate constituents. The slot size depends on the geologic material in which it is placed. For most material a 0.15-mm (0.006-in.) or 0.25-mm (0.01-in.) opening is suggested; a larger slot size of 0.5 mm (0.02 in.) may be used for a coarse sand or gravel environment. Additional information regarding screen slot openings and filter packs can be found elsewhere (Driscoll, 1986). The slot opening should cover approximately 10% of the screen's surface area. Filter cloth should not be wrapped around the well screen (Bureau of Solid Waste Management, 1985). The material placed around the screen is known as filter pack. Properly sized filter pack filters out sediments and increases the effectiveness of the well. The material used should be nonreactive (e.g., clean silica sand) to chemical constituents monitored. The grain size of a filter pack depends on the grain size of the geologic material surrounding the well. Poorly graded (preferably single-sized particle) medium or coarse clean sand or pea gravel is generally used in a sandy environment. Clean fine sand is

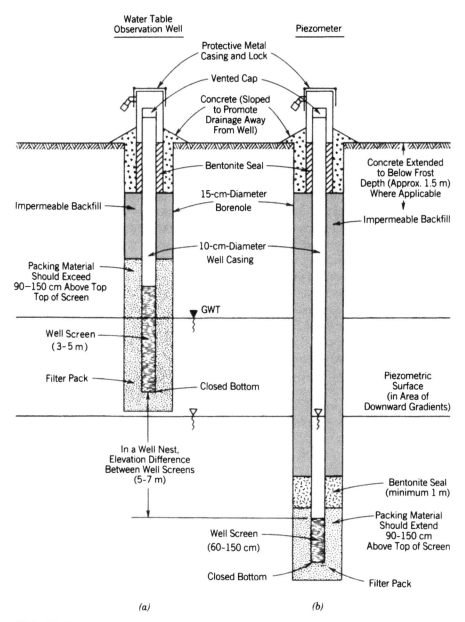

FIG. 22.13 Typical detail of a water table observation well and piezometer. Not to scale.

recommended for a silty and clayey environment. The physical characteristics of the backfill material should exhibit a balance between strength, impermeability, continuity, and chemical compatibility. Bentonite, cement, and polymers are commonly used for backfilling and well sealing. Several types of equipment for drilling well boreholes are available. A drilling method that introduces the least amount of foreign material to the borehole and causes minimum disturbance to the formation should be used.

After installation, a well must be developed (or cleaned) to remove fines accumulated in the borehole during installation and to restore the natural permeability of the surrounding soil. Factors that affect well development include well completion method, open area and slot configuration, slot size, drilling fluid type, filter pack thickness, and type of formation. The best development method is one that causes water to flow rapidly both in an out of the well screen in order to dislodge and remove fine particles. Usually a pump is used to draw the water out of a well. A bailer or surge block may be used for dislodging sediments from the screen. A well should be developed until the water is clear or until the conductivity, pH, and temperature of the pumped water remain constant for a period of time. The various well development methods include overpumping, backwashing, mechanical surging, development using air, high-velocity water, or air jetting, and high-velocity water jetting combined with simultaneous pumping (Driscoll, 1986).

A well must be protected at the ground surface. The log of well installation must be documented properly. The following information should be included in a well installation log: well number; well location (indicating site grid point); data of installation; diameter and type of well casing; mean sea level elevations of the top of the well casing, ground surface near the well, and top of the screen; length and material of the well screen; depth of the bottom of the well; and the type of well (i.e., water level well or piezometer). The quality of groundwater at the site prior to disposing of waste (known as background quality) should be established by sampling on-site wells. The following program for establishing background water quality may be used in the absence of a program set by the regulatory agency: eight rounds of monthly or quarterly sampling of all the wells including the background well for all the parameters for which the groundwater is to be monitored after disposing of waste in the landfill.

Usually groundwater is monitored quarterly, biannually, or annually, depending on the type of waste, size and design of the landfill, aquifer material, and so on. In most cases a quarterly monitoring is undertaken; annual monitoring is undertaken for small landfills located in remote places far away from any groundwater use source.

Collection, preservation, and testing of the groundwater sample are important to obtain representative data. The water level of each monitoring well should be purged by removing four well volumes (internal radius of the well × the height of the water column in the well) of water using a bailer or a pump. Certain parameters (temperature, specific conductance, pH, color, odor, and turbidity) of the water are measured in the field prior to filtering the sample. Field filtering of groundwater samples should be avoided where possible. Low-flow pumps may be used for turbid samples. However, in cold climates field filtering of turbid samples is unavoidable because use of low-flow pumps is not a practical alternative (Connelly, 1994). The collected sample must be preserved using different chemicals for different parameters. Bailer, pumps, and so on must be cleaned after sampling each well to avoid cross-contamination (Lindorff et al., 1987). Groundwater samples should be tested in a laboratory in which quality control measures are good and the lowest possible detection limits are used.

22.5 GAS MONITORING

Gas around a landfill (both above and within ground) needs to be monitored. Although the possibility of the movement of gas through the soil for containment-type landfills is low, it should be monitored on a routine basis. The air on and around landfills should also be monitored to check for hazardous air contaminants injurious to the health of landfill workers and people living around a landfill.

22.5.1 Underground Gas Monitoring

Gas probes similar in design to groundwater wells are installed around landfills. A study of the subsoil stratigraphy must be undertaken prior to selecting monitoring points. Usually migration occurs through sandy deposits, however, highly fractured formations can also serve as a conduit for gas migration. Gas can also migrate through gravel beds of utility lines running close to a landfill. So the possible conduits of migration should be identified before installing a gas probe. Either short-screen or long-screen probes (Fig. 22.14) may be used to monitor gas. The bottom of the probes should be above the highest seasonal elevation of groundwater. Sometimes a nest of three short-screen probes is used to detect gas.

Usually landfill gas is monitored for methane concentration; however, other hazardous air contaminants may be added to the list. Methane is explosive between 5 and 15% volume/volume concentration in air. Landfill gas migration seems to occur in pulses. Figure 22.15 shows a typical concentration variation of methane for an NA-type landfill. Because of the high variability in gas concentration quarterly or even monthly monitoring may not detect the real status of migration because the time and date of sampling may not synchronize with the high concentration (Bagchi and Carey, 1986). It is suggested that gas monitoring be done twice a day (midmorning to early afternoon and late afternoon) for 7–10 consecutive days in the month(s) when migration is most likely to occur. Chances of gas migration are high when the ground is either frozen or saturated.

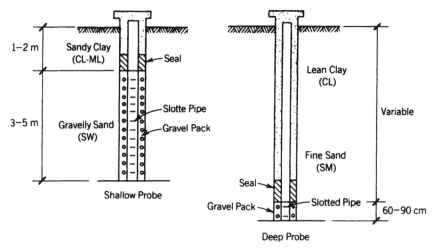

FIG. 22.14   Landfill gas probes. After Bagchi and Carey (1986).

22.5.2   Landfill Air Monitoring

Sample collection and allowable concentration of air contaminants pose special problems for landfill air monitoring. The threshold limit values (TLV) issued by the American Conference of Governmental Industrial Hygienists (1987) are for indoor exposure. In a survey of landfill gas in Wisconsin, Chazin et al. (1987) used source sampling of active or passive gas vents. The sample collection method included a Gillian personal sampling pump and activated charcoal tubes with a sampling time of 60 min. The Wisconsin study included several landfills with different waste types: municipal with industrial, municipal, combined municipal (papermill sludge, refuse, and industrial), and papermill sludge. The study indicated that vinyl chloride (a known carcinogen) is expected to be present in all large municipal landfills (383,000 m^3 or 500,000 yd^3); benzene may also be present in large municipal landfills. In

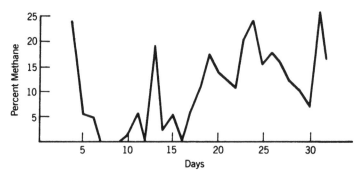

FIG. 22.15   Typical concentration variation of methane observed in landfill gas probes. After Bagchi and Carey (1986).

addition to hazardous air contaminants, dust concentration may also pose a health risk. The following possible causes for the presence of vinyl chloride in municipal landfills are suggested: (1) *in situ* formation by bacteria from other chemicals present in the landfill; (2) a by-product of the chemical decomposition of polyvinyl chloride plastic; and (3) formation of vinyl chloride from trichloroethylene by bacterial action. Of these three, the third is believed to be the most likely mechanism. Thus, discarding of plastic with solvents such as trichloroethylene should be avoided to reduce the chance of hazardous air contaminants in landfill gas.

Several sampling techniques are available. The principal objective of sampling is to collect a polluted air sample to analyze the concentration of pollutants. Gas sampling devices an be divided into three categories: passive, grab, and active. Passive sampling involves the collection of pollutants by the diffusion of gas to a collection medium. Although the passive sampling technique is simple and costs less, a long collection time is required (7–30 days) (Godish, 1987). Presently passive sampling is almost obsolete. An evacuated flask, gas syringe, or air collection bag made of synthetic materials is used to collect grab samples. An active gas sampling technique consists of a train of a pump, a tube, and a sampler, which may be a cassette, impinger, or sorbent tube. Particulate matter may be sampled by a hi-vol sampler (Axelrod and Lodge, 1977). In a hi-vol sampler a collecting glass fiber filter is installed upstream of a heavy-duty vacuum pump with a high air flow rate of 1.12–1.68 m^3/min (40–60 ft^3/min). The filter is placed in a housing to protect it from impact of debris and so on. Paper tape samples draw ambient air through a cellulose filter tape. The particle concentration is measured by optical densiometers usually installed with the tape sampler. Because of difficulties in correlating optical reading with the gravimetric data, their use is limited (Godish, 1987). In impactors, air is drawn and then led through a deflecting surface whereby it can be fractionated to six or more sizes. The hi-vol samplers collect samples of all sizes. Since larger particles do not pose a health risk, data from a hi-vol sampler may not be reliable in assessing particulate density. Impactor samplers are better devices for assessing dust concentrations that pose health risks.

Table 22.2 includes time-weighted average threshold limit values (TLV) for some air contaminants. It may be noted that these are based on a 40-hr/week exposure in a workplace. Usually the allowable concentration in ambient air is much less (e.g., the allowable concentration of vinyl chloride in Wisconsin is 0.0006 ppb for a 24-hr exposure) because a 24-hr exposure is assumed. A health risk assessment should be undertaken to establish allowable concentrations of a contaminant in ambient air.

22.6 LEACHATE TANK MONITORING

A leachate tank should be monitored for level of accumulated liquid and leachate quality. The leachate generation rate is estimated during the landfill

TABLE 22.2 Threshold Limit Values of Selected Air Contaminants[a]

Contaminant	TLV
Dust	1 mg/m^3
Carbon monoxide	50 ppm
Asbestos	0.2–2 fibers/cm^3 (depending on asbestos type)
Benzene	10 ppm
Coal dust	2 mg/m^3
Cotton dust	0.2 mg/m^3
Grain dust	4 mg/m^3
Hydrogen sulfide	10 ppm
Nuisance particulates	10 mg/m^3
Phenol	5 ppm
Vinyl chloride	5 ppm
Wood dust	
Hard wood	1 mg/m^3
Soft wood	5 mg/m^3

[a] Values of TLV obtained from the American Conference of Governmental Industrial Hygienists (1987).

design. Therefore the rate of filling the tank can be anticipated. Based on this rate, the head level in the tank and extracted volume should be monitored daily/weekly/monthly to ensure that overflowing of the tank does not occur. The quality of leachate should be monitored at least annually during the active life of the landfill and 2–5 years after closure. Leachate quality monitoring is needed to interpret groundwater data and revise a groundwater monitoring program. In addition, leachate quality monitoring is helpful in running the treatment plant. Sometimes BOD and volume of leachate are monitored daily prior to discharging the leachate into the intake point of the treatment plant.

22.7 FINAL COVER STABILITY MONITORING

The stability of the final cover should be monitored if a higher than usual waste slope is used for a landfill. Approaches for monitoring a synthetic cover and clay cover are discussed below.

22.7.1 Synthetic Cover Monitoring

Settlement hubs are installed over synthetic covers to monitor settlement. Excessive settlement may lead to shearing of the synthetic membrane. Usually settlement is monitored for sludge-type waste. Settlement hubs should be placed at 30 m (100 ft) grid points (or closer for very unstable sludge) and may be monitored quarterly or biannually. They may be placed on the side slope to monitor its stability. Settlement hubs are usually installed on the top surface of sludge landfills, which is more likely to settle.

22.7.2   Clay Cover Monitoring

Settlement of clay-covered landfills may also be monitored. Usually settlement monuments are used to monitor settlement. Monuments may be installed on the side slope to monitor the stability of the slope. Monuments may be established at 30 m (100 ft) grid points (or closer) on the top surface and monitored quarterly or biannually. Usually slope failure occurs along a circular arc (Fig. 22.16). So if the side slope is to be monitored, then a minimum of three monuments along a slope line should be established. Both horizontal and vertical movement of these monuments should be monitored biannually or annually to judge stability.

22.8   GROUNDWATER DATA ANALYSIS

If an allowable concentration for a parameter is set by a regulatory agency, then it is easy to know whether any of the monitoring wells has been impacted. However, in the absence of such set concentration, interpretation and assessment of groundwater data are difficult tasks. Although several statistical approaches are available (e.g., Student's t test), time versus concentration plots and box plots are two powerful tools for analyzing the data and detecting contamination. Figure 22.17 shows a plan view of an example NA-type landfill and the groundwater monitoring points. The time versus concentration plots of the monitoring points are shown in Fig. 22.18. As is obvious from Fig. 22.17, MW 1 and MW 2 are background wells that were installed to monitor the water quality of the groundwater before getting contaminated by the landfill leachate. Groundwater sampled from MW 3 and MW 4 show an increase in specific conductance with time (Fig. 22.18) indicating impact. The box plots of MW 1, MW 2, MW 3, and MW 4 are shown in Fig. 22.19. Note the spread of data in the box plots for MW 3 and MW 4 whereas box plots of MW 1 and MW 2 shows no spread. It may be mentioned that in many cases contamination of a well cannot be detected easily from a time versus

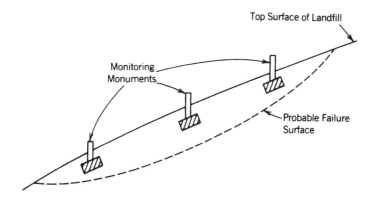

FIG. 22.16   Arrangement of monuments for slope stability monitoring.

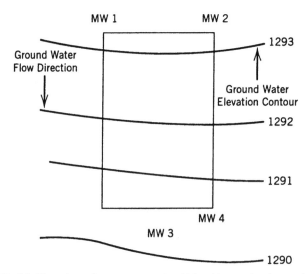

FIG. 22.17   Plan of an NA-type landfill with monitoring points.

concentration plot. From a detailed study on groundwater monitoring wells around landfills in Wisconsin, Fisher and Potter (1989) concluded that a box plot from a contaminated monitoring well will typically show a spread of data compared to the box plot(s) of uncontaminated background well(s), although natural variation with time may exist in both wells.

A box plot is a graphic display of the data that provides the summary information in the quartiles. In a box plot the first and third quartile data are connected by a box with the median indicated by a bar (Johnson and Bhat-

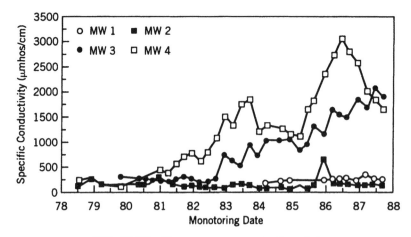

FIG. 22.18   Time versus concentration plot.

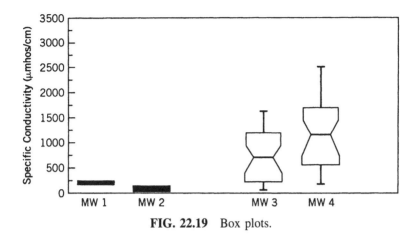

FIG. 22.19 Box plots.

tacharya, 1992). The highest and lowest data are connected to the box by a straight line. Example 22.1 shows how to draw a box plot. Box plots should be drawn for each well for each parameter. Study of box plots for each parameter for the monitoring wells around a landfill site will clearly show whether any well has been impacted.

Example 22.2 Calculate the first, second, and third quartiles for the specific conductivity data shown in Table 22.3 from a groundwater monitoring well down-gradient of a landfill.

The first step is to arrange the data from the smallest to the largest value (already done in Table 22.3). The division of ordered data set into two equal halves is called median; when the division is in quarters, it is called quartiles. The quartiles are 25th, 50th and 75th percentiles. After arranging the data, the quartile points are calculated by multiplying the quartile with the sample size (or number of data). Thus, for the example:

The first (or lower) quartile $= 18 \times 0.25 = 4.5$. Since this is not an integer, round it to the next integer. The value of the first quartile is the 5th data $(= 545)$.

TABLE 22.3 Specific Conductivity Data from a Landfill Monitoring Well

500	600	930
510	640	995
525	697	1030
537	750	1070
545	810	1130
557	850	1195

The second (or median) quartile = $18 \times 0.5 = 9$. Since this is an integer, the average of the 9th and the 10th data is the median [= $(697 + 750)/2 = 723.5$]

The third (or upper) quartile = $18 \times 0.75 = 13.5$. Since this is not an integer, round it to the next integer. The value of the third quartile is the 14th data (= 995).

Use these numbers to draw the box plot as indicated in the text.

23

LANDFILL OPERATION

A simple, well-organized operating plan is the key to the successful operation of a landfill. The cost of operating a landfill may be up to five times the total cost of design and construction of the landfill (Bolton, 1995). An operating plan merely satisfying regulatory compliance is of little use. A good operating plan should provide guidance for day-to-day and year-to-year operation so that landfill volume is efficiently used, a safe working environment is created, and environmental nuisances are not created. No two landfills are alike. The following guidelines may be used to develop a site-specific plan:

1. The highest compaction is obtained by compacting from the base and up of the landfill.
2. End dumping from the side of the landfill should be avoided.
3. All-weather access roads should be built inside and outside of the landfill.
4. The active area should be as small as possible.
5. A maximum of three horizontal to one vertical waste slopes should be maintained on all internal waste faces.
6. The surface water should be diverted away from the landfill. Temporary surface water drainage swales should be constructed whenever possible.
7. Waste contact liquid must not be allowed to run off from the landfill.
8. Access to the landfill should be controlled by a fence and locking gate.
9. Proper care should be exercised during burning of waste (if allowed by the regulatory agency).

10. Clear and visible on-site directional signs should be posted for proper traffic routing. In municipal landfills burnable woods, white goods (refrigerators, etc.), recyclable material (paper, glass, etc.) should be deposited separately so that they may be salvaged.
11. Landfill operator(s) should know about fire control measures and emergency measures to be taken in case of accident leading to physical injury.
12. The operators should take necessary safety precautions when operating a landfill.
13. The operator(s) and other landfill personnel should be conscientious about bacterial and chemical contamination while storing and eating food within the landfill office.
14. The operator(s) should know about monitoring and maintenance requirements for the landfill.
15. The operator(s) should have a basic knowledge about design and construction-related issues of landfills.

The above guidelines are at best partial. Discussions of equipment used, litter and dust control measures, and routine maintenance issues follow.

23.1   EQUIPMENT USED FOR COMPACTION

Proper equipment should be chosen to operate a landfill. Proper attention should be given to the capability of the equipment and its need. In many instances backup equipment may be necessary if severe operational problems are envisioned during equipment downtime. Essentially three major operations are involved in landfilling a waste: spreading waste after dumping, compacting waste, and spreading and the compacting daily or intermediate cover (where necessary). Table 23.1 provides a list of landfill equipment and how each rates in performing the above operational needs. There are essentially two approaches for selecting landfill equipments: basic task based method and comprehensive task based method. In a basic task based method proper equipments are chosen based on tasks in a landfill; the cost for performing each task for each machine is calculated. Alternative equipments are then compared to choose the correct one. In a comprehensive task based method the total tonnage as well as the site features (e.g., topography, soil type) are considered in arriving at the most cost effective equipment. An equipment selection table for a specific landfill for managing various tonnage waste per day (e.g., 100 T/day, 200 T/day) can be developed using the comprehensive task method. Detailed discussion regarding these approaches can be found elsewhere (Bolton, 1995).

Although draglines are normally not used in landfills, they may be useful in disposing of sludge. Consultation with equipment salespersons regarding newly marketed equipment and its use should be undertaken before purchasing any equipment. A machine shed equipped to handle emergency repair and

TABLE 23.1   Performance Characteristics of Landfill Equipment[a,b]

	Solid Waste		Cover Material			
Equipment	Spreading	Compacting	Excavating	Spreading	Compacting	Hauling
Crawler dozer	E	G	E	E	G	NA
Crawler loader	G	G	E	G	G	NA
Rubber-tired dozer	E	G	F	G	G	NA
Rubber-tired loader	G	G	F	G	G	NA
Landfill compactor	E	E	P	G	E	NA
Scraper	NA	NA	G	E	NA	E
Dragline	NA	NA	E	F	NA	NA

After Sorg and Bendixen (1975).

[a]Basis of evaluation: easily workable soil, and cover material haul distance greater than 1000 ft.
[b]Rating key: E, excellent; G, good; F, fair; P, poor; NA, not applicable.

with a stock of accessories should be constructed for larger landfills. Backup equipment should be kept on site or arrangements should be made to borrow equipment whenever necessary. Routine maintenance should be performed as recommended by the manufacturer. Figure 23.1 shows a steel wheel landfill compactor and Fig. 23.2 shows a dozer.

FIG. 23.1   Steel wheel compactor.

FIG. 23.2   Dozer.

23.2   PHASING PLAN

A phasing plan is important, both in terms of operating a landfill and for establishing financial proof for landfill closure (see Section 26.1.2 for details). A phasing plan is necessary for both natural attenuation and containment-type landfills. An ideal phasing plan would be one in which each phase receives a final cover in the shortest possible time. This would mean disposing of waste in one phase until it reaches the final grade. Figure 23.3 shows a phasing plan for a landfill. However, such a simple phasing plan may not be applicable to all landfills. If the landfill height is more than 9 m (30 ft) from the base, usually an intermediate cover is used over parts of the landfill, at 3–4.5 m (10–15 ft) above ground (not above the base). In such cases an intermediate cover consisting of 60 cm (2 ft) of clayey soil and 15 cm (6 in.) of topsoil is used over the area. New phases are started on top of the lower phases after scraping the intermediate cover.

Figure 23.4 shows a phasing plan that follows the above concept. It may be noted that the soil used for constructing the intermediate cover should not be used for constructing the final cover because it gets contaminated with waste. The soil may be reused for daily cover or disposed of in the landfill. However, the topsoil may be reused in the final cover.

The direction of filling should be clearly indicated in the plan to avoid any confusion. The access road should not run over a closed phase. Usually a permanent all-weather access road is constructed parallel to the phases outside of the landfill area and branch roads, leading to the base of the landfill, are

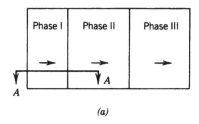

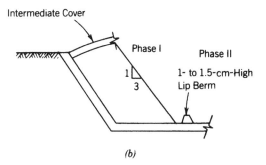

FIG. 23.3 Phasing plan for single-stage landfill: (*a*) plan and (*b*) cross section *A–A*.

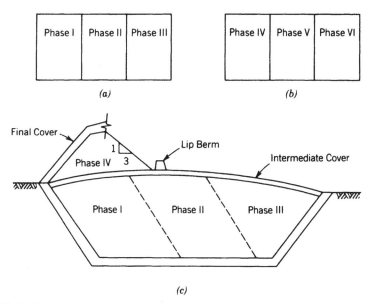

FIG. 23.4 Phasing plan for a multiphase landfill: (*a*) lower phase, (*b*) upper phase, and (*c*) cross section.

joined with it as necessary. Traffic routing should be planned carefully so that all waste is dumped in its final resting place.

23.3   COVERING WASTE

Three types of cover may be used in a landfill: daily cover, intermediate cover, and final cover. The design and construction of a final cover are discussed in Sections 19.8 and 20.9, respectively. Issues related to daily and intermediate covers are discussed in this section. Daily cover serves many functions that are considered essential for municipal landfills. For most municipal landfills 15 cm (6 in.) of daily cover is used, although its use may be warranted in some putrescible sludge landfills. Although a daily cover serves many functions, it also uses up valuable landfill space. (Note: Daily cover occupies about one-fifth to one-sixth the volume of waste.) Depending on the geographical location of a landfill and considering other factors such as odor control, a weekly or monthly cover may also be used if permitted by the regulatory agency. Daily cover improves access, improves aesthetics, reduces windblown debris (e.g., paper, plastics), reduces risk of disease transmittal through vectors (e.g., birds, insects, and rats), reduces odors, reduces fire risk, and provides a media for partial attenuation of leachate. Historically, sandy soils have been the most commonly used daily cover material. If clay is used as daily cover, then the daily cover should be scarified or removed prior to the placement of the next lift to avoid perching of leachate in the future. The other types of materials used for daily cover are: petroleum contaminated soil, synthetic textile, foundry sand, chemical foams and slurries, tire chips, bark and wood chips, and auto shredder fluff. Use of synthetic textiles saves valuable landfill space. The synthetic textile is stretched, either manually or using a dozer, over the lift of refuse at the end of a working day. The synthetic textile is removed prior to disposing waste next day. The same piece of synthetic textile can be reused for several days. If permitted by the regulatory agency, permanent-type daily cover (e.g., tire chips, soil) should be used every 7–10 days.

An intermediate cover is used when portions of a landfill are expected to remain open for a long period of time (2 years or more). Intermediate cover helps in reducing leachate production. The precipitation falling on the intermediate cover is allowed to run off the site. Surface runoff is not allowed from daily cover, so daily cover does not help in reducing leachate volume.

23.4   CO-DISPOSAL OF NONHAZARDOUS SLUDGE IN MUNICIPAL LANDFILLS

Co-disposal of sludge should be done properly so that nuisance conditions do not develop in a landfill. Disposal of slurry-type sludge should be avoided; the solids content of all sludge disposed in a landfill should be at least 40%.

Mixing of sludge with municipal waste is preferred. Mixing can be done by spreading the sludge in thin layers and then covering it with municipal waste. The daily volume of sludge disposal should be restricted to a maximum of 25–30% of municipal waste. A 1:1 daily ratio of sludge to municipal waste may be used with due consideration toward fill stability and change in gas generation rate. It should be borne in mind that each landfill is designed for specific physical and chemical properties of waste. So design-related issues should be revisited prior to changing waste type in an active landfill. The interior access road should be carefully designed and maintained so hauling vehicles can maneuver easily. In some situations a separate disposal area of sludge is preferable for better operational control. Mixing of sludge with dry soil may be necessary to maintain trafficability within the landfill.

23.5 FIRE PROTECTION

Fire hazards exist in certain waste-type landfills. Landfills, where open burning is practiced, are especially susceptible to such fires. Landfill fire can generate toxic fumes injurious to health. The following guidelines should be followed in landfills in which open burning is practiced. (Note: Permission from the regulatory agency may be necessary regarding open burning.)

1. An attendant should be on duty until the fire is extinguished.
2. A minimum of 3–4.5 m (10–15 ft) of open space should be maintained around the burning area.
3. Asphalt, rubber, plastics, and so on should not be burned because they will create air pollution problems.
4. A pit should be used to burn material. Ash generated from burning should be disposed in the landfill.
5. Public access to the burning pit should be restricted.

Usually open burning is allowed in small landfills located in remote places. Air curtain destructors may be used for a cleaner burning operation. A fire extinguisher, sand, and a water wagon should be readily available even in landfills in which open burning is not practiced. Operator(s) should be trained to extinguish small fires. The telephone number and location of the nearest fire station should be displayed in a conspicuous place in the landfill site office. In larger landfills a drill to train operator(s) in fire emergency handling procedures should be arranged once a year.

23.6 LITTER CONTROL

Paper and other lightweight material create littering problems around landfills. Dumping near the landfill base, especially during windy days, can help reduce littering. A small movable screen made of chicken wire may be placed near the active face to catch windblown papers. Regular manual pickup at the end of a working day may become necessary in some landfills. If windblown paper poses a serious problem, then very high fixed wire net may be placed down-gradient of the wind (direction prevailing in most days) or an earthen berm may be constructed up-gradient of the wind direction.

More than one row of trees or shrubs or an earthen berm may be used to screen a site from obvious public view. A well-kept landfill can improve public perception of landfills, which is beneficial to landfill siting both in the short and long term.

23.7   DUST CONTROL

Controlling dust in a landfill is somewhat difficult. The heavy traffic always creates dust problems within a landfill because of dry soil on the road. Although watering of the road can reduce dust, it may increase leachate volume. So judgment has to be used regarding the quantity of water added to the road inside a landfill to reduce dust. Dumping dusty loads in low areas of a landfill or spraying a small quantity of water to a dusty load before dumping will help in controlling dust.

23.8   ACCESS ROAD MAINTENANCE

The interior and exterior roads must be maintained properly at all times. Proper drainage of interior road beds is essential. The temporary interior roads are vital for proper operation of a landfill. Since heavy vehicles run to and from a landfill, the exterior access roads deteriorate significantly. Sufficient funds should be allocated to maintain access roads. Landfill equipment that does not have rubber tires (e.g., steel wheel compactor or crawler tractor) should not be allowed on the paved portion of exterior roads. When necessary, this equipment may be allowed to move along unpaved portions of the road. An even surface, sufficiently wide to accommodate this equipment, should be maintained between the landfill and the maintenance shed.

23.9   LEACHATE COLLECTION SYSTEM MAINTENANCE

All items connected with leachate collection must be maintained properly. The items include mainly the leachate collection pipes, manhole(s), leachate collection tank and accessories, and pumps. The leachate lines should be cleaned once a year to clear out any organic growth. The manhole(s), tank(s), and pump(s) should be inspected visually once a year. Leachate can corrode metallic parts. An annual inspection and necessary repair will therefore prevent many future emergency-type problems such as leachate overflow from the tank due to pump failure. Extreme care must be exercised when entering such confined space. Use of a harness and direct watch by personnel above ground are considered essential when entering manholes. All regulations regarding confined space entry must be followed to ensure the safety of the maintenance crews. A record of all repair activities should be maintained to assess (or claim) long-term warranties on pump(s) and other equipment.

23.10   FINAL COVER MAINTENANCE

The final cover should be maintained properly to reduce infiltration into the landfill. Erosion due to surface runoff, settlement, stress on vegetation due to gas in a landfill, and so on, damage the landfill final cover. Any damaged area should be repaired as soon as possible. The cause of the damage must be investigated so that proper repair measures are implemented. For instance, if the damage is due to stress on vegetation because of gas, then proper gas venting measures should be undertaken. Otherwise, the problem will recur each year. On the other hand, if the damage is due to erosion, then the cause of the erosion must be investigated. The length and steepness of the slope, improper vegetation growth due to bad planting, uneven settlement of the solid waste, and so on, cause erosion. Protection matting should be used in long swales to reduce erosion. Once completed, access to the final cover area should be restricted to avoid damage by vehicular traffic. Every effort should be made to maintain healthy vegetation because vegetation improves the stability of the slopes, reduces surface erosion (Gray and Leiser, 1982), and reduces leachate production by increasing evapotranspiration.

For the first 3–5 years the final cover area should be inspected twice a year: once during a season when the vegetative growth is minimum and once during a season when vegetative growth is maximum. Any problem of erosion can be detected easily when the surface is bare. Any problem of vegetative growth can be detected during the peak season inspection. The inspection frequencies may be reduced to once a year after the initial years and should be continued for another 10–15 years. Special attention should be given to protect the surface from burrowing animals. It is very likely that native vegetative species will invade the final cover in the long run. The root system of large trees may increase infiltration into the landfill. However, realistically very little can be done to stop the invasion of native species. No documentation of this phenomenon is available; this phenomenon is worth researching.

23.11   COMMENTS ON ENFORCEMENT-RELATED ISSUES

It is normally the duty of the landfill manager to ensure that all conditions imposed by the regulatory agency are met. In general, the main areas of regulations are the health and safety of workers and environmental protection. In most cases two different government agencies are responsible for enforcing the relevant laws. It is essential that the landfill manager be familiar with both types of regulations. It is a good idea to meet with relevant officials at least prior to beginning the landfill operations and annually thereafter to discuss regulatory issues. The owner should include a lawyer in the group during the annual meeting who can help in understanding the legal aspects of regulatory requirements. Running an operation in accordance with government regulations costs a lot less than getting involved in litigation. However, if such litigation is unavoidable, then advice from an attorney should be sought at an early stage. The manager must read the conditions of permit and any changes thereof. All unrealistic permit conditions should be resolved at the beginning rather than trying to bypass them. A landfill manager should not knowingly force workers to engage in dangerous operations because the manager not the corporation (if privately owned) may be made personally liable for such action. The manager should be knowledgeable about the steps a regulatory agency takes prior to referring a case for legal action.

Although uniform legal procedures are not followed by all regulatory agencies, in many instances more than one notice is issued regarding a violation prior to taking legal action against a landfill owner. Each notice should be acted upon promptly and the lapse(s) should be rectified within the specified date.

23.12   LANDFILL MINING

Mining of landfill is undertaken to recover valuable landfill space and salvage materials. In landfill mining an old landfill is excavated and the materials are sorted into different categories. Big white goods (e.g., refrigerators) and other metallic items are salvaged and recycled whenever possible. The waste is sieved using trommels (approx. 2.1 m diameter, 7 m long) that separate soil and burnable material. The landfill mining project in Lancaster County, Pennsylvania, reported a 68% recovery of burnable materials and 28% of cover materials; only 4% of excavated material (by weight) was returned to the landfill. Mining of a landfill in Florida has also been reported (U.S. Congress, 1989). According to Vasuki (1988), landfill mining is helpful for reusing of landfill space after a period of decomposition (which can extend the life of landfills), repairing of liners and leachate collection systems, and recovering materials of value. Although landfill mining is not widely practiced now, the idea merits consideration.

24

COMPENSATORY
WETLAND DEVELOPMENT

In many cases, an area proposed for landfill siting includes wetland(s) or it is next to a wetland. Wetlands have many functions and values. In addition to supporting numerous micro and macroorganisms, wetlands support a diverse population of animal species. Wetlands are an important element of the ecosystem and hence should be protected to minimize environmental impacts. There are laws aimed at protecting wetlands. However, as mentioned, there are situations where wetland impact cannot be avoided. In unavoidable circumstances compensatory wetlands are developed.

Wetlands may be crudely defined as wet habitats for a variety of biota (plants, microbes, and animals). Wetlands are subject to permanent or periodic inundation or prolonged soil saturation. Because of this saturation condition, the soil develops anaerobic conditions that are suitable for supporting specific plant species called hydrophytes. Hydrophytes are individual plants that can sustain their roots in anaerobic conditions. Because of variation in climate, hydrologic cycles, and soil types, a vast number of wetland plant species have evolved worldwide. In general, wetlands are not permanently flooded areas but are subject to a cycle of wetness and dryness over a year. However, wetland species can also be found in permanently flooded areas.

It is a difficult task to come up with a definition of a wetland. Many wetland definitions have been forwarded over the last century. It is interesting to note that definition of wetland may focus on a particular aspect depending on the field of study. Thus a hydrologist's definition of wetland may be different from a soil scientist's definition (Tiner, 1999). While wetland can be identified using ecological characteristics, conservationists like to adopt a definition of wetland that will include, in addition to traditional wetlands, all deep-water habitats (e.g., lakes, rivers). The Ramsar convention, an international government body representing 90 countries, defined wetlands as "areas of marsh, fen, peatland, or water, whether natural or artificial, permanent or temporary, with water that is static or flowing, fresh, brackish, or salt, including areas of marine water the depth of which at low tide does not exceed 6 meters. It may incorporate riparian and coastal zone adjacent to wetlands, and islands or bodies of marine water deeper than 6 meters at low tide lying within the wetland" (Ramsar Information Bureau, 1998). This is a broad definition aimed at protecting both natural and human made wetlands (e.g., fish and shrimp ponds, irrigated agricultural lands). A narrower definition adopted by the U.S. Army Crops of Engineers, which is more relevant for landfill-related mitigation projects, is as follows: "Wetlands are those areas that are inundated or saturated by surface or groundwater at a frequency and duration sufficient to support, and that under normal circumstances do support, a prevalence of vegetation typically adapted for life in saturated soils. Wetlands generally include swamps, marshes, bogs and similar areas" (U.S. Government, 1997).

The definitions of wetlands adopted by various countries and various states within United States are different. Because of this variation in official definition of wetlands, it is essential to work closely with proper regulatory agencies for identifying and delineating wetlands. Although every effort should be made to avoid disturbing or encroaching on wetlands during landfill siting, there are situations in which encroaching wetlands for landfill siting becomes unavoidable. In addition to ecology, function of wetland is an important issue that must be considered prior to opting for mitigation. This chapter will provide basic information regarding wetland identification, delineation, and compensatory wetland development.

24.1   WETLAND IDENTIFICATION

Common land forms associated with wetlands that are relevant for landfill sites include: depressed land with or without a drainage, low-lying areas along water bodies (e.g., pond, swamp), flat areas lacking drainage outlets, and areas adjacent to bogs or marshes. Figures 24.1 and 24.2 show typical cross sections of two major wetland types. Identification of wetland is based on a relationship between soil, water, and plants. Wetlands generally form in those portions of the landscape with relatively stable sources of surface and/or subsurface water that can saturate the ground for long periods. The frequency and duration of inundation (flooding or ponding) or soil saturation is a major factor for identifying wetlands. Seasonal variation in surface and/or subsurface water levels is like a hydrologic signature for a wetland. Hydrologic signature is often the most difficult component to establish for a wetland (State of Wisconsin, 1995). Hydrology is the controlling factor in wetland formation. The following factors influence wetland hydrology: landscape, surfacial soil, precipitation patterns, geology and hydrogeology of the area, and surface

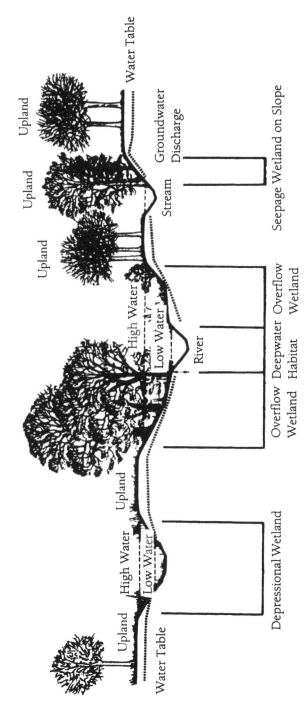

FIG. 24.1 Cross section of typical freshwater wetlands. After Salvesan (1990).

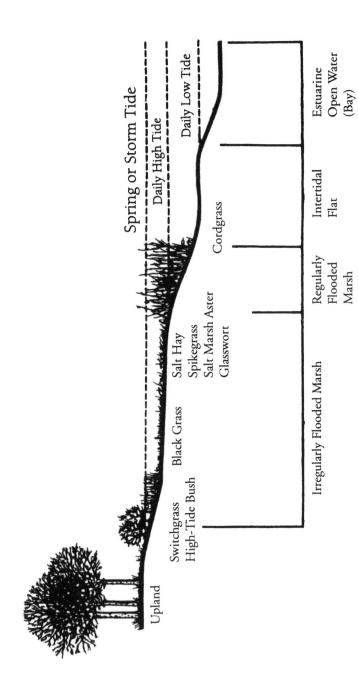

FIG. 24.2 Cross section of a typical salt marsh. After Salvesan (1990).

Spring or Storm Tide

Daily High Tide

Daily Low Tide

Cordgrass

Salt Hay
Spikegrass
Salt Marsh Aster
Glasswort

Black Grass

Switchgrass
High-Tide Bush

Upland

Irregularly Flooded Marsh

Regularly Flooded Marsh

Intertidal Flat

Estuarine Open Water (Bay)

drainage. Although professional judgment is needed to identify a wetland, especially in controversial situations, an area should be assumed to be a wetland based on the following two key indicators: plant types and soil condition.

24.1.1 Wetland Plants

Many wetlands can be readily identified by the types of vegetation present. Flooding and soil saturation cause changes in soil environment that affect plant growth and survival. Wetland plants exhibit a wide range of adaptation in response to frequent flooding. Plants growing in saturated or nearly saturated soil conditions are called hydrophytes. Since flooding causes depletion of oxygen leading to anaerobic conditions, wetland plants develop structural changes to provide oxygen to its roots during flooding. Channels of living roots develop brownish to yellowish red color (due to deposition of iron oxide) because of soil saturation for a significant period during the growing season. This colored channel is known as rizospheres. This zone also aides the plant in oxidizing toxic materials (Tiner, 1999). In order to facilitate wetland identification and delineation, attempts have been made to identify wetland plants (Reed, 1997). Since wetland plant species are different in different parts of the world, experienced plant ecologists familiar with wetland plants for the region in which the landfill is to be located should be consulted.

24.1.2 Wetland Soil

Since wetlands are subject to frequent flooding or saturation conditions, the soil develops unique properties. Wetland soil is called hydric soil (essentially meaning soil that remains saturated for a long time, which cause anaerobic conditions within the soil). A field guide for identifying hydric soils in the USA has been developed by Hurt et al. (1998). Atkinson et al. (1998) has included discussions on several methods for identifying hydric soil.

Blume and Schlichting (1985) have described four types of wetland soil (viz. halomorphic, gypsimorphic, calcimorphic, and redoximorphic) based primarily on oxygen levels and organic matter content in soil. Proper identification of soil is crucial to delineate wetland boundaries. Hydric soil is subdividied into two groups: organic and mineral hydric soil.

24.2 WETLAND DETERMINATION AND DELINEATION

Wetland determination refers to general identification of wetland plants and or soil in an area; delineation refers to drawing an exact boundary of the wetland. After wetland plant, soil, or hydrology is identified within a property, then steps are taken to delineate the boundary. For delineating a wetland area all three criteria—namely plant, soil, and hydrology—are used. As indicated in Section 24.1, hydrologic signature is the most difficult and least exact criteria for delineation purposes. Nevertheless it is a useful tool in the process. Wetland hydrology is recognized through visual observation of inundation, water table in shallow holes (maximum 60 cm deep), saturated surfacial soil, and water marks.

The delineation is dependent primarily on the presence of wetland plants and hydric soil. The boundary thus drawn may be further verified by observing hydrologic criteria, and other plant and soil criteria. Limits of the presence of lichens [a symbiotic combination of alga (aquatic nonvascular plants, e.g. seaweed) and mosses] and mosses may be used to draw the wetland boundary. In addition other indicators such as water marks on trees and bare soil may also be used for wetland delineation purposes. Note that the method of delineation may be different in various countries and in various states within the United States. Note also that the U.S. federal regulatory agency may have jurisdiction over a wetland located on state land. Although regulatory agencies have specific guidelines for wetland delineation (e.g., State of Wisconsin, 1995), the fieldwork involves judgment. So, it is prudent to collaborate with the regulatory agency personnel for determination and delineation of wetland(s) within the property chosen for landfill siting.

24.3 WETLAND FUNCTIONS AND VALUES

There are many functions of wetlands. Wetlands receive and recycle nutrients washed from uplands. The vegetation in wetlands uses these nutrients to convert inorganic chemicals, to organic chemicals, which in turn help support a diverse animal population. It is estimated that in addition to supporting numerous micro and macroorganisms, wetlands in the United States support 190 species of amphibians, 270 species of birds, and over 5000 species of vegetation (Mitsch and Gosselink, 1986). Wetlands help reduce sediments entering a surface water body and also diminish the floodwater damage along river and lake banks. Many wetlands function as groundwater recharge areas. Wetlands are also formed at groundwater discharge areas whereby erosion of the adjoining land is stabilized.

All of the above functions can also be viewed as values of wetlands. The destruction of wetlands has long-term effects on the ecosystem in the immediate areas surrounding it. Since wetlands also support wildlife, bird migration, and provide habitats for numerous living species, destruction of wetlands in an area can affect ecosystems far beyond the immediate surroundings. In addition wetlands can serve as an area of recreatiion and scenic beauty. Marsh-type wetlands support many unique flowering plants (e.g., water lilies). Wetlands are habitats for many endangered species. Thus, the environmental impact of wetland alteration or destruction can be very significant.

It is difficult to put a monetary value on a wetland. A comparison of the ecological value of a wetland and the economic value of proposed activity is made to judge the impact of alteration or destruction of a part or the entire wetland. The following issues are addressed to judge the impact of the proposed action on the wetland: (a) the multiple functions and values of wetlands; (b) impact on public recreation; and (c) the finite nature of commercial values compared to the perpetual values of the wetlands. A detailed discussion on these issues can be found elsewhere (Mitsch and Gosselink, 1986).

24.4   WETLAND PROTECTION

Protection of wetlands should be viewed as an integral part of protecting the environment. In addition, wetlands have recreational and other values as discussed in Section 24.3. Any alteration to wetland hydrology or encroachment on its boundary will impact a wetland, which may lead to severe disturbance to the ecology of a vast area surrounding the wetland. A landfill can impact a wetland in two major ways:

Hydraulic Modifications   The flow of water into and out of an adjacent wetland may be altered due to restricting or diverting surface water flow regimes. This issue must be carefully looked into during landfill design so that the water balance of the wetland is maintained properly. A significant impact may also be caused during construction of a landfill. Chances of an alteration (increase or decrease) of water flow to a wetland is very high during landfill construction. Appropriate measures must be taken to minimize this impact. Construction and maintenance of silt fences and sedimentation pond(s) to trap sediment must be undertaken to minimize sediment load. An increase in sediment load will disturb the aquatic life within the wetland.

Increase in Pollutant Loading   Waste contact liquid from a landfill may find its way into an adjacent wetland. The landfill must be operated in such a way that the waste contact liquid does not flow out of the landfill area. The operational issue is discussed in Chapter 23. If composting or treatment of petroleum-contaminated soil is undertaken outside of the active landfill area, then proper care must be taken to restrict the flow of waste contact liquid into an adjacent wetland.

24.5   WETLAND RESTORATION AND CREATION

As discussed in Section 24.4, the impact of landfills on adjacent wetlands is caused by modification of hydrology and increase in pollutant load. The landfill should be designed and operated in such a way that these impacts can be avoided permanently. Temporary impact during landfill construction comes mainly from an increase in sediment load. Any sediment accumulated during construction should be removed at the end of each construction season or within 3–4 months, whichever comes first. Similar increase in sediment load may also occur during landfill closure. This impact may be somewhat longer term because erosion from landfill final cover continues until the vegetation is well established. Depending on the weather and maintenance program, it may take 2–5 years for the vegetation to cover the entire landfill final cover surface. The wetland should be monitored for increased sediment load during this period. Cleaning of sediment from the wetland must be undertaken as necessary. Restoration of vegetation through planting should also be done as necessary.

Creation of wetland is a long process. Although the structural components can be constructed within one year, establishment of wetland plants will take several years, even several decades. Wetland construction is a relatively simple process. The main issue for constructing a wetland is to create a controlled hydrology whereby there is a balance between inflow and outflow of water. In general, artificial wetlands are constructed essentially as shallow ponds. If the subsoil permeability of the targeted area is high (e.g., sandy soil), then a low permeability liner is constructed over the entire area. The subgrade is excavated to the desired level and then compacted to minimize water loss. Usually a layer of clayey soil is constructed to further minimize water loss. Bentonite may be used with soil to achieve a low permeability layer. If groundwater pollution, due to the chemistry of inflow water is a concern, then plastic liners (e.g., PVC, PE) are used; use of plastic for creation of compensatory wetland is rare. A 18- to 24-in. (45- to 60-cm) layer of a mixture of sandy loam and gravel is laid over the liner, which serves to support the plants (Campbell and Ogden, 1999). The wetland may be surrounded by berm to control the hydrology.

The next step is the establishment of plants. The species most prevalent in the wetlands surrounding the project area should be chosen for planting. The gravel bed should be mulched prior to planting. In cold climates, where temperature goes below freezing, the planting should be done in early to late spring. The success of plant survival depends on proper planting time. The wetland must be flooded immediately after planting. The wetland plants die quickly in dry conditions. The survival of plants must be monitored for several years to ensure long-term viability of the plants. It takes many years for a wetland to mature and function as desired.

24.6   WETLAND MITIGATION

When establishing a landfill, every effort must be made to avoid encroachment of wetland boundaries. The landfill design should be such that it does not alter the hydrology of any adjacent wetlands. However, there are situations where impact and/or encroachment upon wetland boundaries becomes unavoidable. In such cases a compensatory wetland is developed to minimize the environmental impact of landfill siting. In general, regulatory agencies follow a hierarchy of "avoid, minimize, and compensate." So to obtain permission for developing a compensatory wetland, a designer must establish that steps were taken to avoid and minimize impact to adjacent wetland(s).

To avoid a negative impact, one may change the footprint of the landfill. To minimize negative impacts, the landfill design, construction, and operation must be done properly. To compensate for loss of wetland, additional wetland areas need to be developed. The mitigation can be done by creating adjacent wetland areas on site or through mitigation banking. A regulatory agency may specify the distance [usually 1 mile (1.6 km)] within which a compensatory wetland should be developed. To minimize environmental impacts, compensatory wetlands should be developed within the same basin as the landfill site. However, if a suitable area is not found within the same basin, then another area may be used. It is preferable to restore a wetland rather than create a new wetland. In general, an area 1.5–2 times the wetland lost should be compensated. The type of compensatory wetland should be the same as the one lost.

Both short-term and long-term monitoring should be planned for a compensatory wetland. Long-term site management and protection using conservation principles should be included in the planning process. The cost of these activities should be included in the long-term care cost of a landfill (see Chapter 26).

In the United States, a new concept of "mitigation banking" has developed over the years. In mitigation banking, a prospective "wetland banker" develops a compensatory wetland bank after obtaining necessary approvals from the regulatory agency. The banker then sells wetland areas to others who need compensatory wetland. Usually the regulatory agency, the banker, and the buyer enter into a legal agreement to ensure proper long-term management of the wetland by the banker and the buyer. In many states within the United States, the regulatory agency maintains a registry of approved wetland banks. The service area of each bank is specified in the approval from the regulatory agency. The price of the compensatory wetland area is negotiated between the buyer and the banker.

24.7   LONG-TERM EVALUATION

Monitoring of constructed wetland is essential to ensure the success of mitigation. The monitoring frequency should be more initially (e.g., 6 months), which can be reduced (e.g., one year) if wetland development shows maturity. Both field study and analysis of data should be undertaken. At a minimum, the long-term evaluation should include monitoring of hydrology, plant and animal population dynamics, and water quality sampling. The timing and volume of water input, depth, and duration of flooding should also be monitored. Three techniques are employed to characterize wetland vegetation: belt transect, replicate quadrats, and multiple quadrats (Erwin, 1999). Each method essentially involves evaluation of plant species in a marked area within the wetland. The growth and health of vegetation is monitored within the identified areas at regular intervals. The productivity of vegetation is a key criteria to assess the success of created wetlands. The postconstruction monitoring should be recorded for at least 5 years. A panoramic picture of the wetland during a fixed month (usually midsummer) from a fixed point will provide a visual record of the status of vegetation of the wetland each year. An established wetland attracts wildlife. A qualitative report indicating frequency of use of the wetland by wildlife is helpful in assessing the success of created wetlands. Although reptiles and amphibian populations can serve as biological indicators of a wetland (Perry, 1998), birds are also a good biological indicator for wetlands (Casalena, 1998). While short-term monitoring (up to 5 years) should concentrate on hydrology and vegetation, long-term evaluation of a constructed wetland should be done through assessment of the presence of animal habitats and the frequency of use by wildlife, including birds.

25

HEALTH AND SAFETY

Solid waste management creates a number of potential health and safety concerns. Solid waste industry managers, employees, and others must be aware of the possible health hazards from their jobs. Health and safety in the solid waste management industry is not only a worker safety issue, it also involves public health and may have legal implications. For instance, workers involved in medical waste processing may be exposed to a communicable disease (e.g., tuberculosis). If not treated properly, the disease may spread in the community, causing public health concerns. If managers knowingly allow employees to be exposed to hazardous chemicals (without the knowledge of the employees), then the managers are liable to be prosecuted for breaking the law. There are laws regarding health and safety at any worksite that, if not adhered to by management, can have legal consequences.

While all forms of solid waste management have health and safety consequences, landfills in particular require great care. Under normal operations, landfill personnel must safely handle a variety of potentially dangerous materials (e.g., leachate) and operate a number of different types of equipment with inherent health and safety concerns. The landfill work area is often an active place where trucks and compaction equipment are operating while landfill personnel may be inspecting delivered loads.

Many of the design features of a state-of-the-art landfill such as liners and gas collection systems are required to protect the public health and safety. However, special precautions are required in landfill operations since operators are often the people most exposed to the dangers that have led to the stringent environmental protection designs.

Landfill managers and operators should have a clear understanding about occupational safety and health that affects landfills. A solid waste facility should be designed to minimize risk during operation as well as to minimize the long-term risk to the environment and the public. In defining operational health and safety procedures, it is a good idea to work closely with the regulatory agency involved in implementing and enforcing occupational safety and health laws. (Note: Usually this agency is different from an environmental protection agency.) In the United States, the Occupational Safety and Health Administration (OSHA) is the federal agency responsible for this task. Individuals states also have agencies that focus on health and safety issues. While agencies at both the federal and state levels are responsible for enforcement, they are also the source of considerable help in designing and maintaining an effective health and safety program.

25.1 SOURCES OF OCCUPATIONAL HEALTH HAZARDS

There are several types of health and safety hazards in the waste industry. These can range from exposure to the properties of solid waste managed by the facility to the inherent danger of operating heavy equipment required at the facility.

Knowledge about each source of occupational health hazard is essential to avoid injury and health impacts. It should be noted that, in some cases, a prolonged exposure can cause injury (e.g., prolonged exposure to noise may cause loss of hearing), whereas in some other cases a single exposure can lead to severe injury even death (e.g., an accident associated with vehicles or equipment).

25.1.1 Noise

In the context of health hazard, excessive noise may be defined as sound that causes injury to the hearing ability of a person. Effect of sound depends on a number of factors including loudness, frequency, length of exposure, worker's age, and health. While a short exposure to loud sound can cause temporary loss of hearing, a prolonged exposure to sound may cause health problems including permanent loss of hearing. Noise can cause a variety of biochemical reactions including increase in blood pressure, increased heart rate, shallow breathing, abnormal glandular function, and reduced blood flow to the fetus in a pregnant woman. In addition, noisy environments can cause emotional stress and susceptibility to accident (Guinness, 1987).

The ear is a complex organ that carries sound from an external source to the brain. Figure 25.1 shows a cross section of an ear. The process of hearing a sound is complex. Simply put, sound waves cause the eardrum to vibrate. The ear changes these vibrations into impulses that are picked up by the auditory nerves, which in turn send the signal to the brain. Both the frequency and intensity of sound are important factors for determining the effect of noise. High-frequency sound can cause more damage than low-frequency

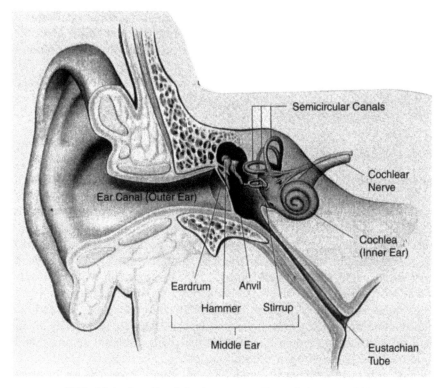

FIG. 25.1   Details of the human ear. From Guinness (1987).

sound. Loudness is measured in decibels (dB). To prevent hearing loss, work-site noise levels should not exceed 85 dB. The OSHA noise standard for 8-hr exposure is 90 dB and 4-hr exposure is 95 dB.

Noise levels in any workplace must be controlled to the allowable levels. The following steps may be taken to reduce noise:

1. Installation of noise-reducing devices or equipment (e.g., mufflers in cars)
2. Placement of machines on shock-absorbing mountings (e.g., rubber pads)
3. Installation of acoustical tiles and blankets
4. Design of machinery to reduce noise level
5. Rearrangement of equipment and machinery to reduce noise exposure
6. Proper maintenance of equipment such as compactors and bulldozers to keep noise levels at a minimum

Even after adopting all of these design and operational recommendations, in many work situations noise levels cannot be maintained below the allowable limit. In these situations, workers should use hearing-related personal protective equipment. Hearing protective devices commonly used in the worksite include earplugs, canal caps, and earmuffs. The selection of the right kind of protective device should be based on the frequency and level of noise, as well as requirements for fit and comfort. All hearing protective devices are labeled with a noise reduction rating (NRR). Effectiveness of protection increases with the NRR number.

25.1.2   Confined Space Entry

A confined space is defined as a space large enough for human entry with restricted means of entry and exit. In addition, the space is not designated for long occupancy. A confined space poses the potential for hazards related to atmospheric conditions such as the existence of toxic fumes, lack of oxygen, presence of methane, and so on. In many cases, accidents due to confined spaces occur due to oxygen deficiency. The minimum safe level of oxygen stipulated by OSHA for confined spaces is 19.5%. A person will feel disoriented at 16% oxygen level and become unconscious at levels between 8 and 12% oxygen. There are a number of potential confined spaces at landfills including structures such as manholes and leachate storage tanks. The air in a confined space may contain toxic gases, which in a landfill setting could include methane and hydrogen sulfide.

The air in a confined space should, at a minimum, be tested for oxygen, hydrogen sulfide, and methane content before entering. Ventilating equipment should be used where necessary to ensure that the oxygen level is above the hazard level. Portable self-contained breathing devices should be used where necessary. Spark-proof tools, explosion-proof fans, and lights should be used within a confined space where there could be an explosion potential.

At a minimum a team of two workers is needed for working in a confined space. One person must remain outside to summon help or offer assistance. The worker inside a confined space should communicate with the person outside the space. Proper equipment such as a full-body harness and lifeline should be used. If workers are routinely required to enter confined spaces to perform a job, then the person serving as the observer should have the necessary rescue equipment and be aware of all contingency procedures that may be required.

25.1.3   Dust and Air Exposure

In landfills, material recovery facilities, and composting facilities, there are a number of substances that may become airborne and cause an exposure with health consequences. Bioaerosols are suspension of particles in the air consisting partially or wholly of microorganisms. These microorganisms can remain suspended in the air for a long period of time. The bioaerosols of concern in the waste industry (particularly in composting facilities) include actinomycetes, bacteria, viruses, molds, and fungi. *Aspergillus fumigatus* is a common fungus present in decaying organic matter. *Aspergillus fumigatus* colonizes easily and can be dispersed from composted waste materials when agitated. The level of this fungus decreases rapidly at a short distance from the source or a short time after the agitation is stopped (Epstein and Epstein, 1989). The spores of this fungus can enter the body through inhalation and breaks in the skin. Although not a hazard to healthy individuals, this exposure can produce fungal infection of the lungs in susceptible individuals. Conditions that predispose individuals to infection by this fungus include a weakened immune system, allergies, asthma, diabetes, tuberculosis, a punctured eardrum, the use of some medications such as antibiotics and adrenal cortical hormones, kidney transplants, leukemia, and lymphoma (USDA and USEPA, 1980; Epstein and Epstein, 1989; USEPA, 1994d). Another health concern from microorganisms is endotoxins. Endotoxins are toxins produced within a microorganism and released upon destruction of the cell in which it is produced. They can be carried by airborne dust particles.

Steps should be taken to minimize dust generation in facilities where bioaerosols and endotoxins are likely to be present (e.g., composting facilities). In addition to measures taken for minimizing dust, the following precautions should be taken for worker safety:

1. Dust masks or respirators should be used under dusty conditions.
2. Workers should remove work clothes and wash their hands before meals and at the end of the work shift.
3. Work clothes and shoes should not be worn home by any employee.
4. Cuts and bruises should be treated and covered immediately.
5. Proper ventilation should be provided for enclosed facilities.
6. Workers performing potential spore-dispersing processes (e.g., turning of compost piles, especially dry piles) should use vehicles with enclosed drivers cabins.
7. Individuals with asthma, diabetes, or a suppressed immune system should not be allowed to work in facilities where chances of the presence bioaerosols and endotoxins are high.

25.1.4 Blood-Borne Pathogens

Blood-borne pathogens are microorganisms present in human blood that can cause disease in humans. Most common of these microorganisms are the hepatitis B virus (HPV) and the human immunodeficiency virus (HIV). Exposure to these microorganisms can be caused through direct contact of eye, mouth, piercing of mucous membrane or skin due to needle stick, human bite, cut, and abrasion. Workers involved in municipal garbage collection and medical waste handling are most susceptible to blood-borne pathogen exposures. However, workers at landfills should be careful not to be exposed to blood-borne pathogens as well. Engineering controls and work practices must be instituted to minimize exposure to blood-borne pathogens. All infectious waste containing blood-borne pathogens must be put in red (or other colored bags as required in the country) bags that are appropriately labeled with a biohazard symbol. While untreated infectious waste should not be brought to a landfill, care needs to be taken by landfill personnel in handling treated infectious waste as well to ensure safety.The following steps should be taken to minimize health hazard due to exposure to blood-borne pathogen:

1. Necessary personal protective equipment (PPE) must be used when the possibility of exposure to blood-borne pathogen exists. (PPE includes gloves, plastic visors, half-face masks, full-body gown, goggles, and so on).
2. Hand gloves must be replaced as soon as possible after they are contaminated or they become torn or punctured.
3. All employees potentially exposed to blood-borne pathogens should be vaccinated against the hepatitis B virus.
4. Equipment and work areas must be cleaned and decontaminated as soon as possible after blood or potentially infectious fluid is spilled.
5. All reported exposure incidents must be evaluated and treated promptly. Follow-up testing and treatment must be undertaken where required.

Employers should arrange for training regarding blood-borne diseases and how they are spread, response plan for emergencies involving exposure, and signs and labels used to identify potential biohazard.

25.1.5 Slips, Trips, and Falls

In general in the waste management industry slips, trips, and falls are a major source of injury. Injuries from falls may include broken bones, back injuries, sprains and strains of muscle, cuts, and bruises.

Slips occur due to a loss of balance caused by lack of friction between feet and the surface in contact with the feet. Slips usually occur due to wearing shoes without adequate friction or because inadequate attention to the surface condition (e.g., icy road). Trips occur whenever a person with a high enough momentum hits a rigid object with foot. Trips usually occur on cluttered work areas or when lighting is poor. Falls occur whenever the body is off of its center of balance. In addition to falls due to slips and trips, falls can also be caused as a result of using makeshift ladders and scaffoldings and accidents while climbing. Falls from a significant height may cause serious injury, even death.

The following safety precautions should be followed to avoid slips, trips, and falls:

1. Short steps should be taken while walking on a slippery surface.
2. All spills of any type of liquid should be cleaned immediately.
3. Shoes with proper friction should be used on work sites particularly on surfaces that can become slippery. Shoes with neoprene soles are good for most slippery surfaces except those where oil has been spilled.
4. If floors are cleaned with soaps or commercial cleaners, then care must be taken to remove all residue from the floor after cleaning.
5. Caution should be exercised while carrying loads. The loads should not block one's front vision.
6. The work area must be kept clean. For example, extension cords should be taped to the floor to avoid trips.
7. Hand rails should be used while walking up or down steps.
8. Jumps from trucks or other equipment must be avoided.
9. Only one person should be on a ladder at any one time.
10. Ladders must not be placed on a stack of blocks or boxes to reach a higher elevation.
11. All necessary safety precautions must be followed while setting up a ladder and working from a ladder.

25.1.6   Thermal Stress

Landfill operators are exposed to both hot and cold temperatures. There are many dangers from exposures to excessive hot and cold temperatures. A good working knowledge about the symptoms of thermal stress is essential to provide protection. Managers of landfills should closely monitor their staff to make sure workers are not exposed to excessively hot or cold temperatures for prolonged periods of time.

Heat exhaustion slows down the circulatory system, which reduces blood flow for removing body heat. Heat exhaustion can cause dizziness, nausea, even fainting. To avoid heat stress during hot summer days, work should be done in short intervals and fluid intake must be increased significantly. The operator's cabin in mobile equipment should be well ventilated or cooled if possible.

Working in extreme cold temperature can also cause health hazards. Blood circulation to feet and hands may be reduced as a result of the cold, causing numbness of fingers. On cold temperature days with high wind, the risk of frostbite is high. Several layers of garments should be worn during winter days. On extremely cold winter days, outdoor work should be done in short intervals.

25.1.7 Exposure to Harmful Chemicals

In the solid waste management industry incidences of exposure to chemicals with health hazards is relatively low. However, small quantities of hazardous chemicals are disposed of in many landfills. In many areas, household hazardous waste collection programs are used to manage these materials. Employees involved in household hazardous waste collection (also known as clean sweep; refer to Chapter 3) may get exposed to a number of common harmful chemicals. Therefore, it is a good practice to have a clear understanding regarding the hazards of chemicals on human health. Flammable or reactive chemicals pose physical hazards. Some chemicals can cause immediate obvious effects while others have health effects that result from repeated long-term exposure. Often, chemicals brought to a collection point or disposed in garbage are not labeled. As a result all unknown chemicals must be handled with care. At a minimum a dust mask and hand gloves should be used while handling small quantities of hazardous waste. Employees must receive proper training for handling flammable liquids and explosives. Many waste materials contain small quantities of harmful chemicals that find their way into landfill leachate. Table 25.1 includes the health effects of various volatile compounds in landfill leachate. Landfill leachate is a highly contaminated liquid and hence must be handled with care. Proper precaution must be taken while pumping leachate out of a holding tank or discharging it to a treatment plant.

25.1.8 Electrical Equipment

Proper precaution must be taken while dealing with electrical equipment such as that associated with landfill gas collection and treatment or leachate pumping. Shocks, burns, and fires are common hazards from electricity. All electrical equipment must be installed and insulated properly. While working on electrical equipment lockout/tagout must be practiced for safety. Lockout is accomplished by installing a locking device, such as a lock or chain, that keeps an electrical or mechanical switch in the off position. Tagout refers to the process of putting a tag or sign (e.g., "do not operate") on the power source so as to provide notification that the device is not to be actuated.

25.1.9 Fire

The risk of fire increases with the storage of flammable materials or chemicals in an unsafe manner. Electrical equipment can also be a source of fire. In landfills, methane accumulated in a vent or manhole can cause an explosion or a fire. Use of an open flame near a manhole at landfills or leachate storage tanks must be avoided. In case of fire, quick action to extinguish the fire and informing the local fire department should be undertaken. It is a good idea to have fire drills at least once a year. A well-managed landfill will have a contingency plan in place to cover incidences such as fires. Personnel should

TABLE 25.1   Health Effects of Various Volatile Organic Chemicals Found in Landfill Leachate

Benzene	Human carcinogen, mutagen, and possible teratogen; central nervous system (CNS), peripheral nervous system, immunological and gastrointestinal effects; blood cell disorders; allergic sensitization; eye and skin irritation
Chloroform	Probable human carcinogen and possible teratogen; CNS and gastrointestinal effects; kidney and liver damage; embryotoxic; eye and skin irritation
1,1-Dichlorethane	Embryotoxic; CNS effects; kidney and liver damage
Ethylbenzene	CNS effects; kidney and liver damage; upper respiratory system, eye and skin irritation
Methylene chloride	Possible carcinogen; CNS, lung/respiratory system, and cardiovascular effects; blood disorders; eye and skin irritation
Tetrachloroethylene	Probable carcinogen; CNS and lung/respiratory effects; embryotoxic; kidney and liver damage; upper respiratory tract and eye irritation
Toluene	Possible mutagen and carcinogen; CNS and cardiovascular effects; kidney and liver damage; upper respiratory tract, eye and skin irritation; and allergic sensitization
Trichloroethylene	Possible carcinogen and teratogen; CNS, kidneys, liver, cardiovascular system, and lung/respiratory system effects; blood cell disorders; skin, eye, and upper respiratory irritation
1,1,1-Trichloroethylene	Carcinogenic; mutagenic; CNS and lung/respiratory effects; kidney and liver damage; eye and skin irritation
Vinyl chloride	Carcinogenic; mutagenic; possible teratogen; CNS effects; kidney and liver damage; blood cell disorders; and skin irritation
Xylene	CNS and cardiovascular effects; kidney and liver damage; upper respiratory and eye irritation

be trained on how they should react should a fire occur. Clearly, this training should include instructions that personnel should not place themselves in danger by fighting a fire that would be better fought by trained local fire department personnel.

25.1.10   Lifting

Improper lifting is the most common cause for back pain and injury. In addition to lifting, poor posture and physical condition (e.g., being overweight) and repetitive minor strains can cause back pain. Although forklifts, hoists, and other types of equipment may be used to lift heavy objects, it is often necessary to lift moderate to heavy objects by hand. Centering the body over the load and bending knees while lifting a heavy object minimizes chances of back injury. Once the object is lifted, the body must not be turned or twisted. In most cases simple bending exercises and bed rest will cure back pain. One should see a doctor if the pain persists or becomes unbearable.

25.2 PERSONAL PROTECTIVE EQUIPMENT

Personal protective equipment (PPE) must be used when necessary. It is an obligation of the employer to provide PPE as needed. However, the employer should not rely on PPE alone to ensure occupational safety but must also implement engineering controls. Employees must be trained in the proper use of PPE. In general, PPE includes protection for eyes, feet, hands, hearing, respiratory systems, and from falls.

25.2.1 Eye Protection

Personal protective equipment for eyes includes safety glasses, goggles, face shields, and absorptive lenses. Each of these is designed for a specific use. Proper assessment of the worksite danger must be done prior to choosing a particular eye protection design. In addition to the PPE, arrangement should be made for eyewash in an emergency. Eyewash stations can be set up for this purpose. Each employee should practice eyewash for proper benefit. If PPE for eyes is needed to perform a job, vision tests should also be done routinely.

25.2.2 Foot Protection

Feet are subject to many injuries such as cuts, punctures, sprains, fractures and so on. The aim of foot protection is to provide protection for toes, ankles, and feet against injury. Sprains and fractures and other injuries to feet can be avoided through safe working practices. There are various types of shoes available to provide foot protection. OSHA requires that all safety shoes meet the requirements of the American National Standards Institute (ANSI) standard for protective footwear. There are shoes that provide protection against static buildup or electrical hazard. Safety shoes and safety boots are commonly used in the waste industry and should especially be required at landfills. These shoes often include steel, reinforced plastic, or rubber for protecting toes. Shoes in which a metal guard extends over the entire foot, to protect the upper foot from impacts, are called metatarsal guards. Safety shoes usually have puncture-resistant soles. Safety boots are made of rubber or plastic for providing protection against oil, acids, and other corrosive chemicals. Depending on the work situation, these boots may be fitted with steel toe caps and puncture-resistant insoles.

25.2.3 Hearing Protection

As indicated in Section 25.1.1, exposure to noise can lead to hearing loss. Earplugs are most common device for hearing protection. Earplugs are devices that fit into the ear canal. There are three primary types of earplugs:

1. Premolded inserts are reusable earplugs made from soft silicone rubber. These inserts are commercially available and can be used by any individual. These plugs must be kept clean to avoid ear infection.
2. Formable plugs are disposable ear plugs made of waxed cotton or acoustical fibers. These can also be used by any individual.
3. Custom-molded earplugs are made to fit a specific individual.

Canal caps seal the external edge of the ear. The caps are held in place by a headband. These caps are designed to reduce noise. These caps are for workers who cannot use earplugs and who enter and leave a noisy worksite frequently during their work.

Earmuffs cover the whole ear to seal out noise. Cups filled with foam or other material are fitted with a cushion filled with liquid, air, or foam. A headband forming the earmuff then joins the cups (J. J. Keller & Associates, 1996).

If use of earplugs is required for any regular noisy work, then audiometric testing should be done under the supervision of a physician. An audiometric test is a procedure for checking a person's hearing and should be done periodically to determine if hearing is deteriorating.

25.2.4 Respiratory Protection

Respirators are used to prevent dust and various gases from entering the respiratory system (Fig. 25.2). A person using a respirator must be trained in its use. Improper use of a respirator can be almost as dangerous as not having a respirator if it is required.

There are various types of respirators. Selection of the type of respirators to use is made on the basis of the hazard present. Medical evaluation is necessary before a respirator may be used by an individual. Individuals with breathing problems such as asthma or those who lack the physical endurance to wear the respirator and perform the necessary work should not be allowed to use it on the job.

There are two principal types of respirators: Air-purifying (also called air-filtering) respirators (Fig. 25.3) and air-supplying respirators (Fig. 25.4). Air-purifying respirators are used when oxygen level is sufficient. Although air-purifying respirators are used to provide protection from gas fumes, dust, and vapors, they cannot filter all dangerous substances. Air-supplying respirators are used in oxygen-deficient situations. Air-supplying respirators may

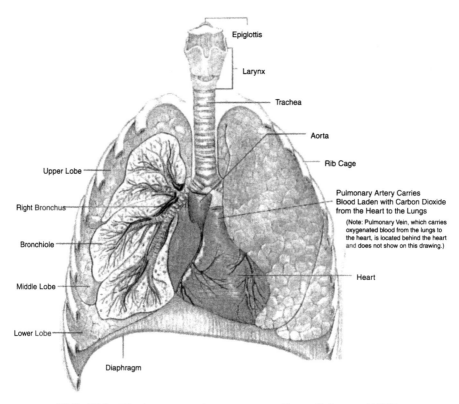

FIG. 25.2   The human respiratory system. From Guinness (1987).

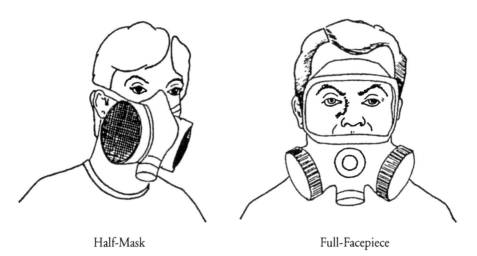

Half-Mask   Full-Facepiece

FIG. 25.3   Air-purifying respirator.

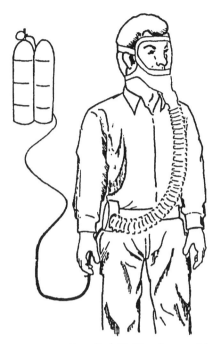
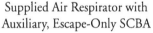

Supplied Air Respirator with
Auxiliary, Escape-Only SCBA

Self-Contained Breathing
Apparatus (SCBA)

FIG. 25.4 Air-supplying respirator.

be connected to a stationary air supply or may include portable air supply
tanks [known as self-contained breathing apparatus (SCBA)]. Since heat cannot escape from a SCBA suit, caution must be exercised to avoid heat stroke
while using an SCBA.

25.2.5 Hand Protection

Hand protection is needed to safeguard against injuries from machines, extreme heat or cold, or chemical or biological agents. Usually equipment is
designed with engineering controls to reduce hand injuries. An example of
these engineering controls is machine guards that protect personnel from moving belts and other such potential injury sources. Several types of hand protecting devices are available including:

1. *Gloves* Gloves provide protection to fingers, and hands. Gloves may
   be longer to also include the protection of wrists and forearms. Gloves
   may be made of canvas, rubber, vinyl, or neoprene.
2. *Mitts* These are similar to gloves but include only two pockets. One
   is for the thumb and the other is for all the four fingers.

3. *Hand Pads*   Hand pads are used to protect the palm against heat or friction. These are not as flexible as gloves or mitts.
4. *Finger Cots*   These are used for providing protection to a single finger or thumb.
5. *Sleeves*   These are used to provide protection to wrists and arms. These are also known as forearm cuffs.

25.2.6   Head Protection

Head protection is important in worksites where there are possibilities of falling objects. In addition head protection should also be provided in worksites involving moving equipment. Hard hats are often used for head protection. Properly sized hard hats should be used so that the hat does not fall off during normal working head movements. Although hard hats are usually made of fiberglass, metallic hard hats are also available.

25.3   ELEMENTS OF A HEALTH AND SAFETY PROGRAM

A health and safety program (HSP) must be well structured and properly implemented. Health and safety programs should be developed based on specific facility criteria because each facility will have its own unique characteristics. A good knowledge regarding occupational hazards at a specific facility is necessary to develop an effective HSP. Hazards involved for each piece of equipment used in a facility and the overall work environment should be taken into consideration while developing a health and safety program. Visitor access should be restricted from areas of a facility that pose a health or safety hazard. For instance, the public should not be allowed to wander around a landfill work area. In addition to employees, representatives of regulatory agencies and the public should be made to wear PPE if needed. The following elements should be included in a well-structured HSP.

25.3.1   Commitment and Involvement

Management and the owners of a facility, whether public or private, must commit to provide a safe environment at the worksite. Apart from developing a proper program through the assistance of a knowledgeable person or consultant, regular employee training and implementation monitoring is needed. Some of this training may be required by existing laws relating to health and safety. Management needs to allocate sufficient resources, both human and financial, for running a successful program.

Employee productivity increases in a safe worksite. Therefore, in addition to the social and human aspects of employee welfare, a safe worksite provides economic benefits. In addition, compliance with occupational safety laws minimizes fines associated with enforcement-related issues implemented by the regulatory agency entrusted with the administration of worksite health and safety (e.g., OSHA).

25.3.2 Worksite Analysis

A thorough understanding of all equipment and employee job duties is necessary for developing an HSP. After an initial analysis of the jobs involved, the regulatory agency entrusted with enforcing health and safety issues at worksites should be consulted. These initial steps help in developing a good understanding of health and safety hazards and the relevant legal obligations related to a worksite. Expert help, either from outside consultants or from in-house health and safety experts, is key to proper worksite analysis. Once an initial analysis is completed a written document summarizing the hazards and the related steps to safeguard employee health is developed. A survey of injuries and illnesses in similar industries would help identify patterns and necessary safeguards. A thorough and complete review of employee health and hazard complaints should be done to improve on the program. A discussion with employees can also be helpful in identifying worksite hazards. All worksite accidents or job-related employee illness should be investigated thoroughly. These investigations can help prevent recurrences of health and safety problems. Employees should be encouraged to report the malfunction of any equipment or the existence of any worksite hazard. One person or a group of persons within the organization should be designated to record and analyze all employee health complaints.

25.3.3 Hazard Prevention and Control

Based on the worksite analysis, safe working procedures should be developed. Each employee must be knowledgeable of all safety procedures. Disciplinary actions, easily understandable and fair to all employees, should be developed. Employees need to be informed about the consequences of failure to follow health and safety procedures. It may be noted that management will be made responsible by the regulatory agency if the procedures are not implemented properly. Employees must be provided with PPE where necessary. It is essential to develop a maintenance program for all PPE. Employees may be given the responsibility for maintenance of PPE or it may be assigned to staff providing maintenance of a larger group. Regular maintenance of all machinery and equipment is a part of the safety program because sudden breakdown may create hazard or lead to accidents.

25.3.4 Staff Training

Proper staff training regarding safe work habits and use of PPE is an integral part of an HSP. All employees, supervisors, and managers must receive training in safe handling of all equipment, worksite hazards, related safety procedures, and the use of PPE if it is required. Training on all these issues should be provided prior to having new employees assume their responsibilities. In addition, refresher training is required to keep all health and safety procedures and practices current. Inclusion of accident data both at the worksite and in similar industries may be included to motivate trainees to follow worksite safety procedures. It is essential to ensure, through test or verbal question and answer sessions, that the trainees have understood the training presentations. While employees have responsibilities regarding safe work procedures and proper use of PPE, supervisors and managers also need to have proper training regarding safe work procedures to be followed by their employees. This knowledge will help in the proper implementation of the HSP.

In addition to the initial training, refresher training is needed. The goal of refresher training is to reinforce the original training and to discuss any updated data or practices. During refresher training, employees should be given opportunities to discuss any changes that they feel will improve safe working conditions.

25.3.5   Emergency Planning

Emergency planning for fires, accidents, illness, spills of hazardous substances, and natural disasters (e.g., tornados) should be an integral part of an HSP. A written emergency planning document detailing procedures, duties, and responsibilities of designated personnel needs to be developed. The contact phone numbers and location of fire, police, hospital(s), and ambulance should be readily available in a designated location within the worksite. Representatives from external organizations (i.e., fire department, hospital, and police) should be made aware of the type of emergency situation that may occur at the worksite. This will help the fire department and the hospitals acquire necessary equipment/medicine and training for handling an emergency situation from the facility. Representatives of the fire department should be invited to the worksite to gain familiarity with the buildings or worksite.

All employees, including supervisors and managers, should know the emergency procedures. Drills dealing with emergency procedures should be undertaken at least once a year. Such drills help in dealing with an emergency situation properly without creating any panic.

After a drill and after observing actual emergency response of employees and the personnel delegated to manage the emergency situation, the entire planning process should be reviewed. The purpose of this review is to check whether the emergency procedures were followed and determine whether any changes to the procedures are necessary.

25.3.6   Record Keeping

Reports of every injury requiring medical treatment beyond first aid should be kept in a file. Each year an annual summary of all injuries and work-related illnesses should be developed. The summary document should be discussed with the employees during the refresher training. The summary report along with the detail report should be retained for 5 years. Such record-keeping practices are very helpful in resolving any litigation brought by an employee. Record-keeping requirements of the regulatory agency, if any, should be followed. The regulatory agency should be given full access to HSP and to the injury files. The person in charge of the HSP should have complete knowledge regarding record-keeping requirements and should maintain liaison with the regulatory agency for updates.

26

ECONOMIC ANALYSIS

Enough money must be made available for not only constructing but also operating, maintaining, and monitoring a landfill. Proper cost analysis must be done to ensure cash flow for performing all the above tasks. Monitoring of a landfill may be required for 30–40 years after closing of the last phase. The total cost of long-term maintenance and monitoring (also known as long-term care) of a closed landfill could be higher than the cost of construction of a landfill. Enough money should be set aside during the active life of the landfill so that closure and long-term care can be performed. Methods for estimating these costs on a future date taking into account the inflation factor are discussed in Section 26.2.

Landfills may be either publicly (e.g., a municipality) or privately (e.g., industry) owned. The methods used to raise money for landfills differ for these two categories. For privately owned landfills the correct cash flow must be ensured by estimating costs for each activity discussed in subsequent sections.

Funds for a publicly owned landfill can be obtained from either general fund revenues or service charges. The disadvantages of using general fund revenues is that local officials are hard pressed to raise taxes. If service charges are used for funding purposes, then updating these charges for inflation and so on must be carefully calculated. Municipalities usually draw from two basic sources for financing a project: capital borrowed funds and current revenue funds. Municipalities may raise money by issuing general obligation bonds, by revenue bonds, or by bank loans. They may also lease equipment necessary for operating the landfill. A detailed economic analysis of the suitability of both these methods of financing must be undertaken before a recommendation on funding is made. A municipality may also sell the operating permit to a private operator to operate the site. In such cases the private operator fixes the dumping fee.

26.1   COST ESTIMATES

As previously indicated, the life-cycle cost can be categorized as follows: construction, operation (including all monitoring), closure, and long-term care (LTC). In many cases the cost of hauling is expected to be high because the landfills are located in remote areas. The cost of road construction for borrowed materials (e.g., clay) must be included in the unit cost.

26.1.1   Cost of Construction

Table 26.1 includes a list of typical construction items for a containment-type landfill; the list is by no means complete. The basis for estimating unit costs

TABLE 26.1   Items for Estimating Site Construction Costs

Item	Basis of Cost
Clearing and grubbing	Clear and grub forest cover
Site access road	Place and compact crushed stone
Strip topsoil	Strip and stockpile topsoil
Drainage swale construction	Rough grade and erosion protection
Sedimentation basin construction	Berm, riprap, and discharge pipe
Excavation to subgrade	Cut for subgrade
Berm construction	Fill for perimeter berms (compaction only)
Collection lysimeter	Excavation, liner, bedding, piping, risers, and storage manhole
Clay liner construction	Hauling, placement, compaction, and borrow restoration
Drainage layer construction	Haul and place
Leachate collection piping	Trenching, bedding, backfill, pipe, and filter fabric
Leachate header pipe to manhole	Double encased, antiseep collars
Leachate header pipe in clay liner	Trench, piping, backfill, and compaction
Leachate cleanouts	Trench, pipe, backfill, and compaction
Leachate collection tank and loading station	Excavation, placement of existing tank, clay backfill, and pump
Leachate pump manhole	Excavation, place MH, backfill, pump
Topsoil placement	Haul and place
Seed, fertilizer, and mulch	Using acreage or similar criteria
Survey, documentation, and technical supervision	Inspection, testing, and report

is also included to provide some idea as to what activities should be included to arrive at unit costs. Items of construction for a natural attenuation landfill may also be arrived at easily from Table 26.1.

26.1.2 Cost of Closure

Table 26.2 includes a list of typical construction items necessary for closing a landfill and the basis of unit cost estimate. The unit cost should be a third-party cost since it is possible that the owner (especially if private) may fail to close the landfill properly due to lack of funds. In that case the regulatory agency may have to close the site. In most cases the regulatory agency will have to hire an outside contractor to perform the job. So the cost of each item must reflect third-party costs. If topsoil is available on site, the cost of purchasing topsoil may not be included in the unit cost but a legal document providing access to the topsoil should be made available to the regulatory agency. A worst-case scenario (e.g., one phase has reached final grade but is not covered and waste is being disposed of in the next phase) should be assumed for estimating closure costs. Closure costs in the example case will include final cover construction in both phases and filling the second phase with soil up to the final grade.

26.1.3 Cost of Long-Term Care

Table 26.3 includes a list of typical items necessary for estimating long-term care costs for a containment-type landfill. For an NA-type landfill there will be no cost related to leachate management; all other cost items will remain the same. Here also the unit cost must be a third-party cost. Damage of 10–20 m^2 of surface area per half a hectare ($\sim$100–200 ft^2/acre) of closed area per year may be assumed in most cases. Erosion loss is high in the first year and decreases steadily in a well-maintained landfill. The cost of land surface

TABLE 26.2   Items for Estimating Site Closure Costs

Item	Basis of Cost
Final cover construction	Construction of barrier layer and other layers, haul and place as required, restoration of borrow source(s), top soil placement, grading
Seed, fertilize, and mulch	Using acreage or similar criteria
Leachate head wells (if designed to be installed after completion)	Boring, well installation
Survey and documentation	Inspection, testing, report preparation, technical assistance, and supervision
Contingency	10–25% of total

TABLE 26.3 Items for Estimating Long-Term Care Cost

Item	Basis of Cost
Land surface care	Erosion damage repair, reseeding, mulching
Leachate and lysimeter pipeline cleaning	Cleaning of pipeline using proper techniques
Groundwater monitoring (includes lysimeter sample)	Sample collection, transportation, and laboratory and field testing
Gas monitoring	Field testing
Leachate monitoring	Sample collection, transportation, and laboratory and field testing
Leachate management	Hauling and treatment of leachate, leachate sump pump, and tank maintenance
Annual assessment	Site inspection and report preparation
Administration and contingency	10–25% (or as appropriate) of total

care can thus be reduced to 10% of the original value in a 7- to 10-year period. Leachate hauling and treatment are high cost items. The leachate production rate is reduced significantly a few years after closure. As mentioned in Section 13.4.2.6, the reduction in leachate production rate after landfill closure is difficult to estimate. A 20–30% infiltration of total precipitation may be used to estimate the leachate generation in the initial years (models mentioned in Section 13.4.2 may also be used to predict postclosure). The funds necessary for leachate management (hauling and treatment) can be adjusted in subsequent years based on actual field data. Third-party costs for hauling and treatment should be assumed even if the landfill owner owns a treatment plant. (Note: Most industries have wastewater treatment plants that are capable of treating the landfill leachate.) Although leachate treatment can be continued in the owner's treatment plant, a third-party cost is to be assumed only for estimating long-term care costs.

26.1.4 Cost of Operation

Table 26.4 includes a list of typical items for estimating the annual operating cost for a containment-type landfill. Costs of leachate management and other irrelevant items (e.g., lysimeter monitoring) should be excluded when estimating the cost of operating an NA-type landfill. A third-party cost need not be used for estimating the annual operating cost. The waste fund is a fund administered in some states to which each landfill owner pays. The money from the fund may be used for closure, remedial action, or long-term care of abandoned landfills. Usually a fee is charged based on tonnage of waste disposed in a landfill. If such a fund does not exist, then the fee need not be taken into consideration when estimating operational costs. In addition to costs of monitoring, necessary money for closure and long-term care should

TABLE 26.4 Items for Estimating Annual Operating Costs

Item	Basis of Cost
Waste placement	Manpower, equipment necessary for compaction of waste
Intermediate cover	Hauling, placement, and compaction of soil
Groundwater, leachate, and gas monitoring	See Table 26.3
Leachate management	See Table 26.3
Equipment	Purchase and maintenance
Annual payment to waste management	Disposal volume/weight

be invested in a fund during the active site life. In addition, money may also be saved for developing future landfill(s) during the active life of a landfill. Thus, once a landfill is constructed, the user fee may be estimated in such a way that it will cover all future activities as related to disposal of waste. In some instances the revenue generated from the user fee may be grossly underestimated if the disposal volume is reduced significantly due to increased recycling activities or wrong estimates of the waste volume generation rate. The chance of overestimating generation rate is high in small communities. So due care should be taken when estimating the waste generation rate. A discussion on cash flow estimates is included in Section 26.4, which will provide additional help in estimating operating costs.

26.2 ESTIMATE FOR PROOF OF FINANCIAL RESPONSIBILITY

Establishing proof of financial responsibility may be required by a regulatory agency. The purpose is to ensure that enough money is available for closure and long-term care of a landfill by the regulatory agency if the owner fails to perform these tasks due to a lack of funds. Not all regulatory agencies require this proof. The proof of financial responsibility (hereafter called "the proof") needs to be established prior to licensing the landfill site. There are two issues related to financial proof: methods of establishing the proof and estimation of the amount of proof. Each is discussed separately in the following sections.

26.2.1 Methods for Establishing Proof of Financial Responsibility

There are several methods by which proof can be established. The regulatory agency should be consulted to determine the acceptability of the chosen method. Since the cost of establishing the proof is different for different methods, the costs for each method should be calculated so that the owner can choose the method that is most suitable. Essentially there are three types of methods: putting money in an interest-bearing account, buying a payment guarantee from an insurance company or bank, and increasing company liability. If necessary, separate methods can be chosen to establish the proof to cover closure and LTC costs.

Money can be deposited in an interest-bearing account (e.g., escrow account) each year so that at the end of the active life of the landfill (hereafter called site life) enough money is available for closure and or LTC. Proof for LTC is required to be in place for several years (usually 30–40 years) after closure, whereas proof for closure is required to be in place until the site is closed. A trust fund may also be created for establishing the proof. The regulatory agency should be named as the joint operator of this account so that no cash can be withdrawn without written permission from the agency.

Forfeiture or a performance bond, a letter of credit issued by a financial institution, or risk insurance may also be used to establish the proof. The issuer of such bonds guarantees the payment for closure and/or LTC if the owner fails to perform the task. The owner has to keep the bond in place for closure until closure is actually performed. Bonds for LTC must be kept in place until the LTC period (usually 30–40 years after closure) is over. The owner has to pay a fee to the bonding company for each year the bond is in place. The fee depends on the amount of the bond. The regulatory agency should be named as the beneficiary of the bond. The bond may be for cash payment to the agency or for performing the task by the issuing company to the satisfaction of the agency.

Large companies may guarantee closure and LTC by increasing their liability for cost of closure and LTC. A financial statement regarding economic soundness commonly known as "net worth of the company" is usually required to be submitted to the regulatory agency each year. Usually companies with a set minimum total asset (e.g., $5 million) and a minimum net asset (total assets minus total liabilities) exceeding the total cost of closure and LTC by a factor (usually 6 or more) are allowed to use the "net worth" method. It is possible that a company's net assets in a year fall below the minimum required by the agency. In such cases the owner has to switch to a different method for establishing the proof. The cost of total LTC is reduced each year after the LTC is performed for the year. Money may be withdrawn, if allowed by the regulatory agency from an interest-bearing account after LTC is successfully performed in a year. If the money withdrawn from the interest-bearing account is saved, then it can be used for LTC in the current year; this means that cash flow is not needed every year for LTC. However, if the bond or net worth method is used for LTC then although the cost of the remaining LTC is reduced and the cost of buying the bond and so on is reduced, no money is really available for performing LTC in the current year. Thus, an interest-bearing account is the best choice for establishing the proof provided enough money is available initially.

26.2.2 Estimation of Financial Responsibility for the Amount of Proof

The costs of closure and LTC are estimated based on Section 26.1. These base costs are then used to calculate the costs of closure and LTC at a future date. These amounts must be available to perform the tasks at the specified time in the future. Inflation reduces the value of money. Thus, if $1000 is required for closure today, the amount required a year from today will be $(1000)(1 + f)$, where f is the percentage inflation rate for the 1-year period. (Note: Usually a constant inflation rate is assumed for each year.) At the end of year 2 the required amount will be $(1000)(1 + f)(1 + f)$. Similarly, the required amount at the end of year n is $(1000)(1 + f)^n$. Now, if the money is invested in an interest-bearing account then less money needs to be kept in that account because of the interest being earned. The money to be invested in an interest-bearing account to yield $(1000)(1 + f)$ at the end of year 1 is $(1000)(1 + f)/(1 + i)$ ($= D$ say), where i is the percentage interest rate. Thus, the amount to be invested in an interest-bearing account to yield the necessary money at the end of year n will be $1000(1 + f)^n/(1 + i)^n$. This is the basic concept used to estimate the amount of financial proof required for closure and long-term care for a landfill. Proofs for both closure and LTC may be required to be established prior to disposing of waste in a landfill or within the active life (also known as site life) of the landfill. The regulatory agency should be contacted as to whether and when the proofs are to be established.

Formulas used for estimating the amount of proof depend on whether any interest is earned. Although interest increases the deposited amount, inflation reduces it. If a bond or net worth is used for proof then, only inflation needs to be accounted for because the additional money earned through interest on the deposited amount is not available to the bond account. The cost of closure and LTC should be increased each year by the inflation rate while using them for the net worth method so that these costs remain current throughout the LTC period. The official inflation factor is obtained only at the end of the year. If the trend indicates that the rate of inflation is increasing each year (e.g., year 1, 3%; year 2, 3.5%; year 3, 4.5%) for the last 2 years then a reasonable increase in inflation rate may be used (i.e., if the actual inflation rate is 4%, then a 5% inflation rate should be assumed). Similarly, if the trend indicates that interest rates will decrease each year then a reasonable reduction in interest rate may be used unless the bank has guaranteed a fixed rate for the entire LTC period. These suggestions regarding increases in the inflation rate and decreases in the interest rate are given to provide additional safeguards against shortages of the funds available to perform the closure and long-term care tasks. A discussion with the financial institution regarding available interest rates for an escrow account (if used) is helpful.

As mentioned in Section 26.1.3, LTC costs may vary over the LTC period (30–40 years after closure of a landfill). Therefore a variable LTC cost may need to be used to estimate the amount of financial proof. The approach for estimating the amount of financial proof for closure for an interest-bearing account is discussed at the beginning of this section. The approach for estimate the amount of financial proof for different cases are included in the following sections.

26.2.2.1 Amount of Proof for Closure   For interest-bearing accounts ($C =$ cost of closure; SL = site life)

$$D = C(1 + f)^{SL}/(1 + i)^{SL} \tag{26.1}$$

For non–interest-bearing accounts (e.g., bonds) and net worth

$$D = C(1 + f)^{SL} \tag{26.2}$$

26.2.2.3 Amount of Proof for LTC

For Non–Interest-Bearing Accounts (e.g., bonds)

EQUAL ANNUAL OUT PAYMENT   The account should be set up in such a way that enough money is available to pay all LTC-related bills at the end of each year of LTC. Thus, LTC money is needed for each year of the LTC period beginning 1 year after landfill closure. The total cost of LTC (A) for the entire LTC period (Fig. 26.1) is given by Eq. (26.4) in which $L =$ the cost estimate of LTC calculated in year zero (i.e., in the year when waste was first disposed in the landfill):

$$A = L(1 + f)^{SL+1} + L(1 + f)^{SL+2} + \cdots + L(1 + f)^{SL+LTC}$$

or

$$A = L(1 + f)^{SL+1}[1 + (1 + f) + (1 + f)^2 + \cdots + (1 + f)^{LTC-1}] \tag{26.3}$$

$1 + (1 + f) + (1 + f)^2 + \cdots + (1 + f)^{LTC-1}$ is a geometric series of the form $1 + a + a^2 + \cdots + a^{n-1}$, the sum of which is expressed as (Tuma, 1979) $(a^n - 1)/(a - 1)$. Thus, Eq. (26.3) can be written as

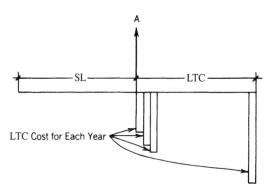

FIG. 26.1 Cash flow pattern used for estimating proof of financial responsibility for long-term care.

$$A = L(1 + f)^{SL+1} \left[\frac{(1 + f)^{LTC} - 1}{f} \right] \qquad (26.4)$$

Example 26.1 Find the amount of financial proof required when a bond is used for a constant LTC cost of $5000/year estimated 2 years after the landfill was opened. The estimated SL for the landfill is 10 years, the LTC period is 20 years, and the constant rate of inflation is found to be 5%.

Note that the LTC estimate was made 2 years after the landfill was opened, which has an SL of 10 years. Thus, the SL to be used in Eq. (26.4) shall be $10 - 2 = 8$ years rather than 10 years.

$$A = 5000 (1 + 0.05)^{8+1} \left[\frac{(1 + 0.05)^{20} - 1}{0.05} \right]$$

$$= \$256,480.74$$

UNEQUAL ANNUAL OUT PAYMENTS   For estimating the amount of proof for variable LTC L_x for year x, A is given by

$$A = L_1(1 + f)^{SL+1} \sum_{\substack{x=2 \\ r=1}}^{\substack{x=n \\ r=LTC-1}} L_x(1 + f)^r \qquad (26.5)$$

For Interest-Bearing Accounts

EQUAL ANNUAL OUT PAYMENTS   In deriving this formula it is assumed that all the money needed for LTC is deposited in an interest-bearing bank account at the end of the site life, but the cost estimate for LTC was made at the beginning of the site life. Therefore, the money earns interest for 1 year before payment is needed for LTC. The money that should be deposited in an interest-bearing account is given by

$$A = L\frac{(1 + f)^{SL+1}}{1 + i} + \frac{L(1 + f)^{SL+2}}{(1 + i)^2} + \cdots + \frac{L(1 + f)^{SL+LTC}}{(1 + i)^{LTC}}$$

$$= L(1 + f)^{SL} \left[\frac{1 + f}{1 + i} + \frac{(1 + f)^2}{(1 + i)^2} + \cdots + \frac{(1 + f)^{LTC}}{(1 + i)^{LTC}} \right] \qquad (26.6)$$

The series within the brackets is a geometric series. Thus, Eq. (26.6) can be written as

$$A = L(1 + f)^{SL}\left[\frac{\left(\frac{1 + f}{1 + i}\right)^{LTC+1} - 1}{\left(\frac{f - i}{1 + i}\right)} - 1\right] \qquad (26.7)$$

Note that the formula is not valid when $f = i$.

Example 26.2   Estimate the amount to be deposited in an interest-bearing bank account for long-term care. Use the following values of various variables: $L + \$10,000$, $SL = 2$, $f + 3\%$, $i = 6\%$, LTC period $= 5$:

$$A = 10,000(1 + 0.03)^2\left[\frac{\left(\frac{1 + 0.03}{1 + 0.06}\right)^{5+1} - 1}{\left(\frac{0.03 - 0.06}{1 + 0.06}\right)} - 1\right]$$

The interest earned and the expenditure for LTC for each year is included in Table 26.5; Fig. 26.2 depicts partial cash flow.

UNEQUAL OUT PAYMENTS   To estimate the amount of proof (A) for variable LTC costs of L_x for year x, Eq. (26.8) is to be used:

$$\begin{aligned} A &= \frac{L_1(1 + f)^{SL+1}}{1 + i} + \frac{L_2(1 + f)^{SL+2}}{(1 + i)^2} + \cdots + \frac{L_n(1 + f)^n}{(1 + i)^n} \\ &= \sum_{x=1}^{x=LTC} \frac{L_x(1 + f)^{SL+x}}{(1 + i)^x} \end{aligned} \qquad (26.8)$$

For Net Worth   (Note: Net worth must be evaluated each year.)

TABLE 26.5   Long-Term Care Cash Flow for Example 26.2

Year	Opening Balance ($)	Interest Earned ($)	LTC Cost ($)	Balance ($)
1	48,707.57	2,922.45	10,927.27	40,702.75
2	40,702.75	2,442.17	11,255.09	31,889.83
3	31,889.83	1,913.39	11,592.74	22,210.49
4	22,210.40	1,332.63	11,940.52	11,602.51
5	11,602.51	696.15	12,298.74	≈0.00

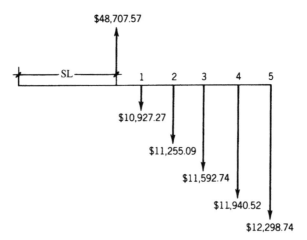

FIG. 26.2   Partial cash flow for Example 26.2.

$$A = L(1 + f) \tag{26.9}$$

The following symbols are used in the above formulas: A = the total inpayment for long-term care per year of site life, i = the estimated annual rate of interest, expressed as a decimal, f = the estimated annual rate of inflation, expressed as a decimal, SL = the estimated life of the facility in years rounded to the lower nonzero whole number, L = the estimated annual long-term care costs, L_x = the estimated unequal annual costs, x = the year of long-term care, and LTC = the period of long-term care.

26.3   USER FEE ESTIMATE

User fees can be calculated by dividing the total of the annual operating cost, the average annual cost for a future landfill, and the average annual cost for establishing financial proof, by the estimated annual disposal weight (or volume). The fee should be updated for inflation and revenue shortfall due to decreased disposal volume and so on. The following example enumerates how to estimate user fees.

Example 26.3   The following items are estimated for construction, operation, and long-term care of a landfill. The site life of the landfill is 3 years. It is estimated that the landfill will receive 25,000 tons of garbage each year. It was decided to raise funds for a future landfill of the same volume when the current landfill is closed. An escrow account will be used to establish proof of financial responsibility for closure and long-term care. The proof for closure is to be established in year 1 of the landfill site life and the proof for LTC is to be established by the end of the site life of the landfill. Cost estimates are included in Table 26.6. Estimate the user fee for the landfill. Assume an inflation rate of 4%, an interest rate of 5%, and a long-term care period of 20 years.

Average Cost for a Future Landfill

Cost for the landfill at the end the 3-year period

$$= (item\ a + item\ b\ of\ Table\ 26.6)(1 + f)^3$$

$$= (70{,}000 + 25{,}000)(1 + 0.04)^3 = \$106{,}862.00$$

The average approximate amount to be made available each year is

$$106{,}862/3 = \$35{,}621.00$$

The amount kept in an interest-bearing account (interest rate: 5% per year) would accrue interest and hence the amount actually needed to be deposited in the first year would be

$$35{,}621/(1.05)^3 = \$30{,}770.76$$

Similarly the cost for the two subsequent years can be estimated as follows:

$$Second\ year\ deposit = 35{,}621/(1.05)^2 = \$32{,}309.30$$
$$Third\ year\ deposit = 35{,}621/(1.05) = \$33{,}924.76$$

TABLE 26.6   Cost Estimates for the Landfill in Example 26.3

Items		Cost ($)
a. Construction of the landfill		70,000.00
b. Site investigation, report preparation, etc.		25,000.00
c. Annual operating cost		
1. Waste placement		20,000.00
2. Intermediate cover construction		2,000.00
3. Groundwater monitoring		5,000.00
4. Lysimeter monitoring		1,000.00
5. Gas monitoring		500.00
6. Leachate management		50,000.00
7. Equipment purchase and maintenance		20,000.00
8. Waste fund payment (at 5¢/ton)		1250.00
	Subtotal	99,750.00
d. Estimated cost of closure at the beginning of site life		10,000.00
e. Estimated cost of land surface care each year		3,000.00

To estimate user fees one may use an average of these three amounts, [(30,770.76 + 32,309.30 + 33,924.76)/3 = $32,334.94] or the maximum annual payment ($33,924.76). In this example the average amount is used.

Average Annual Cost for Establishing Proof of Financial Responsibility

Closure

$$D = \frac{C(1 + f)^{SL}}{(1 + i)^{SL}}$$

$$= \frac{10,000(1 + 0.04)^3}{(1 + 0.05)^3} + \$9716.99$$

The average amount to be made available each year for closure is $9716.99/3 = $3238.99

LONG-TERM CARE   The total cost for long-term care is obtained by adding items c3, c4, c5, c6, and e in Table 26.6.

Long-term care cost = 5000 + 1000 + 500 + 50,000 + 3000

= $59,500.00

$$\text{Total amount of proof} = 59,500(1 + 0.04)^3 \left[\frac{\left(\frac{1 + 0.04}{1 + 0.05}\right)^{20+1} - 1}{\frac{0.05 - 0.04}{1 + 0.05}} - 1 \right]$$

= $1,212,468.26

The average amount to be made available each year is $1,212,468.26/3 = $404,156.09.

AVERAGE ANNUAL COST   The total amount to be made available each year for closure and long-term care is $3238.99 + $404,156.09 = $407,395.08. The amounts to be saved each year are: year 1, 407,395.08/(1 + 0.05)³ = $351,923.19; year 2, 407,395.08/(1 + 0.05)² = $369,519.35; and year 3, 407,395.08/(1 + 0.05) = $387,995.31. The average amount to be used for user fee is

($351,923.19 + $369,519.35 + $387,995.31)/3 = $369,812.62

Money must be accumulated during the site life of a landfill to perform LTC of the landfill for 1 year after closure. Usually money is released from the LTC proof at the end of an LTC period. Therefore from the second year onward money will be available from the financial proof withholding if an escrow account is used. However, if a non-escrow-type method is used for LTC, then money to perform LTC for the entire LTC period must be accumulated in a fund. Money for closure will not be available at the time of performing closure. However, the proof may be released by the regulatory agency once the closure document is accepted. Thus, if an escrow account is maintained for closure, then a lump-sum amount may be available in the second year of the LTC period that may be used for LTC.

Thus, one should estimate which of the following will be the cheapest way of establishing proof: depositing money in an interest-bearing account, a bond, net worth, or a combination of two methods. The total average annual cost to be used for estimating user fee is (Note: The annual operating cost for the landfill is obtained from item c of Table 26.6)

$$\$32,334.94 + \$99,750.00 + \$369,812.62 = \$501,897.56$$

Thus the user fee is

$$\$501,897.56/25,000 = \$20.08/\text{ton}$$

The user fee may be raised by 10–15% for contingency. For privately owned landfills, the user fee should be increased to include profit.

26.4   CASH FLOW ESTIMATE

Cash flow estimate is important for the smooth running of a landfill. As previously indicated, the annual cost for running a landfill should include both the cost of the daily operation and the long-term cost.

The above example will provide some ideas for estimating cash flow for a landfill. The money to be borrowed for running the landfill for the first month can be somewhat high because of equipment purchase and so on, and this should be taken into account. Additional cash is needed at the end of the landfill site life for closure of the landfill and to perform long-term care. A careful study of the cash flow needed should be undertaken for the smooth running of a landfill. It is important to keep a record of waste actually disposed of in a landfill for future planning purposes. A faulty cash flow estimate, especially underestimates, can significantly increase the cost of operation because of emergency borrowing of money.

LIST OF SYMBOLS

A = total cost of long-term care
C = cost of closure
D = required deposit for closure proof
f = inflation rate
i = interest rate
L = cost of long-term care per year
L_x = estimated unequal annual costs of long-term care
LTC = period of long-term care
SL = site life
x = year of long-term care
Σ = sum from year 1 through 11

APPENDIX

CONVERSION OF U.S. CUSTOMARY UNITS TO INTERNATIONAL SYSTEM OF UNITS (SI)

SI units are derived from seven base units [length in meters (m); mass in kilograms (kg); time in seconds (s); electric current in amperes (A); thermodynamic temperature in Kelvin (K); amount of substance in moles (mol); luminous intensity in candelas (cd)] and two supplementary units [plane angle in radians (rad); solid angle in steradians (sr)]. Force is expressed as kg-m/s^2 and is termed a newton (N), and pressure is expressed as N/m^2 and is termed a pascal (Pa). A list of commonly used factors and their names used to form unit multiples and submultiples is given in Table A.1. Prefixes are used closed up to the unit symbol. For example, millimeter is written as mm, kilonewton is written as kN, and so on. Table A.2 lists conversion factors for commonly used engineering units. The following references may be consulted for a detailed guide on the SI system.

TABLE A.1 Name and Factors of Commonly Used SI Units for the Formation of Multiples and Submultiples

Name (Symbol)	Factor
mega (M)	10^6
kilo (k)	10^3
hecto (h)	10^2
deka (da)	10
deci (d)	10^{-1}
centi (c)	10^{-2}
milli (m)	10^{-3}
micro (μ)	10^{-6}

TABLE A.2   Conversion Factors

To Convert From	To	Multiply By
inches	centimeters	2.54
feet	centimeters	30.48
yard	meters	0.914
miles	kilometers	1.609
mils	inches	1×10^{-3}
mils	millimeters	0.0254
square inches	square centimeters	6.452
square feet	square meters	0.0929
square feet	hectares	9.29×10^{-6}
square yards	square meters	0.836
square yards	hectares	8.361×10^{-5}
acres	hectares	0.405
acres	square kilometers	4.047×10^{-3}
acre-feet	cubic meters	1.23×10^{3}
cubic inches	cubic millimeters	16.387×10^{3}
cubic feet	cubic meters	28.317×10^{-3}
cubic feet	U.S. gallons	7.48
cubic feet	liters	28.317
cubic feet	acre-feet	22.957×10^{-6}
cubic yards	cubic meters	0.765
U.S. gallon	liters	3.785
pounds mass	kilograms	0.454
short tons (2000 lb)	kilograms	907.185
slugs	kilograms	14.594
ounces	grams	28.3495
pounds force	newtons	4.448
kilogram force	newtons	9.81
dynes	newtons	1.0×10^{-5}
feet per second	meters per second	0.305
feet per second squared	meters per second squared	0.305
standard gravitational acceleration	meters per second squared	9.807
cubic feet per second	liters per second	28.317
U.S. gallons per minute	cubic meters per second	0.631×10^{-4}
U.S. gallons per minute	liters per second	0.063
acre-feet per day	cubic meters per day	1233.79
pounds force per square inch	kilopascal	6.895
pounds per square foot	kilopascal	0.048
kilopascals	newtons per square meter	1.0×10^{-3}
pounds per foot-hour	pascal second	4.134×10^{-4}
pounds per foot-hour	centipoise	0.413
centipoise	pascal second	1.0×10^{-3}
pounds per foot-second	centipoise	1.488×10^{3}
pounds per cubic foot	kilograms per cubic meter	16.018
pounds per U.S. gallon	kilograms per cubic meter	119.826

$$°C = \frac{°F - 32}{1.8} = K - 273.15$$

1. "ASTM E-380," American Society for Testing and Materials, 100 Barr Harbor Dr., P.O. Box C700, West Conshohocken, PA 19428-2959.

2. T. Goldman and R. J. Bell, eds., *The International System of Units (SI)*, NBS Publ. No. 330. National Bureau of Standards, Washington, D.C., 1981.

REFERENCES

Abichou, T., Benson, C. H., and Edil, T. B. (1998). Database on Beneficial Reuse of Foundry By-Products, Env. Geotech. Report, No. 98-3. Department of Civil and Environmental Engineering, University of Wisconsin, Madison, Wisconsin.

Abou Najm, M., El-Fadel, M., Ayoub, G., El-Taha, M., and Al-Awar, F. (2002). An optimisation model for regional integrated solid waste management I. Model formulation and An optimisation model for regional integrated solid waste management II. Model application and sensitivity analysis. *Waste Mngt. Res.* **20,** 37–45 and 46–54, respectively.

Acar, Y. B., and Ghosh, A. (1986). Role of activity in hydraulic conductivity of compacted soils permeated with acetone. *Proc. Int. Symp. Environ. Geotechnol.* **1,** 403–412.

Acar, Y. B., and Seals, R. K. (1984). Clay barrier technology for shallow land waste disposal facilities. *Hazard. Waste* **1**(2), 167–181.

Alexander, M. (1981). Biodegradation of chemicals of environmental concern. *Science* **211,** 132–138.

Alexander, M. (1999). *Biodegradation and Bioremediation,* 2nd ed., Academic, San Diego, California.

Alkire, B. D., and Morrison, J. M. (1982). Changes in soil structure due to freeze–thaw and repeated loading. *Transport. Res. Rec.* **918,** 15–22.

Allison, F. E., and Klein, C. J. (1962). Rates of immobilization and release of nitrogen following additions of carbonaceous materials and nitrogen to soils. *Soil Sci.* **93,** 383.

Alther, G. R. (1982). The role of bentonite in soil sealing applications. *Bull. Assoc. Eng. Geol.* **19**(4), 401–409.

Alther, G. R. (1983). The methylene blue test for bentonite liner quality control. *Geotech. Test. J.* **6**(3), 128–132.

American Conference of Governmental Industrial Hygienists (ACGIH) (1987). *Threshold Limit Values and Biological Exposure Indices for 1987–1988.* ACGIH, Cincinnati, Ohio.

American Petroleum Institute (API). (1990). *Petroleum Release Decision Framework.* API.

American Public Works Association (APWA) (1966). *Municipal Refuse Disposal.* Public Administration Service, Chicago, Illinois.

American Society of Civil Engineers (ASCE) (1960). *Design and Construction of Sanitary and Storm Sewers,* Manual of Engineering Practice, No. 37. ASCE, New York.

American Society of Civil Engineers (ASCE) (1976). *Sanitary Landfill,* ASCE Manuals and Reports on Engineering Practice, No. 39. ASCE, New York.

American Society of Civil Engineers (ASCE) (1995). *Groundwater Quality: Guideline for Selection of Commonly Used Groundwater Models.* ASCE, New York.

American Society of Civil Engineers/Water Pollution Control Federation (ASCE/WPCF) (1977). *Waste Water Treatment Plant Design,* WPCF Manual of Practice, No. 8. WPCF, Washington, D.C.

American Society for Testing and Materials (ASTM) (1986). *Annual Book of ASTM Standards.* ASTM, West Conshohocken, Pennsylvania.

American Society for Testing and Materials (ASTM) (1989). *Standard Guide for General Planning of Waste Sampling (D 4687).* ASTM, West Conshohocken, Pennsylvania.

American Society for Testing and Materials (ASTM) (1990). In *Geotechnics of Waste Fills—Theory and Practice, STP 1070* (A. O. Landva and G. D. Knowles, Eds.). ASTM, West Conshohocken, Pennsylvania.

American Society for Testing and Materials (ASTM) (1992). *Standard Test Method for Determination of the Composition of Unprocessed Municipal Solid Waste,* D5231-92. ASTM, West Conshohocken, Pennsylvania.

Amiran, M. C., and Wilde, C. L. (1994). Remediation of contaminated soil and sediment using the BioGenesis wasting process. *Hydrocarbon Contam. Soils* **4,** 425–437.

Andersen, J. R., and Dornbush, J. N. (1967). Influence of sanitary landfill on groundwater quality. *J. Am. Water Works Assoc.* **59**(4), 457–470.

Andersland, O. B., and Anderson, D. M. (1978). *Geotechnical Engineering for Cold Regions.* McGraw-Hill, New York.

Andersland, O. B., and Matthew, P. W. (1973). Consolidation of high ash paper mill sludges. *J. Soil Mech. Found. Div., Am. Soc. Civ. Eng.* **99,** No. SMS-5, 365–374.

Anderson D. M., Push, R., and Penner, E. P. (1978). Physical and thermal properties of frozen ground. In *Geotechnical Engineering for Cold Regions* (O. B. Andersland and D. M. Anderson, Eds.), McGraw-Hill, New York, pp. 37–102.

Anderson, D. G., Husmand, B., and Martin, G. R. (1992). Seismic response of landfill slopes. In *Proc. Stability Perform. Slopes Embankments II,* Geotech. Sp. Publ. 31. American Society of Civil Engineers, New York, pp. 973–989.

Andrews, C. A. (1991). Analysis of laboratory and field leachate test data for ash from twelve municipal solid waste combustors. In *Proc. 2nd Annual Int. Conf. Municipal*

Waste Combustion. Air and Waste Management Association, Pittsburgh, Pennsylvania, pp. 739–758.

Andreasen, A. R. (1995). *Marketing Social Change,* Jossey-Bass, San Francisco, California.

Andromalos, K. A., and Pettit, P. J. (1986). Jet grouting: Snail's pace of adoption. *Civil Eng.,* December, pp. 40–53.

Anex, R. P., Lawver, R. A., Lund, J. R., and Tchobanoglous, G. (1996). GIGO: Spreadsheet-based simulation for MSW systems. *J. Environ. Eng.,* **122**(4), 259–262.

Apgar, M. A., and Langmuir, D. G. (1971). Groundwater pollution potential of a landfill above the water table. *Groundwater* **9**(6), 76–96.

Arora, H. S., Cantor, R. R., and Nemeth, J. C. (1982). Land treatment: A viable and successful method of treating petroleum industry wastes. *Environ. Int.* **7,** 285–291.

Asanten-Duah, D. K. (1998). *Risk Assessment in Environmental Management.* Wiley, New York.

Assink, J. W., and Rulkens W. W. (1984). *Proc 5th Natl. Conf. on Mngt. of Uncontrolled Haz. Waste Sites,* November 7–9, Washington, D.C., pp. 576–583.

Atkinson, R. B., Daniels, W. C., and Cairns, J., Jr. (1998). Hydric soil development in depressional wetlands: A case study from surface mined landscapes. In *Ecology of Wetlands and Associated Systems* (S. K. Majumdar, E. W. Miller, and F. J. Brenner, Eds.). Pennsylvania Academy of Science, Easton, Pennsylvania, pp. 182–197.

Axelrod, H. D., and Lodge, J. P., Jr. (1977). Sampling and calibration of gaseous pollutants. In *Air Pollution* (A. C. Stern, Ed.), 3rd ed., Vol. 3. Academic, New York, pp. 145–177.

Babcock, J. D., and Mossien, C. P. (1993). Leachate cleaning test for the Mill Seat landfill. *Proc. 16th Int. Madison Waste Conf.,* U.W. Extn., Madison, Wisconsin, pp. 384–405.

Bagchi, A. (1980). Discussion of leachate generation from sludge disposal area by Charlie et al. (1979). *J. Environ. Eng. Div. (Am. Soc. Civ. Eng.)* **106**(EE-5), 1005.

Bagchi, A. (1983). Design of natural attenuation landfills. *J. Environ. Eng. Div. (Am. Soc. Civ. Eng.)* **109**(EE-4), 800–811.

Bagchi, A. (1984). Closure to "Design of natural attenuation landfills" by Bagchi (1983). *J. Environ. Eng. Div. (Am. Soc. Civ. Eng.)* **110**(EE-6), 1211–1212.

Bagchi, A. (1986a). Landfill geostructure construction using blended soil. *Proc. Int. Symp. Environ. Geotechnol.* **1,** 43–52.

Bagchi, A. (1986b). Simplified analysis of clay liner blow out. *J. Civ. Eng. Pract. Des. Eng.* **5**(7), 533–551.

Bagchi, A. (1987a). Improving stability of a paper mill sludge. *Proc. Purdue Ind. Waste Conf.* **42,** 137–141.

Bagchi, A. (1987b). *Natural Attenuation Mechanisms of Landfill Leachate and Effects of Various Factors on the Mechanisms.* Waste Management and Research, Copenhagen, Denmark, pp. 453–463.

Bagchi, A. (1987c). Discussion on "Hydraulic conductivity of two prototype clay liners" by S. R. Day and D. E. Daniel (1985). *J. Geotech. Eng. Div. (Am. Soc. Civ. Eng.)* **113**(7), 796–799.

Bagchi, A. (1987d). Discussion of "Overburden pressures exerted on clay liners," by J. J. Pierce (1986). *J. Environ. Engl. Div. (Am. Soc. Civ. Eng.)* **113**(EE-5), 1180–1182.

Bagchi, A. (1993). Effect of freeze-thaw on hydraulic conductivity of compacted clay liners in Wisconsin. *Proc. 16th Int. Madison Waste Conf.,* U.W. Extn., Madison, Wisconsin, pp. 583–592.

Bagchi, A., and Carey, D. (1986). More effective methane gas monitoring at landfill. *Public Works* **117**(12), 44–45.

Bagchi, A., and Ganguly, A. (1990). Leachate apportionment in active landfills. *Proc. 13th Madison Waste Conf.,* U.W. Extn., Madison, Wisconsin, pp. 14–27.

Bagchi, A., and Sopcich, D. (1989). Characterization of MSW incinerator ash. *J. Environ. Eng. Div. (Am. Soc. Civ. Eng.)* **115**(EE-2), 447–452.

Bagchi, A., Dodge, R. L., and Mitchell, G. R. (1980). Application of two attenuation mechanism theories to a sanitary landfill. *Proc. 3rd Annu. Madison Conf. Appl. Res. Pract. Munic. Ind. Waste,* U.W. Extn., Madison, Wisconsin, pp. 201–213.

Barbeito, M. S., and Shapiro, M. (1977). Microbiological safety evaluation of a solid and liquid pathological incinerator. *J. Med. Primatol.* **6**(5), 264–273.

Barbeito, M. S., Taylor, L. A., and Seiders, R. W. (1968). Microbiological evaluation of a large volume air incinerator. *Appl. Microbiol.* **16**(3), 490–495.

Barber, E. S. (1946). Application of triaxial test results to the calculations of flexible pavement thickness. *Proc. 26th Annu. Res. Meet., Highway Res. Board,* pp. 26–39.

Barden, L., and Sides, G. R. (1970). Engineering behavior and structure of compacted clay. *J. Soil Mech. Found. Div., Am. Soc. Civ. Eng.* **96**(SM-4), 1171–1200.

Barker, J. F., Tessman, J. S., Poltz, P. E., and Reinhard, M. (1986). The organic geochemistry of a sanitary landfill leachate plume. *J. Contam. Hydrol.* **1**(2), 171–189.

Barlaz, M. A., and Ham, R. K. (1993). Leachate and gas generation. In *Geotechnical Practice for Waste Disposal* (D. E. Daniel, Ed.). Chapman & Hall, London, pp. 113–136.

Barlaz, M. A., Milke, M. W., and Ham, R. K. (1987). Gas production parameters in sanitary landfill simulators. *Waste Manag. Res.* **5**.

Barlaz, M. A., Ham, R. K., and Schaefer, D. M. (1989a). Mass balance analysis of decomposed refuse in laboratory scale lysimeters. *J. Environ. Eng.* **116**(6), pp. 1088–1102.

Barlaz, M. A., Schaefer, D. M., and Ham, R. K. (1989b). Bacterial population development and chemical characteristics of refuse decomposition in a simulated sanitary landfill. *Appl. Environ. Microbiol.* **55**(1).

Barlaz, M. A., Ham, R. K., and Schaefer, D. M. (1990). Methane production from municipal refuse: A review of enhancement techniques and microbial dynamics. *Crit. Rev. Environ. Control,* **19**(6), 557–584.

Bartha, R., and Atlas, R. M. (1977). The microbiology of aquatic oil spills. *Adv. Appl. Microbiol.* **22,** 225–266.

Barvenic, M. D., Hadge, W. E., and Goldberg, D. T. (1985). Quality control of hydraulic conductivity and bentonite content during soil bentonite cutoff wall construction. In *Land Disposal of Hazardous Waste* (N. P. Barkley, proj. officer), 11th Annu. Res. Symp. EPA-600/9-85/013. U.S. Environmental Protection Agency, Cincinnati, Ohio, pp. 66–79.

Bass, J. M. (1985). Avoiding failure of leachate collection systems. *Waste Manag. Res.* **3,** 233–243.

Bass, J. M., Cornish, R. M., Ehrenfield, J. F., Spennenburg, S. P., and Vallentine, J. R. (1983). Potential mechanisms for clogging of leachate drain systems. In *Land Disposal of Hazardous Waste* (D. W. Shultz, Ed.), 9th Annu. Res. Symp., EPA-600/9-83/018. U.S. Environmental Protection Agency, Cincinnati, Ohio, pp. 148–156.

Basu, A. N., Mikherjee, D. C., and Mikherjee, S. K. (1964). Interaction between humic acid fraction of soil and trace element cations. *J. Indian Soc. Soil Sci.* **12,** 311–318.

Battista, J., and Connelly, J. P. (1988). VOC contamination at selected municipal and industrial landfills in Wisconsin—Sampling results and policy implications. *Proc. 11th Annu. Madison Waste Conf. Cations,* U.W. Extn., Madison, Wisconsin, pp. 59–83.

Bear, J. (1969). Hydrodynamic dispersion. In *Flow through Porous Media* (R. J. M. De Wiest, Ed.). Academic, New York, pp. 109–199.

Bear, J. (1972). *Dynamics of Fluids in Porous Media.* Elsevier, New York.

Bedient, P. G., Rifai, H. S., and Newell, C. J. (1994). *Ground Water Contamination: Transport and Remediation.* Prentice-Hall, Engelwood Cliffs, New Jersey.

Benjamin, J. R., and Cornell, C. A. (1970). *Probability, Statistics, and Decision for Civil Engineers.* McGraw-Hill, New York.

Benson, N. (1980). Waste Characterization of Pulp and Papermill Sludges. Unpublished report, Bureau of Solid & Hazardous Waste Management, Wisconsin Department of Natural Resources, Madison, Wisconsin.

Benson, C. H. (1990). Minimum thickness of compacted soil liners. *13th Annu. Madison Waste Conf.,* U.W. Extn., Madison, Wisconsin, pp. 395–422.

Benson, C. H., and Daniel, D. E. (1990). Influence of clods on hydraulic conductivity of compacted clay. *J. Geotech. Eng. (Am. Soc. Civ. Eng.)* **116**(8), 1231–1248.

Benson, C., and Khire, M. (1994). Soil reinforcement with strips of reclaimed hDPE. *J. Geotech. Eng.* **120**(5), 835–855.

Benson, C. H., and Khire, M. V. (1995). Earthen covers for semiarid and arid climates. In *Landfill Closures—Environmental Protection and Land Recovery,* Geotech. Sp. Publ. No. 53 (R. J. Dunn and U. P. Singh, Eds.). American Society of Civil Engineers, New York, pp. 201–217.

Benson, C. H., Daniel, D. E., and Bontwell, G. P. (1999). Field performance of compacted clay liners. *J. Geotech. Geoenviron. Eng.* **125**(5), 390–403.

Bhandari, A., Dove, D. C., and Novak, J. T. (1994). Soil washing and biotreatment of petroleum contaminated soils. *J. Environ. Eng.* **120**(5), 1151–1169.

Bishop, A. W. (1955). Use of slip circle in slope stability analysis. *Geotechnique* **5**(1), 7–17.

Bishop, A. W., and Morgenstern, N. R. (1960). Stability co-efficients for earth slopes. *Geotechnique* **10**(4), 129–150.

Bjarngard, A. B., and Edgers, L. (1990). Settlement of municipal solid waste landfills, *13th Annu. Madison Waste Conf.,* U.W. Extn., Madison, Wisconsin.

Black, W. V., Ahlert, R. C., Kosson, D. S., and Brugger, J. E., (1991). Slurry based biotreatment of contaminants sorbed onto soil constituents. In *On-Site Biorecla-*

mation: Processes for Xenobiotic and Hydrocarbon Treatment (R. E. Hinchee and R. F. Olfenbuttel, Eds.). Butterworth-Heinemann, Stoneham, Massachusetts, pp. 408–422.

Blackwell, R. J. (1959). Laboratory studies of microscopic dispersion phenomena in porous media. In *Joint Symposium on Fundamental Concepts Miscible Fluid Displacement*, Part II. *American Institute of Chemical Engineers*, San Francisco, California.

Blight, G. E., Ball, J. M., and Blight, J. J. (1992). Moisture and suction in sanitary landfills in semiarid areas. *J. Env. Engr.* **118**(6), 865–877.

Blume, H. P., and Schlichting, E. (1985). Morphology of wetland soils. In *Wetland Soils: Characterization, Classification and Utilization.* International Rice Research Institute, Los Banos, Laguna, Philippines, pp. 161–176.

Blumenthal, M. H. (1993). Tires. In *The McGraw-Hill Recycling Handbook* (H. G. Lund, Ed.). McGraw-Hill, New York, pp. 18.1–18.64.

Bolt, G. H. (1955). Analysis of the validity of Gouy-Chapman theory of the electric double layer. *J. Colloid Sci.* **10**, 206.

Bolton, N. (1995). *The Handbook of Landfill Operations.* Blue Ridge Solid Waste Consulting, Bozeman, Montana.

Bonaparte, R., Giroud, J. P., and Gross, B. A. (1989). Rates of leakage through landfill liners. *Proc. Geosynthet.* **89**, 18–29.

Borch, M. A., Smith, S. A., and Noble, L. N. (1993). *Evaluation, Maintenance and Restoration of Water Supply Wells.* AWWA Research Foundation, Denver, Colorado.

Borden, R. C. (1994). Natural bioremediation of hydrocarbon-contaminated groundwater. In *Handbook of Bioremediation* (J. E. Mathews, project officer). Lewis Publishers, Boca Raton, Florida, pp. 177–199.

Bosscher, P. J. (1987). *P. C. Disk of N.Y. State Slope Stability Program.* University of Wisconsin, Geotech. Eng. Div., Madison, Wisconsin.

Bosscher, P., Edil, T., and Eldin, N. (1992). *Construction and Performance of a Shredded Waste Tire Test Embankment,* No. 1345. Transportation Research Board, Washington, D.C., pp. 44–52.

Bosscher, P., Edil, T., and Kuraoko, S. (1997). Design of highway embankments using tire chips. *J. Geotech. Geoenviron. Eng., ASCE* **123**(4), 295–304.

Bouazza, N., and Kavazanjian, Jr. (2001). Construction on old landfills. *Proc. 2nd ANZ Conf. Env. Geotechnics.* Newcastle, NSW, Australia.

Bowders, J. J., Daniel, D. E., Broderick G. P., and Liljestrand, M. M. (1984). Methods for testing the compatibility of clay liners with landfill and leachate. In *4th Annu. Symp. Hazard. Ind. Solid Waste Test.* American Society for Testing and Materials, Arlington, Virginia.

Bouwer, H. (1979). *Groundwater Hydrology.* McGraw-Hill, New York.

Bouwer, E. J. (1994). Bioremediation of chlorinated solvents using alternate electron acceptors. In *Handbook of Bioremediation* (R. D. Norris et al., Eds.). Lewis Publishers, Boca Raton, Florida.

Bower, C. A., and Hatcher, J. T. (1967). Adsorption of fluoride by soils and minerals. *Soil Sci.* **103**, 151–154.

Boyle, W. C., and Ham, R. K. (1974). Biological treatability of landfill leachate. *J. Water Pollut. Control Fed.* **46**(5), 860–872.

Bramble, G. M. (1973). Spray Irrigation of Sanitary Landfill Leachate. Unpublished Master's Thesis, Department of Civil Engineering, University of Cincinnati, Cincinnati, Ohio.

Broadbent, F. E., and Ott, J. B. (1957). Soil organic matter-metal complexes. 1. Factors affecting retention of various cations. *Soil Sci.* **83,** 419.

Brookman, R. S., Kamp, L. C., and Weintraub, L. (1984). PVC/EVA grafts: A new polymer for geomembranes. *Proc. Int. Conf. Geomembr.* **1,** 25–28.

Brown, R. A. (1997). Air sparging: A primer for application and design. In *Subsurface Restorations* (C. H. Ward, J. A. Cherry, and M. R. Scalf, Eds.). Ann Arbor Press, Chelsea, Michigan, pp. 301–341.

Brown, K. W., and Anderson, D. C. (1980). Effect of organic chemicals on clay liner permeability: A review of the literature. In *Land Disposal of Municipal Solid Waste* (D. W. Schultz, Ed.), EPA-600/9-80/010. U.S. Environmental Protection Agency, Cincinnati, Ohio.

Brown, K. W., and Thomas, J. C. (1985). Influence of concentrations of organic chemicals on the colloidal structure and hydraulic conductivity of clay soils. In *Land Disposal of Hazardous Waste* (N. P. Barkley, proj. officer), 11th Annu. Res. Symp., EPA-600/9085/013. U.S. Environmental Protection Agency, Cincinnati, Ohio, p. 272, (abstr.).

Brown, K. W., Green, J. W., and Thomas, J. C. (1983). The influence of selected organic liquids on the permeability of clay liners. In *Land Disposal of Hazardous Waste* (D. W. Schultz, Ed.), 9th Annu. Res. Symp., EPA-600/9-83/018. U.S. Environmental Protection Agency, Cincinnati, Ohio, pp. 114–125.

Brown, R. A., Loper, J. R., and McGarvey, D. C. (1986). In situ treatment of groundwater: Issues and answers. *Proc. 1986 Hazar. Mat. Spills. Cont.,* May 5–8, St. Louis, Missouri, pp. 261–264.

Bruch, J. C., and Street, R. L. (1967). Two dimensional dispersion. *J. Sanit. Eng. Div., Am. Soc. Civ. Eng.* **93**(SA-6), 17–39.

Brunner, C. R. (1991). *Handbook of Incineration Systems.* McGraw-Hill, New York.

Brunner, C. R. (1994). Waste-to-energy conversion, incineration. In *Handbook of Solid Waste Management* (F. Kreith, Ed.). McGraw-Hill, New York, pp. 11.3–11.97.

Buckholz, D. M. (1993). Aluminum cans. In *The McGraw-Hill Recycling Handbook.* McGraw-Hill, New York, pp. 12.1–12.24.

Bumb, A. C., McKee, C. R., Evans, R. B., and Eccles, L. A. (1988). Design of lysimeter leak detector networks for surface impoundments and landfills. *Ground Water Monit. Rev.* **8**(2), 102–114.

Bureau, B. (1981). Design of impervious clay liners by unsaturated flow principles. In *Proc. 4th Symp. Uranium Mill Tailings Manage.* Colorado State University, Ft. Collins, Colorado, pp. 647–664.

Bureau of Reclamation (1978). *Drainage Manual,* 1st ed. Bureau of Reclamation, Eng. Res. Cent., Denver Federal Center, Denver, Colorado.

Bureau of Solid Waste Management (1985). *Guidelines for Monitoring Well Installation.* Wisconsin Department of Natural Resources, Madison, Wisconsin.

Burmaster, D. E. (1996). Benefits and costs of using probabilistic techniques in human health risk assessments—With emphasis on site specific risk assessments. *Human Ecol. Risk Assess.* **2,** 35–43.

Burton, P. E. (1987). An evaluation of three vadose zone monitoring techniques. Unpublished M.S. Thesis, University of Texas, Austin, Texas.

Buscheck, T. L. E., and Peargin, T. R. (1991). Summary of a nation-wide vapor extraction system performance study. In *Proc. Petroleum Hydrocarbons and Organic Chemicals in Groundwater: Prevention, Detection and Restoration,* November, NWWA, pp. 205–219.

Byrne, R. J., Kendall, J., and Brown, S. (1992). Cause and mechanism of failure, Kettleman Hills landfill B-19, Unit IA. In *Proc. Stability Performance of Slopes and Embankment* (R. B. Seed and R. W. Baulanger, Eds.), Geotech. Sp. Publ. 31. American Society of Civil Engineers, New York, pp. 1188–1215.

Cadena, F., and Jeffers, S. W. (1987). Use of tailored clays for selective adsorption of hazardous pollutants. *Proc. Purdue Ind. Waste Conf.* **42**, 113–119.

Cadwallader, M. W. (1986). Selection and specification criteria for flexible membrane liner of the high density polyethylene variety. *Proc. Int. Symp. Environ. Geotechnol.* **1**, 323–333.

Caffrey, R. P., and Ham, R. K. (1974). The role of evaporation in determining leachate production from milled refuse landfills. *Compost Sci.* **15**(2), 11–15.

Calabrese, E. J., and Baldwin, L. A. (1993). *Performing Ecological Risk Assessments.* Lewis Publishers, Boca Raton, Florida.

California Water Pollution Control Board (CWPCB) (1954). *Report on the Investigation of Leaching for a Sanitary Landfill,* Publ. No. 10. Resources Agency of California, Sacramento, California.

California Water Pollution Control Board (CWPCB) (1961). *Effects of Refuse Dumps on Ground Water Quality,* Publ. No. 24. Resources Agency of California, Sacramento, California.

Callahan, M. A., Slimak, M. W., Gabel, N. W., May, I. P., and Fowler, C. F. (1979). *Water Related Environmental Fate of 129 Priority Pollutants,* EPA-440/4-79-029a and b. U.S. Environmental Protection Agency, Washington, D.C.

Cameron, R. D., and McDonald, E. C. (1982). Toxicity of landfill leachates. *J. Water Pollut. Control Fed.* **52**(4), 760–769.

Campbell, G. S. (1974). A simple method for determining unsaturated hydraulic conductivity from moisture retention data. *Soil Sci.* **117**(6), 311–314.

Campbell, C. S., and Ogden, M. H. (1999). *Constructed Wetlands in the Sustainable Lanscape.* Wiley, New York.

Carlson, D. A., and Johansen, O. J. (1975). Aerobic treatment of leachates from sanitary landfills. In *Waste Management Technology and Resource and Energy Recovery.* U.S. Environmental Protection Agency, Washington, D.C., pp. 359–382.

Carman, P. C. (1956). *Flow of Gases through Porous Media.* Academic, New York.

Carpenter, G. W., and Stephenson, R. W. (1986). Permeability testing in the triaxial cell. *Geotech. Test. J.* **9**(1), 3–9.

Carroll, R. G., Jr. (1983). *Geotextile Filter Criteria, Engineering Fabrics in Transportation Construction,* Transp. Res. Rec. No. 916. National Academy of Sciences, Washington, D.C.

Carter, J. N., and Allison, F. E. (1961). The effect of rates of application of ammonium sulphate and gaseous losses of nitrogen from soils. *Soil Sci. Soc. Am. Proc.* **25**, 484.

Casalena, M. J. (1998). Birds as indicators of wetland ecosystems. In *Ecology of Wetlands and Associated Systems* (S. K. Majumdar, E. W. Miller, and F. J. Brenner, Eds.), The Pennsylvania Academy of Science, Easton, Pennsylvania.

Cavalli, N. J. (1992). Composite barrier slurry wall. In *Slurry Walls, Design, Construction, and Quality Control, STP 1129* (D. B. Paul, R. R. Davidson, and N. J. Cavalli, Eds.). American Society for Testing and Materials, West Conshohocken, Pennsylvania, pp. 78–85.

Cedergren, H. R. (1977). *Seepage, Drainage, and Flow Nets,* 2nd ed. Wiley, New York.

Chamberlain, E. J. (1997). Freeze-thaw cycling and hydraulic conductivity of bentonite barriers. *J. Geotech. Geoenviron. Eng.* **123**(3), 229–238.

Chamberlain, E. J., and Blouin, S. E. (1976). *Freeze–Thaw Enhancement of the Drainage of Consolidation of Fine Grained Dredged Material in Confined Disposal Areas,* CRREL Final Report. U.S. Army Cold Regions Research and Engineering Laboratory, Hanover, New Hampshire.

Chamberlain, E. J., and Blouin, S. E. (1977). *Frost Action as a Factor in Enhancement of the Drainage and Consolidation of Fine Grained Dredged Material,* Tech. Rep. No. D-77-16. Department of the United States Army Engineer Water Ways Experiment Station, Dredged Material Research Program, Hanover, New Hampshire.

Chamberlain, E. J., and Gow, A. J. (1978). Effect of freezing and thawing on the permeability and structure of soils. In *Int. Symp. Ground Freez.* Department of the U.S. Army, Hanover, New Hampshire.

Chamberlain, E. J., and Gow, A. J. (1979). Effect of freezing and thawing on the permeability and structure of soils. *Eng. Geol.* **13,** 73–92.

Chamberlain, E. J., Iskander, I., and Hansiker, S. E. (1990). Effect of freeze-thaw on the permeability and microstructure of soils. *Proc. Int. Symp. Frozen Soil Impact on Agricultural Range and Forest Lands,* March 21–22, Spokane, Washington, pp. 145–155.

Chan, K. Y., Davey, B. G., and Geering, H. R. (1978). Interaction of treated sanitary landfill leachate with soil. *J. Environ. Qual.* **7**(3), 306–310.

Chao, T. T., Harward, M. E., and Fang, S. C. (1962). Soil constituents and properties in the adsorption of sulphate ions. *Soil Sci.* **94,** 276.

Charlie, W. A., and Wardwell, R. E. (1979). Leachate generation from sludge disposal area. *J. Environ. Eng. Div., Am. Soc. Civ. Eng.* **105**(EE-5), 947–960.

Chazin, J. D., Allen, M., and Pippin, D. D. (1987). *Measurement, Assessment and Control of Hazardous (Toxic) Air Contaminants in Landfill Gas Emissions in Wisconsin.* EPA/APCA Symp. Meas. Toxic Other Relat. Air Pollut. Research Triangle Park, North Carolina.

Chen, H. W., and Yamamoto, L. O. (1987). Permeability tests for hazardous waste management unit clay liners. In *Geotechnical and Geohydrological Aspects of Waste Management* (D. J. A. VanZyl et al., Eds.). Lewis Publishing, Chelsea, Michigan, pp. 229–243.

Chen, Y. H., Simons, D. B., and Demery, P. M. (1981). Hydraulic testing of plastic filters. *J. Irrig. Drain. Div., Am. Soc. Civ. Eng.* **107**(IR-3), 307–324.

Chen, M., Hindley, R., and Killough, J. (1996). Computed tomography imaging of air sparging in porous media. *Water Res.* **32**(10), 3013–3014.

Cherry, J. A., Barker, J. F., Buszka, P., and Hewetson, J. P. (1981). Contaminant occurrence in an unconfined sand aquifer at a municipal landfill. *Proc. 4th Annu. Madison Conf. Appl. Res. Pract. Munic. Ind. Waste,* U.W. Extn., Madison, Wisconsin, pp. 393–411.

Cheru, H. T., and Bozzelli, J. W. (1994). Thermal desorption of organic contaminants from sand and soil using a continuous feed rotary kiln. *26th Haz. Ind. Waste Conf.* pp. 417–424.

Chesner, W. H. (1994). Waste glass and sewage sludge frit use in asphalt pavement. In *Utilization of Waste Materials* (H. I. Inyang and K. L. Bergensen, Eds.). American Society of Chemical Engineers, New York, pp. 296–307.

Chian, E. S., and Dewalle, F. B. (1975). *Compilation of Methodology for Measuring Pollution Parameters of Landfill Leachate,* EPA-600:3075-011. U.S. Environmental Protection Agency, Washington, D.C.

Chian, E. S., and Dewalle, F. B. (1977). *Evaluation of Leachate Treatment,* Vols. 1 and 2, EPA-600/2-77/186a and b. U.S. Environmental Protection Agency, Cincinnati, Ohio.

Childs, K. A. (1985). Mathematical modelling of pollutant transport by groundwater at contaminated sites. In *Contaminated Land, Reclamation and Treatment* (M. A. Smith, Ed.). Plenum, New York.

Chow, V. T. (1959). *Open-Channel Hydraulics.* McGraw-Hill, New York.

Chow, V. T., Ed. (1964). *Handbook of Applied Hydrology.* McGraw-Hill, New York.

Christel, B. J., Rehm, B. W., and Lowery, B. (1985). Field performance of pressure-transducer equipped tensionmeters in fly ash. *Proc. Natl. Water Well Assoc. Conf. Charact. Monit. Vadose (Unsaturated) Zone,* pp. 182–197.

Collins, R. J., and Ciesielski, S. K. (1992). Highway construction use of wastes and by-products. In *Utilization of Waste Materials* (H. I. Inyang and K. L. Bergensen, Eds.). American Society of Chemical Engineers, New York, pp. 140–152.

Collins, K., and McGown, A. (1974). The form and functions of microfabric features in a variety of natural soils. *Geotechnique* **24**(2), 223–254.

Collins, P., Nag, A. S., and Ramanujam, R. (1998). Superfund success, superfast. *Civ. Eng.,* December, pp. 42–45.

Composting Council (1994). *Recommended Test Methods for the Examination of Compost and Composting, Draft 4.3.* Composting Council, Alexandria, Virginia.

Composting Council (1995). *Suggested Compost Parameters & Compost Use Guidelines.* Composting Council, Alexandria, Virginia.

Connolly, R. A. (1972). Soil burial tests: Soil burial of materials and structures. *Bell Syst. Tech. J.* **51**(1), 1–21.

Connelly, J. (1994). Monitoring Well Sampling. Personal communication, Wisconsin Dept. of Natural Resources, Madison.

Constazo, M., Archer, D., Aronson, E., and Pettigrew, T. (1986). Energy conservation behavior: The difficult path from information to action. *Am. Psychol.* **41**, 521–528.

Cook, E. N., and Foree, E. G. (1974). Aerobic biostabilization of sanitary landfill leachate. *J. Water Pollut. Control Fed.* **46**(2), 380–392.

Cook, C. M., Grefe, R. P., and Kuehling, H. H. (1991). The role of active gas extraction systems in capturing VOCs from municipal landfill waste and leachate: A preliminary assessment, *Fourteenth Annual Madison Waste Conference,* U.W. Extn., Madison, Wisconsin.

CPHEEO (2000). *Manual on Municipal Solid Waste Management,* 1st ed. Central Public Health and Environmental Engineering Organization (CPHEEO), Ministry of Urban Development, Government of India, New Delhi.

Cullimore, R. (1981). *Controlling Iron Bacterial Plugging by Recycling Hot Water.* Canadian Water Well, Agri-Book Publishing.

Daniel, D. E. (1981). Problems in predicting the permeability of compacted clay liners. In *Proc. 4th Symp. Uranium Mill Tailings Manag.* Colorado State University, Fort Collins, Colorado, pp. 665–675.

Daniel, D. E. (1984). Predicting the hydraulic conductivity of compacted clay liners. *J. Geotech. Eng. (Am. Soc. Civ. Eng.)* **110**(2), 285–300.

Daniel, D. E. (1989). In situ hydraulic conductivity tests for compacted clay. *Geotech. Eng. Div. (Am. Soc. Civ. Eng.)* **115**(9), 1205–1226.

Daniel, D. E. (2000). Barriers and containment methods. In *Vadose Zone, Science and Technology Solutions* (B. B. Looney and R. W. Falta, Eds.). Battelle, Columbia, Ohio, pp. 1309–1424.

Daniel, D. E., and Benson, C. H. (1990). Water content-density criteria for compacted soil liners. *J. Geotech. Eng.* **116**(2), 1811–1830.

Daniel, D. E., and Kurtovich, M. (1987). Monitoring for hazardous waste leaks. *Civ. Eng. (N.Y.)* **57**(2), 48–51.

Daniel, D. E., and Shackelford, C. D. (1988). Disposal barriers that release contaminants only by molecular diffusion. *Nucl. Chem. Waste Manag.* **8,** 299–305.

Daniel, D. E., and Trantwein, S. T. (1986). Field permeability test for earthen liners. *Proc. Am. Soc. Civ. Eng. Spec. Conf. Use In-Situ Tests Geotech.* (S. P. Clemence, Ed.), New York *Eng.,* pp. 146–160.

Daniel, D. E., and Wu, Y.-K. (1993). Compacted clay liners and covers for arid sites. *J. Geotech.* Eng., **119**(2), 223–237.

Daniel, D. E., Hamilton, J. M., and Olson, R. E. (1981). Suitability of thermocouple psychrometers for studying moisture movement in unsaturated soils. *ASTM Spec. Tech. Publ.* **STD 746,** 84–100.

Daniel, D. E., Trantwein, S. J., Boyton, S. S., and Foreman, D. E. (1984). Permeability testing with flexible wall permeameters. *Geotech. Test. J.* **7**(3), 113–122.

Daniel, D. E., Shan, H. Y., and Anderson, J. D. (1993). Effects of partial wetting on the performance of bentonite component of a geosynthetic clay liner. In *Proc. Geosynthetics' 93,* Vol. 3. Industrial Fabrics Association International, St. Paul, Minnesota, pp. 1483–1496.

D'Appolonia, D. J. (1980). Soil-bentonite slurry trench cutoffs. *Geotech. Eng. Div. Am. Soc. Civ. Eng.* **106**(GT-4), 399–417.

Dass, P., Tamke, G. R., and Stoffel, C. M. (1977). Leachate production at sanitary landfill sites. *J. Environ. Eng. Div., Am. Soc. Civ. Eng.* **103**(EE-6), 981–988.

Davis, J. J. (1995). The effects of message framing on response to environmental communications. *J. Mass Commun. Q.* **72,** 285–299.

Davis, E. L., (1997). *How Heating Can Enhance In-Situ Soil and Aquifer Remediation: Important Chemical Properties and Guidance on Choosing the Appropriate Tech-*

nique, Groundwater Issue, EPA/540/S-97/502. U.S. Environmental Protection Agency, Cincinnati, Ohio.

Davis, J. L., Sling, R., Stegman, B. G., and Waller, M. J. (1984). *Innovative Concepts for Detecting and Locating Leaks in Waste Impoundment Liner Systems: Acoustic Emission Monitoring and Time Domain Reflectrometry,* EPA-600/S2-84/058. U.S. Environmental Protection Agency, Cincinnati, Ohio.

Dawson, G. W., and Mercer, B. W. (1986). *Hazardous Waste Management.* Wiley, New York.

Day, S. R., and Daniel, D. E. (1985). Hydraulic conductivity of two prototype liners. *J. Geotech. Eng. Div., Am. Soc. Civ. Eng.* **111**(8), 957–970.

Decesare, R., and Plumley, A. (1992). Results from the ASME/US Bureau of Mines investigation—Vitrification of residue from municipal waste combustion. In *Proc. Asme, Natl. Waste Processing Conf.,* Detroit, Michigan.

Deleon, I. G., and Fuqua, R. W. (1995). The effects of public commitment and group feedback on curbside recycling. *Environ. Behav.,* **27,** 233–250.

Demetracopoulos, A. C., and Korfiatis, G. P. (1984). Design considerations for landfill bottom collection systems. *J. Civ. Eng. Pract. Des. Eng.* **3**(10), 967–984.

Demetracopoulos, A. C., Korfiatis, G. P., Bourodimos, E. L., and Nawy, E. G. (1984). Modeling for design of landfill bottom liners. *J. Environ. Eng. Div., Am. Soc. Civ. Eng.* **110**(6), 1084–1098.

Demetracopoulos, A. C., Sehayek, L., and Erdogan, H. (1986). Modeling leachate production from municipal landfills. *J. Environ. Eng. Div., Am. Soc. Civ. Eng.* **112**(EE-5), 849–866.

Dempsey, J. (2000). Getting better with age. *Waste Age* **31**(12), 76–80.

Department of the Navy (1971). *Design Manual, Soil Mechanics, Foundations, and Earth Structures,* NAVFAC DM-7, U.S. Department of the Navy, Washington, D.C., pp. 7-9-8–7-9-9.

Derflinger, B. (2002). How Wisconsin promotes household sharps collection. *J. Am. Pharmaceut. Assoc.* **42**(6, Suppl. 2).

Devine, K. (1994). Bioremediation: The state of usage. In *Applied Biotechnology for Site Remediation* (R. E. Hinchee, D. B. Anderson, F. B. Metting, Jr., and G. D. Sayles, Eds.). Lewis Publishers, Boca Raton, Florida.

Devlin, J. F., and Barker, J. F. (1993). A semi-passive injection system for in situ bioremediation. In *Proc. Natl. Conf. Hydraulic Engr. Hydrology,* July 25–30, San Francisco, CA. American Society of Civil Engineers, New York.

DiMartino, C. (2000). Does pay-as-you-throw pay off? *Waste Age,* **31**(6), 51–55.

Dirksen, C., and Miller, R. D. (1966). Closed system freezing of unsaturated soil. *J. Am. Soil. Sci. Soc.* **50,** 1112–1114.

Doedens, H., and Cord-Landwehr, K. (1989). Leachate recirculation. In *Sanitary Landfilling: Process Technology and Environmental Impact* (T. H. Christenson, R. Cossu, and R. Stegmann, Eds.). Academic, London.

Domenico, P. A., and Schwartz, F. W. (1990). *Physical and Chemical Hydrogeology.* Wiley, New York.

Driscoll, F. G. (1986). *Groundwater and Wells,* 2nd ed. Johnson Filtration Systems, St. Paul, Minnesota.

Druback, G. W., and Artola, S. V., Jr. (1985). Subsurface pollution containment using a composite system vertical cutoff barrier. In *Hydraulic Barriers in Soil and Rock, STP 874* (A. I. Johnson, R. K. Frobel, N. J. Cavalli, and C. B. Petterson, Eds.). American Society for Testing and Materials, West Conshohocken, Pennsylvania, pp. 24–33.

Druschel, S. J., and Underwood, E. R. (1993). Design of lining and cover system sideslopes. *Proc. Geosynthetics '93 Conf.,* Vancouver, B.C., Canada, pp. 1341–1356.

Duncan, J. M. (1992). State-of-the-art: Static stability and deformation analysis. In *Proc. Stability and Performance of Slopes and Embankments-II,* American Society of Civil Engineers, Reston, Virginia, pp. 222–266.

Dunn, R. J. (1986). Clay liners and barriers—considerations of compacted clay structure. *Proc. Int. Symp. Environ. Geotechnol.* **1**, 293–302.

Dunn, R. J. (1995). Design and construction of foundation compatible with solid wastes. In *Landfill Closures,* (J. Dunn and V. P. Singh, Eds.) Geotech. Sp. Publ. No. 53. American Society of Civil Engineers, Reston, Virginia, pp. 139–159.

Dupont, R. R., and Reineman, J. A. (1986). *Evaluation of Volatilization of Hazardous Constituents at Hazardous Waste Land Treatment Site,* EPA-6-/2-86/071. U.S. Environmental Protection Agency, Ada, Oklahoma.

Duranceau, P. E. (1987). US EPA's new leaching test: The toxicity characteristic leaching procedure (TCLP), *Proc. 10th Annu. Madison Waste Conf.,* U.W. Extn., Madison, Wisconsin, pp. 547–560.

Durning, A. (1992). *How Much Is Enough? The Consumer Society and the Future of the Earth.* Norton, New York.

Eagly, A. H., and Chaiken, S. (1975). An attributional analysis of the effect of communicator characteristics on opinion change: The case of communicator attractiveness. *J. Personality Social Psychol.* **32**, 136–144.

Edil, T., and Benson, C. H. (1998). Geotechnics of industrial byproducts. In *Proc. GeoCongress 98* (C. Vipulanandan, Ed.), Geotech Sp. Publ. No. 79. American Society of Chemical Engineers, Reston, Virginia, pp. 1–8.

Edil, T., and Bosscher, P. (1994). Engineering properties of waste tire chips and soil mixtures. *Geotech. Testing J.,* **17**(4), 453–464.

Edil, T. B., Park, J. K., and Berthouex, P. M. (1992). Attenuation and transport of volatile organic compounds in clay liners. *Proc. Mediterranean Conf. Env. Geotech.,* Izmir, Turkey.

Edwards, D. A., Liu, A., and Luthy, R. G. (1994). Surfactant solubilization of organic compounds in soil/aqueous systems and Experimental data and modeling for surfactant micelles, HOCs and soil. *J. Environ. Eng.* **120**(1), 5–22 and 23–41, respectively.

Egosi, N. G. (1992). Mixed broken glass processing solutions. In *Utilization of Waste Materials* (H. I. Inyang and K. L. Bergensen, Eds.). American Society of Chemical Engineers, New York, pp. 71–79.

Ehrenfeld, J. R., Ong, J. H., Farino, W., Spawn, P., Jasinski, M., Murphy, B., Dixon, D., and Rissman, E. (1986). *Surface Impoundments.* Noyes Publications, Park Ridge, New Jersey.

Ehrig, H. J. (1983). Quality and quantity of sanitary landfill leachate. *Waste Manag. Res.* **1**(1), 53–68.

Ehrig, H. J. (1984). Treatment of sanitary landfill leachate: Biological treatment. *Waste Manag. Res.* **2**(2), 131–152.

Ellerd, M. G., Massman, J. W., Schwaegler, D. P., and Rohay, V. J. (1999). Enhancements for passive vapor extraction: The Hanford study. *Groundwater* **37**(3), 427.

Ellis, S. (1994). Air pollution and odor control methods. In *Proc. Northeast Regional Solid Waste Composting Conf.* Composting Council, Alexandria, Virginia, pp. 23–26.

Ellis, B. G., and Knezek, B. D. (1972). Adsorption reactions of micro-nutrients in soils. In *Micro-Nutr. Agric.* Soil Sci. Soc. Am., Madison, Wisconsin, p. 59.

Ellis, W. D., Payne, J. R., and McNabb, G. D. (1985). *Treatment of Contaminated Soils with Aqueous Surfactants,* EPA-600/2-85/129. U.S. Environmental Protection Agency, Cincinnati, Ohio.

EMCON Associates (1975). *Sonoma County Solid Waste Stabilization Study,* EPA-530/SW-65dl. U.S. Environmental Protection Agency, Cincinnati, Ohio.

EMCON Associates (1980). *Methane Generation and Recovery from Landfills.* Ann Arbor Science, Ann Arbor, Michigan.

Epps, J. A., Little, D. N., Holmgreen, R. J., Terrel, R. L., and Ledbetter, W. B. (1980). *Guidelines for Recycling Pavement Materials,* Natl. Coop. Hwy. Res. Program, Transportation Res. Board Report No. 224, Washington, D.C.

Epstein, E., and Epstein, J. I. (1989). Public health issues and composting. *Biocycle* **30**(8), 50–53.

Erwin, K. L. (1999). Wetland evaluation for restoration and creation. In *Wetland Creation and Restoration* (J. A. Kusler and M. E. Kentual, Eds.). Island Press, Washington, D.C.

Evangelou, V. P. (1998). *Environmental Soil and Water Chemistry.* Wiley, New York.

Evans, J. C., Stahl, E. D., and Drooff, E. (1987). Plastic concrete cut off walls. In *Geotechnical Practice for Waste Disposal '87,* Geotech. Sp. Publ. 13 (R. D. Woods, Ed.). American Society of Chemical Engineers, New York, pp. 462–472.

Evans, T. M., Meyers, D. K., Gharios, K. M., Haj-Harmon, T., and Kavazanjian, E., Jr. (1998). The use of a capillary barrier final cover for reclamation of a closed municipal solid waste landfill. In *Proc. 3rd Annual Arid Climate Symp.,* SWANA, Albuquerque, New Mexico, April 12–14.

Evdokimoff, V. (1987). Dose assessment from incinerator of deregulated solid biomedical rodwaste. *Health Phys.* **52**(3), 325–329.

Everett, L. G. (1981). Monitoring in the vadose zone. *Groundwater Monit. Rev.* **1**(2), 44–51.

Everett, L. G., Wilson, L. G., and McMillion, L. G. (1982). Vadose zone monitoring concepts for hazardous waste sites. *Groundwater* **20**(3), 312–324.

Fabbricino, M. (2001). An integrated programme for municipal solid waste management. *Waste Mngt. Res.* **19,** 368–379.

Fahoum, K. (1998). Utilization of lagoon-stored lime in embankment construction. *Sessions of Geo-Congress* (C. Vipulanandan, Ed.), Geotech. Sp. Publ., No. 79. American Society of Chemical Engineers, Reston, Virginia, pp. 115–121.

Fair, G. M., and Geyer, J. C. (1954). *Water Supply and Waste Water Disposal.* Wiley, New York.

Fam, S. (1996). Vapor extraction and bioventing. In *In-Situ Treatment Technology* (E. K. Nyer, S. Fam, D. F. Kidd, F. J. Johns II, P. L. Palmer, G. Boettcher, T. L. Crossman, and S. S. Suthersan, Eds.). CRC Press, Boca Raton, Florida, pp. 101–147.

Fang, H. Y., and Evans, J. C. (1988). Long-term permeability tests using leachate on a compacted clayey liner material. *ASTM Spec. Tech. Publ.* **STP 936,** 397–404.

Farquhar, G. K. (1977). Leachate treatment by soils methods. In *Management of Gas and Leachate in Landfills* (S. K. Banerjee, Ed.), EPA-600/9-77-026. U.S. Environmental Protection Agency, Cincinnati, Ohio, pp. 187–207.

Farquhar, G. J., and Rovers, F. A. (1976). Leachate attenuation in undisturbed and remoulded soils. In *Gas and Leachate from Landfills* (E. J. Genetelli and J. Cirello, Eds.), EPA-600/9-76-004. U.S. Environmental Protection Agency, Cincinnati, Ohio, p. 55.

Fattah, D. N. (1974). Investigation and Verification of a Model for the Dispersion Coefficient Tensor in Flow thru Anisotropic Homogeneous Porous Media with Application to Flow from a Recharge Well through a Confined Aquifer. Unpublished Ph.D. Thesis, University of Wisconsin, Madison, Wisconsin.

Federal Register (2002). *Part V: Environmental Protection Agency, Thursday, May 23.* National Archives and Records Administration, U.S. Government Printing Office, Washington, D.C.

Fenn, D. G., Hanley, K. J., and Degeare, T. V. (1975). *Use of the Water Balance Method for Predicting Leachate Generation from Solid Waste Disposal Sites,* EPA-530/SW-168. U.S. Environmental Protection Agency, Cincinnati, Ohio.

Fernandez, F., and Quigley, M. R. (1985). Hydraulic conductivity of natural clays permeated with simple liquid hydrocarbons. *Can. Geotech. J.* **22**(2), 205–214.

Ferro, A., Chard, J., Kjelgren, R., Turner, D., Montague, T., and Chard, B. (2001). Groundwater capture using hybrid poplar trees: Evaluation of a system in Ogden trees, Utah. *Int. J. Phytoremedation* **3**(1).

Fetter, C. W. (1988). *Applied Hydrology,* 2nd ed. Merrill Publishing, Columbus, Ohio.

Fetter, C. W. (1993). *Contaminant Hydrogeology.* Macmillan, New York.

Ficks, M. (2001). Transfer station economics. *Waste Age* **32**(12), 73–76.

Finley, B., and Paustenbach, D. P. (1994). The benefits of probabilistic exposure assessment: Three case studies involving contaminated air, water, and soil. *Risk Anal.* **14,** 53–73.

Finley, B., Proctor, D., et al. (1994a). Recommended distributions for exposure factors frequently used in health risk assessment. *Risk Anal.* **14,** 533–553.

Finley, B., Scott, P. K., and Mayhall, D. A. (1994b). Development of a standard soil-to-skin adherence probability density function for use in Monte Carlo analysis of dermal exposures. *Risk Anal.* **14,** 555–569.

Finstein, M. S., and Hogan, J. A. (1993). Integration of composting process microbiology, facility structure and decision making. In *Science and Engineering of Composting* (H. A. J. Hoitink and H. M. Keener, Eds.). Renaissance Publications, Worthington, Ohio, pp. 1–23.

Finstein, M. S., and Miller, F. C. (1984). Principles of composting leading to maximization of decomposition rate, odor control, and cost effectiveness. In *Composting of Agricultural and Other Waste* (J. K. R. Casser, Ed.). Elsevier Applied Science, London.

Fishbein, B. (1993). *Making Less Garbage: A Planning Guide for Communities,* INFORM, New York.

Fishbein, B. (1994). *Germany, Garbage, and the Green Dot.* INFORM, New York.

Fishbein, B. K., Geiser, K., and Gelb, C. (1994). Source reduction. In *Handbook of Solid Waste Management* (F. Kreith, Ed.). McGraw-Hill, New York, pp. 8.1–8.33.

Fisher, E. A. (1927). Some factors affecting the evaporation of water from soil. II. *J. Agric. Sci.* **17,** 407–419.

Fisher, S. R., and Potter, K. W. (1989). *Methods for Determining Compliance with Groundwater Quality Regulations at Waste Disposal Facilities.* Wisconsin Department of National Resources, Bureau of Solid and Hazardous Waste Management, Madison, Wisconsin.

Fleming, I. R., Rowe, R. K., and Cullimore, D. R. (1999). Field observations of clogging in a landfill leachate collection system. *Can. Geotech. J.* **36,** 685–707.

Follett, R. H., and Lindsay, W. L. (1971). Changes in DPTA-extractable zinc, iron, manganese and copper in soils following fertilization. *Soil Sci. Soc. Am. Proc.* **35**(4), 600–602.

Folz, D. H. (1991). Recycling program design, management, and participation: A national survey of municipal experience. *Public Admin. Rev.* **51,** 222–231.

Fong, M. A., and Haxo, H. E., Jr. (1981). Assessment of liner materials for municipal solid waste landfills. In *Land Disposal of Municipal Solid Waste* (D. W. Shultz, Ed.), 7th Annu. Res. Symp., EPA/600/9-81/002a. U.S. Environmental Protection Agency, Cincinnati, Ohio, pp. 148–162.

Foose, G. J., Benson, C. H., and Edil, T. B. (2002). Comparison of solute transport in three composite liners. *J. Geotech. Geoenviron. Eng.* **128**(5), 391–403.

Ford, H. W. (1974). Low pressure jet cleaning of plastic drains in sandy soil. *Trans. ASAE* **17**(5), 895–897.

Ford, L. R., Jr., and Fulkerson, D. R. (1962). *Flows in Networks.* Princeton University Press, Princeton, New Jersey.

Foreman, D. E., and Daniel, D. E. (1984). Effects of hydraulic gradient and method of testing on the hydraulic conductivity of compacted clay to water, methanol, and heptane. In *Land Disposal of Hazardous Waste* (N. P. Barkley, proj. officer), 10th Annu. Res. Symp., EPA-600/9-84-007. U.S. Environmental Protection Agency, Cincinnati, Ohio, pp. 138–144.

Forseth, J. M., and Kmet, P. (1983). Flexible membrane liners for solid and hazardous waste landfills—A state of the art review. *Proc. 6th Annu. Madison Waste Conf.* U.W. Extn., Madison, Wisconsin, pp. 138–166.

Foster, W. S., and Sullivan, R. H. (1977). *Sewer Infiltration and Inflow Control Product and Equipment Guide,* EPA-600/12-77/107C. U.S. Environmental Protection Agency, Cincinnati, Ohio.

Foth, H. D., and Turk, L. M. (1943). *Fundamentals of Soil Science,* 5th ed. Wiley, New York.

Fountain, J. C. (1997). Removal of nonaqueous phase liquids using surfactants. In *Subsurface Restoration* (C. H. Ward, J. A. Cherry, and M. R. Scalf, Eds.). Ann Arbor Press, Chelsea, Michigan, pp. 199–207.

Fox, P. J., Rowland, M. G., and Sceithe, J. R. (1998). Internal shear strength of three geosynthetic clay liners. *J. Geotech. Geoenviron. Eng.* **124**(10), 933–944.

Freber, B. W. (1996). Beneficial reuse of selected foundry waste material. In *Proceedings of the Nineteenth International Madison Waste Conference,* U.W. Extn., Madison, Wisconsin, pp. 246–257.

Freeze, R. A., and Cherry, J. A. (1979). *Groundwater.* Prentice-Hall, Englewood Cliffs, New Jersey.

Friedly, J. C., and Rubin, J. (1992). Solute transport with multiple equilibrium-controlled or kinetically controlled chemical reactions. *Water Resour. Res.* **28**(6), 1935–1953.

Friedman, M. A. (1988). *Volatile Organic Compounds in Ground Water and Leachate at Wisconsin Landfills,* PUBL-WR-192-88. Wisconsin Department of Natural Resources, Madison, Wisconsin.

Fuller, W. H. (1977). *Movement of Selected Metals, Asbestos and Cyanide in Soil, Applications to Waste Disposal Problem,* EPA 600/2-77/020. U.S. Environmental Protection Agency, Cincinnati, Ohio.

Fuller, W. H., and Korte, N. (1976). Attenuation mechanisms of pollutants through soils. In *Gas and Leachate from Landfills* (E. J. Genetelli and J. Cirello, Eds.), EPA 600/9-76-004. U.S. Environmental Protection Agency, Cincinnati, Ohio, pp. 111–122.

Fuller, A., and Newenhouse, S. (2002). *Construction Waste Reduction and Recycling Demonstration Project (Interim report #1).* Madison Environmental Group, Madison, Wisconsin.

Fungaroli, A. A., and Steiner, R. L. (1979). *Investigation of Sanitary Landfill Behavior,* Vol. 1, Final Rep. EPA-600/2-79/053a. U.S. Environmental Protection Agency, Cincinnati, Ohio, p. 331.

Garland, G. A., and Mosher, D. C. (1975). Leachate effects of improper land disposal. *Waste Age* **6**(3), 42–48.

Garvin, M. L. (1994). *Infectious Waste Management—A Practical Guide.* CRC Press, Boca Raton, Florida.

Gay, A. E., Beam, T. G., and Mar, B. W. (1993). Cost effective solid waste characterization methodology. *J. Environ. Eng.* **119**(4), 631.

Gear, B. (1988). Leachate head well design (personal communication). Wisconsin Department of Natural Resources, Madison.

Gebhard, A. (1978). The potential of leachate attenuation in soils. In *Disposal of Residuals by Landfilling* (A. Gebhard, proj. manager). Minnesota Pollution Control Agency, Minneapolis, Minnesota, pp. vi-1–vi-110.

Gee, G. W., and Dodson, M. E. (1981). Soil water content by microwave drying: A routine procedure. *Soil Sci. Soc. Am.* **45**(6), 1234–1237.

Gee, J. R. (1986). Predicting percolation at solid waste disposal sites—A direct method. *Proc. 9th Annu. Madison Conf. Appl. Res. Pract. Munic. Ind. Waste,* U.W. Extn., Madison, Wisconsin, pp. 623–645.

Geerdink, M. J., Kleijintjensm, R. H., VanLoosdrecht, M. C. M, et al. (1996). Microbial decontamination of polluted soil in a slurry process. *J. Environ. Eng.* **122**(11), 975–982.

Gemmell, R. S. (1971). Mixing and sedimentation. In *Water Quality and Treatment* (P. D. Haney, Chair.), 3rd ed. McGraw-Hill, New York, pp. 123–157.

Gerhardt, R. A. (1977). Leachate attenuation in the unsaturated zone beneath three sanitary landfills in Wisconsin. *Inf. Circ.—Wis. Geol. Nat. Hist. Surv.* **35**, 93.

Ghassemi, M., Haro, M., Metzgar, J., et al. (1983). *Assessment of Technology for Constructing Cover and Bottom Liner Systems for Hazardous Waste Facilities,* EPA/68-02/3174. U.S. Environmental Protection Agency, Cincinnati, Ohio.

Gifford, G. P., Landva, A. O., and Hoffman, V. C. (1990). Geotechnical considerations when planning construction on a landfill. In *Geotechnics of Waste Fills—Theory and Practice,* ASTM STP 1070. American Society for Testing and Materials, Philadelphia, Pennsylvania, pp. 41–56.

Gilmore, M. W., and Hayses, T. L. (1993). Glass beverage bottles. In *The McGraw-Hill Recycling Handbook* (H. G. Lund, Ed.). McGraw-Hill, New York, pp. 13.1–13.22.

Giroud, J. P. (1982). Filter criteria for geotextiles. *Proc. Int. Conf. Geotext. 2nd, 1982* **1,** 103–108.

Giroud, J. P. (1984). Impermeability: The myth and a rational approach. *Proc. Int. Conf. Geomembr.* **1,** 157–162.

Giroud, L. P. (1996). Granular filters and geotextile filters. In *Proc. Geofilters '96,* Montreal, Canada, pp. 565–580.

Giroud, J. P., and Bonaparte, R. (1989). Leakage through liners constructed with geomembranes—Part 1, geomembrane liners. In *Geotextiles and Geomembranes,* Vol. 8. Elsevier Science, London, England, pp. 27–67.

Girvin, D. C., and Sklarew, D. S. (1986). *Attenuation of Polychlorinated Biphenyls in Soil: Literature Review,* CS 4396, Electric Power Research Institute. Palo Atlo, California.

Glaser, J. A., Tzeng, J. W., and McCauley, P. T. (1995). Slurry biotreatment of organic contaminated solids. In *Proc. 3rd Int. In Situ On-Site Bioreclam. Symp.* (R. E. Hinchee, R. S. Skeen, and G. D. Sayles, Eds.). Battelle Memorial Institute, Columbus, Ohio, pp. 145–152.

Glaub, J. C., Savage, G. M., Tuck, J. K., and Henderson, T. M. (1983). Waste characterization for North Santa Clara County, California. *Proc. 6th Annu. Madison Conf. Appl. Res. Pract. Munic. Ind. Waste,* U.W. Extn., Madison, Wisconsin pp. 74–96.

Glenn, J. (1990). Odor control in yard waste composting. *Biocycle* **31**(11), 38–40.

Godish, T. (1987). *Air Quailty.* Lewis Publishers, Chelsea, Michigan.

Goldbeck, N., and Goldbeck, D. (1995). *Choose to Reuse.* Ceres Press, Woodstock, New York.

Goldman, L. J., Truesdale, R. W., Kingsbury, G. L., Northeim, C. M., and Damle, A. S. (1986). *Design Construction and Evaluation of Clay Liners for Waste Management Facilities,* EPA/530-SW-86/007. U.S. Environmental Protection Agency, Cincinnati, Ohio.

Goldstein, N. (1989). New insights into odor control. *Biocycle* **30**(2), 58–61.

Goldstein, N. (1997). The state of garbage in America. *Biocycle* **38**(4), 60–67.

Goldstein, R., and Spencer, B. (1990). Solid waste composting facilities. *Biocycle* **31**(1), 36–46.

Gordon, M. E., Huebbner, P. M., and Miazga, T. J. (1989). Hydraulic conductivity of three landfill clay liners. *J. Geotech. Eng. Div., Am. Soc. Civ. Eng.* **115**(8), 1148–1160.

Gotaas, H. B. (1956). *Composting: Sanitary Disposal and Reclamation of Organic Wastes,* Monograph Series No. 31. World Health Organization, Geneva.

Gray, D. H., and Leiser, A. T. (1982). *Biotechnical Slope Protection and Erosion Control.* Van Nostrand-Reinhold, New York.

Gray, K. R., Sherman, K., and Biddlestone, A. J. (1971a). A review of composting: Part 1. *Process Biochem.* **6**(6), 32–36.

Green, W. J., Lee, G. F., and Jones, R. A. (1981). Clay-soils permeability and hazardous waste storage. *J. Water Pollut. Control Fed.* **54**(8), 1347–1354.

Greenland, D. J. (1970). Sorption of organic compounds by clays and soils. *SCI Monogr.* **37,** 79.

Griffin, R. A. (1977). *Geochemical Considerations Bearing a Disposal of Industrial Chemicals at Willsanuil, Macoupin City, Illinois,* Doc. SS-2-2, Courthouse No. 77-CH-10 and 77-CH-13. Illinois Environmental Protection Agency, Macoupin City, Illinois.

Griffin, R. A., and Chian, E. S. K. (1980). *Attenuation of Water-Soluble Polychlorinated Biphenyls by Earth Materials,* Final Report, EPA600/2-80-027. U.S. Environmental Protection Agency, Washington, D.C.

Griffin, R. A., and Shimp, N. F. (1976). Leachate migration through selected clays. In *Gas and Leachate from Landfills* (E. J. Genetelli and J. Cirello, Eds.), EPA-600/9-76/004. U.S. Environmental Protection Agency, Cincinnati, Ohio.

Griffin, R. A., Cartwright, K., Shimp, N. F., Steele, J. D., Ruch, R. R., White, W. A., Hughes, G. M., and Gilkenbon, R. H. (1976a). *Attenuation of Pollutants in Municipal Landfill Leachate by Clay Minerals.* Illinois State Geological Survey, Urbana, Illinois.

Griffin, R. A., Frost, R. R., and Shimp, N. F. (1976b). Effect of pH on removal of heavy metal from leachate by clay minerals. In *Residual Management by Land Disposal* (W. H. Fuller, Ed.), EPA-600/9-76/015. U.S. Environmental Protection Agency, Cincinnati, Ohio, pp. 259–268.

Grim, R. E. (1968). *Clay Minerology,* 2nd ed. McGraw-Hill, New York.

Grubb, S. (1993). Analytical model for estimation of steady state capture zones of pumping wells in confined and unconfined aquifers. *Groundwater* **31**(1), 27–32.

Guinness, A. E. (1987). *ABCs of Human Body.* Readers Digest Association, Pleasantville, New York, pp. 206–207.

Hagerty, D. J., Ulrich, C. R., and Thacker, B. K. (1977). Engineering properties of FGD sludges. In *Proc. Conf. Geotech. Pract. Disposal Solid Waste Mater., 1977.* American Society of Civil Engineers, New York, pp. 23–40.

Haith, D. A. (1998). Materials balance for municipal solid-waste management. *J. Environ. Eng.* **124**(1), 67–75.

Hall, T. (1991). Reuse of shredded tire material for leachate collection systems. *Annual Madison Waste Conference.* U.W. Extn., Madison, Wisconsin, pp. 367–376.

Hallberg, R. O., and Martinell, R. (1976). Vyredox-in situ purification of groundwater. *Groundwater* **14**(2), 88–93.

Ham, R. K. (1980). *Decomposition of Residential and Light Commercial Solid Waste in Test Lysimeters,* EPA/SW-190c. U.S. Environmental Protection Agency, Cincinnati, Ohio.

Ham, R. K. (1996). Response. *Warmer Bull. J. World Resour. Found.,* No. 49, p. 4.

Ham, R. K., and Anderson, C. R. (1974). Pollutant production by refuse degradation in test lysimeter. *Waste Age* **5**(9), 33–39.

Ham, R. K., and Anderson, C. R. (1975a). Pollutant production by refuse degradation in test lysimeter. *Waste Age* **6**(1), 30–36.

Ham, R. K., and Anderson, C. R. (1975b). Pollutant production by refuse degradation in test lysimeter. *Waste Age* **6**(2), 38–48.

Ham, R. K., Anderson, M. A., Stegmann, R., and Stanforth, R. (1979). *Background Study on the Development of a Standard Leaching Test,* EPA-600/2-79/109. U.S. Environmental Protection Agency, Cincinnati, Ohio.

Hamblin, G. M., and Hater, G. (1998). *Results from Treating A Million Tons of Soil,* Waste Management Biosite Report. Waste Management, Menomonee Falls, Wisconsin.

Hamilton, N. F. (1988). Antimicrobial controls effects of bioslime. *Mod. Plast.* **65**(5), 166–170.

Hammer, M. J. (1977). *Water and Waste-Water Technology.* Wiley, New York.

Hannapel, R. J., Fuller, W. H., and Fox, R. H. (1964). Phosphorus movement in a calcareous soil: Soil microbial activity and organic phosphorus movement. *Soil Sci.* **97**(6), 421–427.

Hansen, B. P. (1980). Reconnaissance of the effect of landfill leachate on the water quality of Marshall Brook, Southwest Harbor, Hancock County, Maine. *Geol. Surv. Open-File Rep. (U.S.)* **80-120.**

Hargesheimer, E. E., Lewis, C. M., and Yentsch, C. M. (1992). *Evaluation of Particle Counting as a Measure of Treatment Plant Performance.* AWWA Research Foundation, Denver, Colorado.

Harleman, D. R. F., Mehlhorn, P. F., and Rummer, R. R., Jr. (1963). Dispersion permeability co-relation in porous media. *J. Hydraul. Div., Am. Soc. Civ. Eng.* **89**(HY-2), 67–85.

Harper, S. R., and Pohland, F. G. (1988). Landfills: Lessening environmental impacts. *Civil Eng.* **58**, 66–69.

Hart, S. (1996). In situ bioremediation: Defining the limits. *Environ. Sci. Technol.* **30**(9), 398–401.

Haug, R. T. (1980). *Composting Engineering, Principles and Practices.* Ann Arbor Science, Ann Arbor, Michigan.

Haxo, H. E., Jr. (1982). Effects on liner materials of long term exposure in waste environments. In *Land Disposal of Municipal Solid Waste* (D. W. Schultz, Ed.), 8th Annu. Res. Symp., EPA-600/9-82/002. U.S. Environmental Protection Agency, Cincinnati, Ohio.

Haxo, H. E., Miedema, J. A., and Nelson, N. A. (1984). Permeability of polymeric membrane lining materials. *Proc. Int. Conf. Geomembr.* **VI,** 151–156.

Haxo, H. E., Jr., Haxo, R. S., Nelson, N. A., Haxo, P. D., White, R. M., and Dakessian, S. (1985). *Liner Materials Exposed to Hazardous and Toxic Wastes,* EPA/600/2-84/169. U.S. Environmental Protection Agency, Cincinnati, Ohio.

Health Care (2002). Web sites: www.h2e-online.org, www.noharm.org, www.ahrmm.org, www.ciwmb.ca.gov, www.epa.gov/glnpo/bnsdocs/merchealth.

Heckler, S. E. (1994). The role of memory in understanding and encouraging recycling behavior. *Psychol. Marketing Recycling Psychol. Marketing* **11**(Special issue), 375–392.

Hegberg, B. A., Hallenbeck, W. H., Brenniman, G. R., and Wadden, R. A. (1991a). Setting standards for yard waste compost. *Biocycle* **32**(2), 58–61.

Hegberg, B. A., Hallenbeck, W. H., Brenniman, G. R, and Wadden, R. A. (1991b). Municipal solid waste incineration with energy recovery in the U.S.A.—Technologies, facilities and vendors for less than 550 tons per day. *Waste Manag. Res.* **9**(2), 127–142.

Hendricker, A. T., Fredianelli, K. H., Kavazanjian, E., Jr., and McKelvey, J. A., III. (1998). Reinforcement requirements at a hazardous waste site. In *Proc. Sixth Intl. Conf. on Geosynthetics,* Vol. 1, Atlanta, Georgia, pp. 465–468.

Hentges, G. T., Thies, F., and Lemar, T. S. (1993). Leachate extraction well assessments Des Moines, Iowa Metropolitan Park sanitary landfill, Hamilton County Iowa Sanitary Landfill. *Proc. 16th Int. Madison Waste Conf.* U.W. Extn., Madison, Wisconsin, pp. 406–441.

Hewitt, R. D., and Daniel, D. E. (1997). Hydraulic conductivity of geosynthetic clay liners after freeze-thaw. *J. Geotech. Geoenviron. Eng.* **123**(4), 305–313.

Hickey, M. E. (1969). *Investigation of Plastic Films for Canal Linings,* Res. Rep. No. 19. U.S. Department of Interior, Bureau of Reclamation, Washington, D.C.

Hickman, H. L., Jr. (1996). Response. *Warmer Bull. J. World Resour. Found.,* No. 49, p. 4.

Hilf, J. W. (1975). Compacted fill. In *Foundation Engineering Handbook* (H. F. Winterkorn and H. Y. Fang, Eds.). Van Nostrand-Reinhold, New York, pp. 244–341.

Hinkle, R. D. (1990). Landfill site reclaimed for commercial use as a container storage facility. In *Geotechnics of Waste Fills—Theory and Practice,* ASTM STP 1070. American Society for Testing and Materials, Philadelphia, Pennsylvania, pp. 331–334.

Hoekstra, P. (1966). Moisture movement in soils under temperature gradients with the cold-side temperature below freezing. *Water Resour. Res.* **2**(2), 241–250.

Hoffman, M. C., and Oettinger, T. P. (1987). *Landfill Leachate Treatment with the PACT System,* Tech. Report No. HT 602. Zimpro, Rothschild, Wisconsin.

Hokkanen, J., Salminen, P., Rossi, E., and Ettala, M. (1995). The choice of a solid waste management system using the Electre II decision-aid method. *Waste Manag. Res.* **13**, 175–193.

Horz, R. C. (1984). *Gotextiles for Drainage and Erosion Control at Hazardous Waste Landfills,* EPA Interagency Agreement No. AD-96-F-1-400-1. U.S. Environmental Protection Agency, Cincinnati, Ohio.

Houle, M. (1976). Industrial hazard waste migration potential. In *Residual Management by Land Disposal* (W. H. Fuller, Ed.), EPA-600/9-76/015. U.S. Environmental Protection Agency, Cincinnati, Ohio, pp. 76–85.

Howland, J. D., and Landva, A. O. (1992). Stability analysis of a municipal solid waste landfills. In *Proc. Stability and Perform. Slopes and Embankments II* (R. B.

Seed, R. W. Boulanger, Eds.), Geotech. Sp. Publ. 31. American Society of Civil Engineers, New York, pp. 1216–1231.

Huang, C. P., Elliot, H. A., and Ashmead, R. M. (1977). Interfacial reactions and the fate of heavy metals in soil-water systems. *J. Water Pollut. Control Fed.* **49**(5), 745.

Hudel, K., Forge, F., Klein, M., Schroder, H., Jr., and Dohmann, M. (1995). Steam extraction of organically contaminated soil and residue—process development and implementation on an industrial scale. In *Soil Environ. 5, Contaminated Soil 95,* Vol. 2, pp. 1103–1112.

Humphrey, D., Sanford, T., Cribbs, M., and Marion, W. (1993). Shear strength and compressibility of tire chips for use as retaining wall backfill, Transportation Res. Rec. No. 1442. Transportation Research Board, Washington, D.C., pp, 29–35.

Hupe, K., Lueth, J. C., Heerenklage, J., Stegmann, R. (1995). Blade mixing reactors in the biological treatment of contaminated soils. In *Proc. 3rd Int. In Situ On Site Bioreclam. Symp.* (R. E. Hinchee, R. S. Skeen, and G. D. Sayles, Eds.). Battelle Memorial Institute, Columbus, Ohio, pp. 153–159.

Hurt, G. W., Whited, P. M., and Pringles, R. F. (1998). *Field Indicators of Hydric Soils in the United States, a Guide for Identifying and Delineating Hydric Soils (Ver. 4.0).* USDA Natural Resources Conservation Service, Fort Worth, Texas.

Hutzinger, O., and Veerkamp, W. (1981). Xenobiotic chemicals with pollution potential. In *FEMS Symposium No. 12, Microbial Degradation of Xenobiotics and Recalcitrant Compounds* (Leisinger, T., et al., Eds.). Academic, New York.

Hwang, S. (1987). Four Types of Probes for Monitoring the Vadose Zone. Unpublished M.S. Thesis, University of Texas, Austin, Texas.

Iman, R. L., and Helton, J. C. (1988). An investigation of uncertainty and sensitivity analysis techniques for computer models. *Risk Anal.* **8,** 71–90.

Inyang, H. I. (1994). A Weibull based reliability analysis of waste containment systems. *Proc. 1st. Int. Congress Env. Geotech.* Edmonton, Alberta, Canada.

Inyang, H. I., and Myers, V. B. (1993). *Geotechnical Systems for Structures on Contaminated Sites,* EPA 530-R-93-002, PB93-209.419. U.S. Environmental Protection Agency, Washington, D.C.

Irvine, R. L., and Cassidy, D. P. (1995). Periodically operated bioreactors for the treatment of soil and leachates. In *Proc. 3rd Int. In Situ On Site Bioreclam. Symp.* (R. E. Hinchee, R. S. Skeen, and G. D. Sayles, Eds.). Battelle Memorial Institute, Columbus, Ohio, pp. 153–159.

Iwata, V., Westlake, W. E., and Gunther, F. A. (1973). Varying persistence of polychlorinated biphenyls in six California soils under laboratory conditions. *Bull. Environ. Contain. Toxicol.* **9,** 204–211.

James Clem Corporation (JCC) (1992). *Sales Documents for Geosynthetic Clay Liner (GCL).* JCC, Chicago, Illinois.

Javanel, I., and Tsang, C. (1986). Capture zone type curves: A tool for aquifer cleanup. *Ground Water* **24**(5), 616–625.

Jenne, E. A. (1968). Controls on Mn, Fe, Co, Ni, Cu and Zn concentrations in soils and water: The significant role of hydrous Mn and Fe oxides. In *Advances in Chemistry,* Ser. No. 73. American Chemical Society, Washington, D.C.

Jepsen, C. P., and Place, M. (1985). Evaluation of two methods for constructing vertical cutoff walls at waste containment sites. In *Hydraulic Barriers in Soil and Rock,*

STP 874 (A. I. Johnson, R. K. Frobel, N. J. Cavalli, and C. B. Petterson, Eds.). American Society for Testing and Materials, Philadelphia, pp. 45–63.

J. J. Keller & Associates (1996). *Kellers Official OSHA Safety Handbook*. J. J. Keller, Neenah, Wisconsin.

Ji, W., Dahmani, A., Ahlfeld, D., Lin, J., and Hill, E. (1993). Laboratory study of air sparging: Air flow visualization. *Groundwater Monitor. Remediat.*, Fall, pp. 115–126.

Johns II, F. J., and Nyer, E. K. (1996). Miscellaneous in-situ treatment technologies. In *In-Situ Treatment Technology* (E. K. Nyer, S. Fam, D. F. Kidd, F. J. Johns II, P. L. Palmer, G. Boettcher, T. L. Crossman, and S. S. Suthersan, Eds.). CRC Press, Boca Raton, Florida, pp. 289–319.

Johnson, A. I. (1954). Symposium on permeability. *ASTM Spec. Tech. Publ.* **STP 163,** 98–114.

Johnson, R. A., and Bhattacharya, G. K. (1992). *Statistics Principles and Methods,* 2nd ed. Wiley, New York.

Johnson, T. M., and Cartwright, K. (1980). Monitoring of leachate migration in the unsaturated zone in the vicinity of sanitary landfills. *Circ. Ill. State Geol. Surv.* **514,** 82.

Johnson, P. C., and Ettinger, R. A. (1997). Considerations for the design of in situ vapor extraction systems: Radius of influence vs. zone of remediation. In *Subsurface Restoration* (C. H. Ward, J. A. Cherry, and M. R. Scalf, Eds.). Ann Arbor Press, Chelsea, Michigan, pp. 209–216.

Johnson, A. W., and Sollberg, J. R. (1960). *Factors That Influence Field Compaction of Soils,* Bull. No. 272. Highway Research Board, Washington, D.C.

Jordan, R. R., and Crawford, G. L. (1993). Scrap metals and steel cans. In *The McGraw-Hill Recycling Handbook* (H. G. Lund, Ed.). McGraw-Hill, New York, pp. 15.1–15.27.

Jordan, D. L., Mercer, J. W., and Cohen, R. M. (1995). *Review of Mathematical Modeling for Evaluating Soil Vapor Extraction Systems,* EPA/540/R-95/513. U.S. Environmental Protection Agency, Cincinnati, Ohio.

Jumikis, A. R. (1962). *Soil Mechanics.* Van Nostrand, Princeton, New Jersey.

Kalfka, S. (1986). Control Measures for Vinyl Chloride Contained in Landfill Gas. Unpublished Air Management Section Memo. Wisconsin Department of Natural Resources, Madison, Wisconsin, May 30.

Kansas State Highway Commission (1947). *Design of Flexible Pavement Using the Triaxial Compression Test,* Bull. No. 8. Highway Research Board, Washington, D.C.

Kashmanian, R., and Taylor, A. (1989). Costs of composting vs. landfilling yard waste. *Biocycle* **30**(10), 60–63.

Katugampola, P., Vlach, M., and Bandra, J. M. S. J. (1999). Optimal regional planning approach in solid waste management. In *Proc. Forestry and Env. Symp.* Department of Forestry and Environmental Science, University of Sri Jayewardenpura, Sri Lanka.

Katin, R. A. (1995). Operation and maintenance of remediation systems. In *Proc. HAZMACON 95, 12th Haz. Mater. Mgmt. Cont.,* pp. 224–232.

Katsman, K. H. (1984). Hazardous waste landfill geomembrane: Design, installation and monitoring. *Proc. Int. Conf. Geomembr.* **1,** 215–220.

Kauschinger, J. L., Perry, E. B., and Jankour, R. (1992). Jet grouting: State-of-the-practice. In *Grouting, Soil Improvement, and Geosynthetics* (R. H. Bordon, R. D. Holt, and I. Juran, Eds.). American Society of Chemical Engineers, New York, pp. 169–181.

Kayhanian, M., Lindenauer, K., Hardy, S., and Tchobanoglous, G. (1991). *The Recovery of Energy and Production of Compost from the Biodegradable Organic Fraction of MSW Using the High-Solids Anaerobic Digestion/Aerobic Biodrying Process.* Department of Civil and Environmental Engineering, University of California, Davis, California, p. 2-1.

Keen, B. H., Crowther, E. M., and Coutts, J. R. H. (1926). The evaporation of water from soil. III. A critical study of the techniques. *J. Agric. Sci.* **16,** 105–122.

Keet, B. A. (1995). Bioslurping state of the art. In *Applied Bioremediation of Petroleum Hydrocarbons* (R. E. Hinchee, J. A. Kittel, and H. J. Reisinger, Eds.). Battelle Press, Columbus, Ohio, pp. 329–334.

Kester, R. A., and Van Slyke, S. M. (1987). Air Toxic Emissions from Landfill Gas Flares. Presented at the Wisconsin Chapter APCA Meeting, April, Madison.

Keswich, B. H., and Gerba, C. P. (1980). Viruses in ground water. *Environ. Sci. Technol.* **14,** 1290–1297.

Khire, M. V., Benson, C. H., and Bosscher, P. J. (1994). *Final Cover Hydrologic Evaluation Phase III,* Env. Geotechnics Report 94-4. Department of Civil and Environmental Engineering, University of Wisconsin, Madison, Wisconsin.

Khire, M. V, Benson, C. H., and Bosscher, P. J. (1999). Field data from a capillary barrier and model predictions with UNSAT-H. *J. Geotech. Geoenviron. Eng.* **125**(6), 518–527.

Kidd, D. F. (1996). Fracturing. In *In-Situ Treatment Technology* (E. K. Nyer, S. Fam, D. F. Kidd, F. J. Johns II, P. L. Palmer, G. Boettcher, T. L. Crossman, and S. S. Suthersan, Eds.). CRC Press, Boca Raton, Florida, pp. 245–269.

Kim, W. H., and Daniel, D. E. (1992). Effects of freezing on hydraulic conductivity of compacted clay. *J. Geotech. Eng. (Am. Soc. Civ. Eng.)* **118**(7), 1083–1097.

Kim, K. W., Lee, B. H., Park, J. S., and Doh, Y. S. (1992). Performance of crushed waste concrete as aggregate in structural concrete. In *Utilization of Waste Materials* (H. I. Inyang and K. L. Bergensen, Eds.). American Society of Chemical Engineers, New York, NY, pp. 332–343.

Kimmel, G. E., and Braids, O. C. (1974). Leachate plume in a highly permeable aquifer. *Groundwater* **12**(6), 388–393.

Klausmeir, R., and Andrews, C. (1981). Microbial biodeterioration. In *Economic Microbiology* (A. H. Rose, Ed.), Vol. 6. Academic, New York, pp. 432–472.

Kleppe, J. H., and Olson, R. E. (1994). Dessication cracking in soil barrier. In *Hydraulic Barriers in Soil and Rocks* (A. I. Johnson, R. K. Frobel, N. J. Cavalli and C. B. Peterson, Eds.). American Society for Testing and Materials, West Conshohocken, Pennsylvania, pp. 263–275.

Kleven, J., Edil, T. B., and Benson, C. (1998). *Beneficial Re-Use of Foundry Sands in Roadway Subbase,* Env. Geotech. Report No. 98-1. Department of Civil and Environmental Engineering, University of Wisconsin, Madison, Wisconsin.

Klute, A., and Peters, D. B. (1962). A recording tensiometer with a short response time. *Soil Sci. Soc. Am. Proc.* **26**(1), 87–88.

Kmet, P., and Lindorff, D. E. (1983). Use of collection lysimeters in monitoring sanitary landfill performance. *Proc. Natl. Water Well Assoc. Conf. Charact. Monit. Vadose (Unsaturated) Zone.*

Kmet, P., Quinn, K. J., and Slavic, D. (1981). Analysis of design parameters effecting the collection efficiency of clay lined landfills. *Proc. 4th Annu. Madison Conf. Appl. Res. Pract. Munic. Ind. Waste,* U.W. Extn., Madison, Wisconsin, pp. 250–265.

Kmet, P., Mitchell, G., and Gordon, M. (1988). Leachate Collection System Design and Performance—Wisconsin's Experience. Presented at the ASTSWMO National Solid Waste Forum on Integrated Municipal Waste Management, July, Lake Buena Vista, Florida.

Koerner, R. M. (1986). *Designing with Geosynthetics.* Prentice-Hall, Englewood Cliffs, New Jersey.

Koerner, R. M. (1990). *Designing with Geosynthetics,* 2nd ed. Prentice-Hall, Upper Saddle River, New Jersey.

Koerner, R. M. (1998). *Designing with Geosynthetics,* 4th ed. Prentice-Hall, Upper Saddle River, New Jersey.

Koerner, G. R., and Koerner, R. M. (1995). Permeability of granular drainage material. In *Proc. USEPA Biorector Landfill Design and Operation Seminar.* Wilmington, Delaware.

Koerner, R. M., and Soong, T. Y. (2000). Stability assessment of ten large landfill failures, Advances in transportation geoenvironmental systems using geosynthetics, *Proc. Sessions of Geodenver 2000,* Geotech. Sp. Publ. 103 (J. G. Zomberg and B. R. Christopher, Eds.). American Society of Civil Engineers, Reston, Virginia, 1–38.

Kolluru, R. V., Bartell, S. M., Pitblado, and Stricoff, R. S., Eds. (1996). *Risk Assessment and Management Handbook (for Environmental Health, and Safety Professionals).* McGraw-Hill, New York.

Konikow, L. F., and Bredehoeft, J. D. (1978). Computer model of two-dimensional solute transport and dispersion in groundwater. In *U.S. Geological Survey Water Resource,* Investigation B007. Chapter C_2, Middleton, Wisconsin.

Konrad, J. M., and Morgenstern, N. R. (1980). Mechanistic theory of ice lens formations in fine-grained soils. *J. Can. Geotech.* **17,** 473–486.

Korfiatis, G. P., and Demetracopoulos, A. C. (1986). Flow characteristics for landfill leachate collection systems and liners. *J. Environ. Eng. Div. (Am. Soc. Civ. Eng.)* **112**(EE-3), 538–550; errata: 113(EE-6), 1393.

Korfiatis, G. P., Demetracopoulos, A. C., Bourodimas, E. L., and Nawy, E. G. (1984). Moisture transport in a solid waste column. *J. Environ. Eng. Div., Am. Soc. Civ. Eng.* **110**(EE-4), 780–796.

Korte, N. E., Skopp, J., Fuller, W. H., Niebla, E. E., and Alesii, B. A. (1976). Trace element movement in soils, influence of soil physical and chemical properties. *Soil Sci.* **122**(6), 350–359.

Kraus, J. Benson, C., Maltby, V., and Wang, X. (1997). Field and laboratory hydraulic conductivity of paper mill sludges. *J. Geotech. Geoenviron. Eng.* **123**(7), 654–662.

Krauter, P. W. (2002). Using a wetland bioreactor to remediate groundwater contaminated with nitrate (mg/L) and perchlorate (mg/L). *Int. J. Phytoremediation* 3(4).

Kreith, F. (1994). Waste-to-energy conversion, introduction. In *Handbook of Solid Waste Management* (F. Kreith, Ed.). McGraw-Hill, New York, pp. 11.1–11.2.

Kremer, J. G., Lo, M. P., Martyn, P. C., and Directo, L. S. (1987). Regulation of toxic organics in industrial sewer discharges at the sanitation districts of Los Angeles County. *Proc. Purdue Ind. Waste Conf.* **42**, 137–141.

Krishna Raj, S., Dan T. V., and Saxena, P. K. (2000). A fragrant solution to soil remediation. *Int. J. Phytoremediation* **2**(2).

Kristiansen, R. (1981). Sand filter trenches for purification of septic tank effluent. I. The clogging mechanism and soil physical environment. *J. Environ. Qual.* **10,** 53–64.

Kulhawy, F. H., Sangrey, D. A., and Grove, C. S., Jr. (1977). Geotechnical behavior of solvay process wastes. *Proc. Conf. Geotech. Pract. Disposal Solid Waste Mater.*, pp. 118–135.

Kuster, E., and Azadi-Bokhsh, A. (1973). Studies on microbial degradation of plastic films. *Proc. Conf. Degrad. Poly. Plast. Inst. Electr. Eng. Plast. Inst.*, London.

LaGatta, M. D., Boardman, B. T., Cooley, B. H., and Daniel, D. E. (1997). Geosynthetic clay liners subjected to differential settlement. *J. Geotech. Geoenviron. Eng.* **123**(5), 402–410.

Lahti, L. R., King, K. S., Readers, D. W., and Bacopoulos, A. (1987). Quality assurance monitoring of large clay liner. In *Proc. Geotech. Practice Waste Disp.* 87, Geotech. Sp. Publ. 13 (R. D. Woods, Ed.). American Society of Chemical Engineers, New York, pp. 640–654.

Lambe, T. W. (1954). The permeability of fine grained soils. *ASTM Spec. Tech. Publ.* **STP 163,** 56–67.

Lambe, T. W., and Silva-Tulla, F. (1992). Stability analysis of an earth slope. In *Proc. Stability Perform. Slopes and Embankments* II, Geotech. Sp. Publ. 31. American Society of Chemical Engineers, New York, pp. 27–69.

Landmeyer, J. E. (2001). Monitoring effect of poplar trees on petroleum-hydrocarbon and chlorinated-solvent contaminated groundwater. *Int. J. Phytoremed.* **3**(1).

Landva, A. O., and Clark, J. I. (1990). Geotechnics of waste fills. In *Geotechnics of Waste Fills in Theory and Practice*, STP 1070 (A. O. Landva and G. D. Knowles, Eds.). American Society for Testing and Materials, West Conshohocken, Pennsylvania.

Landva, A. O., Valsangkar, A. J., and Pelkey, S. G. (2000). Lateral Earth pressure at rest and compressibility of municipal solid waste. *Can. Geotech. J.* **37**, 1157–1165.

Lange, C. R., Hartman, J. R., Chong, N. M., Weber, A. S., and Matsumoto, M. R. (1987). Constraints of bioaugmentation in enhancing biological treatment process performance. *Proc. Purdue Ind. Waste Conf.* **42**, 275–284.

Larsen, S., and Widowsen, A. E. (1971). Soil fluoride. *J. Soil Sci.* **22**, 210.

Lauber, J. L., Battles, R. D., and Vesley, D. (1982). Decontaminating infectious laboratory wastes by autoclaving. *Appl. Environ. Microbiol.* **44**(3), 690–694.

Lawson, C. R. (1982). Filter criteria for geotextile: Relevance and use. *J. Geotech. Eng. Div. Am. Soc. Civ. Eng.* **108**(GT-10), 1300–1317.

Lechner, P., Lahner, T., and Binner, E. (1993). Reactor landfill experiences gained at the Breitenau Research Landfill in Austria. *Proc. 16th Int. Madison Waste Conf.* U.W. Extn., Madison, Wisconsin, pp. 169–180.

Leckie, J. O., Pacey, J. G., and Halvadakis, C. (1979). Landfill management with moisture control. *J. Environ. Eng. Div., Am. Soc. Civ. Eng.* **105**(EE-2), 337–355.

Lee, J. (1974). Selecting membrane pond liners. *Pollut. Eng.* **6**(1), 33–40.

Lee, G. F., and Jones, R. A. (1990). Managed fermentation and leaching: An alternative to MSW landfills. *Biocycle*, May, pp. 78–83.

Leeson, A., Kittel, J. A., Hinchee, R. E., Miller, R. N., Hass, P. E., and Hoeppel, R. E. (1995). Test plan and technical protocol for bioslurping. In *Applied Bioremediation of Petroleum Hydrocarbons* (R. E. Hinchee, J. A. Kittel, and H. J. Reisinger, Eds.). Battelle, Columbus, Ohio, pp. 335–347.

Lentz, J. J. (1981). Apportionment of net recharge in landfill covering layer into separate components of vertical leakage and horizontal seepage. *Water Resour. Res.* **17**(4), 1231–1234.

Leonards, G. A., Schmednech, F., Chameau, J. L., et al. (1985). Thin slurry cut off walls installed by the vibrated beam method. In *Hydraulic Barriers in Soil and Rock, STP 874* (A. I. Johnson, R. K. Frobel, N. J. Cavalli, and C. B. Petterson, Eds.). American Society for Testing and Materials, Philadelphia, Pennsylvania.

Lewis, P. J., and Langer, J. A. (1994). Dynamic compaction of landfill beneath embankment. In *Settlement '94*, Geotech. Sp. Publ. No. 40, Vol. 1. American Society of Civil Engineers, Reston, Virginia, pp. 451–461.

Liebman, J. C., Male, J. W., and Wathane, M. (1975). Minimum cost in residential refuse vehicle routes. *J. Environ. Eng. Div.* **101**(EE3), 399–412.

Lindgren, E. R., and Brady, P. V. (1995). Electrokinetic control of moisture and nutrients in unsaturated soils. In *Applied Bioremediation of Petroleum Hydrocarbons* (R. E. Hinchee, J. A. Kittel, and H. J. Reisinger, Eds.). Battelle, Columbus, Ohio, pp. 475–481.

Lindorff, D. E., Feld, J., and Connelly, J. (1987). *Groundwater Sampling Procedures Guidelines*, PUBL WR-153-87. Wisconsin Department of Natural Resources, Madison, Wisconsin.

Lindsay, W. L. (1972). *Inorganic Phase Equilibria of Micro-Nutrients in Soils.* Micro-Nutr. Agric., Soil Sci. Soc. Am., Madison, Wisconsin.

Linell, K. A., and Kaplar, C. W. (1959). *The Factor of Soil and Material Type in Frost Action*, Bull. No. 225. Highway Research Board, Washington, D.C., pp. 81–128.

Linsley, R. E., and Franzini, J. B. (1972). *Water Resources Engineering.* McGraw-Hill, New York.

Little, A. L., and Price V. E. (1958). The use of an electronic computer for slope stability analysis. *Geotechnique* **8**(3), 113–120.

Liu, C. W., and Narashimhan, T. N. (1989). Redox-controlled multiple-species reactive chemical transport, 2: Verification and application. *Water Resour. Res.* **25**(5), 883–910.

Liu, M., and Roy, D. (1995). Surfactant induced interactions and hydraulic conductivity changes in soil. *Waste Manag.* **15**(7), 463–470.

Long, F. L. (1982). A new solid state device for reading tensiometers. *Soil Sci.* **133**(2), 131–132.

Longman, F. P. (1990). A treatment concept for leachate from sanitary landfills. Personal Communication, Stork Friesland B.V., Gorredijk, Netherlands.

Loo, W. W. (1994). Electrokinetic enhanced passive in-situ bioremediation of soil and groundwater containing gasoline, diesel and kerosene. In *Proc. 11th HAZMACON. 94, Haz. Mater. Mgmt. Cont. Exhib.*, pp. 254–264.

Lord, A. E., Jr., and Koerner, R. M. (1984). Fundamental aspects of chemical degradation of geomembranes. *Proc. Int. Conf. Geomembr.* **1,** 293–298.

Low, P. F. (1961). Physical chemistry of clay water interaction. *Adv. Agron.* **13,** 269–327.

Lowe, R. K., and Andersland, O. B. (1981). Decomposition effects on shear strength of paper mill sludge. *Proc. Tech. Assoc. Pulp Pap. Ind. Environ. Conf.*, New Orleans, 1981, 239–244.

Lu, J. C. S., Morrison, R. D., and Stearns, R. J. (1981). Leachate production and management from municipal landfills: Summary and assessment. In *Land Disposal of Municipal Solid Waste* (D. W. Shultz, Ed.), 7th Annu. Res. Symp., EPA-600/9-81/002a. U.S. Environmental Protection Agency, Cincinnati, Ohio, pp. 1–17.

Lu, J. C. S., Eichenberger, B., and Stearns, R. J. (1985). *Leachate from Municipal Landfills, Production and Management.* Noyes, Park Ridge, New Jersey.

Lucas, R. E., and Knezek, B. D. (1972). *Climatic and Soil Conditions Promoting Micro-Nutrient Deficiencies in Plants*, Micro-Nutr. Agric., Soil Sci. Soc. Am., Madison, Wisconsin, p. 265.

Lundgren, T. A. (1981). Some bentonite sealants in soil mixed blankets. *Proc. Int. Conf. Soil Mech. Found. Eng. 10th, 1981* **2,** 349–354.

Lustiger, A., and Corneliussen, R. D. (1988). Microscopy shows way to better service in underground PE pipe. *Mod. Plast.* **63**(3), 74–82.

Lutton, R. J., Regan, G. L., and Jones, L. W. (1977). *Design and Construction of Covers for Solid Waste Landfills*, EPA-600/2-79/165. U.S. Environmental Protection Agency, Cincinnati, Ohio, p. 249.

Macintosh, D. L., Sutter, G. W., and Hoffman, F. O. (1994). Uses of probabilistic exposure models in ecological risk assessments of contaminate sites. *Risk Anal.* **14,** 405–419.

Mackay, D. M., Roberts, P. V., and Cherry, J. A. (1985). Transport of organic contaminants in groundwater. *Environ. Sci. Technol.* **19,** 384–392.

Macknight, R. (1990). Controlling the flue-fed incinerator. *J. Air Pollut. Control Assoc.* **10,** 103–109.

Maher, M. H., Gucunski, N., and Papp, W. J., Jr. (1997). Recycled asphalt pavement as a base and sub-base material. In *Testing Soil Mixed with Waste or Recycled Materials.* ASTM 1275. American Society for Testing and Materials, Philadelphia, Pennsylvania, pp. 42–53.

Makdisi, F. I., and Seed, H. B. (1978). Simplified procedure for estimating dam and embankment earthquake induced deformation. *J. Geotech. Eng. Div. (Am. Soc. Civ. Eng.)*, **104**(GT7), 849–867.

Malloy, M. G. (1995a). Medical waste in '95. *Waste Age* **26**(6), 49–62.

Malloy, M. G. (1995b). Plasma arc technology comes of age. *Waste Age* **26**(2), 85–88.

Maltby, V., and Eppstein, L. (1994). A field-scale study of the use of paper industry sludges as hydraulic barriers in landfill cover systems. In *Hydraulic Conductivity and Waste Contaminant Transport in Soils*, STP 1142, American Society for Testing and Materials, West Conshohocken, Pennsylvania, 546–558.

Mar, B. W. (1980). Discussion on leachate generation from sludge disposal area by Charlie and Wardwell (1979). *J. Environ. Eng. Div., Am. Soc. Civ. Eng.* **106**(EE-3), 677–678.

Marcuson, W. F., III, Hynes, M. E., and Franklin, A. G. (1992). Seismic stability and permanent deformation analysis: The last 25 years. In *Proc. Stability Perform. Slopes Embankments II*, Geotech. Sp. Publ. 31. American Society of Civil Engineers, New York, pp. 552–592.

Marley, M. C., and Bruell, C. J. (1995). *In Situ Air Purging: Evaluation of Petroleum Industry Sites and Considerations for Applicability Design and Operation*, Publ. 4609. Am. Petroleum Ins.

Marthaler, H. P., Vogelsanger, W., Richard, F., and Wierenga, P. J. (1983). A pressure transducer for field tensiometers. *Soil Sci. Soc. Am. J.* **47**(4), 624–627.

Martin, R. T. (1958). *Rhythmic Ice Banding in Soil*, Bull. No. 218. Highway Research Board, Washington, D.C., pp. 11–23.

Martin, J. P., Sims, R. C., and Mathews, J. (1986). *Review and Evaluation of Current Design and Management Practices for Land Treatment Units Receiving Petroleum Wastes*, EPA-600/J-86/264, PB 87166339. U.S. Environmental Protection Agency, Cincinnati, Ohio.

Martinson, M. M., and Linck, J. A. (1993). Field pilot-testing for air sparging of hydrocarbon contaminated groundwater. In *Proc. 16th Mad. Waste Conf.*, U.W. Extn., Madison, Wisconsin.

Maser, K. R., and Kaelin, J. J. (1986). Leakage detection of liners using seismic boundary waves. *Barrier Technol.* 362–368.

Mbela, K. K., Sridharan, L., O'Leary, P., Bagchi, A., Mack, D. P., and Mitchell, G. R. (1991). Leachate generation patterns in MSW landfills in Wisconsin and HELP model Assessment. *Proc. Waste Tech'91*, Toronto, Canada.

McBean, E. A., and Anderson, W. A. (1996). A two stage process for the remediation of semi-volatile organic compounds. In *Environmental Biotechnology: Principles and Applications* (M. Moo-Yount, W. A. Anderson, and A. M. Chakrabarty, Eds.). Kluwer Academic, Boston, Massachusetts, pp. 269–277.

McBean, E. A., Poland, R., Rovers, F., and Crutcher, A. J. (1982). Leachate collection design for containment landfills. *J. Environ. Eng. Div., Am. Soc. Civ. Eng.* **108**(EE-1), 204–209.

McBee, W. C., Ward, W. T., Dohner, W. T., and Weber, H. (1992). Utilization of waste sulfur in construction materials as a stabilization/encapsulation agent for toxic, hazardous and radioactive waste. In *Utilization of Waste Materials* (H. I. Inyang and K. L. Bergensen, Eds.). American Society of Chemical Engineers, New York, pp. 116–127.

McBride, M. B. (1994), *Environmental Chemistry of Soils.* Oxford University Press, New York.

McDonald, M. G., and Harbaugh, A. W. (1988). *A Modular Three-Dimensional Finite Difference Groundwater Flow Model*, Techniques of Water Resource Investigations, Book 6, U.S. Geological Survey. Middleton, Wisconsin.

McDougall, F., White, P., Franke, M., and Hindle, P. (2001). *Integrated Solid Waste Management: A Life Cycle Inventory.* Blackwell Science, Oxford.

McElwee, C. D. (1982). Sensitivity analysis and the groundwater inverse problem. *Ground Water* **20**(6), 723–735.

McGinley, P. M., and Kmet, P. (1984). *Formation, Characteristics, Treatment and Disposal of Leachate from Municipal Solid Waste Landfills.* Bureau of Solid Waste Management, Wisconsin Department of Natural Resources, Madison, Wisconsin.

McGuire, W. J. (1989). Theoretical foundations of campaign. In *Public Communications Campaigns* (R. E. Rice and C. K. Atkin, Eds.). Sage, Newbury Park, California, pp. 43–66.

McKee, C. R., and Bumb, A. C. (1988). A three dimensional analytical model to aid in selecting monitoring locations in the vadose zone. *Ground Water Monit. Rev.* **8**(2), 124–136.

McKenzie-Mohr, D., and Smith, W. (1999). *Fostering Sustainable Behavior.* New Society Publishers, Gabriola Island, Canada.

McKenzie-Mohr, D., Nemiroff, L. S., Beers, L., and Desmarais, S. (1995). Determinants of responsible environmental behavior. *J. Soc. Issues* **51**, 139–156.

McKone, T. E., and Daniels, D. (1991). Estimating human exposure through multiple pathways from air, water, and soil. *Regul. Toxicol. Pharmacol.* **13**, 36–61.

McShane, S. F., Montgomery, J. M., Lebel, A., Pollack, T. E., and Stirrat, B. A. (1986). Biophysical treatment of landfill leachate containing organic compounds. *Proc. 41st Purdue Industrial Waste Conf.*, Purdue University, West Lafayett, pp. 167–177.

Meadows, D. H., Meadows, D. L., Randers, J., and Behrens, W. W. III (1972). *The Limits to Growth: A Report to the Club of Rome's Project on the Predicament of Mankind.* Universe Books, New York.

Means, R. E., and Parcher, J. V. (1963). *Physical Properties of Soils.* Charles E. Merrill Book, Columbus, Ohio.

Meegoda, J., Ho, W., Bhattacharjee, M., Wei, C. F., Cohen, D. M., Magee, R. S., and Fredrick, R. M. (1995). Ultrasound enhanced soil washing. *Proc. 27th Mid-Atlantic Ind.* Waste Conf., pp. 733–742.

Meidl, J. A., and Peterson, R. L. (1987). The treatment of contaminated groundwater and RCRA wastewater at Bofors—Nobel Inc. *Proc. 4th Natl. RCRA Conf. Haz. Waste Haz. Materials.*

Melchior, S., and Miehlich, G. (1994). Hydrological studies on effectiveness of different multilayered landfills caps. In *Landfilling of Waste Barriers* (T. H. Christensen, R. Cossu, and R. Stegmann, Eds.). E & FN Spoon, London, pp. 115–137.

Menser, H. A., and Winant, W. M. (1980). Landfill leachate as nutrient source for vegetable crops. *Compost Sci./Land Util.* **21**(4), 48–53.

Mercer, J. W., and Waddell, R. K. (1993). Contaminant transport in groundwater. In *Handbook of Hydrology* (D. R. Maidment, Ed.). McGraw-Hill, New York.

Mercer, J. W., Parker, R. M., and Spalding, C. P. (1997). Use of site characterization data to select applicable remediation techniques. In *Subsurface Restoration* (C. H. Ward, J. A. Cherry, and M. R. Scalf, Eds.). Ann Arbor, Chelsea, Michigan.

Merill, S. D., and Rawlins, S. C. (1972). Field measurement of soil water potential with thermocouple psychrometers. *Soil Sci.* **113**(2), 102–109.

Metcalf & Eddy, Inc. (1979). *Wastewater Engineering: Treatment, Disposal, Reuse,* 2nd ed. McGraw-Hill, New York.

Meyer, C. F., Ed. (1973). *Polluted Groundwater Source: Causes, Effects, Controls and Monitoring*, EPA-600/4-73/001b. U.S. Environmental Protection Agency, Cincinnati, Ohio.

Michaels, A. S., and Lin, C. S. (1954). The permeability of kaolinite. *Ind. Eng. Chem.* **46**, 1239–1246.

Miller, R., and Collins, R. (1976). *Waste Materials as Potential Replacement for Highway Aggregates.* Natl. Highway Res. Program, Report No. 166. Transportation Research Board, Washington, D.C. pp. 1–24.

Miller, C. J., and Mishra, M. (1989). Discussion of "Field verification of HELP model for landfills" by Peyton and Schroedder (1988). *J. Environ. Eng. Div. (Am. Soc. Civ. Eng.)*, **115**(4), 882–884.

Minnesota Extension Service (1993). *Municipal Solid Waste Composting: Is It Right for Your Community.* NR-BU-6179-S. University of Minnesota, St. Paul, Minnesota.

Minnesota Pollution Control Agency (PCA) (1978). *Disposal of Residuals by Landfilling.* Minnesota PCA, Minneapolis, Minnesota.

Mitchell, J. K. (1956). The fabric of natural clays and its relation to engineering properties. *Proc. Highway Res. Board* **35**, 693–713.

Mitchell, J. K. (1976). *Fundamentals of Soil Behavior.* Wiley, New York.

Mitchell, J. K. (1993). *Fundamentals of Soil Behavior.* 2nd ed., Wiley, New York.

Mitchell, J. K., and Madsen, F. T. (1987). Chemical effects on clay hydraulic conductivity. In *Proc. Conf. Geotech. Pract. Waste Disposal, 1987* (R. D. Woods, Ed.) Geotech. Spec. Publ. No. 13. American Society of Civil Engineers, New York, pp. 87–116.

Mitchell, R. A., and Mitchell, J. K. (1992). Stability evaluation of waste landfills. In *Proc. Stability Perfom. Slopes Embankments II*, Geotech. Sp. Publ. 31. American Society of Civil Engineers, New York, pp. 1152–1187.

Mitchell, J. K., Hooper, D. R., and Campanella, R. G. (1965). Permeability of compacted clay. *J. Soil Mech. Found. Div., Am. Soc. Civ. Eng.* **91**(SM-4), 41–65.

Mitsch, W. J., and Gosselink, J. G. (1986). *Wetlands.* Van Nostrand Reinhold, New York.

Moo-Young, H. K., and Zimmie, T. F. (1996). Geotechnical properties of paper mill sludges for use in landfill covers. *J. Geotech. Engr.* **122**(9), 768–775.

Morel, J. L., Colin, F., German, J. C., Godin, P., and Juste, C. (1984). Methods for the evaluation of the maturity of municipal refuse compost. In *Composting of Agricultural and Other Waste* (J. K. R. Casser, Ed.). Elsevier Applied Science, London.

Morgenstern, N. R. (1992). The evaluation of slope stability—A 25 year perspective. In *Proc. Stability Perform. Slopes Embankments II* (R. B. Seed and R. W. Boulanger, Eds.). Geotech. Sp. Publ. 31. American Society of Civil Engineers, New York, 1–26.

Morrison, W. R., and Parkhill, L. D. (1985). *Evaluation of Flexible Membrane Liner Seams after Chemical Exposure and Simulated Weathering*, Interagency Agreement No. DW 14930547-01-2. Hazard Waste Eng. Res. Lab., U.S. Environmental Protection Agency, Cincinnati, Ohio.

Mundell, J. A. (1984). Discussion of "Design of natural attenuation landfills" by Bagchi (1983), *J. Environ. Eng. Div., Am. Soc. Civ. Eng.* **110**(EE-6), 1207–1210.

Murdoch, L. (2000). Remediation of organic chemicals in the vadose zone. In *Vadose Zone*, Vol. II (B. B. Looney, R. W. Falta, Eds.). Battelle, Columbus, Ohio, pp. 34–44.

Naik, T. R., Ramme, B. W., and Tews, J. H. (1994). Use of high volumes of class C and F fly ash in concrete. *Cement, Concrete & Aggregates,* June.

Nash, J. H. (1987). *Field Studies of In Situ Soil Washing.* EPA-600/2–87/110, PB 88146808. U.S. Environmental Protection Agency, Cincinnati, Ohio.

Nasim, K., Meyer, M. C., and Autian, J. (1972). Permeation of aromatic organic compounds from aqueous solutions through polyethylene. *J. Pharm. Sci.* **61**(11), 1775–1780.

National Research Council (1990). *Groundwater Models: Scientific and Regulatory Applications.* National Academy Press, Washington, D.C.

National Research Council (1994). *Alternative for Groundwater Clean Up.* National Academy Press, Washington, D.C.

NCASI (1990). A field-scale study of the use of paper industry sludges in landfill cover systems: first progress report, Tech. Bulletin 595, Natl. Council of Paper Industry, New York, New York.

Newmark, N. M. (1965). Effects of earthquakes on dams and embankments. *Geotechnique* **29**(3), 215–263.

Nichols, R. (1992). Backfill-stiffened foundation wall construction. In *Utilization of Waste Materials* (H. I. Inyang and K. L. Bergensen, Eds.). American Society of Chemical Engineers, New York, pp. 286–295.

Nicholson, R. V., Cherry, J. A., and Reardon, E. J. (1980). *Hydrogeologic Studies on an Aquifer at an Abandoned Landfill*, Part 6. University of Waterloo, Department of Earth Science, Ontario, Canada.

Niessen, W. R. (1995). *Combustion and Incineration Processes, Application in Environmental Engineering*, 2nd ed. Marcel Decker, New York.

Niessen, W. R., and Alsobrook, A. F. (1972). *Municipal and Industrial Refuse, Proc. ASME Incin. Conf. Am. Soc. Mech. Eng.*, New York, p. 319.

Norcell, W. A. (1972). *Equilibria of Metal Chelates in Soil Solution.* Micro-Nutr. Agric. Soil Sci. Soc. Am., Madison, Wisconsin.

Norman, L. E. J. (1958). A comparison of values of liquid limit determined with apparatus with bases of different hardness. *Geotechnique* **8**(1), 70–91.

Norris, R. D. (1994). In-situ bioremediation of soils and groundwater contaminated with petroleum hydrocarbons. In *Handbook of Bioremediation* (J. E. Mathews, proj. officer). Lewis Publishers, Boca Raton, Florida, pp. 17–37.

Norris, R. D., Dowd, K., and Maudlin, C. (1994). The use of multiple oxygen sources and nutrient delivery systems to effect in situ bioremediation of saturated and unsaturated soils. In *Hydrocarbon Bioremediation* (R. E. Hinchee, B. C. Alleman, R. E. Hoeppel, and R. N. Miller, Eds.). CRC Press, Boca Raton, Florida, pp. 405–410.

Norstrom, J. M., Williams, C. E., and Pabor, P. A. (1991). Properties of leachate from construction/demolition waste landfills. *Proc. 14th Annu. Madison Waste Conf.*, pp. 357–366.

Nowatzki, E. A., Lang, R. J., Medelline, M. C., and Sellers, S. M. (1994). Electro-osmotic bioremediation of hydrocarbon contaminated soils in-situ. In *Applied Bi-*

otechnology for Site Remediation (R. E. Hinchee, D. B. Anderson, F. B. Meeting, Jr., and G. D. Sayles, Eds.). CRC Press, Boca Raton, Florida, pp. 295–299.

Nutall, P. M. (1973). The effects of refuse tip liquor upon stream biology. *Environ. Pollut.* **4**, 215–222.

Nyer, E. K. (1991). Biochemical effects on contaminant fate and transport. *Groundwater Monitor. Rev.*, Spring., pp. 80–82.

Nyer, E. K. (1996a). Limitations of pump-and-treat remediation methods. In *In-Situ Treatment Technology* (E. K. Nyer, S. Fam, D. F. Kidd, F. J. Johns II, P. L. Palmer, G. Boettcher, T. L. Crossman, and S. S. Suthersan, Eds.). CRC Press, Boca Raton, Florida, pp. 1–36.

Nyer, E. K. (1996b). Life cycle design. In *In-Situ Treatment Technology* (E. K. Nyer, S. Fam, D. F. Kidd, F. J. Johns II, P. L. Palmer, G. Boettcher, T. L. Crossman, and S. S. Suthersan, Eds.). CRC Press, Boca Raton, Florida, pp. 37–60.

Nyer, E. K., Crossman, T. L., and Boettcher, G. (1996). In situ bioremediation. In *In-Situ Treatment Technology* (E. K. Nyer, S. Fam, D. F. Kidd, F. J. Johns II, P. L. Palmer, G. Boettcher, T. L. Crossman, and S. S. Suthersan, Eds.). CRC Press, Boca Raton, Florida, pp. 61–100.

Nyhan, J., Hakonson, T., and Drennon, B. (1990). A water balance study of four landfill cover design for semiarid regions. *J. Environ. Qual.* **19**, 281–288.

NYSERDA (2003). Sullivan County landfill, Monticello, New York—landfill biostabilization study. Final report 03-03. New York State Energy Research and Development Authority. Albany, New York.

Oakley, R. E., III (1987). Design and performance of earth lined containment systems. *Proc. Conf. Geotech. Pract. Waste Disposal, '87*, R. D. Woods, Ed., Geotech. Spec. Publ. No. 13. American Society of Civil Engineers, New York, pp. 117–136.

Oden, J. T., and Ripperger, E. A. (1981). *Mechanics of Elastic Structures.* Hemisphere Publishing, New York.

Okamp, S., Harrington, M. J., Edwards, T. C., Sherwood, D. L., Okuda, S. M., and Swanson, D. C. (1991). Factors influencing household recycling behavior. *Environ. Behav.* **23**, 494–519.

Olsen, H. W. (1962). Hydraulic flow through saturated clays. *Proc. with Natl. Conf. Clays Clay Miner.*, pp. 131–161.

Olson, R. E., and Daniel, D. E. (1979). Field and laboratory measurements of the permeability of saturated and partially saturated fine-grained soils. *ASTM Spec. Tech. Publ.* **STP 746**, 67.

Olson, R. E., and Daniel, D. E. (1981). *Measurement of the Hydraulic Conductivity of Fine-Ground Soils.* ASTM STP 746. American Society for Testing and Materials, Philadelphia, Pennsylvania, pp. 18–64.

Oosthnoek, J., and Smit, J. P. N. (1987). Future of composting in the Netherlands. *Biocycle* **28**(7), 37–39. In: "Composting of Agricultural and Other Waste" (J. K. R. Casser, ed.), Elsevier Applied Science Publ, London.

Orr, W. R., and Finch, M. O. (1990). Solid waste landfill performance during the Loma Prieta earthquake. In *Geotechnics of Waste Fills Theory and Practice, STP 1070* (A. O. Landva and G. D. Knowles, Eds.). American Society for Testing and Materials, Philadelphia, pp. 22–30.

Osantowski, R. A., Kormanik, R. A., and Huibregtse, G. L. (1989). Leachate treatment system design—flexibility is a must. *Proc. 12th Annu. Madison Waste Conf.* U.W. Extn., Madison, Wisconsin, pp. 542–555.

Oster, J. D., and Ingualson, R. D. (1967). In situ measurement of soil salinity with a sensor. *Soil Sci. Soc. Am. Proc.* **31,** 572–574.

Othman, M. A., and Benson, C. A. (1991). Influence of freeze–thaw on the hydraulic conductivity of a compacted clay. *Proc. 14th Annu. Madison Waste Conf.*, U.W. Extn., Madison, Wisconsin, pp. 296–312.

Otten, A., Alphenaar, A., Pijls, C., Spuij, F., and Wit, H. D. (1997). *In Situ Soil Remediation.* Kluwer Academic Publishers, Boston, Massachusetts.

Oweis, I. S., and Khera, R. P. (1990). *Geotechnology of Waste Management.* Butterworths, London.

Oweis, I. S., Smith, D. A., Allwood, R. B., and Gene, D. S. (1990). Hydraulic characteristics of municipal refuse. *J. Geotech. Engr.*, **116**(4), 539–553.

Pal, D., Weber, J. B., and Overcash, M. R. (1980). Fate of polychlorinated biphenyls (PCBs) in soil-plant system. *Residue Rev.* **74,** 45–98.

Palmer, P. L. (1996a). Vacuum enhanced recovery. In *In-Situ Treatment Technology* (E. K. Nyer, S. Fam, D. F. Kidd, F. J. Johns II, P. L. Palmer, G. Boettcher, T. L. Crossman, and S. S. Suthersan, Eds.). CRC Press, Boca Raton, Florida, pp. 149–194.

Palmer, P. L. (1996b). Reactive walls. In *In-Situ Treatment Technology* (E. K. Nyer, S. Fam, D. F. Kidd, F. J. Johns II, P. L. Palmer, G. Boettcher, T. L. Crossman, and S. S. Suthersan, Eds.). CRC Press, Boca Raton, Florida, pp. 271–288.

Palmer, L. A., and Barber, E. S. (1940). Soil displacement under circular loaded area. *Proc. 20th Annu. Res. Meet., Highway Res. Board*, pp. 279–286.

Palmer, C. D., and Fish W. (1997). Chemically enhanced removal of metals from subsurface. In *Subsurface Restoration* (C. H. Ward, J. A. Cherry, and M. R. Scalf, Eds.). Ann Arbor, Chelsea, Michigan, pp. 217–230.

Papke, C. (1993). Glass recycling and reuse from municipal wastes. In *Recycling Source Book* (T. J. Cichonski and K. Hill, Eds.). Gale Research, Detroit, Michigan, pp. 51–66.

Papp. W. J., Jr., Maher, M. H., Bennert, T. A., and Gucunski, N. (1998). Behavior of construction and demolition debris in base and subbase applications. In *Recycled Materials in Geotechnical Applications* (C. Vipulanandan and D. J. Elton, Eds.). Proc. Sessions of Geo-Congress, Geotech. Sp. Publ. No. 79. American Society of Civil Engineers, Reston, Virginia, pp. 122–136.

Parizek, R., and Lane, B. E. (1970). Soil-water sampling using pan and deep pressure-vacuum lysimeters. *J. Hydrol.* **9**(1), 1–21.

Paruvakat, N. (1993a). Discussion of "Effects of freezing on hydraulic conductivity of compacted clay" by Kim and Daniel (1992). *J. Geotech. Eng. (Am. Soc. Civ. Eng.)* **119**(11), 1862–1864.

Paruvakat, N. (1993b). Deflection analysis of polyethylene leachate collection pipes. *Proc. Geosynthetics-1993*, Vancouver, B.C., Canada, pp. 1413–1424.

Paruvakat, N., Sevick, G. W., and Buechel, L. J. (1990). Freeze–thaw effects on landfill clay liners. *Proc. 13th Annu. Madison Waste Conf.*, U.W. Extn., Madison, Wisconsin, pp. 452–469.

Paschke, N. W. (1982). Mean Behavior of Buoyant Contaminant Plumes in Groundwater. Unpublished M. S. Thesis, Department of Civil and Environmental Engineering, University of Wisconsin, Madison.

Paschke, N. W., and Hoppes, J. A. (1984). Buoyant contaminant plumes in groundwater. *Water Resour. Res.* **20**(2), 1183–1192.

Patrick, W. H., Jr., and Mahapatra, I. C. (1968). Transformation and availability to rice of nitrogen and phosphorus in waterlogged soils. *Adv. Agron.* **29,** 323.

Pearson, W. (1993). Plastics. In *The McGraw-Hill Recycling Handbook.* McGraw-Hill, New York, pp. 14.1–14.32.

Peirce, J. J., Sallfors, G., and Murray, L. (1986). Overburden pressure exerted on clay liners. *J. Environ. Eng. Div., Am. Soc. Civ. Eng.* **112**(EE-2), 280–291.

Penman (1948). Natural evaporation from open water, bare soil and grass. *Proc. R. Soc. London, Ser. A* **193,** 120–145.

Petrov, R. J. Rowe, R. K., and Quigley, R. M. (1997). Selected factors influencing GCL hydraulic conductivity. *J. Geotech. Geoenviron. Eng.* **123**(8) 683–695.

Perkins, T. K., and Johnston, O. C. (1963). A review of diffusion and dispersion in porous media. *Soc. Pet. Eng. J.* **3**(1), 70–84.

Perrier, E. R., and Gibson, A. C. (1980). *Hydrologic Simulation on Solid Waste Disposal Sites*, EPA/SW-868. U.S. Environmental Protection Agency, Cincinnati, Ohio.

Perry, R. H. (1976). *Engineering Manual*, 3rd ed. McGraw-Hill, New York.

Perry, M. C. (1998). Wetland habitats for wildlife of the Chesapeake Bay. In *Ecology of Wetlands and Associated Systems* (S. K. Majumdar, E. W. Miller, and F. J. Brenner, Eds.). Pennsylvania Academy of Science, Easton, Pennsylvania.

Pervaiz, M. A., and Lewis, K. H. (1987). Geotechnical properties of industrial sludge. *Proc. Int. Symp. Environ. Geotechnol.* **2,** 57–76.

Petrasek, A. C., Kugelman, I. J. Austern, B. M., Pressely, T. A., Winslow, L. A., and Wise, R. H. (1983). Fate of toxic organic compounds in wastewater treatment plants. *J. Water Pollut. Control Fed.* **55**(10), 1286–1295.

Pettyjohn, W. A., and Hainslow, A. W. (1983). Organic compounds and groundwater pollution. *Groundwater Monit. Rev.* **3,** 41–47.

Peyton, L., and Schroeder, P. R. (1988). Field verification of HELP model for landfills. *J. Environ. Eng. Div., Am. Soc. Civ. Eng.* **114**(EE-2), 247–269.

Peyton, L., and Schroeder, P. R. (1989). Closure to "Field verification of HELP model for landfills" by Peyton and Schroeder (1988). *J. Environ. Eng. Div. (Am. Soc. Civ. Eng.)* **115**(4), 884–886.

Phaneuf, R. J., and Vana, J. M. (200). Landfill bioreactors. *MSW Manag.* 46–52.

Phene, C. J., Hoffman, G. J., and Rawlins, S. L. (1971a). Measuring soil matric potential in situ by sensing heat dissipation within a porous body: I. Theory and sensor calibration. *Soil Sci. Soc. Am. Proc.* **35**(1), 27–33.

Phene, C. J., Rawlins, S. L., and Hoffman, G. J. (1971b). Measuring soil matric potential in situ by sensing heat dissipation within a porous body. II. Experimental results. *Soil Sci. Soc. Am. Proc.* **35**(2), 225–229.

Phillips, C. R., and Nathwani, J. (1976). *Soil Waste Interactions, a State-of-the-Art Review*, Solid Waste Manag. Rep. EPS 3-EC-76-14m. Environmental Conservation Directorate, Toronto, Ontario, Canada.

Pickens, J. F., and Lennox, W. C. (1976). Numerical simulation of waste movement in steady groundwater flow system. *Water Resour. Res.* **12**(2), 171–180.

Pitrowski, M. R. (1991). Bioremediation of hydrocarbon contaminated surface water, groundwater, and soils: The microbial ecology approach. In *Hydrocarbon Contam-*

inated Soils and Groundwater: Analysis, Fate, Environmental, and Public Health Effects, Remediation (P. T. Kostecki and E. J. Calabrese, Eds.), Vol. 1. Lewis Publishers, Chelsea, Michigan, pp. 203–238.

Pohland, F. G. (1975). *Sanitary Landfill Stabilization with Leachate Recycle and Residual Treatment*, EPA-600/2-75/043. U.S. Environmental Protection Agency, Cincinnati, Ohio.

Pohland, F. G. (1980). Leachate recycle as landfill management option. *J. Environ. Eng.* **106**(EE6), 1057–1069.

Pohland, F. G., Dertien, J. T., and Ghosh, S. B. (1983). Leachate and gas quality changes during landfill stabilization of municipal refuse. *Proc. Third Intl. Symp. Anaerobic Dig.*, August 14–19, Boston, Massachusetts.

Polkowski, L. B., and Boyle, W. C. (1970). *Groundwater Quality Adjacent to Septic Tank Soil Adsorption System.* Wisconsin Department of Natural Resources, Madison, Wisconsin, p. 85.

Pollock, D. W. (1988). Semianalytical computation of path lines for finite difference models. *Groundwater* **26**(6), 743–750.

Pollock, D. W. (1989). *Documentation of Computer Programs to Compute and Display Pathlines Using Results from the U.S. Geological Survey Modular Three Dimensional Finite-Difference Groundwater Model.* USGS Open File Report. U.S. Geological Survey, Middleton, Wisconsin, pp. 89–381.

Polprasert, C. (1989). *Organic Waste Recycling.* Wiley, New York.

Ponnamperuma, F. N. (1972). The chemistry of submerged soils. *Adv. Agron.* **24**, 29–96.

Ponnamperuma, F. N. (1973). *Oxidation-Reduction Reactions, Soil Chemistry*, Vol. 2. Decker, New York.

Potts, J. E., Clendinning, R. A., and Ackart, W. B. (1973). *The Effects of Chemical Structure on the Biodegradability of Plastics, Proc. Conf. Degrad. Poly. Plast.* Inst. Electr. Eng. Plast. Inst., London.

Power, M., and McCarty, L. S. (1996). Probabilistic risk assessment: betting on its future. *Human Ecol. Risk Assess.* **2**, 30–34.

Powers, J. P. (1992). *Construction Dewatering.* Wiley Interscience, New York.

Press, W. H., Teukolsky, S. A., Vetterling, W. T., and Flannery, B. P. (1992). *Numerical Recipes*, 2nd ed. Cambridge University Press, New York.

Preul, H. C. (1964). Travel of Nitrogen Compounds in Soils. Unpublished Ph.D. Thesis, Department of Civil Engineering, No. MN U-D, University of Minnesota, Minneapolis, pp. 64–115.

Prickett, T. A., Naymik, T. G., and Lonnquist, C. G. (1981). *A "Random Walk" Solute Transport Model for Selected Groundwater Quality Evaluations.* Bulletin 65. Illinois State Water Survey, Champaign, Illinois.

Proctor, R. R. (1933). Fundamental principles of soil compaction. *Eng. News Rec.* 3(Aug. 31, Sept. 7, 21, and 28).

Pugh, M. P. (1993). The use of mathematical modes in evaluating resource recovery options. *Resour. Conserv. Recycl.* **8**, 91–101.

Qian, X., Koerner, R. M., and Gray, D. H. (2002). *Geotechnical Aspects of Landfill Design and Construction.* Prentice-Hall, Upper Saddle River, New Jersey.

Quasim, S. R., and Burchinal, J. C. (1970a). Leaching from simulated landfills. *J. Water Pollut. Control Fed.* **43**(3), 371–379.

Quasim, S. R., and Burchinal, J. C. (1970b). Leaching of pollutants from refuse beds. *J. Sanit. Eng. Div., Am. Soc. Civ. Eng.* **96**(SA-1), 49–58.

Quastel, J. H., and Scholefield, P. G. (1953). Arsenite oxidation in soil. *J. Soil Sci.* **75**, 279–285.

Quinn, J. J., Negri, M. C., Hinchman, R. R., Moos, L. P., Wozniak, J. B., and Gatliff, E. G. (2001). Predicting the effect of deep-rooted hybrid poplars on the groundwater flow system at a large scale phytoremediation site. *Int. J. Phytoremed.* **3**(1).

Quiroz, J. D., and Zimmie, T. F. (1998). *Paper Mill Sludge Landfill Cover Construction.* Proc. Sessions of Geo-Congress (C. Vipulanandan and D. J. Elton, Eds.). Geotech. Sp. Publ. No. 79. American Society of Civil Engineers, Reston, Virginia, pp. 19–36.

Quon, J. E., Charnes, A., and Wersan, J. S. (1965). Simulation and analyses of a refuse collection system. *J. Sanit. Eng. Div.*, **91**(SA5), 17–36.

Raimondi, P., Gardner, G. H., and Petric, C. B. (1959). Effect of pore structure and molecular diffusion on the mixing of miscible liquids flowing in porous media. *Jt. Symp. Fundam. Concepts Miscible Fluid Displacement*, Part II. Am. Inst. Chem. Eng., San Francisco, California.

Ram, N. M., Bass, D. H., Falotico, R., and Leahy, M. (1993). A decision framework for selecting remediation technologies at hydrocarbon contaminated sites. *J. Soil Contam.* **2**(2), 167–189.

Ramaswami, S. D., and Aziz, M. A. (1992). Some waste materials in road construction. In *Utilization of Waste Materials* (H. I. Inyang and K. L. Bergensen, Eds.). American Society of Civil Engineers, New York, pp. 153–165.

Ramaswami, S. D., and Tanabooriboon, Y., and Kheok, S. C. (1986). Utilization of steel slag for road surfacing in Singapore. *Proc. 5th Conf. Road Eng. Assoc., Asia and Australia*, Adelaide, pp. 11–15.

Ramaswami, A., Carr, P., and Burkhardt, M. (2001). Plant uptake of uranium: Hydroponic and soil system studies. *Int. J. Phytoremed.* **3**(2).

Ramme, B. W., and Tharaniyil, M. P. (1999). *Wisconsin Electric Power Company Coal Combustion Products Utilization Handbook.* Wisconsin Electric Power Company, Milwaukee, Wisconsin.

Ramsar Information Bureau (RIB) (1998). *What Are Wetlands?* Information Paper No. 1. RIB, Gland, Switzerland.

Rankin, S. (1993). Plastics recycling. In *Recycling Source Book* (T. J. Cichonski and K. Hill, Eds.). Gale Research, Detroit, Michigan, pp. 39–50.

Rao, S. K., Moulton, L. K., and Seals, R. K. (1977). Settlement of refuse landfills, *Proc. Conf. Geotech. Pract. Disp. Solid Waste Mater. Am. Soc. Civ. Eng.*, pp. 574–598.

Raudkivi, A. J., and Callandar, R. A. (1976). *Analysis of Groundwater Flow.* Wiley, New York.

Razvi, A. S., Walsh, P., and O'Leary, P. (1990). Marketing composts. *Waste Age* **21**(3), 137.

Rebhun, M., and Galil, N. (1987). Biotreatment inhibition by hazardous compounds in an integrated oil refinery. *Proc. Purdue Ind. Waste Conf.* **42**, 163–174.

RecycleWorlds Consulting (1994). *Everything You Wanted to Know about Waste Sorts. But Were Afraid to Ask.* RecycleWorlds Consulting, Madison, Wisconsin.

Reed, P. B., Jr. (1997). *Revision of the National List of Plant Species That Occur in Wetlands.* Department of the Interior, U.S. Fish and Wildlife Service, Washington, D.C.

Reeve, R. C., and Tamaddoni, G. H. (1965). Effect of electrolyte concentration on laboratory permeability and field intake rate of a sodic soil. *Soil Sci.* **99**(4), 261–266.

Reinhardt, P. A., and Gordon, J. G. (1991). *Infectious and Medical Waste Management.* Lewis Publishers, Chelsea, Michigan.

Reinhart, D. R., and Townsend, T. G. (1998). *Landfill Bioreactor Design and Operation.* Lewis Publishers, Boca Raton, Florida.

Revah, A., and Avnimeleih, Y. (1979). Leaching of pollutants from sanitary landfill models. *J. Water Pollut. Control Fed.* **51**(11), 2705–2716.

Reynolds, W. D., and Elrick, D. E. (1985). In situ measurement of field-saturated hydraulic conductivity, sorptivity and the a-parameter using Guelph permeameter. *Soil Sci.* **140**(4), 292–302.

Rhoades, J. D. (1979). Inexpensive four-electrode probe for monitoring soil salinity. *Soil Sci. Soc. Am. J.* **43**(4), 817–818.

Rhoades, J. D., and Van Schilfgaarde, J. (1976). An electrical conductivity probe for determining soil salinity. *Soil Sci. Soc. Am. J.* **40,** 647–651.

Richards, L. A., and Gardner, W. O. (1936). Tensiometers for measuring the capillary tension of soil water. *J. Am. Soc. Agron.* **28,** 352–358.

Richards, D. J., and Shieh, W. K. (1986). Biological fate of organic priority pollutants in the aquatic environment. *Water Res.* **20**(9), 1077–1090.

Richardson, G., and Reynolds, D. (1991). Geosynthetic consideration in a landfill on compressive clays, Geosynthetics '91, Vol. 2, Ind. Fabrics Assoc. Intl., St. Paul, Minnesota.

Ridley, W. P., Dizikes, L. P., and Wood, J. M. (1977). Biomethylation of toxic elements in the environment. *Science* **197,** 329–332.

Rifai, H. S. (1997). Natural aerobic biological attenuation. In *Subsurface Restoration.* Ann Arbor, Chelsea, Michigan, pp. 411–427.

Rifai, H. S., Bedient, P. B., Wilson, J. T., Miller, K. M., and Armstrong, J. M. (1988). Biodegradation modeling at aviation fuel spill site. *J. Environ. Eng.* **114**(5) 1007–1029.

Rinne, E. E., Dunn, R. J., and Majchrzak, M. (1994). Design considerations for piles in landfills. *Proc. Fifth Intl. Conf. on Piling and Deep Foundations*, Bruges, DFI, pp. 2.2.1–2.2.5.

Riser-Roberts, E. (1998). *Remediation of Petroleum Contaminated Soils.* Lewis Publishers, Boca Raton, Florida.

Roberts, K. J., and Sangrey, D. A. (1977). Attenuation of Inorganic Landfill Leachate Constituents in Soils of New York, Res. Rep. Contract No. C110367. Presented to N.Y. State Department of Environmental Conservation, Division of Solid Waste Management.

Robinson, H. D., and Maris, P. J. (1985). The treatment of leachates from domestic waste in landfill sites. *J. Water Pollut. Control Fed.* **57**(1), 30–38.

Rogoff, M. J. (1987). *How to Implement Waste-to-Energy Projects.* Noyes Publication, Park Ridge, New Jersey.

Rogowski, A. S., and Richie, E. B. (1984). Relationship of laboratory and field determined hydraulic conductivity in compacted clay soils. *Ind. Waste Proc., Mid-Atl. Conf., 16th, 1984*, pp. 520–533.

Rogowski, A. S., Weinrich, B. E., and Simmons, D. E. (1985). Permeability assessment of a compacted clay liner. *Proc. 8th Annu. Madison Conf. Appl. Res. Pract. Munic. Ind. Waste*, U.W. Extn., Madison Wisconsin, pp. 315–337.

Rollin, A., and Fournier, J. F. (2001). Biogas barrier beneath buildings: Case studies using geomembranes. *Proc. Geosynthetics 2001*, Portland, Oregon, pp. 203–216.

Rossabi, J., Looney, B. B., Eddy-Diley, C. A., Riha, B. D., and Rohay, V. J. (1994). Passive remediation of chlorinated volatile organic compounds using barometric pumping. *Proc. Water Environ. Fed: Innovative Solutions for Contaminated Site Management*, March 6–9, Miami, Florida, pp. 377–388.

Rovers, F. A., and Farquhar, G. J. (1973). Infiltration and landfill behavior. *J. Environ. Eng. Div., Am. Soc. Civ. Eng.* **99**(EE-5), 671–690.

Rowe, P. W. (1972). The relevance of soil fabric to site investigation practice. *Geotechnique* **22**(2), 195–300.

Ruhl, J. L., and Daniel, D. E. (1997). Geosynthetic clay liners permeated with chemical solutions and leachates. *J. Geotech Geoenviron. Eng.* **123**(4), 369–381.

Rutala, W., and Sarubbi, F. (1983). Management of infectious waste from hospitals. *Infect. Waste Manag.* **4**(4), 198–203.

Rutala, W., Stiegel, M., and Sarubbi, F. (1982). Decontamination of laboratory microbiological waste by steam sterilization. *Appl. Environ. Microbiol.* **43**(6), 1311–1316.

Rutherford, K., and Johnson, P. (1996). Effects of process control changes on aquifer oxygenation rates during in situ air sparging in homogenous aquifer. *Groundwater Monitor. Remed.*, Fall, pp. 132–141.

Ryan, C. R. (1987). Vertical barriers in soil pollution contaminant. In *Geotechnical Practice for Waste Disposal, '87*. Geotech. Sp. Publ. 13 (R. D. Woods, Ed.). American Society of Civil Engineers, New York, pp. 182–204.

Ryan, J. V. (1989). *Characterization of Emissions from the Simulated Open-Burning of Scrap Tires*, EPA-600/2-89-054. U.S. Environmental Protection Agency, Washington, D.C.

Ryan, J. V., and Lutes, C. C. (1993). *Characterization of Emissions from the Simulated Open-Burning of Non-Metallic Automobile Shredder Residue*, EPA-600/R-93-044. U.S. Environmental Protection Agency, Washington, D.C.

Sai, J. O., and Anderson, D. C. (1991). In *State-of-the-Art Field Hydraulic Conductivity Testing of Compacted Soils*, EPA/600/S2-91/022. U.S. Environmental Protection Agency, Cincinnati, Ohio.

Salame, M. (1961). An empirical method for the prediction of liquid permeation in polyethylene and related polymers. *SpE Trans.* (Poly. Eng. and Science), October, pp. 153–163.

Salvato, J. A., Witkie, W. G., and Mead, B. E. (1971). Sanitary landfill leaching prevention and control. *J. Water Pollut. Control Fed.* **43**(10), 2084–2100.

Salvesan, D. (1990). *Wetlands: Mitigating and Regulating Development Impacts.* Urban Land Institute, Washington, D.C.

Sangrey, D. A., Noonan, D. K., and Webb, G. S. (1976). Variation in Atterberg limits of soil due to hydration history and specimen preparation. ASTM Spec. Tech. Publ. STP 599, 158–168.

Santillan-Medrano, J., and Jarinak, J. J. (1975). The chemistry of lead and cadmium in solid phase formation. *Soil Sci. Soc. Am. Proc.* **39,** 851.

Savage, G. M., Diaz, L. F., and Golueke, C. G. (1985). Disposing of organic hazardous waste by composting. *Biocycle* **26,** 31–34.

Scharch, J. F., Huebner, P. M., and Mack, D. P. (1985). An improved analytic method for estimating leachate heads within clay lined landfills. *Proc. 8th Annu. Madison Conf. App. Res. Prod. Munic. Ind. Waste*, U.W. Extn., Madison, Wisconsin, pp. 338–354.

Schmugge, T. J., Jackson, T. J., and McKim, H. L. (1980). Survey of methods for soil moisture determination. *Water Resour. Res.* **16**(6), 961–979.

Schnabel, P., Lysmer, J., and Seed, H. B. (1972). SHAKE: *A Computer Program for Earthquake Response Analysis of Horizontally Layered Sites,* Report No. EERC 72-2. Earthquake Engineering Research Center, University of California, Berkeley, California.

Schroeder, P. R. (1985). Personal communication regarding HELP Model experiments, Waterways Exp. Stn., Corps of Engineers, Vicksburg, Mississippi, August 5, 1985.

Schroeder, P. R., Gibson, A. C., and Smolen, M. D. (1984a). *The Hydrologic Evaluation of Landfill Performance* (*HELP*) *Model*, Vol. II, Doc. for Version 1, EPA 530-SW-84-101. U.S. Environmental Protection Agency, Cincinnati, Ohio.

Schroeder, P. R., Morgan, J. M., Walski, T. M., and Gibson, A. C. (1984b). *The Hydrologic Evaluation of Landfill Performance* (*HELP*) *Model*, Vol. 1, User's Guide for Version I, EPA/530-SW-84-009, U.S. Environmental Protection Agency, Cincinnati, Ohio.

Schultz, D. W., and McKias, M. P., Jr. (1980). Assessment of liner installation procedures. In *Land Disposal of Municipal Solid Waste* (P. W. Shultz, Ed.), 6th Annu. Res. Symp., EPA 600/9-80/010. U.S. Environ. Agency, Cincinnati, Ohio.

Schuster, K. A., and Schur, D. A. (1974). *Heuristic Routing for Solid Waste Collection Vehicles.* SW-113. U.S. Environmental Protection Agency, Washington, D.C.

Schwab, G. O., and Manson, P. W. (1957). Engineering aspects of land drainage. In *Drainage of Agricultural Land* (J. N. Luthin, Ed.). Madison, Wisconsin, Chapter 3, pp. 287–344.

Scott, E. M., and Gorman, S. P. (1983). Sterilization with glutaraldehyde. In *Disinfection, Sterilization, and Preservation*, 3rd ed., (S. S. Block, Ed.). Lea & Febiger, Philadelphia, Pennsylvania, pp. 65–88.

SCS Engineers (1976). *The Selection and Monitoring of Land Disposal Case Study Sites*, Vol. I: Project Description and Findings: Contract No. 68-01-2973. U.S. Environmental Protection Agency, Washington, D.C.

Seed, H. B., and Bonaparte, R. (1992). Seismic analysis and design of lined waste fills: Current practice. In *Proc. Stability Perform. Slopes Embankments II*, Geotech. Sp. Publ. 31. American Society of Civil Engineers, New York, 1521–1545.

Seed, H. B., and Chan, C. K. (1959). Structure and strength characteristics of compacted clays. *J. Soil Mech. Found. Div., Am. Soc. Civ. Eng.* **85**(SM-5), 87–128.

Seed, H. B., Woodward, R. J., and Lundgren, R. (1962). Prediction of swelling potential for compacted clays. *J. Soil Mech. Found. Div., Am. Soc. Civ. Eng.* **90**(SM-4), 107–131.

Seed, H. B., Woodward, R. J., and Lundgren, R. (1964a). Clay minerological aspects of the Atterberg limits. *J. Soil Mech. Found. Div., Am. Soc. Civ. Eng.* **90**(SM-4), 107–131.

Seed, H. B., Woodward, R. J., and Lundgren, R. (1964b). Fundamental aspects of the Atterberg limits. *J. Soil Mech. Found. Div., Am. Soc. Civ. Eng.* **90**(SM-6), 75–105.

Shackelford, C. D. (1988). Diffusion as a transport process in fine-grained barrier materials. *Geotechnical News*, June, 24–27.

Shan, C., Falta, R. W., and Javendal, I. (1992). Analytical solutions for steady state gas flow to a soil vapor extraction well. *Water Resour. Res.* **28**(4), 1105–1120.

Shan, H. Y., and Daniel, D. E. (1991). Results of laboratory tests on geotextile/bentonite liner material. In *Proc. Geosynthetics 91*, Vol. 2. Industrial Fabrics Association International, St. Paul, Minnesota, pp. 517–535.

Sharma, H. D. (1992). *Field Performance of a Vertical Cutoff (Slurry) Wall*. American Society of Civil Engineers, New York.

Sharma, H. D., and Lewis, S. P. (1994). *Waste Containment Systems, Waste Stabilization, and Landfills: Design and Evaluation*. Wiley, New York.

Sherard, J. L., Dunnigan, L. P., and Talbot, J. R. (1984a). Basic properties of sand and gravel filters. *J. Geotech. Eng. Div., Am. Soc. Civ. Eng.* **110**(GT-6), 684–700.

Sherard, J. L., Dunnigan, L. P., and Talbot, J. R. (1984b). Filters for silts and clays. *J. Geotech. Eng. Div., Am. Soc. Civ. Eng.* **110**(GT-6), 701–717.

Shimizu, K. (1997). Geotechnics of waste landfills. *Proc. 2nd Intl. Congress Env. Geotech.*, Vol. 3, Osaka, Balkema, pp. 1475–1491.

Shoda, M. (1991). Methods for the biological treatment of exhaust gases. In *Biological Degradation of Wastes* (A. M. Martin, Ed.). Elsevier Science, New York, pp. 31–46.

Silka, L. R. (1988). Simulation of vapor transport through the unsaturated zone—Interpretation of soil-gas surveys. *Groundwater Monit. Rev.* **8**(21), 115–123.

Sims, R., and Bass, J. (1984). *Review of In-Place Treatment Technique for Contaminated Surface Soils*, Vol. 1: *Technical Evaluation*. EPA-540/2-84-0039, U.S. Environmental Protection Agency, Washington, D.C.

Singh, S. (1992). *Response of Clay Liner System to Seismic Loading*, CE Report No. 92-5. Department of Civil Engineering, Santa Clara University, Santa Clara, California.

Singh, S., and Murphy, B. (1990). Evaluation of stability of sanitary landfills. In *Geotechnics of Wastefills—Theory and Practice STP 1070* (A. D. Landva and G. D. Knowles, Eds.). American Society for Testing and Materials, Philadelphia, Pennsylvania, pp. 240–258.

Skempton, A. W. (1953). The colloidal activity of clay. *Proc. Int. Conf. Soil Mech. Found. Eng., 3rd, 1953* **1**, 57–61.

Skempton, A. W. (1964). Long-term stability of clay slopes. *Geotechnique* **14**, 77–101.

Slane, K. O. (1987). An Evaluation of Groundwater Models to Predict Groundwater Mounding beneath Proposed Groundwater Gradient Control Systems for Sanitary Landfill Design. Unpublished M.S. Thesis, Department of Civil and Environmental Engineering, University of Wisconsin, Madison.

Smith, R. L. (1994). Use of Monte Carlo simulation for human exposure assessment at a Superfund site. *Risk Anal.* **14,** 433–439.

Smith, S. A. (1995). *Monitoring and Remediation Wells—Problems, Prevention, Maintenance, and Rehabilitation.* CRC Press, Boca Raton, Florida.

Smyth, D. J. A., and Cherry, J. A. (1997). Sealable joint steel sheet piling for groundwater pollution control. In *Barrier Technology for Environmental Management.* National Academy Press, Washington, D.C., pp. D-144–D-152.

Soil and Conservation Service (SCS) (1971). *National Engineering Handbook, Section 16—Drainage of Agricultural Land.* U.S. Department of Agriculture, Eng. Div., Washington, D.C.

Soil and Conservation Service (SCS) (1972). *Procedure for Computing Sheet and Rill Erosion on Project Areas,* Release No. 51. U.S. Department of Agriculture, Eng. Div., Washington, D.C.

Soil and Conservation Service (SCS) (1975). *Urban Hydrology for Small Watersheds,* Tech. Release No. 55. U.S. Department of Agriculture, Eng. Div., Washington, D.C.

Solseng, P. B. (1978). Determining the moisture content of residual waste. In *Disposal of Residuals by Landfilling* (A. Gebhard, proj. manager). Minnesota Pollution Control Agency, Minneapolis, Minnesota, pp. I-1–I-40.

Somogyi, F., and Gray, D. H. (1977). Engineering properties affecting disposal of red muds. *Proc. Conf. Geotech. Pract. Disposal Solid Waste Mater., 1977,* pp. 1–22.

Sopcich, D., and Bagchi, A. (1988). Discussion on mass burning of MSW with energy recovery, by Beckman and Dragovich (1986). *J. Environ. Eng. Div., Am. Soc. Civ. Eng.* **114**(EE-1), 235–236.

Sorg, T. J., and Bendixen, T. W. (1975). Sanitary landfill. In *Solid Wastes: Origin, Collection, Processing, and Disposal* (C. L. Mantell, Ed.). Wiley, New York, pp. 71–113.

Sowers, G. B., and Sowers, G. F. (1970). *Introductory Soil Mechanics and Foundations,* 3rd ed. Macmillan, New York.

Spangler, M. G., and Handy, R. L. (1982). *Soil Engineering,* 4th ed. Harper & Row, New York.

Spencer, W. F., and Cliath, M. M. (1977). The solid–air interface: Transfer of organic pollutants between the solid-air interface. In *Fate of Pollutants in the Air and Water Environments,* Vol. 8, Part 1: *Advances in Environmental Science and Technology* (I. H. Suffet, Ed.). Wiley, New York, pp. 107–126.

Spitzglass, J. M. (1912). Flow of gas formula, derived, analyzed and checked by experimental data with diagrams for figuring the flow of gas in street mains and services. *J. Am. Gas Light* **96,** 269–315.

Spooner, P., Wetzel, R., Spooner, C. Furman, C., Tokarski, E., Hunt, G., Hodge, V., and Robinson, T. (1985). *Slurry Trench Construction for Pollution Migration Control.* Noyes Publications, Park Ridge, New Jersey.

Sridharan, L., and Didier, P. (1988). Leachate quality from containment landfills in Wisconsin. *Proc. Int. Solid Waste Conf. 5th* **2,** 133–138.

State and Territorial Association (STA) (1994). *Technical Assistance Manual: State Regulatory Oversight of Medical Waste Treatment Technologies.* Report on alternate treatment technologies. STA, Washington, D.C.

Stanczyk, T. F. (1987). The development of treatment alternatives through an understanding of waste chemistry. *Proc. Purdue Ind. Waste Conf.* **42,** 309–320.

Stark, T. I. D. (1999). Stability of waste containment facilities. *Proc. WasteTech'99,* New Orleans, Louisiana.

Stark, T. D. and Eid, H. T. (1998). Performance of three-dimensional slope stability methods in practice. *J. Geotech. Geoenviron. Eng.* **124**(11), 1049–1060.

Starke, J. O. (1989). Effect of freeze/thaw weather conditions on compacted clay liners. *Proc. 12th Annu. Madison Waste Conf.,* U.W. Extn., Madison, Wisconsin, pp. 412–420.

State of Wisconsin (1995). *Basic Guide to Wisconsin's Wetlands and Their Boundaries.* Wisconsin Coastal Management Program, Madison, Wisconsin.

Stegmann, R. (1979). Leachate treatment at the sanitary landfill of Lingen, West Germany—Experiences with the design operation of the aerated lagoons. *Proc. 2nd Annu. Madison Conf. Appl. Res. Pract. Munic. Ind. Waste,* pp. 456–471.

Stein, K. (1997). *Beyond Recycling: A Re-user's Guide.* Clear Light Publishers, Santa Fe, New Mexico.

Steiner, R. L., Keenan, J. D., and Fungaroli, A. A. (1979). *Demonstrating Leachate Treatment,* Report on a Full-scale Operating Plant, EPA SW-758. U.S. Environmental Protection Agency, Cincinnati, Ohio.

Stephens, D. B., Unruh, M., Havelana, J., Knowlton, R. G., Jr., Mattson, E., and Cox, W. (1988). Vadose zone characterization of low permeability sediments using field permeameters. *Ground Water Monitor. Rev.* **8**(2), 59–66.

Stone, R. (1974). *Disposal of Sewage Sludge into a Sanitary Landfill,* EPA-SW-71d. U.S. Environmental Protection Agency, Cincinnati, Ohio.

Stone, R. (1975). Aerobic landfill stabilization. In *Solid Waste: Origin, Collection, Processing, and Disposal* (C. L. Mantell, Ed.). Wiley, New York, pp. 153–183.

Stormont, J. C. (1995). The performance of two capillary barriers during constant infiltration. In *Landfill Closures—Environmental Protection and Land Recovery,* Geotech. Sp. Publ. No. 53 (R. J. Dunn and U. P. Singh, Eds.). American Society of Civil Engineers, New York, pp. 77–92.

Stormont, J. C., and Anderson, C. E. (1999). Capillary barrier effect from underlying coarser soil layer. *J. Geotech. Geoenviron. Eng.* **125**(8), 641–648.

Strack, O. D. L. (1989). *Groundwater Mechanics.* Prentice-Hall, Englewood Cliffs, New Jersey.

Straub, W. A., and Lynch, D. R. (1982a). Models of landfill and leaching: Moisture flow and inorganic strength. *J. Environ. Eng. Div., Am. Soc. Civ. Eng.* **108**(EE-2), 231–250.

Straub, W. A., and Lynch, D. R. (1982b). Models of landfill leaching: Organic strength. *J. Environ. Eng. Div., Am. Soc. Civ. Eng.* **108**(EE-2), 251–268.

Streng, D. R. (1976). The effects of the disposal of industrial waste within a sanitary landfill environment. In *Residual Management by Land Disposal* (W. H. Fuller, Ed.), EPA-600/9-76/015. U.S. Environmental Protection Agency, Cincinnati, Ohio, pp. 51–70.

Stucki, G., and Alexander, M. (1987). Role of dissolution rate and solubility in biodegradation of aromatic compounds. *Appl. Environ. Microbiol.* **53,** 292–297.

Stumm, W., and Morgan, J. J. (1970). *Aquatic Chemistry.* Wiley Interscience, New York.

Suflita, J. M., Gerba, C. P., Ham, R. K., Palmisano, A. C., Rathje, W. L., and Robinson, J. A. (1992). The world's largest landfill—A multidisciplinary investigation. *Environ. Sci. Technol.* **26**(8), 1486–1494.

Sundberg, J., Gipperth, P., and Wene, C.-O. (1994). A systems approach to municipal solid waste management: A pilot study of Goteborg. *Waste Manag. Res.* **12,** 73–91.

Suthersan, S. S. (1996). In situ air sparging. In *In-Situ Treatment Technology* (E. K. Nyer, S. Fam, D. F. Kidd, F. J. Johns II, P. L. Palmer, G. Boettcher, T. L. Crossman, and S. S. Suthersan, Eds.). CRC Press, Boca Raton, Florida, pp. 195–220.

Sutter, G. W., Efroymson, R. A., Sample, B. E., and Jones, B. E. (2000). *Ecological Risk Assessment for Contaminated Sites.* Lewis Publishers, Boca Raton, Florida.

Swaine, D. J., and Mitchell, R. L. (1960). Trace element distribution in soil profiles. *J. Soil Sci.* **11,** 347–368.

Sykes, J. F., Soyupak, S., and Farquhar, G. J. (1969). *Modeling of Leachate Migration and Attenuation in Groundwater below Sanitary Landfill.* University of Waterloo, Department of Civil Engineering, Ontario, Canada.

Tabak, H. H., Quave, S. A., Mashni, C. I., and Barth, E. R. (1981). Biodegradability studies with organic priority pollutant compounds. *J. Water Pollut. Control Fed.* **53**(10), 1503–1518.

Tallard, G. R. (1992). New trenching method using synthetic bio-polymers. In *Slurry Walls: Design, Construction and Quality Control*, STP 1129 (D. B. Paul, R. R. Davidson, and N. J. Cavalli, Eds.). American Society for Testing and Materials, Philadelphia, Pennsylvania, pp. 86–102.

Tatlisoz, N., Benson, C., and Edil, T. (1997). Effect of fines on mechanical properties of soil-tire chips mixtures. In *Testing Soil Mixed with Waste or Recycled Materials*, ASTM STP 1275 (A. Wasemiller and K. B. Hoddinott, Eds.). American Society for Testing and Materials, Philadelphia, Pennsylvania, pp. 93–108.

Taylor, D. W. (1948). *Fundamentals of Soil Mechanics.* Wiley, New York.

Taylor, F. H. (1991). Comparison of potential greenhouse gas emissions from disposal of MSW in sanitary landfills vs. waste-to-energy facilities. In *Proc. 2nd Annual Intl. Conf. Municipal Waste Combustion.* Air & Waste Management Association, Pittsburgh, Pennsylvania, pp. 413–424.

Taylor, G. S., and Luthin, J. N. (1978). A model for couple head and moisture transfer during soil freezing. *J. Can. Geotech.* **15,** 548–555.

Technical Enforcement Guidance Document (1986). *RCRA Ground-Water Monitoring Technical Enforcement Guidance Document*, OSWER-9950.1. U.S. Environmental Protection Agency, Office of Solid Waste and Emergency Response, Cincinnati, Ohio.

Teller, A. (1994). Waste to energy conversion: Emission control. In *Handbook of Solid Waste Management* (F. Kreith, Ed.). McGraw-Hill, New York, pp. 11:127–11:171.

Telles, R. W., Unger, S. L., and Lubowitz, H. R. (1988). *Technical Considerations for De Minimis Pollutant Transport through Polymeric Liners*, EPA/600/2-88/042. U.S. Environmental Protection Agency, Cincinnati, Ohio.

Terzaghi, K. (1936). The shearing resistance of saturated soils. *Proc. 1st Int. Conf. Soil Mech. Found. Eng., 1936* **1,** 54–56.

Tharp, L. (1991). Leachate characteristics for Missouri sanitary landfills. *Proc. 14th Annu. Madison Waste Conf.*, U.W. Extn., Madison, Wisconsin, pp. 313–326.

Thiel, T. J., Focuss, J. C., and Leech, A. P. (1963). Electrical water pressure transducers for field and laboratory use. *Soil Sci. Soc. Am. Proc.* **27**(5), 601–602.

Thornthwaite, C. W., and Mather, J. R. (1955). *The Water Balance*, Publ. Climatol. Lab. Climatol. Drexel Institute of Technology, Lab. Climatol., Centerton, New Jersey.

Thornthwaite, C. W., and Mather, J. R. (1957). *Instructions and Tables for Computing Potential Evapotranspiration and the Water Balance*, Publ. Climatol. Lab. Climatol. Drexel Institute of Technology, Lab. Climatol., Centerton, New Jersey.

Tiner, R. W. (1999). *Wetland Indicators, a Guide to Wetland Identification, Delineation, Classification, and Mapping.* Lewis Publishers, Boca Raton, Florida.

Tisdale, S. L., and Nelson, W. L. (1975). *Soil Fertility and Fertilizer.* Macmillan, New York.

Tobin, W. R., and Wild, P. R. (1986). Attapulgite; a clay liner solution? *Civil Eng.* February, pp. 56–58.

Todd, D. K. (1980). *Groundwater Hydrology*, 2nd ed., Wiley, New York.

Topp, G. C., Davis, J. L., and Annan, A. P. (1980). Electromagnetic determination of soil water content: Measurements in coaxial transmission lines. *Water Resour. Res.* **16**(3).

Torrance, J. K. (1974). A laboratory investigation of the effect of leaching on the compressibility and shear strength of Norwegian marine clays. *Geotechnique* **24**(2), 155–173.

Tortensson, B. A. (1984). A new system for groundwater monitoring. *Groundwater Monit. Rev.* **4**(4), 131–138.

Trainor, D. P. (1986). Moisture and saturation effects on hydraulic conductivity testing. *Proc. 9th Annu. Madison Conf. App. Res. Pract. Munic. Ind. Waste*, U.W. Extn., Madison, Wisconsin, pp. 646–657.

Trombley, J. (1994). Electrochemical remediation takes to the field. *Environ. Sci. Technol.* **28**(6), 289–291.

Tuma, J. J. (1978). *Engineering Mathematics Handbook.* McGraw-Hill, New York.

Turk, M., Collins, H. J., Wittmaier, M., Harborth, P., and Hanert, H. H. (1997). Maintenance of the functioning of landfill drainage systems. In *Advanced Landfill Liner Systems* (H. August, U. Holzlohner, and T. Meggyes, Eds., English. Trans. Ed. T. Meggyes). Thomas Telford, London, pp. 163–181.

Uloth, V. C., and Mavinic, D. S. (1977). Aerobic bio-treatment of a high strength leachate. *J. Environ. Eng. Div., Am. Soc. Civ. Eng.* **103**(EE-4), 647–661.

Uni-Bell Plastic Pipe Association (Uni-Bell) (1979). *Handbook of PVC Pipe Design and Construction.* Uni-Bell, Dallas, Texas.

University of Florida. (1998). Recommended management practices for removal of hazardous materials from buildings prior to demolition. Dept. Env. Engr., Gainesville, Florida.

U.S. Congress (1988). *Issues in Medical Waste Management—Background Paper*, OTA-BP-O-49. U.S. Government Printing Office, Washington, D.C.

U.S. Congress (1989). *Facing America's Trash: What's Next for Municipal Solid Waste?* OTA-0-424. Office of Technology Assessment, U.S. Government Printing Office, Washington, D.C.

U.S. Congress (1992). *Green Products by Design: Choices for a Cleaner Environment.* Office of Technology Assessment, OTA-E-541, Washington D.C.

U.S. Department of Agriculture and Environmental Protection Agency (USDA and USEPA) (1980). *Manual for Composting Sewage Sludge by the Beltsville Aerated-Pile Method*, EPA/600-8-80-022. USEPA, Washington, D.C.

U.S. Department of Commerce (USDC) (1977). *European Waste-to-Energy Systems: An Overview*, CONS-2103-6. Energy Research and Development Administration, USDC, Springfield, Virginia.

U.S. Department of Labor (USDL) (1984). Occupational exposure to ethylene oxide, final standard. *Federal Register*, June 22, 49 (122), pp. 25734–25809.

U.S. Department of Labor (USDL) (1987). Occupational exposure to formaldehyde, final standard. *Federal Register*, December 4, pp. 56168–46312.

U.S. Environmental Protection Agency (USEPA) (1973). *Recommended Methods of Reduction, Neutralization, Recovery or Disposal of Hazardous Waste*, Vol. 3: *Disposal Process Description, Ultimate Disposal, Incineration and Pyrolysis Processes*, USEPA 670/2-73-053C. USEPA, Washington, D.C.

U.S. Environmental Protection Agency (USEPA) (1979). *Hazardous Material Design Criteria*, EPA 600/2-79-198. USEPA, Washington D.C.

U.S. Environmental Protection Agency (USEPA) (1980a). *Source Category Survey: Industrial Incinerators*, USEPA 450/3-80-013. USEPA, Washington, D.C.

U.S. Environmental Protection Agency (USEPA) (1980b). *Sources of Toxic Compounds in Household Wastewater*, EPA 600/2-80-129, Cincinnati, Ohio.

U.S. Environmental Protection Agency (USEPA) (1982). *Draft Manual for Infectious Waste Management.* SW-957. USEPA, Washington, D.C.

U.S. Environmental Protection Agency (USEPA) (1984). *Slurry Trench Construction for Pollution Migration Control*, EPA-540/2-84-001. USEPA, Cincinnati, Ohio.

U.S. Environmental Protection Agency (USEPA) (1985a). *Principles of Risk Assessment: A Non-technical Review*, Office of Policy Analysis, Washington, D.C.

U.S. Environmental Protection Agency (USEPA) (1985b). *Toxicology Handbook.* Office of Waste Programs Enforcement, Washington, D.C.

U.S. Environmental Protection Agency (USEPA) (1985c). *Handbook for Remedial Action at Waste Disposal Sites*, Rev., EPA/625/6-85/006. USEPA, Washington D.C.

U.S. Environmental Protection Agency (USEPA) (1986a). *Test Methods for the Evaluation of Solid Waste*, EPA-SW-846. USEPA, Washington, D.C.

U.S. Environmental Protection Agency (USEPA) (1986b). *Methods for Assessing Environmental Pathways of Food Contamination: Methods for Assessing Exposure to Chemical Substances*, Vol. 8, EPA/560/5-85/008. Exposure Evaluation Division, Office of Toxic Substances, Washington, D.C.

U.S. Environmental Protection Agency (USEPA) (1988). *Superfund Exposure Assessment Manual*, EPA/540/1-88/001. Office of Remedial Response, Washington, D.C.

U.S. Environmental Protection Agency (USEPA) (1989a). *Exposure Factor Handbook*, EPA/600/8-89/043. Office of Health and Environmental Assessment, Washington, D.C.

U.S. Environmental Protection Agency (USEPA) (1989b). *Interim Methods for Development of Inhalation Reference Doses*, EPA/600/8-88/066F. Office of Health and Environmental Assessment, Washington, D.C.

U.S. Environmental Protection Agency (USEPA) (1989c). *Requirements for Hazardous Waste Landfill Design, Construction and Closure*, EPA/625/4-89/022. USEPA, Cincinnati, Ohio.

U.S. Environmental Protection Agency (USEPA) (1990). 40 CFR 300.

U.S. Environmental Protection Agency (USEPA) (1991a). *Risk Assessment Guidance for Superfund, Volume I—Human Health Evaluation Manual, Supplemental Guidance, Standard Default Exposure Factors*, (*Interim Final*), OSWER Directive 9285.6-03. Office of Emergency and Remedial Response, Washington, D.C.

U.S. Environmental Protection Agency (USEPA) (1991b). *Design and Construction of RCRA/CERCLA Final Covers*, EPA/625/4-91/025. Office of Research and Development, USEPA, Washington, D.C.

U.S. Environmental Protection Agency (USEPA) (1992a). *Hazard Ranking System Guidance Manual*, EPA 540-R-92-026. USEPA, Washington, D.C.

U.S. Environmental Protection Agency (USEPA) (1992b). *Draft Guidelines for Controlling Sewage Sludge Composting Odors*. Office of Wastewater Enforcement and Compliance, Washington, D.C.

U.S. Environmental Protection Agency (USEPA) (1993). *Report on Workshop on Geosynthetic Clay Liners*, EPA/600/R-93/171. Office of Research and Development, USEPA, Washington, D.C.

U.S. Environmental Protection Agency (USEPA) (1994a). In situ vitrification, Geosafe Corporation. In *SITE Demonstration Bulletin*, EPA/540/MR-94/520. U.S. Environmental Protection Agency, Washington, D.C.

U.S. Environmental Protection Agency (USEPA) (1994b). Emerging Technology Program of Electrokinetics, Inc. (Electrokinetic Remediation). In *SITE: Technology Profiles*, EPA/540/R-94/526. USEPA, Washington D.C.

U.S. Environmental Protection Agency (USEPA) (1994c). Emerging technology of Battelle Memorial Institute (in situ electroacoustic soil decontamination). In *SITE: Technology Profiles*, EPA/540/R-94/526. USEPA, Washington, D.C.

U.S. Environmental Protection Agency (USEPA) (1994d). *Composting Yard Trimmings and Municipal Solid Waste*. EPA 530-R-94-003. USEPA, Washington, D.C.

U.S. Environmental Protection Agency (USEPA) (1998). *Characterization of Municipal Solid Waste in the United States*, EPA-R-980-007. EPA, Washington, D.C.

U.S. Environmental Protection Agency (USEPA) (1999). *Guide for Industrial Waste Management*, EPA 530-R-99-001. USEPA, Washington, D.C.

U.S. Environmental Protection Agency (USEPA) (2000). State-of-the-practice for bioreactor landfills. In *Proceedings of the USEPA Workshop on Bioreactor Landfills*. USEPA, Washington, D.C.

U.S. Government (1977). *Federal Register*, July 19, Washington, D.C.

Vaccari, P. L., Tonat, K., Dechristoforo, R., Gallelli, J. F., and Zimmerman, P. F. (1984). Disposal of antineoplastic waste at the National Institutes of Health. *Am. J. Hosp. Pharm.* **41,** 87–93.

Velazquez, L. A., and Noland, J. W. (1993). Low temperature stripping of volatile compounds. In *Principles and Practices for Petroleum Contaminated Soils* (E. J. Calabrese and P. T. Kostecki, Eds.). Lewis Publishers, Boca Raton, Florida, pp. 423–431.

Vallejo, L. E. (1980). A new approach to the stability analysis of thawing slopes. *Can. Geotech. J.* **17,** 607–612.

Vallejo, L. E., and Edil, T. B. (1981). Stability of thawing slopes: Field and theoretical investigation. *Proc. 10th Int. Conf. Soil Mech. Found. Eng., 1981* **3,** 545–548.

Van Impe, W. F., and Bouazza, A. (1996). Densification of waste fills by dynamic compaction. *Can. Geotech. J.* **33**(6), 879–887.

Van Olphen, H. (1963). *An Introduction to Clay Colloid Chemistry.* Wiley Interscience, New York.

Varnes, D. J. (1978). Slope movement types and process. In *Landslides: Analysis and Control* (R. L. Schuster and R. J. Krizek, Eds.), Sp. Report 176, Transportation Research Board, National Research Council, Washington, D.C., pp. 11–33.

Varshney, R. S. (1979). *Engineering Hydrology.* NemChand & Bros. Roorkee, U.P, India.

Vasuki, N. C. (1988). Why not recycle the landfill. *Waste Age* November, pp. 165–170.

Veihmeyer, F. J. (1964). Evapotranspiration. In *Handbook of Applied Hydrology* (V. T. Chow, Ed.). McGraw-Hill, New York, pp. 11-1–11-38.

Veihmeyer, F. J., and Henderickson, A. H. (1955). Rates of evaporation from wet and dry soils and their significance. *Soil Sci.* **80,** 61–67.

Velter, R. J. (1987). Low-level radioactive waste: A national disposal problem. *Health Environ. Dig.* **1**(7), 1–3.

Vesiland, P. A., Worrell, W., and Reinharrt, D. (2002). *Solid Waste Engineering.* Brooks/Cole, Pacific Grove, California.

Viessman, W., Jr., Knapp, J. W., Lewis, G. L., and Harbaugh, T. E. (1977). *Introduction to Hydrology.* Harper & Row, New York.

Vipulanandan, C., and Basheer, M. (1998). Recycled materials for embankment construction. In *Recycled Materials In Geotechnical Applications* (C. Vipulanandan and D. J. Elton, Eds.), Proc. Sessions of Geo-Congress, Geotech. Sp. Publ., No. 79. American Society of Civil Engineers, Reston, Virginia, pp. 100–114.

Walker, W. H. (1969). Illinois groundwater pollution. *J. Am. Water Works Assoc.* **61,** 31–40.

Walker, J. M. (1993). Control of composting odors. In *Science and Engineering of Composting* (H. A. J. Hoitanl and H. M. Keener, Eds.). Renaissance Publications, Worthington, Ohio, pp. 185–218.

Walsh, J. J., and Kinman, R. N. (1979). Leachate and gas production. Under controlled moisture conditions. In *Land Disposal: Municipal Solid Waste* (D. W. Shultz, Ed.), 5th Annu. Res. Symp., EPA-600/9-79/023. U.S. Environmental Protection Agency, Cincinnati, Ohio, pp. 41–57.

Walsh, J. J., and Kinman, R. N. (1981). Leachate and gas from municipal solid waste landfill simulators. In *Land Disposal: Municipal Solid Waste* (D. W. Shultz, Ed.), 7th Annu. Res. Symp., EPA-600/9-81/002. U.S. Environmental Protection Agency, Cincinnati, Ohio, pp. 67–93.

Walsh, P., Pferdehrt, W., and O'Leary, P. (1993). Transfer stations and long-haul transport systems. *Waste Age* **24**(12) 57–65.

Wardwell, R. E., and Charlie, W. A. (1981). Effects of fiber decomposition on the compressibility and leachate generation at a combined paper mill sludge landfill area. *Proc. Tech. Assoc. Pulp Pap. Ind. Environ. Conf.*, New Orleans, pp. 233–238.

Water Pollution Control Federation (WPCF) (1980). *Operation and Maintenance of Wastewater Collection Systems.* Moore & Moore Lithographers, Washington, D.C.

Water Pollution Control Federation (WPCF) (1981). *Standard Methods for the Examination of Water and Wastewater.* WPCF, Washington, D.C.

Watkins, R. K. (1990). Plastic pipes under high landfills. In *Buried Plastic Pipe Technology*, ASTM STP 1093 (G. S. Buczala and M. J. Cassady, Eds.). American Society for Testing and Materials, Philadelphia, Pennsylvania.

Wathane, M. (1972). Optimal Routing of Solid Waste Collection Vehicles. Ph.D. Thesis, Johns Hopkins University, Baltimore, Maryland.

Weiler, W. A., Ozarowski, P. P., and Soydemir, C. (1993). Earthquake engineering for landfills. *Waste Age* **24**(8), 53–62.

Weisman, R. J., Falat Ko, S. M., Kuo, B. P., and Eby, E. (1994). Effectiveness of innovative technologies for treatment of hazardous soil. In *Proc. 87th Annual Meeting Exhibit., Air Waste Assoc.*, June 19–24, Cincinnati, Ohio.

Wellings, F. M., Lewis, A. L., and Mountain, C. W. (1974). Virus survival following wastewater spray irrigation of sandy soils. In *Virus Survival in Water and Wastewater Systems* (J. F. Mallina and B. P. Sagic, Eds.), Water Resour. Symp., No. 7, University of Texas, Austin, Texas, pp. 253–260.

Wenger, R. B., and Rhyner, C. R. (1984). Optional service regions for solid waste facilities. *Waste Manag. Res.* **2**(1), 1–15.

Weston, A. F. (1984). Obtaining reliable priority-pollutant analyses. *Chem. Eng. (N.Y.)* **9**(9), 54–60.

Wexell, D. R. (1993). Cold crown vitrification of municipal waste combuster flyash. In *Proc. A & WMA 86th Annual Meeting*, Denver, Colorado.

Wheless, E., and Wiltsee, G. (2001). Demonstration test of the capstone microturbine on landfill gas. In *Proc. 24th SWANA Annual Landfill Gas Symposium*, Dallas, Texas.

White, P. (1996). "So what is integrated waste management. *Warmer Bull.* No. 49, May 6.

Whitman, R. V., and Bailey, W. A. (1967). Use of computers for slope stability analysis. *J. Soil. Mech. Found. Div., Am. Soc. Civ. Eng.* **93**(SM-4), 475–498.

Wigh, R. J. (1979). *Boone County Field Site Interim Report*, EPA-600/2-79/058. U.S. Environmental Protection Agency, Cincinnati, Ohio.

Wigh, R. J., and Brunner, R. D. (1981). Summary of landfill research, Boone County field site. In *Land Disposal of Municipal Solid Waste* (D. W. Schultz, Ed.), 7th Annu. Res. Symp., EPA-600/9-81/002a. U.S. Environmental Protection Agency, Cincinnati, Ohio, pp. 209–242.

Wilbourn, R. G., Newburn, J. A., and Schofield, J. T. (1994). Treatment of hazardous wastes using the Thermatrix treatment system. In *Proc. 13th Int. Conf. Thermal Treatment of Radioactive Haz. Chem., Mixed Munitions, Pharmaceutical Waste.* University of California, Irvine, pp. 221–223.

Williams, P. J. (1986). Pore pressure at a penetrating frost line and their prediction. *Geotechnique* **16**(3), 187–208.

Williams, N. D., and Houlihan, M. (1986). Evaluation of friction coefficients between geomembranes, geotextiles and related products. *Proc. 3rd Int. Conf. Geotext., 1986*, pp. 891–896.

Williams, T. O., and Miller, F. C. (1992). Odor control using biofilters, Part I. *Biocycle* **33**(10), 72–77.

Wilson, L. G. (1980). *Monitoring in the Vadose Zone: A Review of Technical Elements and Methods*, EPA-600/7-80/134. U.S. Environmental Protection Agency, Cincinnati, Ohio.

Wilson, L. G. (1981). Monitoring in the vadose zone. Part I. Storage changes. *Ground Water Monit. Rev.* **1**(3), 32–41.

Wilson, L. G. (1982). Monitoring in the vadose zone. Part II. *Ground Water Monit. Rev.* **2**(1), 31–42.

Wilson, L. G. (1983). Monitoring in the vadose zone. Part III. *Ground Water Monit. Rev.* **3**, 155–166.

Wilson, J. D. (1997). So carcinogens have thresholds: How do we decide what exposure levels should be considered safe?. *Risk Anal.* **17**(1), 1–3.

Wilson, S. A., and Card, G. B. (1999). Reliability and risk in gas protection design. *Ground Eng.*, February, pp. 33–36.

Wisconsin Department of Natural Resources (WDNR) (1993a). *Guidance for Design, Installation and Operation of Soil Venting Systems*, Publ. SW 185-93. WDNR, Madison, Wisconsin.

Wisconsin Department of Natural Resources (WDNR) (1993b). *Guidance for Design, Installation and Operation of In Situ Air Sparging Systems*. Publ. SW 186-93. Madison, Wisconsin.

Wisconsin Department of Natural Resources (WDNR) (1996). *The Burning Question, Open Burning Guidance for DNR Staff and Public Officials*. Publ. AM 194–196, WDNR, Madison, Wisconsin.

Wisconsin Department of Natural Resources (WDNR) (1997). *Beneficial Use of Industrial Byproducts, Chapter 538*. WDNR, Madison, Wisconsin.

Wisconsin Department of Natural Resources (WDNR) (2001). *Site Discovery, Screening, and Ranking, Chapter 710*. WDNR, Madison, Wisconsin.

Wisconsin Department of Natural Resources (WDNR) (2003). *Land Spreading of Solid Waste, Chapter 518*. WDNR, Madison, Wisconsin.

Wisconsin Statutes (1995). Statutes 287-07(7)(c)1.C, Revisor of statutes bureau, state of Wisconsin, Madison, Wisconsin.

Witt, P. A. (1971). Disposal of solid wastes. *Chem. Eng.*, October 4, 67–

Wood, J. A., and Porter, M. L. (1987). Hazardous pollutants in class II landfills. *J. Air Pollut. Control. Assoc.* **37**(5), 609–615.

Wood, D., and Tarman-Ramcheck, B. (2002). *Beyond Recycling: Zero Waste*. ARROW. April, Associated Recyclers of Wisconsin, Portage, Wisconsin.

Wood, J. M., Segal, H. J., Ridley, W. P., Cheh, A., Chudyk, W., and Thayer, J. S. (1975). Metabolic cycles for toxic elements in the environment. *Proc. Int. Conf. Heavy Met. Environ. 1975* **1**, 49–58.

Woodley, R. M. (1988). How to select, install and prevent damage to membrane liners used in settling ponds. *Pulp Paper* **52**(13), 28–32.

Woods, R. (1995). Napa Valley's solid waste: Fine wines rail lines. *Waste Age* **26**(12) pp. 58–68.

Wollak, H. F. (1984). PVC/ethylene interpolymer alloys for plasticizer-free membranes. *Proc. Int. Conf. Geomember.* **1**, 29–32.

Wong, J. (1977). The design of a system for collecting leachate from a lined landfill site. *Water Resour. Res.* **13**(2), 404–410.

World Commission on Environment and Development (WCED) (1997). *Our Common Future.* Oxford University Press, Oxford.

Xanthakos, P. P. (1979). *Slurry Walls.* McGraw Hill, New York.

Yeh, T., Guzman, A., Srivastava, R., and Gagnard, P. (1994). Numerical simulation of the wicking effect in liner systems. *Ground Water* **32**(1), 2–11.

Yeung, A. T. C., and Mitchell, J. K. (1992). Coupled fluid, chemical, and electrical flows in soil. *Geotechnique* **42**(4).

Ying, W., Bonk, R. R., and Sojka, S. A. (1987). Treatment of a landfill leachate in powdered activated carbon enhanced sequencing batch bioreactors. *Env. Progr.* **6**(1), 1–8.

Yoder, E. J. (1967). *Principles of Pavement Design.* Wiley, New York.

Yong, R. N. (1986). Selective leaching effects on some mechanical properties of a sensitive clay. *Proc. Int. Symp. Environ. Geotechnol.* **1**, 349–362.

Yong, R. N., and Warkentin, B. P. (1966). *Introduction to Soil Behavior.* Macmillan, New York.

Yong, R. N., Sethi, A. J., Booy, E., and Dascal, O. (1979). Basic characterization and effect of some chemicals on a sensitive clay from outardes 2. *Eng. Geol.* **14**, 83–107.

Young, J. E. (1991). *Discarding the Throwaway Society.* World Watch Institute, Washington, D.C.

Zeiss, C. (1991). Municipal solid waste incinerator impacts on residential property values and sales in host communities. In *Proc. 2nd Annual Intl. Conf. Municipal Waste Combustion, Air & Waste Management Assoc.*, Pittsburgh, Pennsylvania, pp. 821–839.

Zeng, C. (1990). MT3D: A Modular Three-Dimensional Transport Model for Simulation of Advection, Dispersion and Chemical Reactions of Contaminants in Groundwater Systems. Report to the U.S. Environmental Protection Agency, Ada, Oklahoma.

Zeng, C., and Bennett, G. D. (1995). *Applied Contaminant Transport Modelling.* Van Nostrand Reinhold, New York.

Ziezel, A. J., Walton, W. C., Sasman, R. T., and Prickett, T. A. (1962). Groundwater resources of Dupage County, Illinois. *Ill. State Geol. Surv. Ill. State Water Surv. Rep.*, No. 2.

Zimmerman, R. E., and Perpich, W. M. (1978). Design and construction of pulp and paper mill residue disposal sites. *Proc. 1st Annu. Conf. Appl. Res. Pract. Munic. Ind. Waste*, U.W. Extn., Madison, Wisconsin, pp. 495–508.

Zimmerman, R. E., Chen, W. W. H., and Franklin, A. G. (1977). Mathematical model for solid waste settlement. *Proc. Conf. Geotech. Pract. Disposal Solid Waste Mater., 1977*, American Society of Civil Engineers, pp. 210–226.

Zimmie, T. F., and Laplante, C. (1990). The effect of freeze–thaw cycles on the permeability of a fine grained soil. *Proc. 22nd Mid Atlantic Industrial Waste Conf.*, Philadelphia, Pennsylvania, July 24–27, pp. 580–593.

Zimmie, T. F., Doynow, J. S., and Wardwell, J. T. (1981). Permeability testing of soils for hazardous waste disposal sites. *Proc. 10th Int. Conf. Soil Mech. Found. Eng., 1981* **2,** 403–406.

Zonberg, J. G., Jernigan, B. L., Sanglerat, T. R., and Cooley, B. H. (1999). Retention of free liquids in landfills undergoing vertical expansion. *J. Geotech. Geoenv. Engr.,* **125**(7), 583–594.

INDEX